Tall Buildings and Urban Habitat

Tall Buildings and Urban Habitat

Cities in the Third Millennium

6th World Congress of the Council On Tall
Buildings and Urban Habitat

26 February to 2 March, 2001

COUNCIL ON TALL BUILDINGS AND URBAN HABITAT
MELBOURNE ORGANIZING COMMITTEE
Council Report No. 903.503

CRC Press
Taylor & Francis Group
Boca Raton London New York

CRC Press is an imprint of the
Taylor & Francis Group, an **informa** business
A TAYLOR & FRANCIS BOOK

CRC Press
Taylor & Francis Group
6000 Broken Sound Parkway NW, Suite 300
Boca Raton, FL 33487-2742

First issued in paperback 2019

ISBN-13: 978-0-415-23241-8 (hbk)
ISBN-13: 978-0-367-86668-6 (pbk)

British Library Cataloguing in Publication Data
A catalogue record for this book is available from the British Library

Library of Congress Cataloging in Publication Data
Tall buildings and urban habitat : cities in the third millennium / Council on Tall Buildings and Urban Habitat.
    p. cm.
    Papers from the Council on Tall buildings and Urban Habitat 6th World Congress: Cities in the Third Millennium, Melbourne, 2001.
    Includes bibliographical references and index.
    1. Tall buildings—Congresses. 2. Skyscrapers—Congresses. 3. City planning—Congresses. I. Council on Tall Buildings and Urban Habitat.

TH5 .T35 2001
720'.483—dc21

00-067028

Visit the Taylor & Francis Web site at
http://www.taylorandfrancis.com

and the CRC Press Web site at
http://www.crcpress.com

Sixth World Congress

Tall Buildings and Urban Habitat

Cities in the Third Millennium

February 26 to March 2, 2001
Melbourne Convention Centre
Melbourne, Australia

COUNCIL ON TALL BUILDINGS AND URBAN HABITAT
MELBOURNE ORGANIZING COMMITTEE

MAJOR SPONSOR
Otis Elevators

CONGRESS PARTNERS
Smorgan ARC
BHP
Turner Construction
Melbourne City Council

CONGRESS PARTNERS
Bonacci Group
Norman Disney & Young
Rider Hunt
Davis Langdon Seah International
Skilling Magnusson Barkshire Inc.
Ove Arup

SUPPORTED BY A GRANT FROM
Victorian Department of Infrastructure
Urban Land Institute

AUSTRALIAN HOSTING ASSOCIATIONS
Australian Institute of Steel Construction
Australian Wind Engineering Society
Concrete Institute of Australia
Institution of Engineers, Australia
Royal Australian Institute of Architects
Steel Reinforcement Institute of Australia
Victorian Department of Infrastructure

Council on Tall Buildings and Urban Habitat

SPONSORING SOCIETIES

International Association for Bridge and Structural Engineering (IABSE)
American Society of Civil Engineers (ASCE)
American Planning Association (APA)
International Union of Architects (UIA)
American Society of Interior Designers (ASID)
Japan Structural Consultants Association (JSCA)
Urban Land Institute (ULI)
International Federation of Interior Designers (IFI)

AFFILIATED ORGANIZATIONS

Chicago Committee on High-Rise Buildings
Council for Asian Tall Buildings and Urban Habitat
Los Angeles Tall Buildings Structural Design Council
Council on Tall Buildings and Urban Habitat – Grupo Brasil
Australian Council on Tall Buildings and Urban Habitat
Polish Group on Tall Buildings and Urban Habitat
Dutch Council on Tall Buildings
High Rise Buildings Working Group (Turkey)
Romanian Tall Building Working Group

Council on Tall Buildings and Urban Habitat

Organizational Members

BENEFACTORS
Al-Rayes Group, Kuwait
Invensys plc, London

PATRONS
Zuhair Fayez Partnership, Jeddah
KLCC (Holdings) Bhd., Kuala Lumpur
Hong Kong Land Ltd., Hong Kong
Kuwait Foundation for the Advancement of Sciences (KFAS), Kuwait
Otis Elevator Company, Farmington
Schindler Elevator Corporation, Morristown
Skidmore Owings & Merrill, L.L.P., Chicago
Turner Construction Company, New York

DONORS
Halvorson + Kaye Structural Engineers, Chicago
Hochtief AG, Essen
Hong Kong Housing Authority, Hong Kong
Jaros, Baum & Bolles, New York
Mori Building Company, Ltd., Tokyo
Multiplex Constructions (NSW) Pty. Ltd., Sydney
Nishkian-Menninger, San Francisco
Ove Arup Partners, London
Leslie E. Robertson Associates, R.L.L.P., New York
Saudi Consulting Services, Riyadh
Takenaka Corporation, Chiba
Thornton-Tomasetti/Engineers, New York
Tishman Speyer Properties, New York
Wing Tai Construction & Engineering, Hong Kong
Wong & Ouyang (HK) Ltd., Hong Kong

CONTRIBUTORS

American Iron and Steel Institute, Washington, D.C.
Buro Happold Partnership, Bath
China Jingye Construction Engineering Contract Company, Beijing
Draka Elevator Products, Inc., Hickory
The Durst Organization, New York
EUROPROFIL SA, Luxembourg
INTEMAC, Madrid
Kone Corporation, Hyvinkaa
Joseph R. Loring & Associates, Inc., New York
Meinhardt Australia Pty. Ltd., Melbourne
Walter P. Moore & Associates, Atlanta
Mueser Rutledge Consulting Engineers, New York
Norman Disney & Young, Melbourne
Pei Cobb Freed & Partners, Architects, L.L.P., New York
Projest Consultoria e Projetos S.C. Ltd., Rio de Janeiro
Shenzhen University, Institute of Architectural Design & Research, Shenzhen
Skilling, Ward, Magnusson, Barkshire, Inc., Seattle
Teng & Associates, Chicago
Werner Voss & Partner, Ascheberg
WHL Consulting Engineers, Los Angeles
Nabih Youssef and Associates, Los Angeles

CONTRIBUTING PARTICIPANTS

ACTEC Consulting Engineers, Bangkok
Allied Environmental Consultants Ltd., Hong Kong
American Institute of Steel Construction, Chicago
Arab Consulting Engineers, Cairo
Architectural Services Department, Hong Kong
Aronsohn Consulting Engineers BV, Rotterdam
Barker Mohandas, L.L.C., Bristol
Lynn S. Beedle, Hellertown
Bonacci Group Consulting Engineers, Melbourne
Botti Rubin Arquitetos Associados S/C Ltda., Sao Paulo
Boundary Layer Wind Tunnel Laboratory (U. Western Ontario), London, Canada
Bovis Lend Lease Engineering, Middlesex

Brandow & Johnston Associates, Los Angeles
Callison Architecture, Inc., Seattle
Cantor Seinuk Group, New York
CBM Engineers Inc., Houston
Cermak Peterka Petersen, Inc., Fort Collins
CMA Architects & Engineers, San Juan
Connell Mott MacDonald, Neutral Bay
Construction Technology Labs, Inc., Skokie
Crane Revolving Door Co., Inc., Lake Bluff
Crone Associates Pty. Ltd., Sydney
D3BN Civil & Structural Consultants, Amsterdam
Davis Langdon & Everest, London
DeSimone Consulting Engineers, New York
DeStefano + Partners, Chicago
Dorma Group North America, Reamstown

Dorma Sistema de Controles P/Portas Ltda., Barueri
Drahtseilerei Gustav Kocks GmbH & Co., Muelheim
Draka Elevator Products, Inc., Hickory
Edgett Williams Consulting Group, Mill Valley
Egis Consulting Australia Pty. Limited, Sydney
Bob Evans, General Superintendent, Atlanta
Flack + Kurtz, Inc., New York
Foster and Partners, London
Fox & Fowle Architects, New York
T. R. Hamzah and Yeang Sdn. Bhd., Selangor
Haynes Whaley Associates, Inc., Houston
Heller Manus Architects, San Francisco
Hellmuth, Obata & Kassabaum Inc., San Francisco
Z. Hemly Engineers Ltd., Tel-Aviv
Hijjas Kasturi Associates, Sdn., Kuala Lumpur
The Hillier Group, New York
HL-Technik AG, Munich
Hornberger + Worstell, San Francisco
Horvath Reich CDC Inc., Chicago
Housing & Development Board, Singapore
IGH Ingenieurgesellchaft Hopfner MBH, Cologne
Ingenhoven Overdiek und Partner, Dusseldorf
Ingenieur Buro Muller Marl GmbH, Marl
Institute Sultan Iskandar, Johor
Raul J. Izquierdo Ingeniero Civil, Mexico
Japan Structural Consultants Association, Tokyo
Jeon and Associates, Seoul
JR3 Management Services, P.C., Allentown
KEO International, Kuwait
Konig und Heunisch, Frankfurt
Land Development Corporation Hong Kong
Dennis Lau & Ng Chun Man, Architects & Engineers (HK) Ltd., Hong Kong
Lerch Bates & Associates Ltd., Littleton
Lim Consultants, Inc., Cambridge

Stanley D. Lindsey & Associates, Nashville
LPC Arquitetura Ltda., Rio de Janeiro
Martin & Huang International Inc., Pasadena
Enrique Martinez-Romero, SA, Mexico
McNamara/Salvia, Inc., Boston
Meinhardt (HK) Ltd., Hong Kong
Middlebrook & Louie, San Francisco
Mischek Ziviltechniker GesmbH, Vienna
MPN Group Pty. Limited, Milsons Point
Murphy/Jahn, Inc., Chicago
I.A. Naman + Associates, Inc., Houston
Nikken Sekkei Ltd., Tokyo
N V Besix SA, Brussels
Obayashi Corporation, Tokyo
Omrania & Associates, Riyadh
Ove Arup & Partners, Melbourne
Cesar Pelli & Associates, New Haven
Petzinka Pink und Partner, Dusseldorf
John Portman & Associates, Inc., Atlanta
PSM International, Chicago
Rahulan Zain Associates, Kuala Lumpur
Rocco Design Ltd., Hong Kong
Rowan Williams Davies & Irwin Inc., Guelph
Sato & Boppana, Los Angeles
SOBRENCO SA, Rio de Janeiro
Southern African Institute of Steel Construction, Johannesburg
Steel Reinforcement Institute of Australia, Crows Nest
Chris P. Stefanos Associates Inc., Oak Lawn
STS Consultants, Ltd., Vernon Hills
The Stubbins Associates, Inc., Cambridge
D. Swarovski + Company, Wattens
Swire Properties Ltd., Hong Kong
Taylor Thomson Whitting Pty. Ltd., Crows Nest
Tooley & Company, Los Angeles
United States Gypsum Company, Chicago
Villa Real Ltd/SA, Brussels
VIPAC Engineers & Scientists, Ltd., Port Melbourne
Weidlinger Associates, New York
Wind Engineering Institute Company, Ltd., Tokyo

Council on Tall Buildings and Urban Habitat

STEERING GROUP

EXECUTIVE COMMITTEE

Chairman
R. Shankar Nair   Teng & Associates, Chicago

Vice-Chairmen
Sabah Al-Rayes (Middle East)	Al-Rayes Group, Kuwait
Joseph P. Colaco (N. America)	CBM Engineers, Inc., Houston
Henry J. Cowan (Oceania)	Consultant, Mosman, Australia
Fu-Dong Dai (N. Asia)	Tongji University, Shanghai
Edison Musa (S. America)	Edison Musa Arq., Rio de Janeiro
Syd Parsons (Africa)	Parsons & Lumsden, Kloof
Ryszard M. Kowalczyk	Universidade da Beira Interior, Portugal
Kenneth Yeang (S. Asia)	T. R. Hamzah & Yeang Sdn. Bhd., Malaysia

Director Emeritus
Lynn S. Beedle   Lehigh University, Bethlehem

Past Chairman
Gilberto M. do Valle   Projet Consultoria e Projectos Sc. Ltd., Rio de Janeiro

Members (Ex-Officio)
Steven Lampert	Turner Construction Co., New York
Ron Klemencic	Skilling Ward Magnusson Barkshire, Seattle
Ignacio Martin	Consulting Engineer, San Juan
George von Klan	Edgett Williams Consulting, Mill Valley
Kenneth Yeang	T. R. Hamzah & Yeang Sdn. Bhd., Malaysia

GROUP LEADERS (Ex-Officio Members)

Mir M. Ali	University of Illinois, Champaign
Frans Bijlaard	TNO Building & Constr. Res., Delft
William Gene Corley	Construction Technology Labs, Skokie
Thomas K. Fridstein	Tishman Speyer Properties, Chicago
James G. Forbes	Scott Wilson Irwin Johnston & Partners, Sydney
Peter Lenkei	Pecs Polytechnic, Pecs
Tom Losek	RP+K Projekt, Warsaw
Thomas J. McCool	Turner Steiner International, Istanbul
Moira M. Moser	M. Moser & Assoc., Arch., Hong Kong
Lee Polisano	Kohn Pedersen Fox Assoc., New York
Gary Pomerantz	Flack & Kurtz, New York
Jerry R. Reich	Horvath Reich CDC, Inc., Chicago
Mark P. Sarkisian	Skidmore Owings & Merrill, Chicago

STEERING GROUP

Members-at-Large

Ali A. Al-Shamlan	Kuwait Fdn for Advancement of Sciences, Kuwait
Georges Binder	Buildings & Data, Brussels
Joseph Bittar	Consultant, Farmington
Daniel Cerf	Value Asset Management Sdn. Bhd., Kuala Lumpur
John E. Chapman	Schindler Elevator Corp., Morristown
Charles DeBenedittis	Tishman Speyer Properties, New York
Zuhair Fayez	Zuhair Fayez & Assoc., Jeddah
Michael Fletcher	Walter P. Moore & Associates, Atlanta
Chandra K. Jha	PSM International Corp., Chicago
A. Eugene Kohn	Kohn Pedersen Fox Assoc., New York
David Morris	David Morris & Assoc., Hong Kong
Ishwar B. Patel	Mangat I. B. Patel, & Ptnrs., Nairobi
Wayne Petrie	HPA Architects, Milton, Queensland
Leslie E. Robertson	Leslie E. Robertson Assoc., New York
Werner Voss	Werner Voss & Partners, Braunschweig

Sponsoring Society Representatives

Finley Charney (ASCE)	Schnabel Engineering, Denver
Lee Polisano (ULI)	Kohn Pedersen Fox Assoc., London
Jeffrey Soule (APA)	American Planning Assoc., Washington
Koichi Takanashi (IABSE)	Chiba University, Chiba
Akira Yamaki (JSCA)	Nihon Sekkei, Inc., Tokyo

Affiliated Organization Representatives

Ian Chin (CCHRB)	Wiss Janney Elstner Assoc., Chicago
Albert K. H. Kwan (CATBUH)	The Univ. of Hong Kong, Pokfulam
Marshall Lew (LATBSDC)	Law/Crandall Engineering & Environ. Services, Los Angeles
Edison Musa (CTBUH-GB)	Edison Musa Arq., Rio de Janeiro
David Morris (ACTBUH)	McWilliam Consulting Engrs., Spring Hill
Jerzy Skrzypczak (PGTBUH)	WARCENT, Warsaw
Jan N.J.A. Vambersky (DCTB)	Corsmit Consulting Engrs., Amsterdam

Contents

BUILDING SYSTEMS AND CONCEPTS

BUILDING SERVICE SYSTEMS

TALL CONCRETE AND MASONRY BUILDINGS

COUNTRY REPORTS

Foreword

The Council on Tall Buildings and Urban Habitat began life in 1969 as the "Joint Committee on Tall Buildings," a group formed by the International Association of Bridge and Structural Engineering and the American Society of Civil Engineers. Today, civil/structural engineering – indeed, all of engineering – represents only a small part of the Council's field of activity.

The First World Congress, held in Bethlehem in 1972, was titled "Planning and Design of Tall Buildings" and focused primarily on technical aspects of skyscrapers. The Sixth World Congress, 29 years later and half a world away, is subtitled "Cities of The Third Millennium" and focuses equally on tall buildings and the urban habitat of which the buildings form an integral part.

These changes reflect the ways in which, at the start of the twenty-first century, the "why" of tall buildings has become as important as the "how." Technology, especially structural technology, plays an increasingly small role today in determining how tall a skyscraper can be, or whether it should be built at all. The more important consideration is that the building be compatible with the way people choose to live and work.

Another important development is that tall buildings are no longer an American preserve: Thirteen of the twenty tallest buildings in the world were built in the last 20 years; eleven of the thirteen are in Asia; only one is in North America. Many of the Asian buildings in the current top-twenty used American technology and expertise, but this is just a passing phase. Knowledge in the twenty-first century does not recognize geographic or national boundaries.

The Council on Tall Buildings and Urban Habitat has worked throughout its existence to lower the barriers to the free flow of knowledge between disciplines and nations. The Council is, today, the only organization in the world in its field that is both multidisciplinary and truly international.

The Council's 60 topical committees are organized into eight Topical Groups that cover Urban Systems, Development and Management, Planning and Architecture, Building Systems and Concepts, Building Service Systems, Design Criteria and Loads, Tall Steel Buildings, and Tall Concrete and Masonry Buildings. The Council's general membership is drawn from 77 countries; the majority of the governing Steering Group are from outside North America; there are Vice Chairmen representing all regions of the world.

The Sixth World Congress in Melbourne represents the clearest celebration yet of the Council's interdisciplinary and international character. There is equal emphasis on "Tall Buildings" and the "Urban Habitat." A broad range of disciplines is covered, from the social sciences to engineering. Speakers and case studies are drawn from around the world.

People everywhere who care about tall buildings and the urban habitat owe a debt of gratitude to the volunteers and staff who have labored for five years to organize this event, under the able leadership of Ron Klemencic, Congress Chairman and member of the Council's Executive Committee, and Jamie Learmonth, Chairman of the Melbourne Organizing Committee.

R. Shankar Nair
Chairman, Council on Tall Buildings and Urban Habitat

Preface

Each of the Council's congresses gets closer and closer to the "urban habitat" that is part of its title. The organizers of this, the Council on Tall Buildings and Urban Habitat's 6th World Congress, have succeeded in that notably, placing urban habitat as the first of its two main streams.

This volume contains the papers submitted in time to meet the publisher's deadline. Following the Congress a second volume will be published – papers that are submitted after the deadline and those that were available at the time of the Congress (see the announcement in the back of this book). The arrangement of these Proceedings corresponds to the Council's topical groups, an arrangement that is fortuitous since it is consistent with the thrust and theme of the Congress. Perhaps that arrangement shows that the Council's leadership truly believes that people – and the "Life and Work" aspects are the driving force for the buildings.

The theme title, "Cities in the Third Millennium" shows a bold stroke by the Melbourne organizers. If one looks back at the changes in the Second Millennium – and what brought them about – one would come to the conclusion that, while plans for the immediate future are essential, those for the millennium are quite impossible. How could one have planned in the year 1000 for a city in the 1990's that was driven by the Internet?

It's even difficult to make plans for the short term, but a review of the papers show that the congress promises to achieve the goal of a vision for a meaningful future. Actually it would be interesting to see what kinds of projections were made (or implied) in our prior congresses. How good were they? For how long a time period? Were the suggestions realized? And what will be the reaction of those making plans for our cities of the Fourth Millennium that was said in Melbourne at the beginning of the Third?

We both commend and thank Jamie Learmonth, Chairman of the Melbourne Organizing Committee and Ron Klemencic, Chairman of the Congress Committee of the CTBUH. We would also like to thank the CTBUH Headquarters staff for their part in making the Congress a reality: Geri Kery, manager of Financial Operations, Abdullah Rahim Abdussalam and the staff of undergraduate and graduate students. We are indebted to the Council's organizational members and the many sponsors of the Congress – identified earlier.

Finally, we thank Council Chairman R. Shankar Nair for his guidance and deep commitment to the Council's mission.

Lynn S. Beedle
Director Emeritus

Recent Infrastructure Developments in Hong Kong – the Background, Current and Future Developments

Raymond W. M. Wong

1 INTRODUCTION

To accommodate somewhat 7 million people in a piece of land of size slightly bigger than 1050 sq km like Hong Kong is not an easy task. Needless to mention the 240 outlying islands and hilly topography that contributes to about 62% of the total land area which cut the territory into bits of almost disconnected lands.

The overall population density per sq km in Hong Kong is about 6300 in 1999. The figure conceals wide variations among different areas in the territory. The density in the metro areas is about 28000 per sq km, while in the New Territories it is around 4100. Thanks to the continual development of new towns outside the metro areas since the 70's, the difference in population density is gradually dropping in the recent years.

For the last 3 decades, growth of population is quite steady roughly at a rate of 1 million every 10 years. However, as Hong Kong has returned to the sovereignty of the People's Republic of China in 1997 that created more active social and economical interactions between people of the two places, it is expected a faster growth will be envisaged in the coming decades. As a result, more lands with acceptable infrastructure facilities have to be provided to cater for the expected growth as well as to improve existing quality of living inside the territory. This paper aims to provide a summary of what has been done in the recent years as a means of strategic improvement to the territory of Hong Kong.

2 THE PHYSICAL ENVIRONMENT

The territory of Hong Kong can be sub-divided into 4 main geographical regions, namely the Island of Hong Kong, Kowloon Peninsula and New Kowloon, New Territories and the outlying islands. Map as shown in Figure 1 provides a rough idea of the physical environment of Hong Kong.

Figure 1   Physical Map of Hong Kong Special Administration Region, Public Republic of China.

The urban or metro areas are located on the northern and southern sides of the Victoria Harbour. North of the harbour is the Kowloon Peninsula, it stretches northward about 5.5 km until it reaches the east-west running Kowloon peaks (averaged 320 m high). The peninsula is a piece of relatively flat land of about 36 sq km in size.

South of the harbour is the Island of Hong Kong, which is a hilly island of about 75 sq km in size, with continual mountain ranges occupying most of the island, leaving only a narrow strip of the coastal areas being practically developable. Within what are called the metro areas with flat land less than 58 sq km altogether, it housed about 49% of the total population of Hong Kong.

North of the Kowloon peaks is the New Territories. Before the 60's, there were only some thinly populated village towns scattered all over the 650 sq km of land. The only piece of flat land (Yuen Long-Kam Tin Plain, area about 55 sq km) situated on the north-western corner of the New Territories, the rest of the areas are practically only narrow patches of flatter land located on the foot of hills and mountains. The 960 m-high Tai Mo Hill and the outward stretching mountain ranges lying almost in the middle, making access from the metro areas into the New Territories very difficult. Table 1 shows the growth in population throughout the past decades.

End of Year	Total Population	Population in Metro Areas*	Population in New Territories
1910	457 000	**	**
1920	625 000	**	**
1930	840 000	**	**
1960	3 133 000	**	**
1965	3 716 000	**	**
1970	3 948 000	**	**
1975	4 439 000	**	**
1980	5 021 000	**	**
1985	5 495 000	3 551 000	1 944 000
1990	5 674 000	3 381 000	2 293 000
1995	6 218 000	3 305 000	2 913 000
2000	6 720 000	3 110 000	3 610 000

* excluding Kwai Tsing and Tsuen Wan districts
** statistic not available under regional sub-group

Table 1   Population Growth in the New Territories.

3   THE METRO AREAS

The over-saturated development and the need to thin out the metro areas in Hong Kong is unquestionable. However, as the majority of the community facilities as well as job opportunities are within the metro areas, purely extending the living room from the city into the northern territories cannot provide solutions to everything.

Due to the lack of long-term city planning strategy since the 1950's when Hong Kong was just recovering from World War II and the civil wars of China, quite a lot of developmental constraints have been undermined during the restructuring process as to improve the urban environment as a whole by today's standards. For instance, the Kwun Tong and Tsuen Wan districts, which were two "satellite towns" being developed in the early 60's, were originally designed as industrial towns where very labour-intensive manufacturing industries would be located. It was so planned that the neighbouring areas, provided with densely packed low-cost public housing, would supply the required manpower for the nearby factories. As the metro development expanded rapidly in the decades that came, these districts became a kind of cancer to the metro environment and cannot meet any loose quality standard even in development countries.

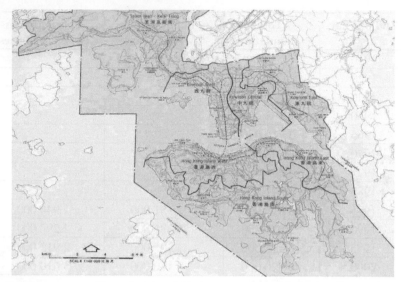

Figure 2   Map of the Metro Area of Hong Kong.

The pace and coordination in development is another important issue. A typical example can be found in the eastern part of Hong Kong Island. Since the 50's, the eastern part of Hong Kong Island stretching from Causeway Bay to North Point, was settled with residents coming from middle-class families, or with new settlers migrated from Mainland China who had brought with them capitals after the 1949 revolution. During the economic take-off in the late 70's, the demand for housing increased drastically and the areas were rapidly developed. For a period of more than 15 years, the pace of development much out-rated the growth in public transport, thus resulting in constant congestion and serious pollution. Needless to mention the poor living environment faced by the 560,000 residents living in the district.

There is not much better in the central part of Kowloon, like the district of Tsim Sha Tsui, Yau Ma Tei, Mong Kok and Sham Shui Po. In these areas, a great amount of old buildings, the majority of which were built in the early 60's, are deteriorating rapidly and causing quite a lot of problems to a modern society. Inappropriate commercial activities like food-stalls, restaurants, retail shops, small-scale entertainment facilities, or even mini red-light zones are scattered all over the district. Owing to the complication in the property ownerships, or limitation and insufficiency in the existing control regulations, the government can do nothing to have the situation easily rectified. Urban restructuring and renewal plans on a strategic level is thus an essential process to improve the existing urban environment.

The government of Hong Kong is at present trying to achieve the urban renewal exercise through several major means. One is to thin-out the urban density by introducing more lands in closer proximity to relieve population and essential facilities in the metro area. This can be achieved by reclamation such as the West Kowloon reclamation, which was carried out between 1994 to 1997, that provided about 3.8 sq km of newly reclaimed land for various development purposes. The second way is by the continual development of new towns with large amounts of public housing provisions to attract people in the urban areas to move in. And at the same time amendments of planning regulations will gradually be made, such as to lower the plot ratio (gross floor areas allowed to build as per unit land area) or by the introduction of other incentive provisions, in order to attract investors to redevelop the old land lots. The Land Development Council, which is a commission that represents the government on a commercial basis to redevelop deteriorating properties in the metro areas, also helps to play an important role in the urban renewal processes.

4 RESTRUCTURE THE ENTIRE TERRITORY OF HONG KONG

In views of achieving sustainable growth to meet with the modern world and to enhance the living standard of the majority, the government of Hong Kong realizes that there must be an up-to-date physical planning framework to guide the development and investment in order to ensure the efficient use of resources and to promote a high quality living and working environment. The process of strategic planning has been established in Hong Kong for many years, involving a number of reviews to take account of changing circumstances. The attitude of the government was becoming more pro-active especially after the 80's when Hong Kong had founded her prosperity in trade and finance that led her to be one of the four dragons in Asia.

The Territorial Development Strategy (TDS), in particular, provided a new framework for urban and territorial growth in Hong Kong. TDS is the highest tier in the hierarchy of town plans in Hong Kong. It provides a board, long-term framework on land use, transport and environmental matters for the planning and development of the territory. The following are the objectives as set out in the TDS.

Objective 1: To enhance the role of Hong Kong as an international city and a regional centre for business, finance, information, tourism, entrepôt activities and manufacturing.

Objective 2: To ensure that adequate provision is made to satisfy the land use and infrastructure needs arising from sectoral policies on industry, housing, commercial, rural, recreation and other major socio-economic activities.

Objective 3: To conserve and enhance significant landscape and ecological attributes, and important heritage features.

Objective 4: To enhance and protect the quality of the environment with regard to air, water quality, noise, solid waste disposal and potentially hazardous installations by minimizing net environmental impacts on the community.

Objective 5: To provide a framework within which to develop a multi-choice, high capacity transport system that is financially and economically viable, environmentally acceptable, energy efficient and makes provision for the safe and convenient movement of people and goods.

Objective 6: To formulate a strategy that can be carried out both by the public and private sectors under variable circumstances, particularly with respect to the availability of resources and significant changes of demand.

Figure 3 Squatter area located in the metro area – the Dae Hom Village at Diamond Hill.

Figure 4 New skyscrapers and aged buildings co-existed in the urban centre of Wanchai in the Island of Hong Kong.

Figure 5 Illegal structure constructed on the roof top of some post-war buildings in Yaumati District of Kowloon.

After years of studies, some of which were done by independently appointed consultants, and the carrying out of thorough public consultations and debates, the TDS has arrived at a generally accepted direction, which comprises with various scenarios and goals that may cater different development situations and rates of growth.

The main scope and major studies in the TDS covers the general development principles, scenarios on economical and population growth, overall review studies, land use and development patterns, formation of new land and new townships, urban restructure and renewal, distribution of socio-economical centres, rural and environmental preservation, port and airport development strategies, as well as land transportation systems etc. Targeted guidelines have been set for each studied areas according to short, medium and long-term needs and each strategic area is going to be fine-tuned according to the condition and pace of development in Hong Kong as a whole under appropriate legislative and executive procedures.

5 STRATEGIC DEVELOPMENT SUB-REGIONS

To have the development strategies implemented in a more efficient manner, the territory of Hong Kong is divided into five sub-regions according to their geographical positions, sectoral functions or development history and patterns etc. These regions are further detailed as below.

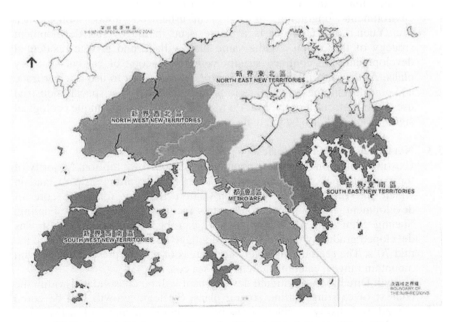

Figure 6 Sub-Regions according to the Strategic Development of Hong Kong.

5.1   Metro Area

Existing condition – It includes the Island of Hong Kong; east, central and west Kowloon; and the Kwai Tsing and Tsuen Wan districts. Present population is about 3.24 million. Except for the southern part of Hong Kong Island where natural coastline is basically preserved, it is a relatively high-density development zone with mixed urban land uses.

Growth direction – The prime objective in the development strategy is to thin-out some of the high-density zones and to restructure some of the old districts such that social and commercial facilities can be better and more conveniently allocated. New highway and mass transit system will continually be provided to ease off the transportation needs between major sub-districts. The development to be a focus point as principal tourist centre and cores for main government and cultural functions will be continued.

5.2   North-West New Territories

Existing condition – It houses a population of about 1.1 million and majority of which are situated along the Tuen Mun-Yuen Long Corridor and the Yuen Long-Kam Tin Plain. Present developments are quite unstructured except around new towns which have been developed since the mid 70's.

Growth direction – Except to support the new towns with significant amount of population, the region is also intended to develop as a "flow corridor" linking the border with Mainland China; new centres for mass housing will be provided at Tin Shui Wai, Yuen Long South and areas around main transport nodes. In order to improve the existing environment, upgrading of the semi-urbanized areas along Tuen Mun/Yuen Long Corridor is a prime issue in the overall development strategy of the region. At the same time, village and private residential developments are confined strictly within the scope of current statutory plans. At the west of Tuen Mun, it will be intensified to use as river trade and port development as well as to provide land for other special industrial uses. The protection of the Mai Po Marshes which are of unique ecological significance will also be intensified.

5.3   North-East New Territories

Existing condition – Existing population is about 1.02 million. Majority of the population is gathered within a south–north running strip of land in which the East Rail of Kowloon-Canton Railway form the core of development. Major population centres include Shatin, Tai Po, Fanling, Sheung Shui and Ma On Shan, all can be regarded as new towns developed gradually from the original village-base rural township since the mid 70's. The region also covers large areas of natural preserved land with mountain ranges, catchment areas and sea coasts.

Growth Direction – Continual development will be consolidated within the context of existing outline zoning plans. Strategic growth will be cored around the railway corridor to receive for another increase of at least 0.15 population within the coming decade. Emphasis will continually be made on conservation of the environment, and provision of low-density housing, tourist and recreation facilities in selected locations.

5.4 South-West New Territories

Existing condition – This part of the territory mainly comprises of the inhabited outlying islands of Hong Kong. These islands include the Lantau, Peng Chau, Cheung Chau and Lamma islands. Within an area of about 140 sq km, it has a population of about 0.11 million. Except for the Fixed Lantau Crossing, the Lantau Express and the North Lantau Expressway opened in 1998 leading to the new airport in Chek Lap Kok at Lantau, there is no other land-based transport leading to these outlying islands.

Growth Direction – Major growth will be clustered in the new towns along the north coast of Lantau Island. The new town of Tung Chung was developed in 1996 and intended to be a working base during the construction of the new airport as well as to house another 0.15 million people in the coming decade. With the signing of an agreement with Walt Disney in 1999 to develop a new theme park in the north-east corner of Lantau at Penny's Bay, there is a strong tendency to develop Lantau into a resort and recreational centre subject to detail investigation by the government. While the central and southern portion of Lantau will be retained as a semi-preserved zone such that the green areas, natural beaches and some old monasteries that were built at least half a century ago can be protected.

5.5 South-East New Territories

Existing condition – The territory is one of the best preserved areas in Hong Kong. The Sai Kung Country Park, the High Island Reservoir and countless uninhabited islands and coastal beaches are lying within this region of the territory. There is about 0.36 million of the population living in this 80 sq km of land. Within which about one-third are living in the scattered low-density houses around the old village town of Sai Kung, and the rest are living in the new town of Tseng Kwan O.

Growth Direction – Major growth will only be confined in the new town of Tseng Kwan O. With the opening of the Mass Transit Railway Tseng Kwan O Extension Line in 2003 and the full development of the vicinity by the end of the decade, population is expected to escalate to about half a million. The rest of the areas will be retained as a preserved zone with limited low-density development under stringent control.

6 REVIEW OF EXISTING INFRASTRUCTURE DEVELOPMENT

It could be assumed that there was practically no planned infrastructure provision for Hong Kong on a territorial scale before the 60's. Every future planning was then on a rather short-term basis depending on almost ad-hoc development opportunities. As the population grew dramatically in the early 60's and the gradual maturity of the manufacturing industries of Hong Kong, the needs to provide more land for the industry and to accommodate the growing population were becoming obvious. The first two satellite towns of Tsuen Wan and Kwun Tong on the western and eastern ends of the Kowloon Peninsula were then developed to relieve such needs.

As population centres grew in a somewhat arbitrary manner in the 70's, bottlenecks of various natures appeared. Common problems such as traffic jams, over development, environmental degradations, under provision of the required urban facilities etc., caused great nuisances to the efficient operation of the city.

Large-scale coordinated infrastructure projects aimed to improve the traffic and other supporting systems within the territory started in the late 70's. These projects can be summarized into 2 main categories according to their geographical sub-region.

6.1   Major Link from Metro into the New Territories (NT) and the Lantau Island

Year Between   Major Projects
1910   • Introduction of the 33.7 km single-tracked Kowloon-Canton Railway serving from Tsim Sha Tsui (South Kowloon) to Lo Wu (North NT).
                      • Other linkage into the NT mainly through pedestrian footpaths and an unpaved single lane roadway on the western side of NT.
1910–1940   • Situation remains unchanged but with an additional roadway from central Kowloon to Central NT.
1950–1960   • Roadway paved and extended to single lane for both ways.
                      • Introduction of a simple circular road network in NT with a single-lane both directions roadway crossing the 960m Tai Mo Hill in Central NT.
                      • Development of the first new town (Tsuen Wan) in the south-western part of NT.
1960–1970   • Opening of the Lion Rock Tunnel linking central Kowloon more directly to the Central NT (1967).
                      • Introduction of the first 2-lane 2-way highway linking western Kowloon to south-western part of NT (Kwai Chung Road).
1970–1980   • Opening of the 2-lane 2-way Tuen Mun Highway that provided swifter traffic for the western part of NT.
                      • Development of two more new towns (Tuen Mun and Shatin) in the western and central-southern part of NT.
                      • Opening of the second Lion Rock Tunnel (1978).
1980–1990   • Completely double-tracked and electrified of the Kowloon-Canton Railway system (1983).
                      • Opening of the 3-lane 2-way Tolo Harbour Highway serving the Central NT from south to north (1985).
                      • Opening of the Light Rail Transit with a route length of 32 km serving the north-western part of NT along the Tuen Mun-Yuen Long Corridor (1988).
                      • Large scale traffic improvement and road extension to the north-western and central part of NT.
1990–1995   • Opening of the 2.8 km Shing Mun Tunnel linking Tsuen Wan to Shatin in the south-central part of NT (1990).
                      • Development of the Tseng Kwan O New Town (with a linking tunnel) at the eastern-end of New Kowloon.

- Opening of the Tate's Cairn Tunnel linking eastern Kowloon with Central NT (1991).
- Development of Ma On Shan New Town on the central-eastern part of NT.
- Development of Tin Shui Wai New Town on the north-western part of NT.
- Development of the north-western NT (Tuen Mun, Yuen Long and Kam Tin) and the central-northern NT (Tai Po, Fanling and Sheung Shui) corridors.
- Completion of the NT Circular Road (in 4 main stages) that provides high speed traffic link for the central, northern and western part of NT.
- Large scale traffic improvement and road extension to enhance existing highway network throughout the NT.

1995–2000
- Development of Tung Chung New Town at Lantau Island.
- Opening of the new international airport at Chek Lap Kok.
- Opening of the Lantau Links comprising a series of long-span channel crossing bridges.
- Opening of the Tung Chung Line and the Airport Express Line which underlie the future large-scale development of Lantau Island.
- Opening of the Route 3 comprising of a long-span cable-stayed bridge (Ting Kau Bridge) and a 5.4 km long country park crossing tunnel (Tai Lam Tunnel).

2000–2005
- Opening of the West Rail that provides a swifter traffic link for the north-western part of NT.
- Opening of the Mass Transit Railway Tseng Kwan O extension to provide a direct link to the new town of Tseng Kwan O with the metro.
- Opening of the Kowloon-Canton Railway Ma On Shan Line to provide the necessary linkage of the new town to the East Rail of the railway network.

6.2   Major Infrastructure Development within Metro Areas of Hong Kong

Years between   Major Projects

1960–1965
- The first highway linking between Wanchai and Central in Hong Kong Island (about 1.8 km in length).

1965–1970
- The first highway linking central to eastern part of Kowloon (about 6 km from Kowloon City to Kwun Tong).

1970–1975
- The first highway linking the eastern to western part of Kowloon peninsula along the foot of Kowloon peaks (8.5 km in length from Lai Chi Kik to Diamond Hill on northern part of Kowloon).
- The first cross-harbour tunnel joining Hong Kong Island and Kowloon (1972).
- Opening of the first Tsing Yi Bridge triggering the large-scale development of the Tsing Yi Island (1974).

1975–1980	•	Opening of the first phase of Mass Transit Railway (1979).
1980–1985	•	Opening of the first Ap Lei Chau Bridge linking the 2 sq km island to Aberdeen at the southern part of Hong Kong Island.
	•	Opening of the second phase of Mass Transit Railway (1982).
	•	Opening of the Aberdeen Tunnel linking the northern and southern part of Hong Kong Island (1982).
	•	Opening of a 1.8 km tunnel under the Kai Tak Airport to provide a direct link between the central and eastern part of Kowloon (1982).
	•	Opening of the first phase of island eastern corridor (part of the Hong Kong Island northern bypass).
1985–1990	•	Opening of the West Kowloon Corridor (a 4.5 km expressway/inter-change system on the western side of Kowloon up to Kwai Chung).
	•	Completion of the final phase of the island northern bypass (totally 13 km expressway/interchanges system from Chai Wan on the east to Kennedy Town on the west, 1989).
	•	Opening of the second cross-harbour tunnel (Eastern Harbour Crossing, 1989).
	•	Final completion of the Mass Transit Railway with the second cross-harbour linkage through the Eastern Harbour Crossing (1989).
	•	Large-scale road improvement throughout the metro areas.
1990–1995	•	Opening of the Tseng Kwan O Tunnel to accelerate the development and moving into the Tseng Kwan O new town.
	•	Opening of the Kwun Tung Bypass (a 2.6 km elevated expressway/inter-change system to link up the Tate's Cairn Tunnel and to ease the traffic of the eastern part of Kowloon).
	•	Large-scale road improvement throughout the metro areas.
1995–2000	•	Opening of the third cross-harbour tunnel (Western Harbour Crossing, 1997).
	•	Opening of the West Kowloon Expressway (part of a 34 km expressway connecting the cross-harbour tunnel to the new airport in Lantau Island, 1997).
	•	Opening of the Hung Hom Bypass (a 1.2 km expressway/interchange to ease the traffic of the southern part of Kowloon, 1999).

7 OVERALL INFRASTRUCTURE PROVISION

Stepping into the mid 80's, infrastructure developments in Hong Kong were introduced at a much organized, long term and coordinated manner. Reasons for this positive shift can be explained by a number of factors. Besides the actual needs of the community as well as the rapid growth in economy due to the success of Hong Kong in securing her position as a regional business and finance centre in South East Asia that provided the capital for development, one guaranteed step was that a joint declaration to return Hong Kong to Mainland

China was made in 1984 between the British and Chinese governments that clarify the future of Hong Kong. With that agreement, more definite investments and resources planning could then be arranged.

Infrastructure developments in Hong Kong in general are aimed at several directions and levels, with rail and highway networks, population centres and new towns, airport and ports, urban restructuring, and environment as the major strategic cores. These cores are integrated and interrelated to form an overall strategic development that aims to restructure, improve and achieve the objectives as stated in previous pages. Detail implementation is somewhat based on the territorial sub-regions for the effectiveness in planning and achieving the set targets. This paper tries to pay more concentration on the parts related to rail and highway networks for it is the backbone of all developments as well as it calls for a huge amount of resources and capitals.

7.1 The New Airport at Chek Lap Kok in the Lantau Island

The old Kai Tak Airport was located in the centre of Kowloon since the 20's. Her existence was quite awkward when compared to other achievements in terms of population and economic growth of Hong Kong. Kai Tai Airport was both dangerous as it is located in the heart of a high-density city and at the same time could not satisfy the demands from all needs.

Figure 7 The new International Airport of Hong Kong and the new town of Tung Chung (foreground).

Figure 8 Layout of the airport seeing the Passenger Terminal Building and the first runway (the second runway on the north along the seawall, was under construction).

After numerous studies and debates, Chek Lap Kok at the north-eastern side of Lantau Island was selected as the site for the new airport. Besides providing a new airport 3 times larger than Kai Tak, it also serves the following functions:

- To relocate the source of air traffic noise to areas with lower population density.
- To relieve huge amount of buildable space in the metro areas by releasing the air traffic height restriction.
- Strengthen the air cargo handling capacity of Hong Kong.
- Provide the land reserve to develop a new commercial, conference and exhibition centre at the threshold of the airport.
- Provide a transportation link from the metro area to Lantau, the largest outlying island of Hong Kong. This is an initial investment for the development of Lantau.
- Trigger the development of the first new town (Tung Chung) in Lantau.
- Provide part of the highway traffic improvement strategic backbone for the north-western part of New Territories.

To have the new airport made deliverable in mid 1998, the following related infrastructure works have been carried out starting from 1994.

- Formation of the new airport platform by the levelling of the original Chek Lap Kok Island (430 Hectares) and to reclaim the nearby sea to form a 1250 hectares airport island site (contract cost US$1.16 billion)
- Construct the Passenger Terminal Building (contract cost US$1.95 billion).
- Construct two 3.8 km long runways and the associated taxiways (contract cost US$0.95 billion).
- Construct other ancillary and transportation facilities within the airport island (contract cost US$0.58 billion).

7.2   The Airport Core Projects

The new airport at Chek Lap Kok is located more than 30 km from the city centre on an originally uninhabited island. The provision of the new airport in fact serves several purposes and sets an important developmental pattern in the overall restructuring of the Hong Kong territory. To implement the plan, 10 related projects, known as the 10 Airport Core Projects (ACPs), were carried out at the same time together with the construction of the new airport. The 10 ACPs can be summarized as follows:

- The new airport at Chek Lap Kok – to form the airport island and provide the required facilities including the terminal building, runway, taxiway, ground transportation and other ancillary facilities to enable the construction and operation of the new airport to replace the old one at Kai Tak.

- Development of the Tung Chung new town – to form and develop a new town from the old village town of Tung Chung (2.5 km on south-eastern side of the terminal building), which is to be used as a support base during the construction of the new airport, and become a population centre to house a population of about 200,000 by the end of 2015.

Figure 9 Layout of the Tung Chung New Town as seen in mid 1997.

- North Lantau Expressway – a 12.5 km long 3-lane both directions expressway constructed along the north coast of Lantau Island to provide direct linkage from the new airport to the metro areas. Several interchanges are provided to allow for the future development of population centres at appropriate locations.

Figure 10 The formation of the North Lantau Expressway at Yam O Section as seen in mid 1996.

Figure 11 The completed section of Expressway as seen in late 1997.

- Lantau Link – this is a series of bridges and elevated expressway linking two sea channels with a total distance of about 2.9 km. It comprises the Tsing Ma Bridge (suspension bridge with main span 1377m, Ma Wan elevated expressway (viaduct, about 600m long) and the Kap Shui Mun Bridge (cable-stayed bridge, main span 430m).

Figure 12 The Lantau Link, with the Kap Shui Mun Bridge in the foreground and the Tsing Ma Bridge in the background.

- Route 3, Kwai Tsing Section – a 7.5 km, 4-lane both ways expressway, provide direct linkage from Lantau Link to western Kowloon. It serves also as an interchange to the Route 3 Country Park Section, another 14 km expressway leading to the north-western part of the New Territories and the border areas.

Figure 13 Route 3 Kwai Chung Section along the very busy Kwai Chung highway.

- West Kowloon Reclamation – to reclaim along the original seawall of western Kowloon to form a strip of new land about 340 hectares in size to accommodate a new expressway, the airport railway, two railway stations and other supportive land for necessary future development. This land forms an important part for the strategic renewal and restructuring of the high-density districts at western Kowloon.

Figure 14 West Kowloon Reclamation on the southern tip of Kowloon.

Figure 15 West Kowloon Expressway on the West Kowloon Reclamation site at Mei Foo District.

- West Kowloon Expressway – a 4.5 km expressway linkage between Hong Kong Island (through the third harbour crossing tunnel) in the south to Route 3 in the north. Three major interchanges are provided to allow traffic to enter/exit into the expressway from West Kowloon.
- Western Harbour Crossing – this is the third harbour crossing tunnel joining the western part of Hong Kong Island with west Kowloon. Interchange provision at the Hong Kong side is also serving an important role for the continual development of the western and southern part of Hong Kong Island.

Figure 16 Coupling arrangement and approach ramp for the Western Harbour Crossing on the south-most tip of the West Kowloon Reclamation.

Figures 17 and 18
Central Reclamation as
seen in mid 1995
(above) and early 1998
respectively. The 20-
hectare reclaimed area
is used as the Terminal
for the Airport Railway,
ferries for the outlying
islands, as well as
reserved land for the
Central-Wanchai By-
pass and other private
developments.

- Central and Wanchai Reclamation – to form a piece of land of about 20 hectares in size at the geographical, financial and political centre of Hong Kong. It accommodates the new passenger terminal of the Airport Railway, and to provide extra land for the Central-Wanchai Bypass and other commercial developments.
- The Airport Railway (Airport Express Line and the Tung Chung Line) – a 34 km-long railway line serving the new airport and the new town of Tung Chung to the metro. There are 7 stations/terminal along the line at this stage, most of these stations are designed as transport nodes to support developments of community, residential or commercial nature.

7.3 Future Strategies in Transport Systems and the Related Infrastructure Development

According to the TDS review, principal areas of concern from a strategic planning point of view with regard to transport systems in Hong Kong can be summarized as follows:

- To improve the distribution of population and jobs in a way that helps to minimise the need for travel and, at the same time, concentrates development in a well designed way within the walking distance from major public transport nodes.
- To provide a comfortable, efficient, safe, affordable, multi-model and environmentally acceptable public transport system that gives good levels of accessibility to residential areas and work places.
- To provide a hierarchy of highways and freight railway services, including the provision of adequate cross border links, to facilitate the movement of goods between centres of demand and supply as to sustain key economic activities.
- To provide comprehensively designed pedestrian networks that facilitate the safe and convenient movement of people within the catchments of transport nodes and interchanges.
- To stress the need for greater provision for high-capacity passenger mover systems within the metro areas as well as between the metro and new population/job centres. Parallel efforts to be made for the upgrading and expansion of public bus services to meet both intra and inter-zonal travel needs in areas beyond the walking distance of railway stations.

7.3.1 Highway Strategies

With regard to the strategic highway network, a number of new links and extensions that aim to create a denser web of North–South and East–West high-capacity expressways within the territory will be developed in appropriate stages.

In the metro areas where room for expansion is indeed highly limited, major highway projects are confined to segmental, short-sectioned roadwork acting mainly as an extension to existing highway networks rather than major development in a large-scale strategic manner. The following are the major projects in the strategic planning within the metro areas:

- The Central Wanchi Bypass and a section of link roads into the existing Island Eastern Corridor – with the provision of this bypass, a continual expressway with at least 3-lane both ways can be provided along the northern coast of Hong Kong Island. However, the execution of the project is subject to the approval of the reclamation in the Wanchai area.
- The Route 7 – a 2-lane both ways highway linking the Belcher Bay on the western end to Aberdeen on the southern part of Hong Kong Island. This is important for the strategic development of the south-western part of the island, which is at present only relatively moderately-populated and has potential to develop further.
- The Central Kowloon Route – to provide an east–west running 3-lane both ways highway link, mainly in the form of subway, that cuts across the densely-developed areas in the central part of Kowloon Peninsula.
- The Route 9 – a 7 km long expressway with tunnel and a 1100m span bridge linking Tsing Yi Island to Stonecutters Island and reach Lai Chi Kok. This is an improvement project to enhance the entire traffic flow in the western part of Kowloon.

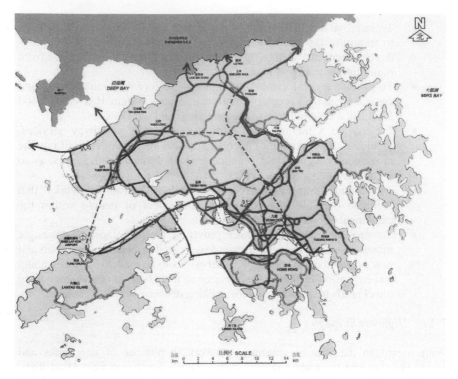

Figure 19   Strategy Highway Network under the 1996 Territorial Development Strategy Review.

For non-metro areas, majority of the highway development are concentrated on the north-western part of the New Territories which is the only sub-region within the territory with major strategic growth in population and other communal-economical activities. The following major projects are prioritized and would be carried out in the medium-term highway strategic plan.

- The Route 10 – a 3-lane both ways expressway serves as the Hong Kong Western Corridor to provide direct link from the western part of Hong Kong Island through a series of bridges and tunnels, via Lantau Island, Yuen Long, and finally reach the western border between Hong Kong and the Mainland. This would provide an additional strategic link between the north-western part of New Territories and North Lantau, relieve pressure on Route 3 Country Park Section and cater for cross-border traffic in the future.
- A 3-lane both ways highway link connecting the Western Corridor of Shenzhen at the Mainland side of the border.
- A 3-lane both ways coastal highway running along the west coast of Tuen Mun, join the new town of Tin Shiu Wai, and reach the Yuen Long plain at the end. This is important for the development of the area into a river trade centre as well as with land reserve for special industries. The highway network will also cater for necessary

interchanges for possible traffic linkage to the Zhuhai Special Economic Zone of Mainland China on the west bank of the Pearl River.

- Widening of the Yuen Long Highway and to provide major interchanges to the existing and newly introduced highway network within the areas.
- Widening of the Tolo Highway and Fanling Highway, which is the only highway system serving the central-eastern part of the New Territories.
- Provide a new 3-lane both ways coastal highway from the eastern tip of Kowloon into the new town of Tseng Kwan O.

7.3.2 Railway Strategies

The Kowloon-Canton Railway Corporation (KCRC) and the Mass Transit Railway Corporation (MTRC) are two government-based corporations at present operating the railway network system of Hong Kong. The present railway networks running under the two corporations are:

- KCRC – At present only the KCR 35 km-long East Rail is serving the central part of Kowloon and New Territories from Hung Hom in the south to Lo Wu in the north, passing through 10 intermediate stations.

  However, by the end of 2004, two more railway lines will be put into operation. They are the 31 km-long West Rail (US$8.2 billion) serving the north-western part of New Territories from Sham Shui Po in the south to Tuen Mun in the north (with 8 intermediate stations), as well as the 11.5 km-long Ma On Shan Line serving the new town of Ma On Shan on the east bank of the Tolo Harbour (with 7 intermediate stations).

- MTRC – At present there are 5 underground railway lines operated by the corporation serving mainly within the metro areas. They are the Island Line running along the northern coast areas of Hong Kong Island; the Quarry Bay Line running from the eastern part of Hong Kong Island through a submerged tube across the harbour to the central-eastern part of Kowloon. The Tsuen Wan Line running from Central of Hong Kong Island, then across the harbour and serves the western portion of Kowloon until it reaches Tsuen Wan. The Tung Chung Line runs from Central to the new town of Tung Chung at Lantau Island. The last one is the Airport Express Line providing direct link from Central to the new airport at Chek Lap Kok. The 5 lines altogether have a length of about 77 km and serve about 35% of the total population in the territory.

  Before the end of 2003, the 12 km-long Tseng Kwan O Line with 5 intermediate stations shall come into operation and serve the new town of Tseng Kwan O.

In order to cater for the continual population growth of Hong Kong up to 2015 and the expected increasing cross border social and economic activities, the expansion of the railway network is ranked as the priority in the government's strategic planning in recent years. The forthcoming railway network will be so designed to fulfil the following objectives:

- To relieve the bottlenecks in the existing railway systems.
- To provide rail service to strategic growth areas for housing and economic developments.
- To meet cross border passenger and freight demands.
- To increase the share of rail in the overall transport system to reduce reliance on road-based transport.

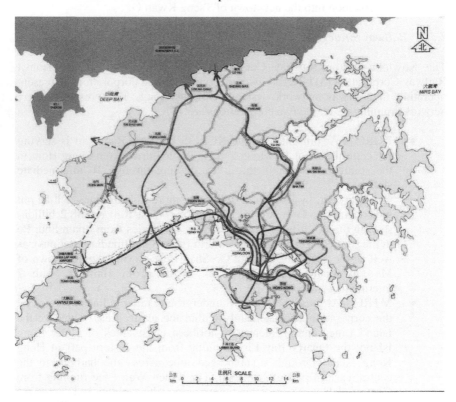

Figure 20   Strategy Railway Network under the 1996 Territorial Development Strategy Review.

To achieve the intended objectives, a number of specific railway schemes are short-listed for the formation of the basic network options. These include:

- The North Hong Kong Island Line – the existing Island Line of MTR will be extended and re-aligned to form a continual line of about 24 km along the north shore of the Island of Hong Kong. The sections to be extended will be the North Point-Central link (4 km) in the island centre and the Sheung Wan-Kennedy Town extension (3.8 km) serving the western part of Hong Kong Island.

- East Kowloon Line – this will be a new line connecting the MTR Diamond Hill Station to the KCR Hung Hom Station. It will serve the eastern part of the Kowloon Peninsula in which no railway service exists at the present moment. In a longer term, the line may be further extended to Hong Kong Island by the fourth harbour-crossing tunnel in the south and to the North-eastern New Territories through a new tunnel to Tai Wai station of the KCR East Rail in the north.
- Kowloon Southern Link – this is an extension of the West Rail to link with East Rail of KCRC through the southern tip of the Peninsula at Tsim Sha Tsui. This 4 km-long railway linkage is considered to be very critical in the overall network system, for it can physically connect two very important railway lines to provide convenient interchange for passengers between the East Rail to the West Rail.
- The Northern Link – similar to the Southern Link, the 12 km-long Northern Link will connect the East Rail to the West Rail on the north-western portion of the New Territories, such that the railway network will be able to have a completed circuit serving the majority of the strategic development sub-regions.
- The Regional Express Line – the purpose of this line is to provide an express rail service that links the urban area with the borders between Hong Kong and Mainland China, in addition with limited stops providing fast domestic service to the northern portion of the New Territories. This 35 km-long express line will be running mainly through tunnels crossing the hilly central part of the New Territories. The urgency to construct this express line is supported by the average 18% annual growth of cross-border passengers in recent years, which will soon over-saturate the already very busy cross-border check point at Lo Wu.

The strategic railway network as proposed above costs about US$12 billion, its implementation will of course depend on the actual economical and population growth, as well as engineering and environmental studies, that will be carried out from time to time in conjunction with the implementation and coordination of other railway projects.

7.4 New Towns

To cope with the increase in population and to improve the living environment by decentralizing the population from the over-crowded metro areas, Hong Kong has developed nine new towns since the initiation of her New Town Development Programme in the early 70's. The new towns by now accommodating a population of about 2.7 million and the programme is aimed to provide housing for at least 3.8 million people by the end of 2010.

The new towns development can be grouped roughly under three generations. Tsune Wan, Shatin and Tuen Mun developed in the early 60's up to mid 70's being the first generation of new towns; Tai Po, Fanling/Sheung Shui and Yuen Long in the late 70's being the second generation; and Ma On Shan, Tseng Kwan O and Tin Shui Wai in the mid 80's onward being the third generation of new towns. The basic concept in developing these new towns is to provide a balanced and self-contained community as far as possible in terms of provision of infrastructure, community facilities and other basic needs.

The development process for new towns, rural townships and new major developments in the metro areas has exhibited a lot of teething problems and detail planning is thus called for in a carefully coordinated manner. In general, for rural townships, the major concerns are for township improvement, provision of the needed infrastructure and community facilities, as well as sometimes flood protection. For new developments in the metro areas, new land will be formed to allow growth and to decant existing population, and to provide or upgrade facilities to enable the redevelopment of old and run-down areas. The functional, environmental and aesthetic aspects of the developments are given priority in consideration in the development process.

Figure 21 The Tuen Mun-Yuen Long Corridor on the North-Western New Territories.

Figure 22 The New Town of Tseng Kwan O as seen in early 2000.

Figure 23 The old urban centre of Kwun Tong and her neighbourhood.

Figure 24 View of the external façade of some old buildings in Kwun Tung constructed in the mid-60's.

Though the development of new towns may have significant differences in their development history, specific geographical environment and constraints, or a particular set of policies and criteria standards applied to their design according to the value and mind-set of the town planners and decision makers by the time the new towns were to be developed, it is beyond doubt that Hong Kong has obtained valuable experience in the development processes of these new towns.

Dating back to the early 60's when Tsuen Wan was being developed, where the town was just provided with the very basic urban facilities that could enable the subsistence of low-income settlers as factory workers. Then came the new town of Shatin in the mid-70's, where the concept of large areas of community spaces and bicycle paths were introduced linking between most public estates and communal centres. Finally to the new town of Tin Shui Wai or Tseng Kwan O in the early 90's, where a mass transit system is employed and the emphasis on a green and energy efficient township design becomes an appealing slogan. It is hard to comment whether there are drastic improvements exhibited in these generations of new town development due to incomparable background. It is still fair enough to conclude that planners and decision makers by now can at least avoid errors that we have frequently made in the past and gained the advantage of having a better chance to provide new township, which is more efficient and operational, and with higher quality in terms of environment and living standards.

8   CONCLUSION

Hong Kong can hardly be regarded as a self-sufficed system in developing herself into a modern and efficient city within the present reality of topography, territorial and population size. Hong Kong can only afford to work according to stringent constraints and limits. Within a short development history of 4 to 5 decades after the Second World War, and based on quite a lot of odd backgrounds, from unstructured ad-hoc schemes to the adaptation of a well-structured strategic planning process in development, it is a big step forward.

The process of development for sure is not a smooth exercise for Hong Kong. Experience, knowledge and technological know-how is only the least of the problems. Physical or economical limitations are the second level of concern. In fact, the key issue may come from the political environment which includes whether there is an effective decision making mechanism based on recognized legislative procedures acceptable by the general public, or whether it can balance the interests of major parties such as from investors, political parties, pressure groups or other voices from ordinary people represented by the media. In the past years, decisions were made in a closed-door manner where public consultation did not exist in practice. People had no way to input their ideas or raised comments to major policies or actions taken by the government.

Figure 25 Views of the border between Hong Kong and the Mainland near Lok Ma Chau.

Figure 26 Border at Lo Wu. The background in the photo is the Shenzhen Special Economic Region of Mainland China. Note the wetland on Hong Kong side which is by now mainly zoned as restricted areas due to internal security reasons.

However, the way to make important decisions has changed, the government nowadays has to carry out a lot of studies, produce piles of technical reports and proposals, conduct public consultation on various levels, prepare work schemes and implementation plans, or to lobby numerous concerning parties to obtain understanding and acceptance in an informal manner in order to pass through all the required legislation or executive procedures, most of which are controlled under independent councils or committees. Except that it takes a much longer time to have a plan implemented, the controlling exercise and function does ensure a high-quality, rationalized and balancing decision be made.

After the return of the sovereignty of Hong Kong back to Mainland China, interaction between the two places has been intensified, not necessarily on the political but on social and economical levels, that makes predictions for making future decisions become more difficult. For instance, should Hong Kong be acting more independently as she was without making necessary coordination with the Chinese government in territorial planning in case issues where regarding cross border matters are concerned? This is extremely important where transportation, environmental, or in the joint development of border areas are taken into consideration so that the two places may have a chance to merge sensibly together to form and function as a megacity as a whole. Hong Kong and her neighbouring city Shenzhen, which is only a few hundred metres north of the border, will have a total population of more than 12 million in the decade to come. We cannot afford to ignore this urging reality. The existing way Hong Kong is using the northern part of her territory with about 90 sq km of land being restricted areas and fenced-off from development is highly questionable. Hong Kong is not being tightly restrained, there is still plenty of room for innovative ideas to help shape the future of Hong Kong in a much more creative way.

Anyway, Hong Kong is somewhat on the right track heading for a more reasonable direction under relatively reliable vehicles. What we need to do is to make the necessary fine-tuning at the required moment, provided that the political environment within Hong Kong can maintain stability and rationality.

9   REFERENCES

Planning Department, the Government of the Hong, 1992
    METRO PLAN – THE SELECTED STRATEGY EXECUTIVE
    SUMMARY.

Planning Department, the Government of the Hong, 1996
    FINAL TECHNICAL REPORT ON TERRITORIAL DEVELOPMENT
    STRATEGY REVIEW 1995, PART I AND II.

Highways Department, the Government of the Hong Kong Special
    Administrative Region, 1998
    HONG KONG HIGHWAYS.

China Trend Building Press, 1998
    THE NEW HONG KONG INTERNATIONAL AIRPORT, edited by Martin
    Choi & Stefan Hammond.

W. M. Lo and L. T. Chan, 1999
    PRINCIPLES OF TOWN PLANNING PRACTICES IN HONG KONG.
    (Joint Publishing HK Ltd.)

Planning Department, the Government of the Hong Kong Special
    Administrative Region, 2000
    URBAN DESIGN GUIDELINES FOR HONG KONG.

Transport Bureau, the Government of the Hong Kong Special Administrative
    Region, 2000
    RAILWAY DEVELOPMENT STRATEGY 2000.

Transportation in the Built Environment: Cities in the Third Millennia

Robert Prieto

1.0 PREPARING FOR THE FUTURE

Transportation in the third millennia will undoubtedly evolve in ways few can even imagine today. Whether the visions embodied by the Jetson's personal spacecraft or the physics stretching concepts of the Star Trek's transporter beams become reality, none here today will ever know. But what we do know is that the challenge in meeting our transportation needs even in just the next century will require a combination of creativity and commitment seldom seen in peaceful endeavors.

The survival of cities and perhaps humanity itself is increasingly linked to our ability to move people and goods in an environmentally, socially and economically sustainable manner. This notion of sustainability or sustainable development must be the central goal of our development and redevelopment of our cities and the transportation networks that bind them together in the 21st century.

In this paper I will briefly look at:
  a. Transportation in the 21st century – A range of challenges
  b. Drivers of transportation form
  c. Opportunities transportation form creates
  d. Forgotten links
  e. A global roundup
  f. The 21st century opportunity

2.0   TRANSPORTATION IN THE 21ST CENTURY – A RANGE OF CHALLENGES

Transportation in the 21st century will be required to meet a range of challenges on a scale never faced by mankind before. These challenges run the gamut from:

a. Economic – Assembling the capital for new transportation development as well as replacement, renewal, maintenance and expansion of the transportation systems of today.

b. Social – Mitigating the impacts of construction while avoiding unintended impacts such as spurring development once construction is complete.

c. Political – Whether considered "political pork," "sacred cows" or "hot potatoes," transportation projects seem to engender the same degree of political attention as their agrarian namesakes.

d. Technological – Making transportation "smarter" is essential to providing the efficiency and flexibility our growing and changing cities of the 21st century will require. New technology brings with it new challenges. Do we understand what they mean?

Simply put, the transportation challenge of the 21st century, as shown in my presentation *21st Century Challenge: Infrastructure Reconstruction,* will be focused on achieving the goal of sustainable development. This is exactly the goal that will drive the redevelopment of today's existing urban centers, the cities spanning the millennia, as well as the development of the new cities and new towns the world's growing and changing population will demand. Transportation corridors are 100+ year assets and in the third millennia we will find that many of the corridors for the next millennia have already been set.

This intimate link between physical infrastructure and urban form will necessitate tomorrow's transportation systems to incorporate the requirements of tomorrow's cities in their planning, implementation and operation. Livable communities and open space will be transportation "issues" in the 21st century as well as transportation's ability to shape urban density and promote the efficient use of energy through mass transit solutions. These issues, the very challenges that the cities of the third millennium will face, will bind the development of cities and infrastructure, particularly transportation infrastructure in unprecedented ways.

Figure 2.1 shows this intimate linkage between transportation and the challenges tomorrow's cities face. More importantly, it recognizes the special opportunity created by the creation of the transportation corridors of the third millennia. This opportunity is to create broader-based infrastructure corridors that allow the needs of the cities' other linear infrastructure to be met. Perhaps as we move through the third millennium we will see these new linear corridors become the logistic backbone of increasingly "smart" urban environments.

The transportation challenges our cities will face in the 21st century will depend heavily on whether they are one of the cities spanning the millennia or the new towns and new cities that changing economic models and demographics will demand. Let us look at each in turn.

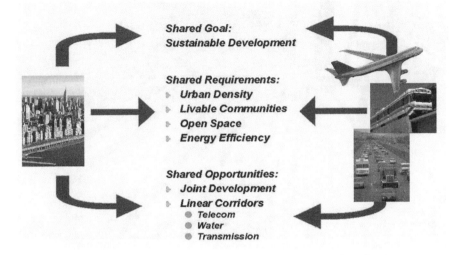

Figure 2.1   The Linkage Between Transportation and Tomorrow's Challenges.

2.1   Rebuilding the Past – A Generational Opportunity

The cities of yesterday and today will be many of the cities of tomorrow. Barring those destroyed by war, rising sea levels or cataclysmic earthquakes these cities will face the dual challenge of developing a sufficiently overarching vision of their future as well as implementing this vision while meeting the demands of the city as it undergoes the required transformations.

2.1.1   Central Artery – The Challenge Met

One only needs to look at Boston's Central Artery to understand the enormity and complexity of this challenge (Figure 2.2). Likened to performing open-heart surgery on someone who is running the Boston marathon, this project has in many ways served as the test-bed for the challenges that lie ahead for all of our cities as they move into and through the third millennium.

Figure 2.2   Central Artery Under Construction; Boston, Massachusetts, USA.

In its efforts to create the next significant step in an enduring urban transportation network, it has to address the significant issues all of our cities will face as they undergo similar transformations. At each step in the construction process, we had to maintain and enhance the livability of the community. The depression of the Central Artery will create new open space tying downtown Boston and a number of its urban communities to the waterfront which has been a key ingredient in that city's vitality since its founding. Preserving that newly created open space and avoiding the temptation to increase urban density are part of the balancing process that the cities of tomorrow must successfully address (Figure 2.3). In meeting this challenge, Boston seized on a generational opportunity to begin the development of more broadly based and coordinated infrastructure corridors. Undertaken in many instances to clear the way for construction, these coordinated "utility corridors" are small scale models for the opportunities available as we build tomorrow's new city infrastructure (Figure 2.4).

Figure 2.3 Open Space Created by the Central Artery/Tunnel Project; Boston, Massachusetts, USA.

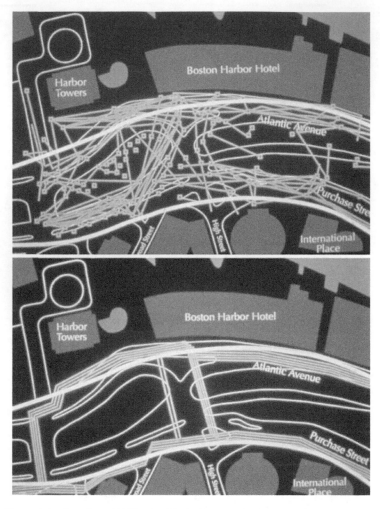

Figure 2.4   Coordinated Utility Corridor in the Central Artery/Tunnel Project; Boston,
Massachusetts, USA.

This project's recognition of the generational opportunity went well
beyond the immediate needs of the transportation project, with several billion
dollars of other "enhancements," some to satisfy the needs of the current population, a price paid, if you will, for undertaking this long-term endeavor, but
time will show many of these enhancements to be demonstrations of prudent
foresight.

2.1.2 Reconstruction, Replacement, Expansion – A Different Kind of Challenge

By no means is the Central Artery unique in its response to this generational opportunity. Other recent and current models of meeting the generational opportunity created by rebuilding the transportation infrastructure of tomorrow span the globe and include:

a. Frankford El – Where the need to keep Philadelphia "running" while a key element of its transportation system was substantially rebuilt, necessitated limiting construction activities to nights and weekends. Through the entire 7-year construction period, this "temporal" separation resulted in only one rush hour period being adversely impacted. The rebuilding of this 5.2-mile double track rail system built circa 1919 was accomplished without disrupting business below (Figure 2.5).

Figure 2.5 Frankford El, Philadelphia, Pennsylvania, USA.

b. Fort Washington Way – Spatial separation, combined with a fully coordinated and comprehensive approach to economic revitalization of this downtown area, have resulted in enhanced vehicular access and pedestrian movement during the construction period and major commitments of private capital for new economic development such as a new stadium for the Cincinnati Reds. This 6–8 year project was fast-tracked to 36 months and $80 million in project savings put into additional urban improvements (Figure 2.6).

Figure 2.6   Fort Washington Way, Cincinnati, Ohio, USA.

c. Melbourne City Link – The need to enhance transportation infrastructure in this established urban area presented a generational opportunity to create a "gateway" to the city while at the same time spatially separating a portion of the city's infrastructure from the immediate environment of its inhabitants (Figure 2.7).

Figure 2.7   City Link, Melbourne, Australia.

2.1.3   Scale – Challenge and Opportunity

The Central Artery project demonstrates what must be accomplished on a "city scale." However, the cities of the third millennia will be perhaps less distinct in their boundaries as today's cities grow together. This blurring of city boundaries will shift the challenge – and opportunity – to a regional or national scale. This is exactly the challenge emerging today in the United Kingdom where 30 years of disinvestments in that nation's rail infrastructure has created a political, economic and technical challenge the scale of which has been seldom seen. It has, however, also created an unprecedented opportunity to reshape not only that nation's rail network but also to redefine the very way in which people live and economic activity is carried out.

To fully appreciate the scale of the challenge ahead, one needs to only look at one of its many major programs; The West Coast Route Modernization (Figure 2.8). The WCRM program, as discussed in my presentation *Renewing British Rail: West Coast Route Modernization* presented at the New York Academy of Sciences, includes:

- 10 Major Subgroups
- 86 Major Projects
- 500 Individual Construction Projects
- Over 5000 Line Items to be Tracked

Figure 2.8 West Coast Route Modernization; United Kingdom.

Together, reconstruction of the WCRM infrastructure will cost approximately $9 billion or about 15% of the total a full upgrade of Railtrack's infrastructure would cost.

These costs exclude the costs of new train sets which will be borne by operating companies such as Virgin and further exclude the investments that will occur at the principal opportunity areas created by the project, namely at its principal city stations. Improved transportation service, coupled with farsighted investments at major stations, will provide a generational opportunity to redefine the core urban areas of the major cities of an entire nation.

That opportunity is still ahead. The challenge today is to rebuild the railroad during possessions as short as a few hours, all the while keeping the railroad operating.

Scale is both the principal challenge that building the transportation infrastructure of the 21st century must overcome as well as the principal opportunity successfully meeting this challenge will create.

2.2 Building the Future – Our Best Chance To Get It Right

The new cities and new towns of the third millennia will develop along either existing intercity transportation links or will cause new such links to be built. The issues and challenges associated with new development are dramatically different and in some ways more complex from those facing existing cities that will span the millennia.

These challenges are being faced in diverse locations from Southern California to the Philippines to Taiwan and elsewhere. The range of issues varies widely from project to project, but the opportunity remains the same – the opportunity to get it right the first time.

It is worthwhile to look at these projects to better understand the challenges and opportunities ahead.

2.2.1 Not In My Backyard

One of the principal challenges the new transportation links of tomorrow will face is getting past a series of environmental hurdles perhaps best described as "not in my backyard." A good example of the hurdles tomorrow's transportation links will face is represented by SR125, the so-called San Miguel Mountain Parkway. Begun in 1991 following award of a franchise under innovative state legislation to promote private development of new transportation projects, this project has so far survived a grueling 10-year environmental review and permitting process.

The SR125 project would stretch 9.5 miles connecting State Route 905 near the border crossing at Otay Mesa to a connection with the regions freeway system. Figure 2.9 shows the general alignment of the project while Figure 2.10 provides a brief chronology of the project development and permitting challenges, the project has faced.

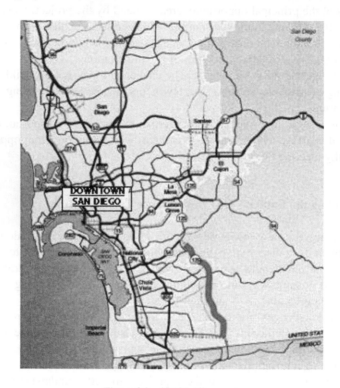

Figure 2.9   SR125 Alignment.

Completion of SR125 would open a new area of Southern California to development while enhancing NAFTA driven transportation links. These opportunities for new town creation and enhanced access to other, more recently completed developments are the very points of challenge in both the political and environmental processes. The absence of a broader, long-term development vision, clearly addressing the intimate links between development, transportation and other infrastructure needs creates a vacuum which raises project risk profiles. In order for the new cities and towns of the third millennia to efficiently develop these visions, long-term integrated plans showing long-term growth and urban evolution will need to be articulated and the necessary political, economic and social frameworks and consensus put in place. "Banks" of developable transportation projects should be created in conjunction with master plan approvals lest someone seem surprised later when the required road or rail network is to be built. Simply put, we must protect and preserve corridor right-of-way.

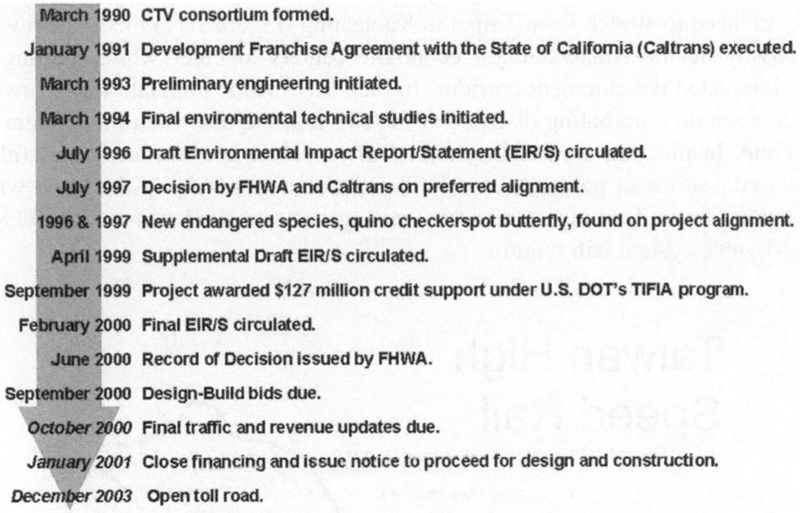

March 1990	CTV consortium formed.
January 1991	Development Franchise Agreement with the State of California (Caltrans) executed.
March 1993	Preliminary engineering initiated.
March 1994	Final environmental technical studies initiated.
July 1996	Draft Environmental Impact Report/Statement (EIR/S) circulated.
July 1997	Decision by FHWA and Caltrans on preferred alignment.
1996 & 1997	New endangered species, quino checkerspot butterfly, found on project alignment.
April 1999	Supplemental Draft EIR/S circulated.
September 1999	Project awarded $127 million credit support under U.S. DOT's TIFIA program.
February 2000	Final EIR/S circulated.
June 2000	Record of Decision issued by FHWA.
September 2000	Design-Build bids due.
October 2000	Final traffic and revenue updates due.
January 2001	Close financing and issue notice to proceed for design and construction.
December 2003	Open toll road.

Figure 2.10 SR125 Project Milestones.

2.2.2 Show Me the Money

If you are able to clear the environmental hurdle, you have in reality only reached the starting block. Mobilizing the money required to design, construct and operate the project until cash flows from project revenues turn positive is the next project challenge. In many ways, these new transportation links face even more risks than those associated with reconstruction and renewal in today's existing urban centers. Demand for the new facility is often closely tied to the success of the very new developments it is designed to serve, creating in effect, an indirect market risk, which is way beyond the capability of the transportation links builders and financiers to assess or mitigate. The true mark of success in not opening day traffic or ridership, but rather what has happened to the planned development 25 to 50 years out!

These long time scale development risks, as discussed in my presentation to the APEC Roundtable on *Best Practices in Infrastructure Development*, significantly raise the bar on public funding support for such projects and in effect raise the cost of risk capital.

One current example of a major new transportation link tying existing cities together while opening up the potential for substantive new development along its route is Taiwan's new high speed rail system.

Planned to stretch from Taipei to Kaohsiung (Figure 2.11) this system will effectively tie the island's major economic centers together while creating a new, integrated development corridor. In fact, 95% of the population of Taiwan will live within commuting distance of the two major cities when this system is complete. In meeting its challenge, the Taiwan HSR consortium successfully addressed a series of political, economic and technical hurdles to be faced with perhaps its most daunting challenge – raising the estimated $14 billion (U.S.) this 207-mile system will require.

Figure 2.11   Route for the Taiwan High Speed Rail System.

Figure 2.12 shows the selected technology for Taiwan's new high speed rail system.

Figure 2.12   Taiwan High Speed Rail Car.

2.2.3   A Road Paved With Dreams

Opening up the new towns of the third millennium will not generally begin with the construction of new major metro or inter-city rail systems. Rather, it will start with initial high capacity road links that carry with them the dreams and vision of a new tomorrow; a chance to avoid the mistakes and problems that today's cities face. The Metro Manila Skyway is a good example of a development linked infrastructure project (Figure 2.13).

The Metro Manila Skyway links the Makati business district to the commuter suburbs of Sucat and Alabany as well as the southern area of Cavile in Luzon. The new 12.6 km, 3-lane toll road is both elevated and at-grade and is essentially a substantive upgrade of an existing 20-year-old at-grade toll road. We should not lose sight of the advantage created by the revision of an existing linear corridor and we should reflect on the difficulty in creating newly assembled similar corridors in the future.

Figure 2.13   Metro Manila Skyway in the Philippines.

Development of the project was government supported and in many ways, government assisted in overcoming many of the early phase hurdles such projects face.

The role such projects play in helping create the new cities and towns of the third millennia can be seen in the first stage of project development where access from the commuter suburbs was reflected in floor space values as well as in residential values. Integration of the toll plaza into the Belle Makati development helped focus value enhancement and represents a clear demonstration of the symbiotic links between new transportation systems and tomorrow's new cities and towns.

Other examples exist worldwide and the impact of projects such as the 90 km, 4 lane Chengdu–Mianzany Expressway which links upwards of 20 million people in China bear watching.

3.0   DRIVERS OF TRANSPORTATION FORM

Transportation form in the third millennia will be driven by a combination of requirements, with sustainable development as an overarching goal and requirement. This overarching goal of sustainable development will directly impact the urban form and by extension, the transportation systems required to serve the cities of the third millennia.

However, sustainable development and its concomitant emphasis on fostering an attractive quality of life will not be the only driver of tomorrow's transportation needs. Other principal drivers will include:

a. Speed and agility to implement and respond to the demands of more dynamic knowledge based processes.

b. Technology that will enable "smart" transportation links in the future. Improved efficiency and safety will be matched by a blurring of modal distinctions.

c. Evolving economic and social models will capitalize on distributed knowledge and customer bases tied together with transportation enabled integrated logistics.

d. Changing competitive requirements of third millennia cities driven by the globalization of all human endeavors.

New fully integrated, multimodal transportation links will become integral requirements of the cities of the 21st century as well as principal drivers in the evolution of those cities. In 1999, Fortune ranked the best metropolitan areas for business, based on comparative statistical data and a survey of 1,200 executives worldwide. The Dallas metropolitan area ranked first because of its air transportation hub, which links the city not only internally, but on a national and global basis (Figure 3.1). Similar efforts at the recently connected Austin Bergstrom Airport (Figure 3.2) attempt to recreate this city linked transportation hub concept on a regional basis. New York City's East Side Access project (Figure 3.3) seeks to provide enhanced links between that city's major rail networks while San Francisco's BART system is being extended to the airport (Figure 3.4). In England, the Channel Tunnel Rail Link (CTRL) helps tie England's transportation network into the major transportation destinations in Europe.

Figure 3.1   Dallas/Fort Worth Airport; Texas, USA.

Figure 3.2   Austin-Bergstrom Airport, Texas, USA.

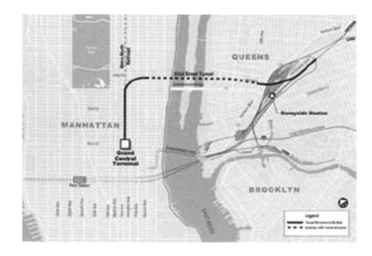

Figure 3.3   East Side Access Project, New York City, USA.

Figure 3.4   BART, San Francisco, California, USA.

High-tech clusters in locations such as Austin, Raleigh-Durham and Portland will require transportation solutions more in balance with their urban form as well as linkages to adjacent clusters (Figure 3.5).

Figure 3.5   Light Rail System, Portland, Oregon, USA.

In other city settings, technology will provide for enhanced transportation access with improved electronic tolling on recently completed or ongoing projects such as those in New Jersey, California and Massachusetts (Figure 3.6). Traveler information systems such as those proposed by TravInfo, a proposed ITS system serving the San Francisco Bay area provides a foundation for the agility the transportation networks of the 21st century will require. These systems also open the door to other forms of user changes to provide better equity in paying for highways.

Figure 3.6   E-ZPass Toll Booths in New Jersey.

4.0   OPPORTUNITIES TRANSPORTATION FORM CREATES

The form of tomorrow's transportation link will be shaped by the cities and towns they must serve. However, in defining these links we must not lose sight of the value created at major transition points (traditional joint development opportunities) as well as the value created <u>along</u> the link itself. Before focusing on this last opportunity, it is worth reiterating that significant access-enhanced development opportunities are created by each major transportation project. In many instances, these transportation projects are the best indicators of future opportunity. This has been discussed in my presentation, *Infrastructure Improvements as Early Indicators of Development.*

Returning to the opportunities created along the link itself, linear transportation corridors, in fact, linear corridors in general, represent a property form that will be increasingly difficult and expensive to assemble. This is particularly true of cities spanning the millennia, but will be equally true in the new cities and towns of tomorrow as well as with regard to their links to established economic centers. Opportunities such as those afforded the original railroad companies in the western United States (generous right-of-way and mineral rights) will be few and far between.

Having assembled such a linear corridor, we are afforded the opportunity to maximize its utilization in a coordinated and planned way. This foresightedness is unfortunately absent in many of our master plans today, although recent trends are promising.

Some examples of the opportunity linear transportation corridors can provide for the development of cities include the Central Artery utility corridors (see Section 2.1.1) and Fort Washington Way comprehensive redevelopment (see Section 2.1.2).

Other examples, not resulting from pre-planning but rather driven by the needs of other utilities or transportation facilities include the wide array of new fiber optic network systems currently being developed around the world as well as additional transportation facilities retrofit into existing corridors. A clear example of the opportunity these linear corridors represent can be seen in the 15,000 mile Level 3 fiber optic network in the U.S. where thousands of miles of fiber were laid either in existing rail or road right of way. Where parcel-by-parcel assemblage of right of way was required, the process can best be described as time consuming. Even with the advantage presented by being on existing transportation facility right of way, progress was paced by the need to obtain different permits as the corridor moved from jurisdiction to jurisdiction. The major transportation corridors of tomorrow need to be approached in a more coherent manner. The need for unified standards and regulatory approvals is becoming increasingly evident as we strive to tie our cities together.

The opportunity for joint transportation development of a corridor can be seen in many places around the world, but let me use an example close to home to illustrate the value these pre-assembled linear transportation corridors can have.

The JFK Airport Access Program (Figure 4.1) will provide access between the various terminals at JFK and both the Long Island Railroad and New York City transit systems. This $1.5 billion, 8.4 mile, 10 station system will leave airport property and travel along the median of the Van Wyck Expressway on its way to its Jamaica Station terminus. People mover structures will utilize the existing Van Wyck Road median during this 2-mile leg of the system.

Figure 4.1   Route for the JFK Airport Access.

5.0   FORGOTTEN LINKS

Modern highways, state-of-the-art metro systems tying a city and its suburbs together, light rail for emerging cities and towns of the third millennia and inter-city high speed rail are all-important transportation links as we move into the 21st century. But the list is not complete. More are required, including:

a. Airport links, responding to the changing roles of airports as both city gateways and shopping centers with runways. Intermodalism in the third millennia will consider air travel just another mode of moving people and goods to and from a growing mega polis.

Recent airport and rail expansions at San Francisco, New York (JFK), Washington, D.C. and Orlando, and a host of other airport systems worldwide, have recognized not only the value of airport people mover systems (Figure 5.1) but also that of direct links to the downtown metropolitan area (Figure 5.2).

b. Ferry links, revitalizing a historical transportation mode in many of the world's port cities. Links in Boston tie the downtown and airport together by water shuttle while more recent links helped by providing an additional altern-ative during the rebuilding of major portions of that city's infrastructure. Major facilities such as those in Hong Kong and Macao (Figure 5.3) will grow in importance as all alternatives are explored in the 21st century.

Figure 5.1 People Mover at Orlando International Airport.

Figure 5.2 WMATA Station at Ronald Reagan Washington National Airport.

Figure 5.3   Macau Ferry Terminal, Hong Kong.

c. <u>Environmentally friendly alternatives</u> such as those typified by the Des Plaines Bike Trail (Figure 5.4) and the sky bridges and pedestrian walkways found in major cities and growing towns worldwide.

Figure 5.4   Des Plaines Bike Trail, Lake County, Illinois, USA.

6.0   A GLOBAL ROUND-UP

The scale of change in the transportation sector has never been greater. Throughout the world major new transportation links tying cities together are underway at unprecedented levels. Transportation investment is on the rise in both developing and developed countries and the prognosis for continued investment in the medium term remains good. More importantly, the links between these projects and broader sustainable development goals seems to be firmly entrenched in many areas of the world and the trend is growing.

Each of the projects listed in Appendix A creates a host of city shaping opportunities ranging from the immediate opportunities created by joint development at key modal transition points to the longer term ones associated with fundamental rearrangement of the city's economic and social fabric.

The lists are by no means comprehensive nor do I attempt to put a price tag on the planned improvements, but needless to say, the scale of the challenge as well as the opportunities it creates are enormous.

7.0   THE 21ST CENTURY OPPORTUNITY

Today's challenges must be addressed for the opportunities of the 21st century and the balance of the third millennia to be realized.

7.1   Some Viewpoints To Consider

As I addressed in my presentation *Next Steps: Implementing Solutions for Easing Congestion Points in the APEC Region*, best practices for meeting tomorrow's transportation needs while avoiding congestion, can be described from the viewpoints of the user, operator, regulator and provider. Each of these is an important viewpoint:

- The <u>user's</u> concern of transit time, cost and reliability cannot be ignored.
- The <u>operator's</u> viewpoints on productivity, revenue and profit are essential.
- The <u>regulator's</u> viewpoint on safety and environmental and social impact provide an important and necessary framework, and
- The <u>infrastructure provider's</u> requirements for asset utilization and return on capital employed – more simply put – <u>profit</u> are necessary prerequisites.

Let me suggest, however, that "best practices" need to recognize two other views:

a. The first of these is the view of long-term capital markets. There is a compelling need for efficient use of capital.

Demand driven by the growth rates we anticipate and desire in the third millennia are too high to tolerate mistakes.

While short-term market forces will drive solutions, reward success and punish failure – too many failures, especially early on, will drive up the cost of capital or just plain make it unavailable for projects down the road.

This highlights the need for long-term system wide planning in each economy and particular attention on a regional basis to the interface points and interaction between the various regional systems.

b. A second view to further look at "best practices" is to recognize the strategic relationship between infrastructure and development, a view of particular interest at this conference. The transport network is the backbone of infrastructure. Congestion results when investment in development and investment in infrastructure are mismatched. The call for coordinated infrastructure and economic development planning is loud and clear in my opinion. We must not lose sight of their interrelationship. We must not lose sight of a simple fact – they are competing for the same sources of capital. There can be no winners or losers. There must be balance.

Coordinated infrastructure and development planning, as discussed in *Engines of Growth? Asia and the Future of World Business*, can work in tandem to maximize overall economic efficiency – not just efficiency in the use of capital in the transport network.

7.2   Realizing the Future

The future is ours to define and achieve. To realize the opportunity that a strong link between transportation and building of the cities of the third millennia can create, we must begin now. We must remember that our opportunity in the 21st century is to:

a. Get it right or live with it for the next 30–50 years and beyond.

b. Recognize urban form is not static, so plan for the evolution of urban form.

c. Clearly recognize the symbiotic linkage between transportation and development and to plan for it.

d. View deferred maintenance for what it is – a generational transfer of cost. A simple analogy would have you eating at the best restaurant in Melbourne for your entire life and leaving the bill to your kids. Deferred maintenance must be funded as you consume it!

e. Mobilize capital from all sources. Focus on the risk/reward balance and government's role in helping get the balance point right, whether through "developable" project banks or project bundling for risk mitigation.

The builders of tomorrow's cities and tomorrow's transportation links must have a <u>shared</u> vision, with a clear focus on what it means and what it will take to achieve it.

Together we must drive this shared vision to fruition.

REFERENCES

Prieto, R., 1998
INFRASTRUCTURE IMPROVEMENTS AS EARLY INDICATORS OF DEVELOPMENT.

Prieto, R., 1997
NEXT STEPS: IMPLEMENTING SOLUTIONS FOR EASING CONGESTION POINTS IN THE APEC REGION, presented at the 11th APEC Transportation Working Group Meeting.

Prieto, R, 2000
21ST CENTURY CHALLENGE: INFRASTRUCTURE RECONSTRUCTION, presented at the IMPACTS Conference.

Prieto, R., 2000
RENEWING BRITISH RAIL: WEST COAST ROUTE MODERNIZATION, presented at the New York Academy of Sciences.

Prieto, R., 1996
BEST PRACTICES IN INFRASTRUCTURE DEVELOPMENT, presented at APEC Roundtable.

Prieto, R., 1998
ENGINES OF GROWTH, ASIA AND THE FUTURE OF WORLD BUSINESS, Back to the Future: Creating Long Term Growth.

Social Infrastructure in Tall Buildings:
A Tale of Two Towers

Carolyn Whitzman, MA, MCIP

Two towering housing complexes, each famous in their respective cities, with different origins and very different fates, illustrate both the challenges and the possibilities of high rise housing, and the importance of social infrastructure in tall buildings. One is St. James Town, designed in downtown Toronto in the 1960s as swinging singles apartments for the newly affluent youth market. It is now a byword throughout Canada for overcrowding, as the highest density neighbourhood in the nation, and is also considered a sinkhole of crime and poverty – although, as I will discuss, this cliche obscures as much as it illuminates. The second example is Trellick Tower in central London, constructed as public housing in 1972, and still the tallest housing block in England. While it was once seen as the perfect example of the perils of modernism in the 1980s, when drug dealing and petty crime was rife in its corridors, it is now listed as architecturally significant, and a two bedroom flat there can be bought for a mere 170,000 pounds, or approximately $400,000 Australian dollars.

1   WHAT IS SOCIAL INFRASTRUCTURE?

But before I discuss these two case studies, we should start off with discussing 'social infrastructure', since it is a term that many people use, but few people bother to define. 'Infrastructure' itself means "the basic framework of a system or organization", while the prefix 'infra' itself, means "below or beneath". So infrastructure is what is below or beneath a structure, in this case high rise buildings. Looking again to the dictionary, 'social' is defined as "of or pertaining to society and its organization" (Funk and Wagnalls, 1980; my 1971 edition of the *Oxford English Dictionary* contains no separate entry for infrastructure, a sign, perhaps, of it being a relatively new catchphrase). So "social infrastructure" can be seen as the basic framework of services pertaining to the organization of society, underlying any human settlement.

What do these services include? Almost anything pertaining to people's socialization – the way they become an active and productive part of society; their social welfare – their emotional and physical well-being; and their social life – what happens apart from the basic survival functions of work, eating, and sleeping. So social infrastructure includes schools, child care centres, libraries, and other places that children learn how to use skills they will need as adults. It also incorporates adult education facilities, literacy centres, multicultural and multilingual centres, services for people with disabilities: places where adults continue the life long process of learning. Hospitals and health clinics, parks, playgrounds, sports and recreation centres – places that help keep people healthy – are vital components of social infrastructure. But so are community gardens, community kitchens, parent-child drop-in centres: places that promote physical and mental health in their broadest meanings. Social infrastructure includes the theatres, art galleries, music halls, and movie houses that bring people together for arts and entertainment. It can also include seniors' centres, home work clubs, computer and Internet cafes, houses of worship, workshops, laundromat ... anywhere which brings people together. Even the common bench is a vital piece of social infrastructure (Marcus and Sarkissian, 1986).

Social infrastructure is often given the term 'soft' services, to be contrasted with the 'hard' services underlying a settlement, such as roads, water, sewers, electricity, and heating pipes. 'Hard' and 'soft' are suggestive terms, hinting at the 'hard-headed' reality of the importance of providing sewers, roads, and water, in contrast to the 'soft', 'wet', 'bleeding heart' or weak arguments in favour of providing adequate schools, recreation, and health care. In reality, social infrastructure is the poor cousin of physical infrastructure. While there are building codes which state clearly to developers how much sewage capacity or electrical outlets must be provided in a given building, there is no corresponding 'social code' which would state, for instance, how much common meeting space should be provided in an apartment building, or guaranteeing a certain amount of affordable grocery stores per hundred units of offices. While there are certified professionals, such as engineers, trained to create and inspect physical infrastructure, planners, architects, and other relevant professionals are rarely given a solid background in social infrastructure before being entrusted with providing it. Whereas, in advanced industrialized countries such as Australia and Canada, we do not now let people move into housing before there are sewers, roads and water in place, we often let people move into new housing years before the necessary schools, recreation centres, and health care facilities are built. There are entire districts of high rise office buildings where the thought that people might have to shop for groceries before, during, or immediately after their work day was not even considered (for instance, St. Kilda Road in Melbourne, Huxley, 1994). As a society, we know far less about social infrastructure than we do about physical infrastructure. Certainly, from a legislative standpoint, we seem to care much less as a society about these 'soft' services.

Yet, time and time again, housing – and the settlements that contain them – succeed or fail because of the quality and quantity of social services. What is more, consumers in the advanced industrialized countries are placing an ever greater emphasis on 'quality of life' concerns, such as recreation, tourism, education, and health care, leading to increased jobs in that sector. Businesses in key growth sectors of the economy are basing decisions on where to locate business on 'quality of life' issues largely determined by social infrastructure (UNDP, 1996). My hometown, Toronto, is often named one of the best cities in the world to live and work in by corporate analysts and business magazines (Fortune Magazine, 1996; Corporate Resources Group, 1995). This is because the city is considered safe, stable, clean, and green, not because of the traffic (which is terrible) or taxes (which are high compared to cities in England and the United States). What this means is that there are tremendous economic opportunities related to promoting social infrastructure, and terrible fiscal, as well as human, costs related to dismissing or downplaying it.

Before I return to the question of how we – as architects, planners, engineers, social and physical scientists – might work together to promote high quality social infrastructure in tall buildings, I would like to turn to two housing-related case studies to illuminate some dos and don'ts.

2 TWO TOWERS

2.1 St. James Town, Toronto Canada

St. James Town was built between 1959 and 1976 in central Toronto as a "city within a city" to house affluent young singles. It was a classic urban renewal project, involving the erasure of a 19th century working class neighbourhood that was felt to have become degraded in order to build housing for "the way we live now" (Whitzman, 1991; Collins and Silva, 1995; Doucet and Weaver, 1991). The 8.2 acre site eventually held almost 7,000 units in eighteen buildings, each between 16 and 33 stories. Even as singles housing, this would result in a high-density neighbourhood. With its present 'legal' population of 12,000 residents, as well as countless apartments whose occupancy exceeds the building code, the density is at least 400 persons per acre.

The story of the development of St. James Town typifies a cosy relationship between private developers and local government that is, unfortunately, quite common in not only Toronto but many cities. It also illustrates assumptions about families, housing types, and social life, that are also typical, in this case of society fifty years ago. In the 1950s, Toronto's city government became concerned about housing the demographic bulge now known as the "baby boom". It was able to forecast a potentially large market of "upwardly mobile singles and professional couples" that would be seeking housing in the next 10 to 20 years. The prediction was that young people who had grown up in suburbs would want to live downtown during their single years, then presumably move back to the suburbs to marry, buy a house, and bear 2.4 children. In a neighbourhood of aging 19th century houses, many converted to rooming houses, the city government and a large local developer saw a prime opportunity for 'urban renewal'. City government assisted the developers by taking over the management of 250 'doomed houses' and relocating the tenants to new public housing in the neighbourhood. At one part of the site, the approval of three 24 storey buildings was obtained for $22,500 and the promise of a parkette in the vicinity, which was then quietly built over. The federal government funded the construction of the first two buildings in St. James Town, and provided low-interest mortgages for the next nine buildings, constructed between 1964 and 1969, in return for a rather vague promise of "recreational facilities". Finally, the provincial housing authority contracted the developers to construct four further buildings in 1969-70, for use as public housing.

It should be noted that a considerable amount of 'social infrastructure' was planned into St. James Town. It was developed as a classic modernist "city within a city". There was retail space including space for ground floor grocery shopping, banking, and medical facilities. Comprehensive recreation facilities included indoor swimming pools, saunas, TV and record rooms, and raised jogging tracks. There was an extensive network of underground garages, and enough open space for several football fields. The school was literally the only structure in the nine square blocks of St. James Town that was not torn down. Surely it had everything a swinging single or couple could want, including proximity to downtown jobs and nightlife.

But by the time the complex was finished, tastes had changed among this demographic bulge. 'Swinging singles' found the lack of convenient shopping and street life in St. James Town a distinct 'turn off'. The surrounding working class neighbourhoods, in no better or worse shape than the houses in St. James Town condemned as unliveable, were being bought up by the same young upwardly mobile professionals in a process then known as 'white painting' and now called 'gentrification'. The apartments, especially those constructed as public housing, became known as accommodation of last resort for working class and immigrant families, including those who had been driven out by the construction of St. James Town and the gentrification of the surrounding neighbourhoods.

By the late 1980s, the physical infrastructure in St. James Town had become completely degraded. The elevators were broken down more often than not, the trash collection system was dysfunctional, the front doors were unguarded, locks were left broken for weeks, and the underground garages were strewn with abandoned cars. The social infrastructure was in no better shape. The swimming pools and community rooms had been closed for years. There was no real space for children to play. The remaining shops were overpriced, and the banks had moved away. Not surprisingly, tenants were concerned about crime, drugs, and prostitution, although the top complaint was of poor management and maintenance (Rahder, 1988). Tenant activists complained of systemic harassment, including the windshields of tenant leaders' cars being bashed in, and sexual favours being asked in return for apartment repairs.

Now this is an extreme example, but it does point to the problems that the private sector has in providing social infrastructure and public goods to low-income tenants. The initial provision of social infrastructure was good. But after the area became predominantly low-income, the private sector – the developers themselves, the banks and the shops, all cleared out. With the loss of commercial enterprise, there were few local employment opportunities. A classic vicious circle was created: the kind of middle-class residents who would attract local business would never be attracted to a place without such goods and services.

There were also clear problems with the planning and design of the project. The most obvious is the assumption that only one age group – young adults – and one income group – the upwardly mobile middle class – would live in the housing. As long as the residents were middle class, they could pay for the upkeep of the privately owned recreational facilities, and support the large commercial units within the shopping centre. As long as they were childless adults, it didn't matter that the developer did not improve the local school, build a promised parkette, or assign the swathes of open green space to specific purposes. Once poor people of varying ages moved into St. James Town, however, the stage was set for erosion of recreational and shopping facilities, and tensions over use of open space between teens with nothing to do but hang around and seniors who were intimidated by the youths' presence.

The quite common practice of privatizing the public streets in St. James Town, and consolidating them into a traffic-free 'monster block' also had negative repercussions for the development. There was an attempt in the late 1980s to buy back street right-of-ways from the developers, in order to improve lighting and garbage collection, combat perceived isolation of the neighbourhood through re-establishing through streets, and add some social mix through the construction of street-side townhouses on vacant land. But this would have cost several million dollars per block, and was quickly abandoned in favour of more pressing concerns. All levels of government had given enormous financial concessions to the developers of St. James Town in order to get it built. But once problems appeared, governments had virtually no say over the way the buildings were managed. The owners said that if the city enforced repairs, it would pass on the bill to the tenants, which would eliminate what had become a large source of low-income housing. The owners also refused to sell the privately-owned 'common' space to the local government for much-needed social and recreational uses. The public and non-profit sectors were forced to focus on what little publicly owned land was left. A new recreation centre has been constructed as part of the redevelopment of Rose Avenue School. There has been extensive renovation of the four public housing buildings in St. James Town, including new playgrounds. But, other than the occasional "inspection blitz" in the privately owned buildings, there have been few successful attempts to address the management and social issues which were of most concern to tenants.

Despite high density and problematic infrastructure, there is much about St. James Town that works. First and foremost, St. James Town provides 7,000 units of low cost downtown housing to many of the people who need it most. New immigrants, seniors, people with disabilities, and families with children are all close to schools, work, and downtown community services. There are a number of long time residents, who have given some stability to the building and its tenants' association. The high proportion of Tamil and Filipino newcomers means that services in these languages – English as a Second Language course, cultural activities – are provided in the newly constructed meeting rooms in the school, or in nearby community centres. Stores providing specialized groceries have sprung up on nearby Parliament Street, where smaller and less expensive commercial units are available. Drug dealing and other crimes have declined in the publicly owned buildings, where the underground garages, entrances, and playgrounds have been redesigned, and community involvement in crime prevention has been promoted by the local and national governments.

At the height of St. James Town's problems in the late 1980s, there were those who said that the only solution was to tear down St. James Town and start again. I would now like to turn to another case study, this time in London, England. There, an apartment building underwent the same diagnosis at approximately the same time. But Trellick Towers has undergone an even more remarkable transformation in the past 20 years.

2.2   Trellick Tower, London England

Trellick Tower was designed by a renowned architect and won numerous design awards, which is often a recipe for disaster. In the words of Short (1989), "It is a common belief that buildings which win architectural awards begin to fall apart after two years and send their occupants crazy after four" (see Carroll, 1999 and Gibson, 1996).

After the second world war in England, chronic underhousing of the working poor assumed crisis proportions. According to Alan Gibson's short history of Trellick Tower, "around four million homes were destroyed during the war – some 35 percent of the housing stock". The 1951 Census found 5 million people across the country still dependent on public washrooms for baths.

In response, a massive public housing program was embarked upon by the national government, in conjunction with local councils. Between 1952 and 1976, an average of 145,680 council homes were built each year. Post-war public housing had private baths and central heating, virtually unheard of in working class homes. While most housing estates were large, London County Council also commissioned competitions and hired leading architects.

One such architect, Erno Goldfinger, was hired in 1966 to create a 31 story slab block in North Kensington, at the edge of the then, working class Notting Hill neighbourhood. The building was completed by 1972, and almost immediately fell victim to funding cuts. Goldfinger had requested a concierge for the front door, but the Greater London Council considered concierges too expensive and overly "paternalistic".

By the 1980s, Trellick Tower was known to newspapers as 'The Tower of Terror'. There were only 3 elevators for the 175 unit building, at least one of which always seemed out of order. One elderly resident collapsed and died after broken elevators forced him to climb six flights of stairs. Rapes, muggings, and burglaries were commonly reported. Residents described the stench of urine, beer and stale sweat in the lobby, broken bottles and syringes in the hallways, graffiti, used condoms and vagrants in the elevators. According to an article in *The Guardian* newspaper on the transformation of the building, "one Christmas, vandals opened the fire hydrant and unleashed thousands of gallons of water into the lifts, blowing fuses and leaving the block without electricity, water, or toilet facilities".

The Greater London Council belatedly attempted to fix the front door problem by buying an expensive entryphone system, which was repeatedly vandalized. But the real answers to the problems in Trellick were found by a strong and active residents' association, which was formed in 1984. The residents' association formed a management group, developed surveys and reports on tenants' concerns, and kept after their local council until changes started to happen. One of the first breakthroughs was when the council changed its policy in 1986, and only assigned flats to people who wanted to live in the tower. Surprisingly, many people did want to live at Trellick, and many people also wanted to buy their units when that option became available.

Improvements continued, including the long-awaited concierge, who along with 24-hour security and closed-circuit television, improved the security of the building. New elevators, an improved hot water system, and a playground added to the amenity and comfort of the units.

Now there is a mix of long time residents and newcomers. There are examples of people who brought up their children in Trellick Tower and eventually bought their flat. There are also young families buying. One resident gushed in a recent article: "On this floor there's a girl from the BBC, a woman in a bank, a social worker, pensioners, a market stall holder. They're all great. They don't mind when the kids race up and down on their bikes or whatever. There's no way we'd have as much space anywhere else." Residents also rave about the views, and have developed a strong social network. For instance, in 1997, a light show, visible for miles, was orchestrated by residents, who operated special coloured lights on their balconies in a sequence co-ordinated with the passing of trains below. Indeed, the greatest social concern at Trellick may now be that the units are too expensive, along with most other property in London, for the working poor and lower middle class to buy into.

3 LESSONS FROM ST. JAMES TOWN AND TRELLICK TOWER

St. James Town and Trellick Tower are different in several important ways. St. James Town is a classic modernist 'city within a city', with 18 buildings and almost 7,000 units over a total of 9 former city blocks. It was planned for isolation from the surrounding community, which was then working class, although it has since become gentrified. Trellick, in contrast, is one individual 175 units building in a mixed and thriving neighbourhood. St. James Town was also constructed over 15 years, by a variety of undistinguished architects, under a number of cost constraints. Trellick Tower is one building within a heterogeneous community. Although separated from part of the community by an elevated expressway, it partakes of the same social infrastructure: the famous Portobello Road market, the schools, the excellent parent-child drop-in centres and recreation activities provided by the local council. It was, by and large, the vision of one architect, and although cost constraints were imposed by the local authority, the vision – for better or worse – remained intact.

Yet St. James Town and Trellick have a number of similarities, as well. Both were designed as ground-breaking modernist developments in the 1960s. Both have been faced with challenges related to crime, social isolation, and maintenance. There have been serious suggestions that both St. James Town and Trellick be torn down, that they were simply too expensive to maintain, that they were design and social experiments that had gone horribly wrong. Both St. James Town and Trellick have almost mythic status in their respective cities and countries: they act as symbols of the perils and the pleasures of high rise living. Finally, the social infrastructure has been a key to livability in both places.

St. James Town is considered symbolic of "the nature of problems that can be encountered in many other high rise buildings in the city and elsewhere" in Canada, according to Toronto's Planning and Development Department.

Throughout not only Canada, but many developed nations, there is a legacy of high rise housing from the second half of the last century. Much of it is poorly designed and built, and residents face problems ranging from physical deterioration to criminal activity. Although much attention has focused on publicly owned housing, the public is also left "holding the bag" for private housing which does not work. Problems are particularly acute when high-rise housing is lived in by low-income tenants, where maintenance may easily lead to higher rents pricing out occupants. In St. James Town, this was a concern even before the abolition of rent controls under the present neo-conservative provincial government.

St. James Town is also used in Toronto and in Canada as an example of why high rise housing does not work, period. Both the general public and certain sectors of the "chattering classes" have damned modernism in general, and high rise housing in particular, over the past 50 years. This almost atavistic dislike of modernism peaked in the 1980s, when Alice Coleman, the guru of British crime prevention through environmental design, contended that high rise housing in and of itself, led to "crime, fear, anxiety, marital breakdown, and physical and mental disorder that would largely be avoidable in more socially stabilizing environments", whereas most people could "cope perfectly well with life in more traditional houses" (Coleman, 1985).

Trellick Tower has also become a symbol of both subtle and paradigmatic shifts which have been occurring over the past decade in England, and in other English-speaking advanced industrialized nations. The subtle shifts are towards a partial acceptance of architectural modernism, including high-rise housing towers, and what I believe to be a related acceptance of the principles behind public housing.

In the 1980s, modernism stunk of the social and political experimentation of the sixties, and it was a truism that it did not work. Modernist buildings, and high rises in particular, were ugly, stupid and made people go bad. Prince Charles said so, and he must be right. Now the eighties are history, and modernism, to a certain extent, is back. In 1997, for example, English Heritage listed Trellick Tower as "worthy of preservation for posterity", along with a number of other modernist buildings, including Coventry Cathedral and several public housing estates.

Trellick Tower is also used as an example of the best principles and practices of public housing. In the 1980s, national governments in Great Britain, Canada, the United States, Australia and New Zealand began to step away from their previous commitments to provide housing for low-income people. One reason given was the lack of success of public housing: its tendency to 'ghettoize' residents, the brutality of much of its design, its often dehumanizing bureaucracy. And indeed, public housing has often concentrated on poor people within large areas with few jobs, poorly funded schools, banks unsympathetic to community economic development, and in general, a pretty shabby state of social infrastructure. On the other hand, private housing has also received large subsidies from various levels of government, and has often done just as lousy a job of housing poor people decently, with St. James Town being a perfect example of this sad state of affairs.

Trellick is an especially good spokesperson for both modernist design and public housing. Trellick Tower was well designed and well built from good materials. The local council was responsive to tenants' concerns, although the cynical might well note that the council's improvements coincided with the first offering of the units for sale. Many high rise housing blocks, both privately and publicly owned, were the result of overly cosy relationships between unethical developers, corrupt local councils, and incompetent architects and contractors. Many public housing blocks have been torn down, especially in England and the United States, because they were unsafe or so hated by their residents there seemed no hope of improvement. But as in the case of destroying privately-run tenements 50 years before, the alternative was not necessarily an improvement. It is important to recognize and learn from public housing buildings and communities that work.

Trellick Tower is also an example of a more profound paradigmatic shift. Whether you call it post-modernism or post-industrialism, it goes beyond design labels to encompass the tremendous social and economic changes of the past half century. Who would have predicted 50 years ago, the rate of divorce, single parenthood, and 'non-traditional' families that now characterize western societies? Who would have known that rapid increases in longevity would lead to a tremendous increase in empty nesters and single seniors? And especially, who would have thought, in the triumph of the car and in an age of rapidly changing technology, that these innovations would have led to centrality, to the desire to live downtown, becoming ever more fashionable?

Even and especially in the age of television, video players, the Internet, and working out of the home, living downtown has become ever more popular. Commercial and industrial buildings are being converted into central city 'lofts', which are inhabited not only by the spiritual descendants of the swinging singles, but also by older people and families. Gentrification has priced out ever larger segments of households from formerly declining downtown neighbourhoods. All of the generalizations that used to be made about cities in Canada, the United States, England, Australia and New Zealand are now under revision: that suburbs would continue to draw the middle class from the centre city, that people's ideal accommodation is the single family detached home, that in an increasing era of privatization, public education, health, and other elements of social infrastructure would matter less.

In the London neighbourhood of Notting Hill, Trellick Tower's success has occurred concurrently with the transformation of this traditionally working class immigrant community to one of the most fashionable districts in London. Trellick's success in attracting buyers was not solely a function of location – there are still public housing estates that are not popular with the middle class in the neighbourhood. But Trellick Tower, unlike St. James Town, has been embraced by its surrounding community. It has become to be seen as an icon of 'cool'.

Despite its problems, it is quite possible that St. James Town be fashionable as well in 20 years. As mentioned before, the surrounding neighbourhoods have become quite gentrified. The current neoconservative government in Ontario has ended rent controls and plans on selling off public housing, two key Thatcherite policies which helped to create the sort of supercharged demand that exists in London. Well-to-do residents could buy the kind of social infrastructure – the parent-child drop-in centres, the after-school programs, the recreation centres – which presently barely survive in the face of social service cuts. And of course, shopping and banking choices would follow in the wake of a critical mass of middle-class residents.

And then where will the housing of last resort be? Quite possibly in those quiet allegedly stable suburbs, chock full of single family houses, where the baby boomers grew up. We are seeing this already in London, where virtually the entire central area is now priced out of the hands of the lower middle class, the nurses, teachers, and retail workers, never mind the working poor. The suburbs have generally housed the industrial workers and the immigrants in other parts of Europe and the world. Even countries as wedded to the single family home and the suburban dream as Canada, the United States, New Zealand and Australia will be faced by the challenge of this societal shift in the coming century. The challenges of providing adequate social infrastructure in high rise housing will be especially acute in suburbs, where a more dispersed pattern of settlement, poor public transportation, and a legacy of non-interventionist local governments have left these areas particularly underserved.

4   CONCLUSION: SIX MAXIMS ON SOCIAL INFRASTRUCTURE IN TALL BUILDINGS

The principles behind the provision of social infrastructure are more vague, and the importance of social infrastructure less understood, than is the case for physical infrastructure. Obviously and understandably, local governments in the industrializing countries of the nineteenth century, and local governments in industrializing countries of today, have shown more concern about typhoid epidemics and buildings falling down than about people not reaching their full human potential because of their housing. But beginning in the nineteenth century, local governments began to invest in education, health care, and recreational facilities because these services were understood to be an essential ingredient for the good life in cities.

It is important to remember that modernism did consider social infrastructure: this network of relationships and services that together make people a community. The much maligned Le Corbusier did intend high rise housing to include shops, nurseries, community centre, doctors' surgeries, and even roof gardens. He intended a mix of people, including people who needed to work in the neighbourhood, seniors, and children. Like garden cities, towers in the sky have been poorly served by their alleged followers. And it was the anti-modernists, like Margaret Thatcher, who contended that community did not exist, and did their utmost to try to make it so.

The two case studies I have used suggest questions about ownership and accountability in relation to social infrastructure. The problem with the social infrastructure within St. James Town is that it was almost entirely privatized. The recreation facilities and shopping centre were, at least at first, spanking new and thus unaffordable to the tenants who began to move there. As for public facilities, there was no library, the school was old and neglected, and the local government did not press to retain land for a playground, or indeed any public land for future use. **It is always a mistake to assume that all necessary social infrastructure needs can be provided to people at market prices, without government intervention.**

"There is no simple relationship between density and satisfaction" (Marcus and Sarkissian, 1986), and high density is not necessarily a bad thing. Density can create a demand for social services, like literacy classes or parent-child drop-ins, that would not necessarily be available in a more dispersed community. Sometimes people need to be with people like themselves, which is why there needs to be enough flexible space to accommodate people's varying needs for social interaction. Teenagers often need space to be apart from adults, with minimal supervision. Parents and dog-owners both like to meet suitable play companions for their charges. Seniors have particular needs, as do people who share language and culture. But there also needs to be larger spaces, where a tenants' association or special meeting can take place. Folding walls in larger rooms can sometimes help. **But in order to have the flexibility to respond to these changing needs, developers need to provide meeting rooms and small store-fronts that can be used by community organizations.**

High rise housing can be alienating, especially for people who are used to other forms of living, or who have experienced a traumatic break from their home community, as is true for many newcomers to a city. If there is one thing we have learned from the failures of modernism, it is that apartment corridors do not function as "streets in the sky". Children can't play in them, without disturbing people living inside units. People cannot informally 'hang out' in corridors, the way they often do on streets. **It is necessary to provide a range**

of opportunities in outdoor and indoor space for informal social interaction. That means providing benches, in playgrounds, near dog runs, in front of stores selling food, in front of schools so parents can talk as they drop off and pick up their children. It also means that pubs, low cost cafes, and clubhouses should not be zoned out of high rise housing communities. It means providing space for play centres, advice centres, homework clubs, community gardens, and workshops. It can mean a children's play area or book exchange in the laundry room. There is a remarkable amount of 'dead space' in and around many high rise buildings: much of that space can be used in ways that promote social interaction. I know of a cigar-smoking room in one high rise building that not only took care of a perceived loitering problem, but became a valuable informal social centre for immigrant men from Latin America.

The previous few maxims lead to an obvious point, which always needs to be stressed: **residents have both the right and the ability to suggest social infrastructure improvements. The people who live in communities are the experts.** They not only come up with wacky and wonderful ideas like the light show at Trellick, or the cigar-smoking room, but they can put in the energy to make things happen.

In Trellick, the tenants knew they wanted a concierge. They were right. Concierges are not only less expensive than closed circuit television and complicated entryphone systems, they are more effective in preventing crime. Concierges, and good staff in general, are an essential element of the social infrastructure in tall buildings. They pass on messages, assist residents and non-residents, and help maintain social control in lobbies. They know the people who live there (Bright et al., 1986). On-site superintendents can often perform some of the same functions, although they can also be intimidating to tenants in buildings with poor management practices. **In high rise buildings, there should be resident staff people providing supervision of meeting spaces and laundry rooms, which preferably should be sited near the building's management office.**

There is no fundamental reason why people should not live high up. High rise housing has problems inherent in its design. It is dependent on technology. When lifts, lighting, and garbage disposal don't work, or when communal spaces such as underground garages, entrances, and hallways are inadequately supervised, high rises can be dangerous and depressing places to live. It is sometimes difficult to maintain privacy and restrict noise. But we should also remember that there are problems with the design of low density, low rise housing: lack of sufficient density to support services and public transit, for instance. There are even problems with "the golden mean" of medium density, medium rise housing, so called 'traditional' and 'neo-traditional' housing: namely, not everyone wants to live there for all of their lives. **It is always a**

good idea to plan for a range of housing types across all incomes and ages, in all parts of the city centre and in suburbs. Homogenous neighbourhoods tend to work only for the very wealthy and stable.

Perhaps the greatest lesson from the changing fortunes of Trellick Tower and St. James Town is that no part of the city and no kind of architecture should be seen as finite, when the possibilities of social change are infinite. Architects, planners, building managers, and social service providers all play a part in providing social infrastructure. Together with the people who live and work in tall buildings, ideas and imaginations can be pooled to meet the challenges of the twenty-first century.

5 REFERENCES

Bright, Jon *et al.*, 1986
     AFTER ENTRYPHONES: IMPROVING MANAGEMENT AND
         SECURITY IN MULTI-STOREY BLOCKS, Safe
         Neighbourhoods Unit, London.

Carroll, Rory, 1999
     HOW DID THIS BECOME THE HEIGHT OF FASHION? A
         RENAISSANCE IN CONCRETE, in *The Guardian,* March 11.

Coleman, Alice, 1985
     UTOPIA ON TRIAL: VISION AND REALITY IN PLANNED
         HOUSING, Hilary Shipman, London, pp. 2–3.

Collins, Stuart and Silva, Fatima, unpublished
     SOCIAL HOUSING EXPERIMENTATION, American Planning
         Association Annual Conference, 1995, School of Urban
         and Regional Planning, Ryerson University, Toronto.

Corporate Resources Group, 1995
     TORONTO RANKS AS FOURTH-BEST CITY TO LIVE IN,
         Toronto Star, January 18.

Doucet, Michael and Weaver, John, 1991
     HOUSING THE NORTH AMERICAN CITY, McGill-Queens
         University Press, Montreal and Kingston.

Fortune Magazine, 1996
    TORONTO IS NO. 1 ON FORTUNE'S INTERNATIONAL
        LIST OF BEST CITIES FOR WORK AND FAMILY,
        October 21.

Funk and Wagnalls, 1980
    FUNK AND WAGNALLS STANDARD DICTIONARY.

Gibson, Alan, 1996
    HIGH HOPES, in *Socialist Review*, issue 201, October.

Huxley, Margo, 1994
    PLANNING AS A FRAMEWORK OF POWER, *Beasts of Suburbia:
        Reinterpreting Cultures in Australian Suburbs*, Ferber, Healy, and
        McAuliffe, eds., Melbourne University Press, Melbourne.

Marcus, Clare Cooper and Sarkissian, Wendy, 1986
    HOUSING AS IF PEOPLE MATTERED: SITE DESIGN GUIDELINES
        FOR MEDIUM-DENSITY FAMILY HOUSING, University of
        California Press, Berkeley.

Rahder, Barbara, 1988
    ST. JAMES TOWN REVITALIZATION: SOCIAL ANALYSIS, City of
        Toronto Planning and Development Department, Toronto.

Short, John, 1989
    THE HUMANE CITY, p. 39.

United Nations Development Programme, 1996
    HUMAN DEVELOPMENT REPORT 1995, Economica, Paris.

Wekerle, Gerda and Whitzman, Carolyn, 1995
    SAFE CITIES: GUIDELINES FOR PLANNING, DESIGN, AND
        MAINTENANCE, Van Nostrand Reinhold, New York.

Whitzman, 1991
    COMMUNITY AND DESIGN: AGAINST THE SOLUTION IN
        ST. JAMES TOWN, *The Canadian City,* Kent Gerecke, ed.,
        Black Rose Books, Montreal, pp. 164–174.

Sustainability and Cities: The Role of Tall Buildings in this New Global Agenda

Peter Newman

SUMMARY

Sustainability in cities is defined in terms of the reducing of resource inputs and waste outputs whilst simultaneously improving livability. This definition is traced through the recent history of global politics and is then applied to two case studies: Australian settlements and a squatter settlement in Jakarta. Some key principles are developed that show sustainability is improved when cities get bigger and when they are denser. The data for this is provided and the implications discussed.

INTRODUCTION

Sustainability or sustainable development, is the great global agenda of our time. Despite attempts by some elements of the business sector to define it as 'sustainable profits' the concept continues to provide the central challenge to our age: how we can redefine growth so that we can simultaneously improve the environment, the economy and the community. The overlapping circles approach has been adopted around the world by thousands of local governments, hundreds of national governments, and scores of global NGO's and major companies (Figure 1).

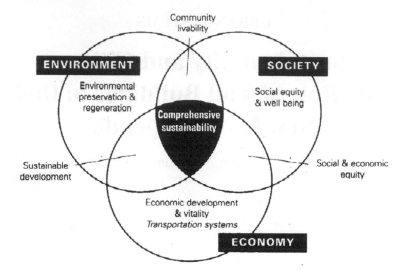

Figure 1   The overlapping areas of Economic Development, Community Development and
Ecological Development which are simultaneously required for sustainable development. Source:
Newman and Kenworthy (1999).

     In this paper I will trace how the concept of sustainability came to be
developed and how it is now being applied to cities. I will then try to show how
tall buildings fit into this agenda. I will do this with two case studies and two
key principles.

SUSTAINABILITY AND GLOBAL POLITICS

In whatever way sustainability is defined and analysed, it is important to see that
its roots did not come so much from academic discussion as from a global
political process.

    The first elements of sustainability emerged on the global arena at the
1972 UN Conference on the Human Environment. At this conference 113
nations pledged to begin cleaning up the environment and most importantly to
begin the process of tackling environmental issues on a global scale. The
problems of air pollution, water pollution and chemical contamination do not
recognize borders. It was acknowledged that it is not possible to allow DDT or

PCBs or radioactive materials to be released anywhere without it affecting everyone. Natural resource depletion was also discussed as awareness had grown that depletion of forests, groundwater, soils, and fishstocks had impacted across national boundaries.

Concern about the global environment was very high. Evidence was presented at Stockholm that the scale of the human economy was now significant relative to the natural environment. For example, the flow of human energy (mostly in settlements) was now roughly equal to the flow of solar energy through ecosystems, with inevitable impacts from the wastes (see Newman, 1974).

This sense of limits was not new for many nations. In the 19th century a similar sense of limits drove Americans to set aside the first national parks as they realised their apparently limitless new frontier had reached the west coast. George Marsh's book 'Man and Nature: Or Physical Geography as Modified by Human Action' first published in 1864 analysed the environmental impacts of US urban and rural development. One hundred years later this sense of limits had become a global phenomenon as the last frontier lands were developed.

The effects of human activity on this biosphere were also beginning to negatively impact human welfare. The spectre of Malthus was raised as a global phenomenon but focussed on the rapidly growing areas of the third world where it was thought that much of the world's future growth and impact would occur (e.g. Ehrlich and Ehrlich, 1977).

However, this environmental sensitivity is only one side of sustainability. The third world was not so impressed by this new environmental globalism. The new agenda was rapidly turning to one of anti-growth as environmentalists saw the rapacious consumption of natural resources as inevitably linked to economic development. The third world saw it as just another way to prevent them from attaining their goals for development. The 1 billion people living in abject poverty with not even enough food to eat, did seem to have some legitimate claim on a little more of the world's resources. Thus the UN established the World Commission on Environment and Development in 1983 to try and resolve this fundamental conflict. In 1987 they published "Our Common Future" or the Brundtland report, which launched into common parlance the phrase 'sustainable development'. This was then given form, as shown below, at the 1992 Earth Summit in Rio.

Sustainability was presented as an agenda to simultaneously solve the global environmental problem and to facilitate the economic development of the poor, particularly those in the third world. Whereas in 1972 the environment was placed on the global *political* agenda, in 1992 the environment was placed on the global *economic* agenda. Thus the characteristics of sustainability can be distilled into four broad policies which have since become the basis of much global action.

CHARACTERISTICS OF GLOBAL SUSTAINABILITY

The following four characteristics are derived from the Brundtland Report and are the fundamental approaches to global sustainability that need to apply simultaneously to any approach to the future.

(1) The elimination of poverty, especially in the third world, is necessary not just on human grounds but as an environmental issue.
Evidence was presented by the Brundtland report from across the globe that poverty is one of the factors degrading the environment because populations grow rapidly when they are based on subsistence agriculture or fishing or plant collection. In the past, the population of subsistence communities was controlled by high death rates but the globalisation of health care has meant that there is no way forward to a new equilibrium but to reduce birth rates. This seems only to occur sustainably when families want fewer children not more and in subsistence economies children are a source of wealth and security (United Nations, 1987).

Where economic and social development do not occur and populations continue growing, the environment inevitably suffers. This feeds back in a poverty cycle, for example much of the Rwanda tragedy has been traced to this process (UN Centre for Human Settlements, 1997). Grass roots economic and social development (particularly women's rights) are necessary to break this cycle (UN, 1987). The alternative is a constant degradation of the 'commons' as more forest is cleared, more soil is overgrazed, more fisheries are destocked (Hardin, 1968). **Thus third world economic and social development are a precursor to global sustainability.**

(2) The first world must reduce its consumption of resources and production of wastes.
The average American (or Australian) consumes natural resources 50 times that of the average Indian, and the poorest groups in abject poverty across the world consume 500 times less. By raising the standard of living of the global poor from 1/500th to 1/50th would not be a huge extra strain on resources. The primary responsibility for reducing impact on global resources lies in the rich part of the world.

Such a goal cannot be achieved without economic and social change. For example, industry cannot be frozen with 80's machinery, it needs to develop new technology for replacing CFC's, for using less energy, for switching to new renewable fuels and more efficient materials; such change requires economic and social development. Cities will not be less energy intensive if they are frozen in their sprawling 80's structures and they can only rebuild in more compact, transit-oriented forms if they are growing economically and socially. **Thus first world economic and social development are precursors to global sustainability but in the future they must be much less resource-intensive.**

(3) Global co-operation on environmental issues is no longer a soft option.
Hazardous wastes, greenhouse gases, CFC's, and the loss of biological diversity are examples of environmental problems that will not be possible to solve if some nations decide to hide from the necessary changes. The spread of international best practice on these issues is not some management fad, nor is it a conspiracy for world domination from certain industries or advanced nations, it is essential for the future of the world. **Thus a global orientation is a precursor to understanding sustainability.**

(4) Change towards sustainability can only occur with community-based approaches that take local cultures seriously.
Most of the debate on sustainability has been through UN conferences and high level international events. However, it is recognised that this can only create the right signals for change, it cannot force the kind of changes which are discussed above. These will only come when local communities find how to resolve their economic and environmental conflicts in a way that creates simultaneous improvement of both. **Thus an orientation to local cultures and community development is a precursor to implementing sustainability.**

Academic discussions on the meaning of sustainability need to build from this base of four principles. The definition most people have picked up on from Brundtland is that "sustainable development is development that meets the needs of the present without compromising the ability of future generations to meet their own needs" (WCED, 1989, p. 43). Undoubtedly the sustainability agenda is about future generations but it is not trying to create some infinitely durable means of managing society so it can be sustained indefinitely. This is particularly important when it comes to discussing sustainable cities which can become a diversion into ideal city forms or on the impossibility of creating eternal cities.

Sustainability has come from a global political process that has tried to bring together, simultaneously, the most powerful needs of our time:

- the need for **economic development** to overcome poverty,
- the need for **environmental protection** of air, water, soil and biodiversity upon which we all ultimately depend, and
- the need for **social justice and cultural diversity** to enable local communities to express their values in solving these issues.

Thus when I refer to sustainability, it will be the simple idea that means the simultaneous achievement of global environmental gains in any economic or social development.

The sustainability movement is first and foremost a global movement that in particular is forcing economists and environmentalists to find mutually beneficial solutions.

The sustainable development process has been proceeding at many different levels:

- In academic discussions (e.g. how ecological economics can be defined and formulated, Daly and Cobb, 1989, Jeroen et al, 1994).
- In laboratories, in industry and in management systems as they try to be innovative within the new parameters of reduced resource use and less waste (e.g. 'the clean production' agenda), and
- Within governments at all levels and in community processes.

These approaches are usually called 'green economics', 'green technology', 'green planning' etc. When they are no longer called 'green' perhaps we can begin to say that we are becoming more sustainable. However, sustainability is never likely to be a state that we reach but one towards which we need to constantly strive. Sustainability is a vision and a process, not an end product.

GLOBAL GOVERNMENT RESPONSES TO SUSTAINABILITY

Most countries, particularly European nations, began to respond to the Brundtland report in the late 80's. One of the first responses was when Canada established their Round Table on the Environment and the Economy which began mapping out what the new agenda meant. In Australia, the Ecologically Sustainable Development process was begun in 1990 involving government, industry, conservation groups, unions, social justice groups and scientists. And in New Zealand the Resource Management Law Reform process began its re-examination of all aspects of government from an environmental perspective. In the US a private sector organisation, the National Commission on the Environment published a report in 1993 entitled 'Choosing a Sustainable Future' which states:

> "The economy and the environment can no longer be seen as separate systems, independent of and even competing with each other. To the contrary, economic and environmental policies are symbiotic and must be molded to strengthen and reinforce each other." (p. 21)

The US Clinton administration set up the President's Council on Sustainable Development in 1996 as the first US government response to sustainability.

At a global level, after 3 years of preparatory meetings involving thousands of the world's scientists and administrators, the UN Conference on Environment and Development was convened in Rio in 1992. The 'Earth Summit' drew together more heads of government than any other meeting in history and its final resolutions were signed by 179 nations representing 98% of the world. This is about as global as is ever likely to be possible.

The documents agreed to were: a statement on sustainability called the **Rio Declaration**, a 700 page action plan for sustainability called **Agenda 21**, a **Convention on Climate Change**, a **Convention on Biological Diversity**, and a **Statement on Forests** (Keating, 1993).

Such plans are still working their way through governments, industries and communities. International treaties are being developed each year to put some substance into the global sustainability agenda including the CFC agreement and the late 90's climate change agreements. In 1997 the 'Rio plus five' Earth Summit occurred in order to report on how well nations were doing on the sustainability agenda. For many environmentalists the 'Rio plus five' summit was a dismal performance as so many issues seemed to be no closer to a solution. Nations seemed to be failing to deliver on so many fronts. However, a more perceptive look would see that significant progress has occurred at the global or international level and at the level of the local community, between Rio and New York which give hope for a continuing global process.

At the global level the following landmarks were achieved:

- The Biodiversity Treaty, which emerged from the Rio Summit, took effect in 1993 and now requires all nations to keep better inventories of their biodiversity and to ensure protection and sharing in profits from the world's life forms. This is now part of international law.
- The Law of the Sea, which took 40 years for agreement, is also now part of international law since the required number of nations finally agreed in 1993. It now means for example that fishing of migratory ocean species is regulated.
- The Ozone Layer treaty took the world's governments a mere 10 years to develop and implement, so that 1996 saw the end of most global production of ozone-depleting chemicals.
- 1997 saw the start of a new global treaty which prevents the transport of hazardous waste to developing countries, and

- 1997 also saw the first serious attempts at setting targets on greenhouse gases begin to be negotiated.

The global agenda in terms of international environmental treaties and laws is quietly and slowly changing how the world does business. There are now over 200 such international agreements.

At the local level the sustainability agenda began to be taken seriously in over 2000 local governments who have implemented Local Agenda 21 Plans or Sustainability Plans since the Rio conference. The stories of hope are rich and diverse when examined at the grass roots level (e.g. Pathways to Sustainability Conference, 1997). The reason for this is that at the local level it is possible for government to more easily make the huge steps in integrating the economic, environmental and social professions, in order to make policy developments that are sustainable. They are also closer to concerned people and more distant from the single issue powerful lobbies like the fossil fuel and road lobbies, who are so obvious in shaping national priorities.

The local sustainability agenda and the global sustainability agenda are beginning to make more sense when the focus is shifted away from nation-states to settlements. This is the theme of this paper and it is partly a plea to do more, partly an attempt to help define how we can be more sustainable in our settlements and partly a celebration of those cities and towns who are showing us what can be done (Newman and Kenworthy, 1999).

APPLICATION OF SUSTAINABILITY TO CITIES

The principles of sustainability outlined above can be applied to cities though the guidance on how this can be done was not very clear in Agenda 21 or the other UN documents. It is probably true to say that the major environmental battles of the past were fought outside cities but that awareness of the need to come back to cities is now universally recognised by environmentalists, government and industry. The OECD, the European Community and even the World Bank now have sustainable cities programs. In 1994 the Global Forum on Cities and Sustainable Development heard from 50 cities (Mitlin and Satterthwaite, 1994) and in 1996 the UN held Habitat II, the Second United Nations Conference on Human Settlements in Istanbul. At the 'City Summit' the nations of the world reported on progress in achieving sustainability in their cities (UN Centre for Human Settlements, 1996).

Anders (1991), in a global review of the sustainable cities movement, pointed out:

"The sustainable cities movement seems united in its perception that the state of the environment demands action and that cities are an appropriate forum in which to act." (p. 17)

In fact others such as Yanarella and Levine (1992) suggest that all sustainability initiatives should be centred around strategies for designing, redesigning and building sustainable cities. In this global view they suggest that cities shape the world and that we will never begin the sustainability process unless we can relate it to cities.

Indeed there is a case to be made that urbanisation to cities is a major step forward in terms of sustainability. The rural issues of land degradation would be greatly exacerbated if the extra population continued to live in rural areas. It is a well known phenomenon that people have much fewer children when they move to cities than if they continued to live in a rural context where children provide farm labour. The total area occupied by the globe's cities occupy between 3000 and 5000 square kilometers of land, roughly two to three times the size of Tasmania. If spread thinly in semi-rural settlements it would have far greater ecological impact than in the dense cities that are the traditional form of urban settlement. As shown below the bigger cities have less ecological footprint per person than small cities. Yet this perspective on cities is not widely understood, perhaps because of the way we think about cities.

AN EMERGING FRAMEWORK – THE CITY AS AN ECOSYSTEM

Throughout this century the city has been conceived by sociologists, planners and engineers as a "bazaar, a seat of political chaos, an infernal machine, a circuit, and more hopefully, as a community, the human creation par excellence" (Brugmann and Hersh 1991, cited in Roseland 1992). One of the strongest themes running through the literature on urban sustainability is that if we are to solve our problems we need to view the city as an *ecosystem*. As Tjallingii (1993) puts it:

"The city is (now) conceived as a dynamic and complex ecosystem. This is not a metaphor, but a concept of a real city. The social, economic and cultural systems cannot escape the rules of abiotic and biotic nature. Guidelines for action will have to be geared to these rules." (p. 7)

Like all ecosystems, the city is an open system, having inputs of energy and materials. The main environmental problems (and economic costs) are related to sustaining the growth of these inputs and managing the increased outputs. By looking at the city as a whole and by analysing the pathways along which energy and materials (and pollution) move, it is possible to begin to conceive of management systems and technologies which allow for the reintegration of natural processes, increasing the efficiency of resource use, the recycling of wastes as valuable materials and the conservation of (and even production of) energy. Such approaches enable us to envision more sustainability in cities.

There may be on-going academic debate about what constitutes sustainability or an ecosystem approach (Slocombe, 1993), but what is clear is that many strategies and programs around the world have begun to apply such notions both for new development and redevelopment of existing areas.

SUSTAINABILITY GOALS FOR CITIES

How does a city define its goals in a way that is more sustainable? How do you make a systematic approach that begins to fulfil the global and local sustainability agenda? The approach adopted here is based on the experience of the Human Settlements Panel in the Australian State of the Environment Reporting process (see Newman et al., 1996).

It is possible to define the goal of sustainability in a city as the reduction of the city's use of natural resources and production of wastes while simultaneously improving its livability, so that it can better fit within the capacities of the local, regional and global ecosystems.

This is set out in Figure 2 in a model that is called the 'Extended Metabolism Model of the City'. Metabolism is a biological systems way of looking at the resource inputs and waste outputs of settlements. The approach has been undertaken by a few academics over the past 30 years, though it has rarely if ever been used in policy development for city planning (Wolman, 1965, Boyden et al., 1981, Girardet, 1992). Figure 2 sets out how this basic metabolism concept has been extended by us to include the dynamics of settlements and livability in these settlements.

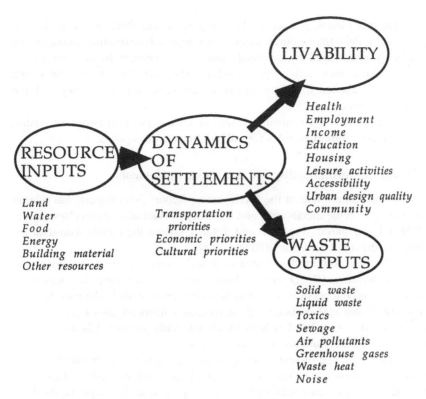

Figure 2 Extended metabolism model of human settlements.

In this model it is possible to specify the physical and biological basis of the city, as well as its human basis. The physical and biological processes of converting resources into useful products and wastes is like the human body's metabolic processes or that of an ecosystem. They are based on the laws of thermodynamics which show that anything which comes into a biological system must pass through and that the amount of waste is therefore dependent on the amount of resources required. A balance sheet of inputs and outputs can be created. It also means that we can manage the wastes produced, but they require energy in order to turn them into anything useful and ultimately all materials will eventually end up as waste. For example, all carbon products will eventually end up as CO_2 and this is not possible to recycle any further without enormous energy inputs that in themselves have associated wastes. This is the entropy factor in metabolism.

What this means, is that the best way to ensure there are reductions in impact, is to reduce the resource inputs. This approach to resource management is implicitly understood by scientists but is not inherent to an economist's approach which sees only 'open cycles' whenever human ingenuity and technology are applied to natural resources. However, a city is a physical and biological system.

The model of sustainability will now be applied to two case studies which both show important roles for tall buildings in this agenda.

Case Study (a): Sustainability and Australian Settlements

The 1996 Australian State of the Environment Report (SoE Report) has a large amount of data on the metabolism and livability of Australian cities (Newman et al., 1996). For example, Figure 3 and Table 1 set out the metabolism data for Sydney and the trend between 1970 and 1990.

The SoE Report shows that in global terms and even in comparison with other wealthy OECD cities, Australian cities are high consumers of resources and producers of wastes. Furthermore, most trends in the past 20 years are not sustainable as they indicate these metabolic flows are growing per capita. Livability on the other hand is high by all standards (though it is not uniform across all settlements) and has also been growing.

The challenge of sustainability is to find out how this livability can be improved for everyone in Australian cities whilst simultaneously reducing the metabolic flows of resources and wastes. Three particular challenges highlighted in the SoE Report will be outlined to show how Australians might proceed with meeting this overall challenge and because they throw light on a number of important issues for tall buildings.

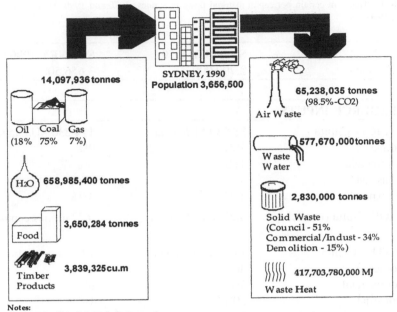

Notes:
Waste water data do not include stormwater
Timber products and food data derived from national per capita data

Figure 3 Resource inputs consumed and waste outputs discharged from Sydney, 1990.
Source: Newman et al., 1996.

Table 1 Trends in certain per capita material flows in Sydney, 1970 and 1990. Source: Newman et al., 1996.

	SYDNEY 1970	SYDNEY 1990
Population	2,790,000	3,656,500
RESOURCE INPUTS		
ENERGY/Capita	88 589 MJ/capita	114 236 MJ/capita
Domestic	10%	9%
Commercial	11%	6%
Industrial	44%	47%
Transport	35%	38%
FOOD/Capita (intake)	0.23 tonnes/capita	0.22 tonnes/capita
WATER/Capita	144 tonnes/capita	180 tonnes/capita
Domestic	36%	44%
Commercial	5%	9%
Industrial	20%	13%
Agricultural/Gardens	24%	16%
Miscellaneous	15%	18%
WASTE OUTPUTS		
SOLID WASTE/Capita	0.59 tonnes/capita	0.77 tonnes/capita
SEWAGE/Capita	108 tonnes/capita	128 tonnes/capita
HAZARDOUS WASTE		0.04 tonnes/capita
AIR WASTE/Capita	7.6 tonnes/capita	9.3 tonnes/capita
CO_2	7.1 tonnes/ capita	9.1 tonnes/capita
CO	204.9 kg/capita	177.8 kg/capita
SO_x	20.5 kg/capita	4.5 kg/capita
NO_x	19.8 kg/capita	18.1 kg/capita
HC_x	63.1 kg/capita	42.3 kg/capita
Particulates	30.6 kg/capita	4.7 kg/capita
TOTAL WASTE OUTPUT	324 million tonnes	505 million tonnes

AUSTRALIAN SUSTAINABILITY CHALLENGE 1: THE BIG CITIES

Big cities can be more sustainable. Contrary to many popular conceptions, there is a worldwide pattern within nations that bigger cities (over 1 million) have smaller metabolic flows *per capita* (energy, water, land inputs and waste outputs) than smaller cities and these are lower than the flows in small country towns per capita. Below I will pursue this question of city size and sustainability. The data from the SoE Report show this to be true in Australian settlements based on a number of metabolism parameters such as energy, water and waste. At the same time, the bigger cities have uniformly higher levels of livability (health, income, employment, housing . . .) than in smaller settlements. The reduction in metabolic flows can be explained in terms of the economies of scale and density, producing greater efficiency in technology, more access to markets for recycling, better public transport, and generally a more efficient use of land. The greater livability is due to the greater diversity of opportunities provided in the bigger cities.

So it is possible to see even in Australian settlements that higher incomes and other features of livability do not necessarily mean higher metabolic flows of natural resources. Such subtle differences are usually washed out when just national averages are compared.

Big cities are reaching local capacity limits. Despite the bigger cities consuming less per capita energy, water, land, forest products, and building matcrials and producing less per capita air, water and solid waste, the bigger cities are likely to sooner meet natural environment capacity constraints in their local areas. These are the capacity of the air shed to absorb air pollutants, the capacity of the surrounding water bodies to absorb liquid wastes and the capacity of the land to be used for solid waste landfill.

In all these local capacity issues there are significant questions being asked about the major cities in Australia. The air sheds are at capacity in Sydney, Melbourne, Perth and Brisbane. Water problems are being experienced as well, e.g. Adelaide is not able to cxpand much further with its water supply, Sydney has reached the limit on the capacity of the Hawkesbury-Nepean river system to absorb stormwater waste, and all cities have questions about the ecological impacts of ocean disposal of sewage.

The response to this is often to try and assert that Australia needs to develop more small towns, to decentralise development away from the big cities with their capacity problems. This may not however be the most sustainable solution. Based on the evidence collected, the new small towns will more than likely increase per capita metabolic flows. And many of these flows are significant in a global context, e.g. oil and greenhouse gases. It is not sustainable if a big city's growth is stabilised only to transfer the problem elsewhere at a higher impact.

The best solution is for each city to reduce its overall metabolic flows whilst simultaneously improving its livability. No growth, no development should occur in Australia's big cities, unless it can be shown that the overall metabolic flows of the cities are reducing. This is the challenge for Australia's big cities. It may seem like an impossible task and it is if we only look at how we have undertaken urban development in the past. But the challenge is also an opportunity for innovation in a world looking for sustainable solutions. The SoE Report thus outlines a few areas where it is possible to see some hope in responding to the sustainability challenge.

Waste is reducing. Sydney grew 30% in its solid waste output per capita between 1970 and 1990 but in the years 1993 to 1995 there was a decrease as recycling systems began to be adopted by local authorities and new markets were created where previously it was not considered possible. Examples now exist where innovative municipalities have reduced their solid waste by 50%. As big cities are at the cutting edge of the global economy they are where the greatest innovation in sustainability ought to be occurring and has the chance to occur. Thus not only do the big cities need sustainability most, they have the best track record (in per capita terms) and the most opportunity to show other cities how it can be done.

Cars and reurbanisation. Transportation seems to be an area where past trends seem impossible to alter and the growth in automobile use seems inevitable. However, even here there are trends which offer some hope. The most recent data from a study we have been conducting on cities around the world, shows that although the Australian city is a heavy car user, the trend was for a slowing down of the increase in per capita car kilometres between 1961 and 1991. The same pattern can be found in Canadian cities but not US cities which continued to grow out of control, between 1960 and 1990. Why is this? It is possible that the trend is connected to another broader trend in Australian cities which was also documented and detailed perhaps for the first time in the SoE Report. This trend is a process of reurbanisation, which is where present suburbs are redeveloped rather than new suburbs on the urban fringe. Although most US cities continued to show a decline of the inner city at least up to 1990, there are also signs here of change, however suburban growth is still the dominant process. The reurbanisation process is well underway in Australian cities; data suggest that 40% to 60% of all new housing is following this pattern and a higher proportion of commercial development is going to the inner city and sub-centres. Melbourne's locus of development has begun to move back towards the city centre after centuries of expansion. Sydney has dramatically increased its density (see Figure 4).

Reurbanisation has an inherently smaller transport requirement as well as being more suitable for other modes of transport. Thus patterns of car use are changing.

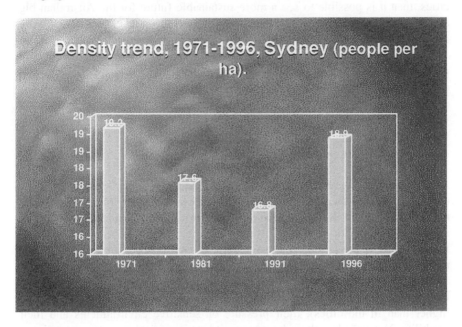

Figure 4   Sydney density trend.

Much should be made of this process of reurbanisation as it appears to be a critical part of urban sustainability. Central to this process is a rapid increase in the number of denser, taller buildings which are now attracting a high market premium due to their locational advantages. The question will be asked as to why this process is occurring so rapidly. Briefly, the pattern in Australian cities appears to be part of a bigger economic process (linked to globalisation of the economy) as the city turns into a series of nodal information-based sub-centres. The new economy seems to need dense, walkable centres where people can meet face-to-face. The SoE Report also shows that livability in these redeveloping areas is equal to or higher than on the fringe.

Thus the processes of reurbanisation can have most of the elements of sustainability: less resources, less waste and higher livability. If governments and the private sector are able to facilitate such a process in all parts of their cities, then it is possible to see a more sustainable future for the Australian big city. On the other hand, it is possible to develop policies and programs that work against this process. For example, urban policies in the late 1990s are not clear whether to facilitate transit-oriented urban villages or to build freeways and push urban development further out. The sustainability agenda in Australia's big cities is however gaining momentum as policy makers begin to recognise that reurbanisation is a major tool that can help overcome automobile dependence whilst improving the economy and increasing opportunities for community.

Australian Sustainability Challenge 2: dispersed, exurban development

At the same time as reurbanisation appears to be accelerating in Australia's big cities, there is another process of exurbanisation or periurbanisation which is taking development away from the cities to semirural areas, particularly coastal regions. Such development is partly related to the city (with people commuting large distances), but is mostly lifestyle related where people are attracted to the rural environment.

These exurban locations are expanding rapidly and are less sustainable than other settlements. This is evident from their metabolic flows (which are higher than in the cities), their pressure on sensitive environments and their livability. Not only has there been a rapid loss in coastal ecosystem resilience and diversity, as well as problems with sewage and loss of rural productivity, there has been little economic development associated with the trend. The economic opportunities and availability of services is much reduced in these areas and yet governments continue to subsidise and promote such scattered, highly car-dependent development. This subsidised exurban development is leading to unsustainable development in all the senses meant by that word.

The exurbanisation process has many causes but it too can be linked to the globalisation process as many of those leaving the city are not participating in the new information-oriented global economy and are attracted to cheaper houses and land as well as the rural environment.

It is not inevitable that this continues. Plans can be made to ensure large areas of the coastal environment are not developed; where development is preferred, then it can be planned around innovative, clustered villages with reduced metabolic flows and greater livability. There are examples in Australia (Lingstrom and Tap, 1994) and across the world the Eco Village network is showing how it can be done in all cultures (Register and Peeks, 1997). The process of bringing these exurban-semirural communities into the sustainability agenda is aimed at making the countryside more rural and cities more urban.

Small rural-oriented towns can be created around community-based technologies, permaculture and community arts with links to the global economy through electronic communication. But the process will not work while development is allowed to scatter through coastal areas without any concern for the impacts.

Australian Sustainability Challenge 3: remote aboriginal settlements

Remote indigenous communities in Australia grew from 88 in 1966 to 1385 in 1992. They have low metabolic flows but are experiencing some capacity issues e.g. from untreated wastes or inadequate water supplies. They also have extremely low livability on all indicators, particularly health. Powerful cultural forces have driven this 'return to country', which has only just begun to be assisted with appropriate technology. The communities lack basic infrastructure, as well as social and economic development to improve livability.

The interesting aspect for sustainability is that these remote aboriginal settlements have become test beds for renewable energy and small scale environmental technologies. These are likely to have a large export future and are becoming more viable for developed cities each year, especially as the sustainability agenda appears to favour more small scale technological systems. It is more than likely that these innovative technologies will move into remote mining towns and then slowly move into larger towns and eventually big cities.

This process can be facilitated by governments removing subsidies on fossil fuel and using it instead to bring in the new technologies where a city or region has a competitive advantage. This will be essential as the world responds to the critical question of declining world oil output as production is expected to peak around the early part of the new millennium. The sustainability agenda will be even more of an imperative then and Australia in part will be turning to its outback technology for assistance in this transition. But in the cities the biggest challenge will be how we can build denser cities to reduce the need for oil in transport (see below).

Case Study (b): Sustainability and Traditional Settlements

The way that some critics of high rise go on, it is possible to think that tall buildings were invented in the 1960's for public housing projects and bear little relationship to historic urbanism. In fact the unusual thing in historical terms is the sprawling car-based suburb. The origins of tall buildings go back into the oldest cities and are still evident in many older cities today. For example Shibam in Yemen is a third century mud brick city with a density of around 300 per hectare. Most buildings are 8 storeys high. Such areas show a strong urbanism with tall buildings clustered close together around walkable public spaces.

In third world cities today there has been a burst of high rise building as well as traditional clusters of housing as well as new squatter settlements. It is difficult to assess these various forms in terms of sustainability but the extended metabolism model provides a solution as shown in the case study below.

Jakarta, in Indonesia, is a huge megalopolis of over 12 million people; like many rapidly growing cities in developing nations, it has a major problem with informal housing or squatter settlements which form on public land along river banks. One such settlement on the Ciliwung River is subject to annual flooding with much loss of property and some loss of life. The city of Jakarta has developed a set of high-rise buildings as replacement housing for squatters. The extended metabolism model was used by Arief (1997) to evaluate the sustainability of the new high-rise housing area in comparison with the squatter area.

Table 2 summarises the per capita data collected from 50 households in both areas, with comments.

Table 2 Comparison of resource use, waste produced and livability in squatter and high-rise housing in the Ciliwung River area of Jakarta, 1994. Source: Arief, 1997.

INDICATOR	CILIWUNG SQUATTERS	HIGH-RISE SETTLERS	COMMENTS
RESOURCE INPUTS			
Water (litres/day)	2,142	2,386	Similar water use.
Energy (MJ/yr)			Reduced household energy use for high-rise
-Household	9,288	8,731	settlers as they must pay for power (squatters use
-Transport	2,270	2,935	informal power connections). High-rise
Total Energy	11,558	11,666	settlers travel a little more.
Land (m²/person)	4.57	0.91	High-rise much more efficient.
Building materials	Mostly wood, bamboo, tiles and tin roofs, some brick.	Bricks, ceramic floor, tile roofs.	Much better quality materials in high-rise.
Food	Inadequate protein.	More balanced but minimal intake.	Both groups remain poor.

WASTE OUTPUTS			
Solid waste (kg/day)	0.42 82% throw solid waste into river.	0.42 100% solid waste collected.	Same waste but much better managed in high rise.
Liquid waste	Direct into river.	Septic tank.	Same waste but much better managed in high-rise.
Air waste (gCO$_2$/day)	23,875	14,675	Reduced levels in high-rise due to lower electricity use.
LIVABILITY			
Health	Environmental health situation poor. 58% no sickness in past 3 months.	Environmental health situation better. 60% no sickness in past 3 months.	Despite better conditions no big difference in health of high-rise settlers. Both groups are still quite poor.
Housing quality	Poor.	Relatively good.	Physical housing quality improved in high-rise.
Employment	Mostly informal economy participants: 55% street traders 19% private business 0% home industries 21% freelance workers.	Mostly formal economy participants: 6% street traders 40% private business 29% home industries 2% freelance workers.	Formal address allows entry to formal economy by high-rise settlers.
Income (av/month)	Rp 151,000	Rp 252,000	Significantly better income for high-rise settlers.
Education	94% primary school and below.	44% primary school and below.	Those who moved to high-rise tend to be more upwardly mobile.
Community	High level of community: 92% know >20 neighbours by name 90% happy to live there 100% trust their neighbours 100% felt safe and secure 100% borrow tools from neighbours 100% borrow money from neighbours.	Not so high level of community: 44% know >20 neighbours by name 76% happy to live there 52% trust their neighbours 4% felt safe and secure 70% borrow tools from neighbours 22% borrow money from neighbours.	Distinct problems in the design of high rise as far as community-building is concerned.

What the data shows is that there are either similar or reduced levels of resource inputs and waste outputs in the high-rise compared with the squatter area, probably because the high rise dwellers have to pay rent and for power. However, land efficiency in the high rise is considerably better as is the management of the wastes, both solid and liquid. Thus in terms of metabolic flows the high-rise is better as a physical development.

In terms of livability, the situation is much more mixed:

- in health there is little difference though housing quality is significantly improved in the high-rise,
- employment, income and education are all better in the high rise as the formal address allows entry to the formal economy,
- community is, however, significantly reduced on all the parameters measured. In the high rise people do not trust each other, they don't know their neighbours and hence don't feel safe and secure. In simple terms the social capital of the new development has significantly been reduced.

Thus as a sustainability model, the high rise is a distinct improvement as a physical development over the squatter settlement, it is a more efficient way to manage the metabolic flows of the settlement and is able to create economic opportunity. However, as a community, the high-rise is a failure – there is little trust, communication or security in comparison with the squatter settlement.

Other approaches to managing such squatter settlements tend to emphasise how to upgrade the housing and its infrastructure in a more organic way. In the Kampung Improvement Program in Indonesia there is a local example of how well this organic approach to urban development has worked and how popular it is (Silas, 1993). The approach is to build in a much more communal way, still with reasonable heights but only by building from the site and with the communities heavily involved. It also uses the local layout and public spaces rather than creating modernist uniform dead space between buildings. It could not be seriously considered on this site on the Ciliwung as it is too flood prone. However, the principle of organic design is something that must become more influential in such high-rise development or else the full sustainability potential of the project can be seriously questioned. This is the main failure of modernist tall buildings in their third world expression and probably their global expression.

Two key concepts have come out of the case studies above in relation to sustainability: increases in urban density and city size. They can both be positive elements of sustainability in cities. These two concepts will be expanded to explain how this can be so.

Urban density and sustainability

Density is clearly inversely related to gasoline use per capita as set out in Figure 5. Density reduces transport energy through several mechanisms: it shortens distances for all modes and it makes transit and bicycling and walking more viable as alternatives to the car; it also makes many journeys redundant as when transit is used many journeys are combined e.g. going to shops on the way to or from the train. Data from a 1996 study by Dunphy and Fisher show a 21% decrease in daily driving between central Manhattan and outer suburbs in New York but transport energy is 500% less in Manhattan on a per capita basis.

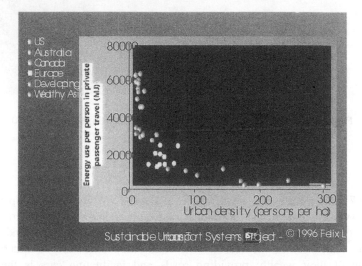

Figure 5   Energy use per capita in private passenger travel and density, in 38 global cities, 1990.

As well as having obvious benefits to cities' environmental sustainability, and especially with the crucial resource of oil, there are also clear economic sustainability gains when cities are able to reduce their car dependence. Figure 6 shows that cities with better public transport (the three middle group of cities from Canada, Europe and the wealthy Asian cities of Singapore, Hong Kong and Tokyo) have much reduced transport costs as a proportion of city wealth. The car dependent cities of the US and Australia have much higher costs, as well as the newly developing cities of Asia like Bangkok and Jakarta, where they are investing heavily in new roads and putting little into quality public transport.

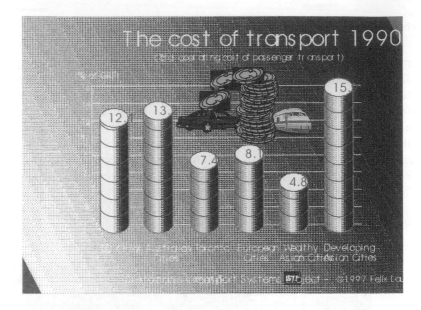

Figure 6   The proportion of city wealth on transportation in global cities, 1990.

Figure 7 demonstrates the significance of density in this relationship to city wealth (via the link to car dependence). Figure 7 shows that the lower the density of the city the more it wastes of its wealth on transport. And this is probably the basis of the reason why road-based cities are so much more wasteful of their wealth: providing roads and facilitating cars is the basic mechanism for sprawling a city and this is an expensive way to build a city. It is less efficient in terms of infrastructure and if a city is constantly sprawling rather than reurbanising then there is less capital available for investment in productive innovative aspects of the city (see Frost, 1991 and Jacobs 1984). Therefore in

economic and environmental terms the sustainability agenda is clear – contain
sprawl, reurbanise and go for sustainable transportation.

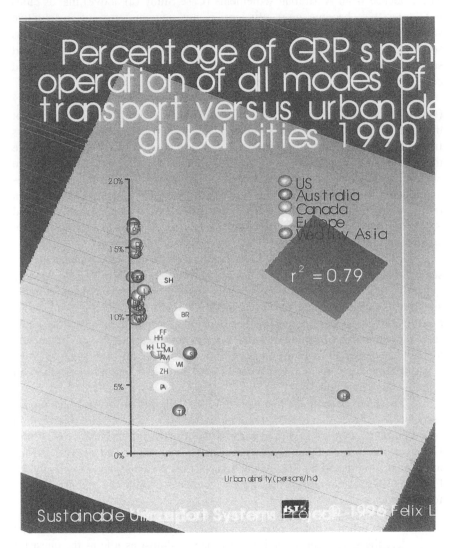

Figure 7   The total operating costs of passenger transport versus urban density in developed
cities, 1990.

City Size and Sustainability

The evidence from Australian settlements (Case Study (a) above) that as cities get larger they become more efficient, is not new in terms of quantitative studies, nor should it surprise us in theoretical terms. But it does. There are thus some important issues that need to be discussed about city size and sustainability. The issue will be looked at from the perspective of economics and ecology.

The data on city size were clear in our original global cities study (Newman and Kenworthy, 1989). This 32-city survey showed the same pattern: transport energy use per capita generally reduces with city size. Peter Naess from the Norwegian Building Research Institute wanted to try and investigate this phenomenon by eliminating the major variable of cultural difference that so clearly interferes with a global survey of cities. He chose 22 Scandinavian cities and found a very clear relationship between the size of the city and its per capita transport energy use as well as with density. The cities of Copenhagen, Oslo and Stockholm were significantly lower in transport energy use per capita than in smaller provincial towns (Naess 1994).

Urban economists have been pointing to the efficiency advantages of scale for decades and also the efficiency advantages of density (the two are generally linked). There have been many studies of cities which have found significant economic benefits from scale and density (Hoch 1976, 1979, Sternlieb 1973, Richardson 1973). The benefits in terms of sustainability come from the same kind of economic efficiencies which are applied to environmental technologies, e.g. public transport systems become more efficient as cities grow, waste treatment and recycling systems become more efficient as cities grow (all other factors being equal).

There are economists like Neutze (1981) who point to the diseconomies associated with size due to the growth in externalities. Others point to growth in social problems that are said to increase with city size and density (Troy, 1996). However, the data on this question are usually very sparse and the issue seems to be more dominated by ideological stances (see Newman and Hogan, 1982). Fischer (1976) summarises the elusive search for optimal city size in the following way:

> "Most urban scholars seem convinced, to quote a British economist, that '... the search for an optimal city size is almost as idle as the quest for the philosophers' stone' (Richardson, 1973: 131). The entire area of

speculation is misconceived on several grounds. First, there are no substantial empirical findings pointing to city size at which any "good" – income or innovation or governmental efficiency – is maximized, or any "bad" – crime or pollution – is minimized. In fact, some data suggest that for economic purposes an optimal city size would be larger than any we now have. We have certainly not identified an optimum size for any social-psychological variable in this book. Even if such ideals could be found, they would probably not be the same for a wide variety of social products. The size that maximizes personal incomes would differ from that which maximizes artistic creativity, or that which minimizes pollution, and so on. And it would surely be a vain task to try to sum up all these various "goods" and "bads" into a single measure'". (p. 250).

In ecological terms, it should come as no revelation that as cities grow and become more complex and diverse, they begin to create more efficiencies. Ecosystems grow from being simple systems with a few pioneering species to more mature ecosystems with diversity and interconnection. Thus after a fire or a flood or some other disturbance, a cleared piece of land will begin developing the structure of its ecosystem with an emphasis on rapid and simple growth. After a period it becomes more diverse and more efficient as it establishes a more complex network of interactions and begins to grow upward. As Table 3 shows, this succession process involves a series of changes which can be paralleled with the processes of urban development.

Critical to the process of succession becoming more efficient in its use of the natural environment is its increase in information. This is both the raw genetic information of its individual components and the complex interactions and networks that hold it together as a system.

All analogies from nature need to be carefully considered before they are taken as principles for human society – Social Darwinism caused wars and the rise of such socially disastrous theories as the 'master race'. But the analogy of the ecosystem and its obvious application to the urban system at least shows us that it is quite understandable when human activity in cities seems to parallel natural activity.

Ecological patterns thus help us to see that big cities can be moving, like a maturing ecosystem, towards a more efficient use of resources and a higher level of information, organisation and environmental control.

Table 3 Characteristics of young and mature ecosystems and their application to sustainable development in cities. See: Newman (1975).

Note: young ecosystems pioneer newly cleared or burned sites and progressively develop into mature ecosystems.

ECOSYSTEM *YOUNG ECOSYSTEM*	SUCCESSION *MATURE ECOSYSTEM*	SUSTAINABLE CITY *YOUNG CITY*	DEVELOPMENT *MATURE CITY*
ENERGY and MATERIALS			
• High gross photosynthetic activity, low efficiency.	• Reduced gross photosynthetic activity, high efficiency.	• High energy, low efficiency.	• Reduced energy, high efficiency.
• Wastage of nutrients.	• Recycling of nutrients.	• Wastage of nutrients and materials.	• Recycling of nutrients and materials.
ECONOMIC DIVERSITY			
• Producers mainly.	• Balance of producers, consumers, decomposers and integrative species.	• Emphasis on producers, less on manufacturers, little on services.	• Balance of producers, manufacturers, and services.
• Few functional niches – generalists.	• Many functional niches – specialists.	• Low functional diversity.	• High functional diversity.
SPATIAL EFFICIENCY			
• Low spatial efficiency – dispersed.	• High spatial efficiency – compact.	• Low spatial efficiency – dispersed.	• High spatial efficiency – compact.
• Low structural diversity – small, lateral, little variety.	• High structural diversity – small and large, lateral and vertical, large variety.	• Low structural diversity – small, lateral, little variety.	• High structural diversity – small and large, lateral and vertical, large variety.
INFORMATION and ORGANISATION			
• Low species and community diversity.	• High species/ community diversity.	• Low community diversity.	• High community diversity.
• Low community organisation – little interconnection.	• High community organisation – much interconnection.	• Low community organisation – few networks.	• High community organisation – many networks.

ENVIRONMENTAL CONTROL			
• Low environmental control – resource availability external to biotic system, climate unbuffered.	• High environmental control – resource availability controlled within biotic system, climate buffered.	• Weak protection from environmental perturbations – resources poorly managed, vulnerable to changes in the physical environment.	• Strong protection from environmental perturbations – resources tightly managed, more able to buffer and cope with changes.
• System instability.	• System stability.	• System instability.	• System stability.

Cities can of course choose not to become more efficient as they grow. Cultural forces discussed in this book can always be used to deny the opportunities created by scale and density of activity. However, the evidence suggests that smaller cities find the process of achieving efficiency in economic and resource terms harder than larger cities.

The strong emotional appeal of 'smallness' to our age is not, however, without its basis. EF Schumacher's book 'Small is Beautiful' with its attack on modern gigantism, developed the message that all technology needs to be at the appropriate scale of the community that it is meant to be serving. Community scale technology is emerging as communities begin to assert their role – whether it be in villages in the third world or modern large cities.

The thrust of the New Urbanism and the 'urban village' movement is that communities need to be physically designed for and given infrastructure at the scale of the community. Vast endless suburbs can be given new coherence when focused around a sub-centre.

The recognition of the need for a community scale does not deny the realities associated with a big city. It tries to bring the social dimension into the city in a meaningful way. To try and suggest that all big cities should be dispersed into small ones is to deny some very major forces – both economic and ecological. And the evidence is that this would be counter-productive in terms of sustainability. However to ensure that communities work within big cities is a major policy for achieving sustainability.

This is also the approach often taken to growth management, trying to bring order and focus into the megalopolis around community goals. This is not anti-city but is a part of asserting the importance of community and the need for environmental and economic responsibility within the city.

In the US there has been a longterm belief in the value of small communities and small towns. The December 15, 1997 issue of *Time* magazine talked about the Rise of Small Towns, based on the move of some (generally wealthy) people to small towns after finding that even the move to suburbs from the city was not meeting their expectations. This is not statistically a significant movement. Nor can it be separated from the continuing problem of US inner cities. But even the US city is growing larger and is now becoming denser, along with most other cities around the world.

The driving force behind the growth of cities is human opportunity. The diversity of opportunity in cities continues around the world to be their main magnetism. We can try to deny these opportunities by attempting to artificially constrain city size, as the former USSR tried to do (unsuccessfully) with Moscow. Or we can even have experiments in de-urbanisation like those conducted by Pol Pot and Mao. Indeed some environmental philosophy becomes very anti-urban and verges on suggesting this as a policy framework (see Trainer, 1985, 1995, 1996). The idea pursued in this paper suggests that the human quest, the process of civilisation, the development of human society, is all about the growth of cities.

There is no question that cities are now heading in many wrong directions. They are not sustainable, as outlined. But to suggest that sustainability means the systematic dismantling of cities is neither realistic nor does it have an historical or theoretically sensible basis to it. The urban adventure needs to be grasped and pursued, not denied.

Cities need to change, but the historic quest for human achievement through urban civilisation, will go on. The great challenge for our cities is that they must now take seriously the quest for sustainability, that cities can be more livable, more human, more healthy places, but they must learn how to do this by simultaneously using fewer natural resources, creating less waste and thus impacting less on the natural world.

CONCLUSIONS

The sustainability agenda is a major global and local issue. Communities across the world are trying to see how they can simultaneously reduce their impact on the earth whilst improving their quality of life. This is happening to a greater or lesser extent in all cities (Newman and Kenworthy, 1999).

The Extended Metabolism model is an effective way for cities to try to create a holistic picture of their sustainability agenda. It enables urban managers to create a series of sustainability indicators which together can give a sense of whether they are reducing or growing in their resource inputs and waste outputs and at the same time reducing or growing in livability. These indicators can be used to show the world how well they are contributing to global issues such as greenhouse and oil depletion, and to show their local citizens how well they are managing sustainability issues that impact on them directly. However, if a city concentrates only on the local issues it is likely to miss major components of the sustainability agenda as outlined above.

REFERENCES

Anders, R., 1991
> THE SUSTAINABLE CITIES MOVEMENT. Working Paper No 2, Institute for Resource and Security Studies, Massachusetts, USA.

Arief, A., 1997
> A SUSTAINABILITY ASSESSMENT OF SQUATTER REDEVELOPMENT IN JAKARTA, Thesis, ISTP, Murdoch University.

Boyden, S., Millar, S., Newcombe, K. and O'Neill, B., 1981
> THE ECOLOGY OF A CITY AND ITS PEOPLE. ANU Press, Canberra.

Brugmann, J. and Hersh, R., 1991
> CITIES AS ECOSYSTEMS: OPPORTUNITIES FOR LOCAL GOVERNMENT. ICLEI, Toronto.

Dunphy, R. and Fisher, K., 1996
> TRANSPORT, CONGESTION AND DENSITY: NEW INSIGHTS, Transportation Research Record, No. 1552, TRB, Washington DC.

Daly, H. E. and Cobb, J. B. Jnr., 1989
> FOR THE COMMON GOOD: REDIRECTING THE ECONOMY TOWARD COMMUNITY, THE ENVIRONMENT AND A SUSTAINABLE FUTURE. Beacon Press, Boston.

Ehrlich, P. R. and Ehrlich, A. H., 1977
> POPULATION, RESOURCES AND ENVIRONMENT. W. H. Freeman, San Francisco.

Fischer, C. S., 1976
THE URBAN EXPERIENCE. Harcourt, Brace, Jovanovich, New York.

Girardet, H., 1992
THE GAIA ATLAS OF CITIES. Gaia Books, London.

Hardin, G., 1968
THE TRAGEDY OF THE COMMONS. Science, 162(3), 1243–1248.

Hoch, I., 1976
CITY SIZE EFFECTS: TRENDS AND POLICIES. Science, 193, 856–863.

Hoch, I., 1979
SETTLEMENT SIZE, REAL INCOME AND THE RURAL TURNAROUND. American Journal of Agricultural Economics. 61(5), 953–959.

Keating, M., 1993
THE EARTH SUMMIT'S AGENDA FOR CHANGE: A PLAIN LANGUAGE VERSION OF AGENDA 21 AND THE OTHER RIO AGREEMENTS, Center for Our Common Future, Geneva.

Kenworthy, J., Laube, F. et al., 1999
AN INTERNATIONAL SOURCEBOOK OF AUTOMOBILE DEPENDENCE IN CITIES 1960–1990. University Press of Colorado, Boulder.

Lindegger, M. and Tap, R., 1989
CONCEPTUAL PERMACULTURE REPORT: CRYSTAL WATERS PERMACULTURE VILLAGE. Nascimanere Pty Ltd., Nambour.

Marsh, G., 1864
MAN AND NATURE: OR PHYSICAL GEOGRAPHY AS MODIFIED BY HUMAN ACTION. Lowenthal, D. (ed.), 1965, Harvard University Press, Cambridge, Massachusetts.

Mitlin, D. and Satterthwaite, D., 1994
CITIES AND SUSTAINABLE DEVELOPMENT. Discussion paper for UN Global Forum, 1994, Manchester, IIED, London.

Naess, P., 1993
> ENERGY USE FOR TRANSPORT IN 22 NORDIC TOWNS. NIBR
> Report No. 2, Norwegian Institute for Urban and Regional
> Research, Oslo.

National Commission on the Environment, 1993
> CHOOSING A SUSTAINABLE FUTURE. Island Press, Washington
> D.C.

Neutze, M., 1978
> AUSTRALIAN URBAN POLICY. George Allen and Unwin, Sydney.

Newman, P., 1974
> ENVIRONMENTAL IMPACT PART I AND PART II. Journal of
> Environmental Systems, 4(2), 97–108 and 109–117.

Newman, P., 1975
> AN ECOLOGICAL MODEL FOR CITY STRUCTURE AND
> DEVELOPMENT. Ekistics, 40(239), 258–265.

Newman, P. W. G. and Hogan, T. L. F., 1981
> A REVIEW OF URBAN DENSITY MODELS: TOWARDS A
> RESOLUTION OF THE CONFLICT BETWEEN
> POPULACE AND PLANNER. Human Ecology, 9(3),
> 269–303.

Newman, P. W. G., et al., 1996
> HUMAN SETTLEMENTS. In Australian State of the Environment
> Report, Department of Environment, Sport and Territories,
> Australian Government Publishing Service, Canberra.

Newman, P. and Kenworthy, J., 1989
> CITIES AND AUTOMOBILE DEPENDENCE: AN
> INTERNATIONAL SOURCEBOOK, Gower, Aldershot,
> England.

Newman, P. W. G. and Kenworthy J. R., 1999
> SUSTAINABILITY AND CITIES: OVERCOMING AUTOMOBILE
> DEPENDENCE, Island Press, Washington DC.

Pathways to Sustainability Conference, 1997
    UN CONFERENCE, City of Newcastle, NSW, Australia.

Register, R. and Peeks, B., 1997
    VILLAGE WISDOM: FUTURE CITIES. Ecocity Builders, Oakland,
        California.

Richardson, B. M., 1973
    THE ECONOMICS OF CITY SIZE. Saxon House, London.

Roseland, M., 1992
    TOWARDS SUSTAINABLE COMMUNITIES. Canadian National
        Round Table on the Environment and the Economy, Ottawa.

Silas, J., 1993
    SURABAYA 1293–1993: A CITY OF PARTNERSHIP. Municipal
        Government of Surabaya.

Slocombe, D. S., 1993
    ENVIRONMENTAL PLANNING, ECOSYSTEM SCIENCE AND
        ECOSYTEM APPROACHES FOR INTEGRATING
        ENVIRONMENT AND DEVELOPMENT. Environmental
        Management, 17(3), 289–303.

Sternlieb, G., 1973
    HOUSING DEVELOPMENT AND MUNICIPAL COSTS. Centre for
        Urban Policy Research, Reutgers University, New Jersey.

Tjallingii, S. P., 1995
    ECOPOLIS: STRATEGIES FOR ECOLOGICALLY SOUND
        URBAN DEVELOPMENT. Backhuys Publishers, Leiden.

Trainer, T., 1985
    ABANDON AFFLUENCE. Zed Books, London.

Trainer, T., 1995
    THE CONSERVER SOCIETY: ALTERNATIVES FOR
        SUSTAINABILITY. Zed Books, London.

Troy, P. N., 1996
    THE PERILS OF URBAN CONSOLIDATION. The Federation Press,
        Leichardt, Sydney.

UN Centre for Human Settlements, 1996
      AN URBANISING WORLD: GLOBAL REVIEW OF HUMAN
           SETTLEMENTS 1996. Habitat, Nairobi.

UN Centre for Human Settlements, 1997
      CHANGING CONSUMPTION PATTERNS IN HUMAN
           SETTLEMENTS. A Dicussion Paper, UN Centre for Human
           Settlements, Nairobi.

United Nations, 1987
      FERTILITY BEHAVIOUR IN THE CONTEXT OF
           DEVELOPMENT: EVIDENCE FROM THE WORLD
           FERTILITY STUDY. Population Studies, 100(2).

Wackernagel, M. and Rees, W., 1996
      OUR ECOLOGICAL FOOTPRINT: REDUCING HUMAN IMPACT
           ON THE EARTH. New Society Publishers, Philadelphia.

Wolman, A., 1965
      THE METABOLISM OF THE CITY. Scientific American, 213, 179

Yanarella, E. J. and Levine, R. S., 1992
      DOES SUSTAINABLE DEVELOPMENT LEAD TO
           SUSTAINABILITY? Futures, 24(8), 759–774.

World Commission on Environment and Development, 1989
      OUR COMMON FUTURE, Oxford University Press, Oxford.

Shaping the Future of Human Settlements: Globalisation, and the Leadership and Management of Change

Peter Ellyard

ABSTRACT

Most of our human settlements: cities, and urban and rural communities are over managed and under led. As the interconnectedness of globalisation grows, market forces are increasingly shaping the forms and functions of human settlements. Vision and strategic intervention have lessened in importance and management has grown at the expense of leadership. Leadership has not been reinvented to operate in an interconnected globalisated planet. However the tools of leadership that have worked in the past are increasingly ineffective in a globalising interconnected world and marketplace of the 21st century. There has not been enough consideration about how the development of human settlements can be shaped and their habitability improved in an era of open global markets and trade, and of increasingly unempowered governments.

INTRODUCTION

The future of our cities and urban areas is interconnected to the future of rural and regional communities. At the beginning of the new Millennium, there are many global cities but very few global rural communities. If one looks at the industrial structures of most rural areas in 2001 one discovers that the industrial structure is remarkably similar to their industrial structure in 1901. It is as if the whole of the 20th century bypassed rural communities and all the new job categories and new industries of the 20th century were born in the cities. Seventy percent of the job categories and products and services of the year 2020 have yet to be invented. Now in an era of globalising markets, universal connectivity and wider access to eduction and knowledge, it is now possible to ensure that many of the new job categories and new industries born in the 21st century can be founded in rural areas.

If the world is to come to grips with continuing urban drift then a new strategy that can work in the conditions of the 21st century must be devised. This the purpose of the MyFuture Foundation, which aims to promote the sustainable prosperity of human settlements in rural/regional areas and developing countries in a globalising 21st century.

I have been asked to discuss the management of change. In this paper I want to discuss the issues of both management and leadership of change. It is my view that most of our problems we have in dealing with rapid change in general and in human settlements in particular are related to the fact that we are trying to manage when we should be leading it.

To be effective in shaping and responding to change in human settlements or elsewhere one also must be able to understand how the various causal elements which shape change converge and interact to create coherent outcomes. This in turn requires one to be both a strategic and systems thinker.

The big changes the 21st century: the Planetist future.

Change of course is constant. Those who do not regard change as a normal part of life will be condemned to become victims of change. However it is useful to examine the change process itself because one must try to understand the causes and consequences of change if one is to successfully deal with it. Let me illustrate how I deal with the change process, and how I turn and understand of change to empower the ability to shape and respond to change.

As an example I would like to discuss global change and how we can work to make sense of the forces which are so dominant in all our lives at the beginning of the 21st century. There are three major global forces which are shaping our future. These forces can be described in three core words: Globalisation, Tribalisation and Technological Change. In my book "Ideas for the New Millennium" I outline in detail how these forces are creating a new global culture/paradigm in the 21st century. The consequences of these processes I predict will be a new paradigm which I call Planetism

The dominant paradigm of the 20th century has been Modernism. This is so deeply entrenched that we have taken it for granted. Modernism meant the triumph of the western European way over everything else. It crushed cultural diversity through the forces of colonialism, religious evangelism and the power of western science and technology. A component of Modernism has been the concept of progress, which for most of this century has been something which we felt we shouldn't or couldn't stop. As the century proceeded, the attitude accompanying the utterance that "we can't stop progress" changed from unbridled enthusiasm in the 1950s to increasing scepticism, even cynicism and sarcasm, in the 1970s. By the 1980s, we needed to modify Modernism, as its dark side had become too significant to ignore, so we invented Post Modernism, in which we borrowed deconstructed parts of previous eras and built them into the new. We had recognised that some forms of progress involved throwing out babies with bath water.

Post-Modernism can be regarded as a process which is responsible for the deconstruction of Modernism tradition and its replacement by a collage of both Modern and Pre-Modern forms and traditions. Until the late 1990s the deconstruction has dominated. The deconstruction has taken place so that a reconstruction of a new paradigm appropriate for the 21st century can begin. Post-Modernism provides a global intercultural mixing bowl, in which Modern, Pre-Modern and many other cultural concepts and viewpoints are being synthesised into a new paradigm. Now the reconstruction of a new paradigm to replace Modernism or Post-Modernism is under way. In the Post-Modern era the value of many of the old ways of doing things, and of viewing the world are being reaffirmed. Post-Modernism fosters a view that the world would be a better place if we paid more respect to Pre-Modern, including indigenous, wisdom and knowledge. It is implicit that Post-Modernism aspires to integrate the best of the old with the best of the new in order to prepare for what follows Post-Modernism.

Most debate about the consequences of the paradigm shift from Modernism to Post-Modernism has been focused in the realms of culture and art and in professional fields with strong links to these areas, such as architecture and urban planning. There has been much less exploration of what this paradigm shift means for the creation of wealth or poverty, for the increase or decrease in equity or opportunity, or for social, economic and industrial development.

In the late 1990s humanity was mid-way through the transition from the disappearing cowboy culture to the emerging spaceship culture. The cowboy culture was dominated by Modernism. Modernism is the paradigm of the cowboy. The paradigm of the transformation from cowboy culture to spaceship culture is Post-Modernism. Post-Modernism does not stand for anything except that Modernism must be deconstructed and replaced with something new. It is a paradigm embodying a process of transformation and reconstruction of something to replace Modernism, but it does not itself replace Modernism. What follows Post-Modernism? What will be the true successor to Modernism? What paradigm will embody a world view which expresses the 21st century just as Modernism has expressed the late 19th and the first two-thirds of the 20th century? This transformation from Modernism began in the 1960s and will most likely be completed by the year 2020. During this Post-Modernist period a new paradigm embodying the synthesis and hybridisation of the old and the new is emerging. This is the paradigm of the spaceship culture, the paradigm of the cosmonaut. This is Planetism. It will be to the 21st century what nationalism was to the 19th and 20th centuries. In the 19th and 20th century people gave their first allegiance to their nation and tribe. In the 21st century they will give their first allegiance to the Planet and their shared humanity, while at the same time celebrating their difference by treasuring their tribal and community roots.

Kenneth Boulding introduced the idea that the Earth needed to change from a "cowboy economy" to a "spaceship economy" if life on the planet was to survive. Today at the mid point of the 1990s humanity is mid way through a transition between what can be recognised as a disappearing Cowboy Culture and an emerging Spaceship Culture in the 21st century. The Cowboy Culture is a paradigm for an unsustainable society while the Spaceship Culture is a paradigm for a sustainable society. The Cowboy and Spaceship Cultures have the following characteristics:

From 1960	To 2020
The Cowboy Culture	The Spaceship Culture
Individualism	Communitarianism
Independence	Interdependence
Autocracy	Democracy
Humanity against nature	Humanity part of nature
Unsustainable lifestyles and behaviour	Sustainable lifestyles and behaviour
Patriarchy	Gender Equality
Intercultural & interreligious intolerance	Intercultural & interreligious tolerance
Conflict resolution through confrontation	Conflict resolution through negotiation
Reliance on Defence	Reliance on Security

The values of 2020 will determine what people regard as valuable in 2020. What they regard as valuable they will want more of. What they want more of they will create markets and products and services for. Those who then become Planetists first and best will get to the future first and thrive by creating the products and services demanded by emerging Planetist markets.

LEADERSHIP AND MANAGEMENT FOR DEALING WITH CHANGE

All of us are part leader, part manager. It is important that all people learn the difference between them and to utilise both of these roles in their lives, not just one of them. Australia is currently an over managed and under led country. It constantly puts managers in positions where leaders are needed. Many Australians fail to understand this critical difference between leadership and management. The first job of good leadership is the leadership of self. Therefore what I am now going to say, refers as much to how we plan our lives as it does to planning and responding to the world around us.

- Managers respond to change and problems, whilst leaders envision, create and shape change.
- Managers are concerned about doing the thing right, while leaders are concerned about doing the right thing.
- Managers reflect about fate. Managers reflect about destiny.

Henry Ford said: "The whole secret of a successful life is to find out what is one's destiny to do, and then do it."

Destiny has two qualities, what you are good at and what you love doing; it is about aptitude and passion. Most people leave school without having any real idea about these qualities because they have not been encouraged to look within. Fulfilling one's destiny is about growing your aptitude and passion in the context of emerging possibilities and opportunities. Destiny is different to fate. You might like to contemplate the difference between these. In Australia we contemplate fate too much and destiny not enough.

- Managers control, Leaders facilitate;
- Managers work in the organisation, while leaders work on the organisation;
- Managers and Leaders also have different kinds of visions and ask different questions about the future.

The <u>manager</u> is most comfortable asking "What <u>will</u> the future of our community be?" I call this the <u>Probable Future</u> question. In over-managed and under-led Australia it is the question most Australians are comfortable with.

An alternative question is "What <u>should</u> our community be like in the year 2020?" I call this the <u>Preferred Futures</u> question. It is the question <u>leaders</u> tend to ask. It is also the question that the average Korean, Japanese, Malaysian or Chinese will ask. It is the question for people who let their dreams play a role in how they think about the future.

You will appreciate that there is a big difference between these questions.

- Managers are Probable Futurists, and contingency plan for Prospective Futures. Leaders are Preferred Futurists and contingency plan for Possible Futures.

What people who ask the probable future question are really indicating is that they have very little influence on the future of Australia. It is a fatalistic view, based on the thought that the future will just happen, and that one cannot shape the future, merely not get run over by it, or if one is smart, make a dollar out of it.

A second way people look at the future involves the process required to get there. The current approach of most people most of the time is a Problem Centred one. <u>This is usually the way of the manager</u>. This involves working towards a future where present problems are lessened or removed. The aim is to remove or lessen present "bads" from the future rather than positively create "goods". The alternative way is to take a Mission Directed approach to the future: to create "goods" in the future, to set out to create a positive future. <u>This is the way of the leader.</u> As it is with the imbalance between probable and preferred futures, there is an imbalance between Problem Centred and Mission Directed approaches. The excessive weight given to Problem Centred approaches makes it very difficult to achieve anything like an optimal result. Again all of this involves using management skills where leadership skills are needed, or because we appoint managers to positions of leadership.

- Managers are Problem Centred people, Leaders are Mission Directed people.

There are many other examples of imbalances between Problem Centred and Mission Directed approaches. The emphasis of medical approaches to health over health promotion is one, the emphasis on pollution control over pollution prevention is another. In the area of structural adjustment of our economy, most of the concentration goes to the Problem Centred repairing the old (modernising the existing industrial structure and infrastructures), rather than the Mission Directed creating the new (designing and building new industrial structures, industries and enterprises appropriate for the 21st century).

Many of the so-called economic and unemployment problems of Australia have neither economic causes nor economic solutions. They are cultural problems with economic, social, cultural and environmental consequences, and the solutions must be found at the cultural level. The biggest problem is the imbalance between Probable and Preferred Futures thinking, and between Problem Centred and Mission Directed approaches to the future.

People change their behaviour for two basic reasons: fear and hope, and their more extreme soul mates, desperation and inspiration. Fear and desperation are the tools used too often by the manager to create change. Hope and inspiration are the tools of the leader.

Too many of our cities and towns are being managed rather than led. A mission–directed preferred future mindset, the way of the leader will help all human settlements from large already global cities to small non-global rural communities, thrive in a globalising 21st century. The way of the manager the problem-centred probable future mindset will not produce a city or community which can manage the change process.

The leader uses the following process to create change: they develop vision, which is used to create hope, hope is used to create inspiration, and inspiration is used to create commitment to change.

If young people enter adulthood after being given the opportunity to grow their leadership capabilities as well as their management capabilities they will be able to first determine their own destinies and shape their own future in a changing world where government's are less important, but communities and individuals are more important.

THE CORE QUALITIES OF LEADERSHIP: THE EIGHT Cs:

Leadership involves what I call "neck down" components including the heart, as well as the intellect The leader embodies six qualities which come from the heart rather than the head. The leader should be:

- confident: having self belief but without hubris;
- courageous: going where others dare not, overcoming self interested opposition;
- committed: doing what must be done, being assertive not aggressive;
- considerate: listening and responding to the opinions and views of others;
- courteous: showing respect in conversation;

- compassionate: responding with empathy to victims and the disadvantaged.

Management on the other hand is largely "neck up", an intellectual exercise. It does not seek to engage the emotions in work. To these six Cs we can add two more Cs which are necessary for effective leadership. This is the ability to:

- conciliate: building and nurturing interdependence and relationships by facilitating compromises which realise win-win outcomes in negotiations.
- communicate: articulating with both head and heart, ensuring both non verbal and verbal forms of expression convey the same message. Leadership therefore embodies eight qualities and capabilities: confidence, courage, commitment, consideration, courtesy, compassion, compromise and communication.

THE LEADER IN ACTION: THE SIX Vs:

I have already mentioned the eight Cs of leadership – the internal characteristics of good leadership: confidence, courage, commitment, consideration, courtesy, compassion, conciliation and communication. This is what the leader is. However it is also the performance of a leader, what the leader does, which is equally critical. A leader who does not get results, or only gets them in ways which alienate the crew or undermine the long term capacity and capability of the crew, is a poor leader.

I want to describe the leader in action by using a metaphor: the leader as a commander of a spaceship leading a culturally diverse and interdependent crew on a mission to a chosen destination

The leader is the facilitator of mission-directed/preferred-future strategies, of mission building. The critical components of mission building are:

- Vision

- Values

- Virtues

- Venturers

- Voyages

- Vehicles

Vision: this involves choosing one's destination. The first task is to develop the preferred future destination: "Where do we want to go and when do we want to get there?" The journey is being made in an environment of change that is driven by both external processes and internal mindsets and behaviours. The second task is to understand and recognise that the spacecraft is already journeying to the probable future, which will be the destination if current trends ("business as usual") are pursued and an alternative destination is not chosen. The leader will also recognise, however, that the leader and the crew are not omnipotent and do not have unlimited resources. Although the leader will undertake measures to improve organisational capacity (financial resources available) and capabilities (human resources available), it may not be possible to achieve the preferred futures goal within the required time-frame. The leader and crew might, therefore, need to select a possible future destination based on a recognition of limitations imposed by the capacities and capabilities as well as by the competition.

Values: this involves asking the questions of "What are the shared beliefs of the crew?" and "What are the rules which determine how the crew behaves and relates to one another during the mission?" Some of these values will be desirable for the completion of the mission and some will not be desirable. It is important that the leader and the crew explore the core values (both desirable and undesirable) which define the ethos of the crew and the rules which govern relationships between crew members and between the crew and the external world. For example, "tall poppyism," "cargo cultism" or "cultural cringism" would not be desirable. Without a clear understanding of shared values, a culture of mutual trust and interdependence will not develop between the crew members, and commitment by the crew to the realisation of the mission will not follow. The crew and leader should examine core values (both desirable and undesirable) and develop strategies to nurture the desirable ones and transform the undesirable into more desirable. Before a crew can determine what virtues it will actually practice, it must first determine what its core beliefs are and assess them in terms of both ethics and performance. Values which undermine or elevate ethical behaviour or which will diminish or improve performance, should be specifically identified. Then strategies for nurturing the desirable and transforming the undesirable would then become part of the capability building strategy to be discussed below under "Vehicles." It is the role of the leader to stretch and challenge values which are no longer appropriate for future needs or which are undermining desirable transformations in behaviour, and to advocate value shifts which can enhance future thrivability. A leader who merely reflects current values, or even a time warp of past ones, and encourages people to be "relaxed and comfortable" with them, such as Australian Prime Minister John Howard will not earn respect. People know that rapid change is always challenging and stretching values and know that they need guidance making value shifts when necessary. They know it is a delusion to sit comfortably holding on to old and dated value systems in a rapidly changing world. The leader must challenge and stretch value systems.

Virtues: these are desirable values that are practised unquestioningly and automatically. They must be desirable and appropriate for the development of a successful mission. Behaviour can be perceived by others but inner beliefs cannot be seen; other people cannot see one's values but they can see one's virtues. Virtues involves behaviour rather than beliefs. It is possible that a leader can harbour an undesirable value such as racism, but this is not of consequence unless this value influences the leader's behaviour. Virtues are values which are actually practised: even if they are not believed. It is important to take core desirable values and ensure that these become practised virtues. The leader must be a practicing exemplar of organisational virtues. A leader who does not "walk the talk" or "practice what the leader preaches" will earn disrespect. This is the flaw of President Clinton: he is a good leader in terms of the seven Cs but is not virtuous: he fails in practice. The leader must be a practitioner/advocate of these virtues, and work to ensure that desirable values become practised virtues and undesirable values remain unpractised. In the context of ensuring prosperity and thrivability in the 21st century these virtues should include the values of Planetism.

Venturers: answer the question "Who will participate in the mission or support the purposes of our mission as allies?" People change behaviour or direction or commit themselves to undertaking a mission for two basic reasons: fear and hope. Fear is the tool of the manager and is over-used; it undermines trust. The creation of an environment of hope is the major tool of the leader. Provided there is trust, vision leads to hope, and hope leads to inspiration. Inspiration will lead to the making of a commitment to the mission. Commitment comes from the heart not from the head, therefore the leader must be able to move the hearts of his crew and this, in turn, delivers the energy required for commitment. The head then comes into the commitment process by bringing intelligence to assess the wisdom of making a commitment. So both head and heart are involved. Management involves the head but not the heart. It is therefore not surprising that the majority of mission statements developed by management gather dust on shelves. They have been developed without an attempt to gain the commitment which comes from the heart.

The leader should promote a climate of organisational interdependence between venturers. There are two particularly important groups of venturers who are critical to the success of any mission. The first group consists of champions who are the "true believers" among the crew, the most strongly committed to the mission and those who will try to ensure other crew members become equally committed. The second group of venturers are external to the crew but are equally committed to the mission. These are allies of the mission who are committed to working towards the same or a similar destination. They will provide external support, intelligence and knowledge and wisdom for the mission. They are also a group who have vested interests in the mission and can be used to counteract vested interests against the mission who might try to stop it or slow it down.

Voyages: involve asking the question "What course should the spaceship take and through what environment?" Many events occur during the voyage. Futurists often use a process called "backcasting". This process describes a journey into the future but it is written as a history from the perspective of the future. It is the opposite of forecasting which identifies major events and their timing and weaves them into a narrative. In this "Future History" of the voyage there are four kinds of events: obstacles, initiatives, improvements and heritages.

They are described in the past tense as events which have occurred and are being detailed after the mission is completed:

- obstacles are constraints and barriers which stopped, slowed down or side-tracked the mission and which were overcome. Descriptor words which can be used include reduce, abolish, overcome, annul, cancel, negate, retard and extinguish;

- initiatives are new infrastructures and actions which were developed and implemented, and qualities, opportunities and facilities which were created during the mission. The descriptor words include establish, initiate, organise, found, increase, encourage, achieve, attain, negotiate and elevate.

- improvements are the changes we make to existing infrastructure, qualities, facilities and opportunities to improve performance and outcomes. Descriptor words include improve, redesign, renew, revitalise, better, enhance, enrich, amplify, fortify and strengthen.

- heritages are priceless elements and qualities relating to the mission, the crew and its culture which must be nurtured and treasured during the mission and for the future. We need to do this so that we do not throw out babies with bath water while we are changing everything else. Descriptor words include protect, defend, nourish, enhance, bolster, support, care for and sustain.

Vehicles: involve answering the question "Which vehicle(s) do we use to reach our destination?" To answer the question we must decide the means by which the mission will be achieved, including the development of new innovations to provide those means. Many of the means will already exist, but it would be foolish to assume that they are all that will be available to the crew. New means will be created in order to realise the mission and these, in turn, will create new opportunities. The Apollo mission, for example, led to the creation of many new innovations. One of the most exciting aspects of mission building is to recognise that mission building is a process of design and innovation. There are two kinds of innovation:

- there are innovations to the crew itself. These are called capacities and capabilities. They serve to improve the ability of the crew and their allies to complete the mission. Capacities refer to additional resources such as financial and technological resources which are identified and utilised to improve the success of the mission. Capabilities involve improving the human resources element, the skills, knowledge and experience of the crew, so that the crew – both as individuals and as a collective – are able to perform at a higher level. A combination of improving capacity and capability will assist the crew to arrive at the preferred future destination, or to move a possible future destination closer to a preferred future destination. The development and maintenance of learning and innovation cultures are important factors relating to capacities and capabilities. Without these cultures the crew will not have the capability to renew and reinvent itself, to adapt, to develop new tools, new means and new resources to fulfil the mission.

- there are innovations which need to develop to realise the mission. The two major vehicles are ways and ware. Ways includes the values, virtues, ethics, beliefs, paradigms, behavioural patterns, customer preferences and professional practices necessary to complete the mission. Many of these ways can be developed through learning and the most effective way to do this is to develop a learning culture. Ways can also be shaped or limited by laws which promote, permit or prohibit actions or things. Incentives and disincentives, both financial and non-financial, are other means of promoting appropriate ways and discouraging inappropriate ways. Ware includes designs, products, services and technologies which will be needed to realise the mission. The development of ware will be most successful if an innovation culture is nurtured. Different ways and ware can be developed for different strategic purposes and their development will provide opportunities for the innovative and enterprising. For example, the ways required to realise a sustainable future can be called "green" ways, while the ware for the realisation of a sustainable future can be called "green" ware. Likewise, we can have health ways and health ware, learning ways and learning ware, and so on.

The way of the leader therefore involves embodying the eight Cs internally and implementing the six Vs externally. It involves creating a climate of hope as the major causal agent of change, instead of utilising a climate of fear. It involves adopting mission-directed, preferred future strategies instead of problem-centred, probably future strategies, or the way of the manager. Those who thrive will be those who fully understand the opportunities provided by the change process initiated by globalisation, tribalisation and technological change, how these trends will develop in the next few decades and who understand the nature of, and practice the values of, a Planetist future. However, they must also be leaders not managers. The world needs management and managers – but not as commanders of spaceships.

WALKING INTO THE FUTURE

We cannot work to create a future which we do not first imagine. Another characteristic of leaders is that they are able to visit the future in their imagination, visualise what is preferred or possible, note their reactions to this and draw conclusions which they can then bring back to the present. Those who are interested in building the industries of the 21st century and designing and innovating the products and services needed will have to walk out into the future in their imaginations in order to appreciate the nature of twenty-first markets. Those who are able to do this will thrive in a Planetist future.

Scenario: the year is 2010 and you are going to buy a new car. There are two cars on the showroom floor. This is a petroleum-powered car and it is a fine piece of European technological excellence. It is quite environmentally friendly and emits only sixty percent of the carbon dioxide that was emitted by cars manufactured ten years previously. It is also very safe.

The second car is a hydrogen-powered vehicle produced by a smaller manufacturer in Australia. It has an Australian-owned majority because all of the former overseas-owned motor vehicle manufacturers decided to withdraw from Australia as part of their global rationalisation programs. The Australian Government intervened to facilitate the formation of an Australian motor vehicle manufacturing consortium. At the same time, it was decided that to be globally competitive this consortium should specialise in hydrogen-powered vehicles because so many Australian journeys involved long distances which meant that solar/electric vehicles were less viable.

Australian-designed and built solar/electric systems produce the hydrogen used as a fuel in these cars from solar energy and water. Thus, hydrogen-powered cars are also partly solar-powered. If everybody drove one of these new hydrogen-powered vehicles all the cities of the world would have much cleaner air and the "greenhouse" effect would be lessened. In another ten years the improvements in both solar/electric vehicles and in solar/hydrogen conversion will mean the virtual elimination of photochemical smog in our cities and a very significant reduction in climate change. End of scenario.

Which car will you buy? Many would buy the hydrogen-powered car or the solar/electric vehicle from Europe because in the year 2010 they would want to be a good Planetary citizen when making such a purchase.

Traditional market analysis emphasises trend projections into the future: a form of probable futurism. This assumes that the future will mostly be composed of a continuation of past trends, whether this is desirable or not, and that one must be a manager and respond to these probable-future projections, rather than be a leader set out to realise a preferred-future. This "walking into the future" scenario operates in a totally different way. You are asked to walk out into the future and to utilise two facts which you know to be true. The first fact is your recognition that values change slowly. You can take your own values, intelligence and intuition a little more than a decade into the future and know that your values will be only slightly different. The second fact is that you recognise that technology changes fast, so it is possible to take your present values ten years ahead and confront a totally different technological scenario. Thus, if one uses imagination and vision in this way one will be able to understand the true nature of a Planetist future and to imagine the nature of a Planetist market-place. One will then be able to create new innovations to service this future market-place.

PREPARING CITIES AND COMMUNITIES FOR SUCCESS IN PLANETIST FUTURE

If we are to be effective managers/leaders of change in our cities and human settlements we must be good systems thinkers. This also means that some of our work to help our cities to thrive might be conducted elsewhere than in our own patches.

The future of our cities and urban areas is interconnected to the future of rural and regional communities. At the beginning of the new Millennium, there are many global cities but very few global rural communities.

While many cities are booming most rural and regional communities are not. Those communities that are genuinely part of the globalising trading and production system are thriving, while those which are not are either just holding their own or are in decline. This includes most rural and regional communities and most developing countries.

If one looks at the industrial structures of most rural areas in 2001 one discovers that the industrial structure is remarkably similar to their industrial structure in 1901. It is as if the whole of the 20th century bypassed rural communities and all the new job categories and new industries of the 20th century were born in the cities. Seventy percent of the job categories and products and services of the year 2020 have yet to be invented. Now in an era of globalising markets, universal connectivity and wider access to education and knowledge, it is now possible to ensure that many of the new job categories and new industries born in the 21st century can be founded in rural areas.

In an article published in *The Weekend Australian* on 1 January 2000, Nelson Mandela, in his role as a Nobel Peace Laureate said:

"Together we live in a global neighbourhood and it is not to the long term benefit of any if there are islands of wealth in a sea of poverty. We need a globalisation of responsibility as well. Above all that is the challenge of the new century."

In these remarks Mandela expresses his concern that globalisation is so far not delivering sufficient equity between the Planet's peoples. This concern can also be seen in the increase in public opposition to the establishment of some of the new institutions of global governance, such as the postponement of negotiations on the Multilateral Agreement on Investment (MAI) in 1998. This was followed by a series of street protests in 1999 and 2000 against meetings of World Trade Organisation (WTO) in Seattle, the World Economic Forum (WEF) in Davos and Melbourne, and the International Monetary Fund (IMF) and the World Bank in Prague. There are also some positive responses to these same concerns such as the Jubilee 2000 campaign, which is demanding the forgiveness of the indebtedness of developing countries. Some of those concerned are opponents of globalisation itself and like King Canute's courtiers have to be persuaded that some things are indeed unstoppable. Others are more realistic, and seek to modify or "civilise" globalisation into a process which will deliver more equitable outcomes, such as providing a better deal for developing countries, safeguarding the global environment, and protecting workers rights in developed countries.

The current juggernaut of globalisation, tribalisation and technological change is delivering both increased global prosperity and many victims. The protestors are drawing global attention to these victims. Any mission to civilise globalisation must include measures to transform the psychological state of victims, and provide them with tools which will empower their capacity to help themselves. It should encourage them to see and take the opportunities in the changes wrought by globalisation as well as the threats, and not only expect others to come to their rescue. This involves encouraging them to exit a world of dependence and to chart the pathway towards independence and then to interdependence.

The traditional ways of delivering social justice have been the social welfare programs of government and the bargaining power of trade unions. Both of these are slowly weakening. The nation state is being unempowered by globalisation, and this functional disempowerment will continue. With the exception of the United States, and possibly China and Japan, most individual governments cannot play a critical role in determining outcomes. The delivery of social justice is one of the major traditional roles of governments. The functional disempowerment of nation states is now limiting the ability of governments to protect its own people from becoming victims of globalisation. They are now not able to introduce a tariff barrier to protect a declining industrial base, except in a short-term way as part of a transitional arrangement.

Once we recognise that national governments are losing their capacity to deliver social justice it is incumbent on us to derive new mechanisms appropriate for the new conditions. Many of these will be based around community empowerment. While national governments are being disempowered by globalisation, communities and corporations are potentially becoming empowered. Whether this empowerment becomes real or remains potential, however, depends on the mindsets and skills of the corporations and communities themselves. Thus far corporations have been much more successful in reinventing themselves for success in an interdependent planetary society than have communities.

There is very little which can prevent a community from charting a new course in a globalised world if they decide to do this. Governments are no longer acting to discourage local self-help initiatives as they once did. The worst excesses which discouraged initiative such as the command economy, and some of the more extreme policies of the welfare state, are things of the past. These encouraged dependence, and discouraged independence and the chance to ever become interdependent. Governments will continue to be willing to be the catalyst of change, provided it does not engender long-term and open-ended commitment. This is the difference between helping out and handing out.

There are a number of steps which a community must take if it is to successfully reinvent itself for prosperity in the 21st century. The first step is to develop a vision, to consciously choose its future. The future is part chance part choice, and too many communities which are in decline have let the balance tip toward chance. Henry Ford's comment about destiny (chapter 3) is apposite. If rural and regional communities want to thrive in a generation's time, they must ensure that their strategic plans are more mission-directed, creating the new, and less problem-centred, repairing the old. Most of the job categories and products and services of the year 2020 have yet to be invented. There is no reason that many of these cannot be created first and best in disadvantaged communities if there is a mission to make this happen. In my work with rural communities in Australia over a number of years, the development of a preferred-future strategic vision is quite easily undertaken. Creating a vision is the first step towards self-empowerment and self -realisation.

Rural communities, and many other disadvantaged communities as well, often tend to suffer from three mindsets which limit the ability of rural/communities to reinvent themselves. The first of these is the "tall poppy' syndrome, the undermining of those people in the community who are the enterprising and the visionaries. The second of these is the "colonial cringe", a belief that rural communities are not sufficiently sophisticated to provide or undertake high knowledge and skill components of the value adding stream. All too often they see themselves only as a commodity producer only. The final one is the "cargo cult" syndrome, the belief that rescue from their predicaments will come from outside, rather than by self-help. They hope that "the cavalry will arrive".

A community committed to collective self-help can be most successful. I am involved with a new international not-for-profit foundation called the MyFuture Foundation. This foundation is responsible for two new programs called MyTown and MyCountry. The goal of MyTown is to "enable communities (or nations) to choose their future and thrive in an interdependent Planetary 21st century". MyTown will work primarily with rural and regional communities to assist them to create and implement strategic visions for sustainable prosperity in the 21st century. MyTown will provide the envisioning and strategic planning tools on-line to enable communities to create this sustainable prosperity for themselves.

In envisioning its preferred-future the community makes choices about its future industrial base, and in the context of both its destiny and of the emerging 21st century planetist market place. To realise this preferred-future on the ground the community will need to collaborate with both governments and corporations. The role which governments can play in these processes is well understood. However the way which communities can realise their envisioned future through collaboration with private sector corporations requires some explanation. Globalisation is changing the potential roles of both governments and corporations. Governments are becoming less influential and corporations more so.

In an era of globalisated relationships, the loyalty of customers to suppliers is lessening and is increasingly valued by a supplier. People have more choices about who can provide them with products and services, and are increasingly able to shop around for them, including on the Internet. In an era of open competition corporate enterprises are looking for long-term relationships with their suppliers and their customers: they are searching for more certainty in a less certain world. To respond to this challenge corporations want to enter long term commercial relationships with their customers and are willing to invest much more to obtain this kind of customer loyalty. The frequent flier programs of airlines are examples of this. Communities can now act as a single aggregated purchaser and negotiate with suppliers for goods and services, and for financial investment in their communities, in return for their loyalty. They get a better price and they provide a larger aggregate market for a supplier. A community which enters a long-term contract with a corporation as part of its strategic plan to help itself can create a strategic alliance, an interdependent relationship which is mutually beneficial. However to do this the community first needs to decide what it wants. It needs a shared vision of its future, and it can then negotiate with corporations and the external world generally to obtain the means to realise its vision of the ground. The community operating as a single collective can enter the global trading system just as nations and individuals can.

In all communities a significant proportion of the income spent by the community, leaves it. Over time, this has tended to increase, and in rural communities a very high proportion of the community's financial resources is used to pay for products and services provided externally. When a local business closes and is replaced by a transnational, money which used to stay in the community leaves the community. Therefore over time the power of communities has tended to decrease as there has been a decline in the capacity of communities to decide how the capital resources they have created are expended. Over time, more and more money leaves the community, never to return. This financial "leakage" has been a major factor in the decline of rural and regional communities.

A community which collectively develops a preferred-future vision and a strategic plan to realise this vision, can now implement policies to ensure that this leakage is minimised in order to increase its ability to finance its own development, from its own resources. In the MyTown initiative each community establishes a community development fund, which can then be used to finance the implementation of the strategy to realise its preferred-future vision. The community can collectively bargain with external providers of community services, in the same way a trade union bargains with an employer. In any community a significant proportion of the community's financial resources leaves the community as payment for services provided to that community. The community can restructure itself organisationally to minimise this "leakage" and retain more of its own money in the community. MyTown is providing expertise to show communities how to achieve this.

This process recognises the reality that the 21st century power will increasingly reside with communities and corporations, and less with governments. This is a completely different model to the traditional model whereby governments provide the machinery and finance for the provision of social justice. In chapter 8, I will be talking about another aspect of community empowerment wherein social welfare is not directed from government to individual welfare recipients but to the communities who provide social welfare rights through a program of mutual responsibility.

Typical of the questions which are asked of communities during the envisioning phase of MyTown are:

- The year is 2020, and this community is world famous for "X". "X" is a product or service, unknown at the turn of the new century, which is now sold on world markets. What is "X"?

- The year is 2020, and this community is now thriving. So much so that the most ambitious and best educated of its young now return to spend most of their lives in the community after some years away undertaking education and work experience. Name a quality, facility or opportunity which has been added to the community to make this difference.

Communities which get to the future first, and together, will become the successful communities in the 21st century. This process can work for communities in rich and poor countries alike, all kinds of communities including indigenous communities. It can also work for whole nations, and this is the purpose of the MyCountry program.

There are many large transnational corporations who are willing to be involved in the MyTown and the MyCountry initiative. It is in their enlightened self interest to do so, and it also affirms their Planetist credentials. No government funding is needed for this model to work, but of course this is welcome. However the model for community development which is used is the one based on the 21st century reality of stronger communities and corporations, and weaker state and national governments.

Some communities will continue to blame governments for not coming to the rescue if nothing is done. This is because many people in rural areas and in disadvantaged groups continually believe that their sole pathway to social justice is through the redistributive role of government. Of course, some of this will and should continue but it is an unwise community to expect this as the major means of financing its reconstruction. Governments should be relied upon for topping up rather than for funding the basic program. Some will find it incredible that I am proposing that communities and corporations can work together to solve their problems, and even without the cooperation of the nation state. However, this is the reality of the emerging global society

While my comments have focused on rural and regional communities, they apply equally to indigenous communities or any community which is currently being disadvantaged. Community initiated programs, based on collective decision-making about their preferred-future, and collective bargaining with service providers, together with the development of community financial resources, including through the lessening of financial leakage, can provide the mechanism for the revitalisation of communities.

The programs which I have described for the revitalisation of communities can also work for disadvantaged nations as well. This is the purpose of the MyCountry program. Developing countries can bargain with the rest of the world in a similar manner. The process has been sucessfully trialled in Papua New Guinea. In my work there I assisted the government to develop a strategic preferred-future vision of itself for the year 2020. This vision and strategic plan is called KUMUL 2020. In a globalised environment developed nations have the best opportunity to succeed if they seek to market their special qualities, their differences. As in communities, however it is important that preferred-future mission-directed mindsets predominate.

For too long we have considered the world as being divided into rich countries and poor countries. In fact, many rich people live in poor countries and many poor people live in rich countries. The significant difference is the proportion of these in various countries. In a globalising world where nations are becoming less significant and individuals, communities and corporations more significant, we must rethink our mindsets on this issue. This mistaken mindset about poor and rich countries reinforces the idea that the major source of redevelopment capital for poor countries should continue to be foreign aid and international capital. Of course this is a part of the equation, but it also ignores the fact that rich people in poor countries could and should be playing a much more significant role. As with communities the issue often is one of "leakage" of capital out of poor nations. Often the mechanisms do not exist to enable these people to reinvest in their own country. The old ways of controlling the export of capital are now not possible in an open global capital market. New means to prevent capital leakage must be found and a renegotiated Multilateral Agreement on Investment (MAI) should incorporate a new set of rules to achieve this. MyCountry aims to assist developing countries to minimise the leakage of their own financial resources, including promoting communitarian mechanisms to encourage local investors to invest more at home.

Some of the steps that all communities should take if they want to find sustainable prosperity in an era of globalisation, and at the same time contribute to the process of civilising globalisation, are:

- becoming Planetist first and collectively so that communities are able to understand the requirements of the Planetist market place, and create a thriving 21st century community. This requires that communities develop mission-directed strategies and that they are committed to "creating the new and emerging" rather than "propping up the old and declining". Accepting assistance from governments or others which is aimed at propping up declining industries is, in essence, accepting a form of palliative care.

- consider their destiny: what they are good at (aptitude) and what they love doing (passion), and consider this destiny in terms of new emerging opportunities for the provision of products and services for Planetist markets. Insight should precede foresight.

- helping themselves by developing preferred-future visions based on the community's destiny and initiating, through the development of strategic action plans, new ways to fund their new initiatives, such as by the minimisation of the financial leakage from the community.

- installing high capacity connectivity, such as via the Internet, to connect their community with potential customers elsewhere who will be interested in their products and services, and with collaborators and allies who can help the community restructure itself. Ensure that this connectivity is available to all, irrespective of capacity to pay.

- seeking to grow the community's intellectual property in the areas which it has chosen to promote for its future success, by investing resources in research and development, and innovation in these areas of chosen excellence, so that the community can ensure that it remains an industry leader in these areas. Ensure that tertiary education programs are the best possible ones in terms of both quality and access. The education system should excel at nurturing the capabilities needed to realise the chosen industrial future. Each community should aim to ensure that new transnational organisations, which specialise in these same areas, are founded and incubated and that the leadership of these organisations remains based in the community. All organisations should nurture appropriate brand names and other ways of differentiating its products and services so they can maximise their chances of being a price maker, rather than a price taker in a globalised market place.

- continuously promoting leadership, learning and innovation. Communities should restructure their leadership from a traditional "control" form to an "empowering" form. In the "empowering" form, the traditional leadership adopts a mentoring/elderhood mode, and empowers the young and emerging leadership to assist the community to reinvent itself. This most effectively combines the wisdom of the old and the energy of the young. It is a form often used in Japanese corporations to ensure that old leadership, which is naturally conservative, does not prevent innovation and the implementation of "creating the new" strategies in the corporation. In rural and regional communities the leadership is often both conservative and closely tied to the old and declining industrial order. It often struggles to embrace a newer industrial order. As well as leadership, the other main cultural issue which needs attention is the promotion of learning and innovation cultures in the community. In particular, entrepreneurship should be promoted and affirmative action programs in these areas should be introduced for disadvantaged members of the community.

- encouraging the rich and successful to show greater loyalty and responsibility to their communities by investing more and providing mentoring to their own communities. Successful communities often have a culture of community philanthropy and community responsibility is upheld by the community's most successful members, including those who have become expatriates from the community and have become successful during their time of expatriation. In my work with communities I often suggest that the community invites its successful expatriates to return to the community as "mentors in residence" for a few months to assist the community reinvent itself for future success.

- ensuring that the NGO sector which is the ethical and moral watchdog of the behaviour of the global trading system, and of social justice at the national and community level, is sufficiently empowered to be effective in this role. This includes community and global NGOs such as Greenpeace International, Amnesty International, Transparency International, religious social justice organisations, international aid NGOs and the like. This will ensure that new international institutions such as the International Criminal Court, and UN agencies are able to be more effective in their critical work. It will be NGOs who will increasingly provide the people who will oversee the work and development of the global system of governance. They can operate as they see fit, and are not constrained by the limitations posed by the consensus or "least common denominator" based negotiations that constrain the effectiveness of negotiations between governments.

Dr. Peter Ellyard is a futurist and strategist who lives in Melbourne. A graduate of the University of Sydney and of Cornell University, he is currently Executive Director of Preferred Futures and Chairman of the Universal Greening Group of companies and of the MyFuture Foundation. A former Executive Director for the Australian Commission for the Future, he held CEO positions in a number of public sector organisations over 15 years. He is Adjunct Professor of Intergenerational Strategies at the University of Queensland, and is a Fellow of the Australian College of Education, the Environment Institute of Australia and the Australian Institute of Management. He has been a Senior Adviser to the United Nations system for 25 years.

The Sustainable Tall Building of the Third Millennium

Harry Blutstein and Allan Rodger

ABSTRACT

Governments, industry and the communities are developing a consensus that the current pattern of human activity is ecologically unsustainable, and that this must change.

The paper explores issues that will need to be resolved for a successful sustainable tall building of the Third Millennium. This will not just incorporate sustainable features, like energy efficiency, but a more fundamental change in the systems – technical, economic, organisational – involved in delivering the project, not to mention re-defining the project itself. The role of externalities also needs to be addressed.

Any project is shaped by constraints, which are part of the creative process. In a sustainable building the constraints need to be identified to allow the imaginative solutions to be developed.

To achieve fundamental changes in delivering a tall building project will require strong drivers. These do not exist, as the development of a project is fragmented between the developer and end user. As companies with a deep commitment to sustainable development begin to appreciate that their accommodation represents a significant component of their ecological footprint, then they will need to look more carefully at the supply chain that delivers that accommodation, and seek to influence it to ensure a sustainable outcome.

INTRODUCTION

High buildings and high density living have gone together since the earliest days of urbanisation, though the driving forces leading to such urban form have varied from time to time and place to place. For much of human history security has been a dominant issue shaping the urban environment. Defence of civil society (and its livestock and artefacts) against a hostile and lawless countryside saw the formation of defensible walled cites and sometimes within these there were hostile competing forces (see the tower palaces of northern Italy[1]). It could be that the corporate fortresses of the present day are the symbolic successors of these imperial fiefdoms[2].

At modest densities, of up to a few hundred persons per hectare, low rise constructions can serve. As density increases, however, or as the expectation for space per person increases so too does the demand for higher buildings.

Physical security is no longer the determining influence of urban form. Within the urban context, the growth in popularity of tall buildings (both number and height) is now mainly driven by economics and demographics, as business location is a strong determining factor as is the scarcity of land in urban centres.

The evolution of the tall building has been shaped by a range of restraints. Principally these are economic, technological, social and basic environmental constraints, with the last mentioned often governed by regulatory requirements, such as over-shadowing and more recently with the need to maintain a high level of internal air quality[3]. Architects have worked within these constraints to generate innovative and imaginative solutions. The main changes over the last hundred years have been improvements in technology, which have allowed sky-scrapers to reach new heights.

The tall building of the twenty-first century will need to address new environmental constraints of an order not yet fully realised, which, like previous challenges, will stimulate new creative solutions.

Systems such as financial analysis, procurement methods and project management. have been introduced to efficiently address current constraints and deliver a profitable product. In addressing new environmental constraints, these institutionalised systems also need to be reviewed to see how they can be adapted to produce a sustainable tall building of the Third Millennium.

THE NEED FOR A SUSTAINABLE BUILDING

In many areas of development, particularly since the Industrial Revolution, the main challenges in delivering tall buildings have been seen in technological terms, and working within an economic paradigm that assumes continuous growth. However natural systems provide a boundary to such growth, which is only now being appreciated[4].

There is now a better understanding that the main input into economic growth is natural capital[5]. Growth is therefore limited by the ability of natural systems to provide ecological service in a sustainable manner. It is conservatively estimated that the economic value of ecological services produced by the Earth is of the order of $US33 trillion per year[6].

Like any business that depletes its capital to generate profit, if humans continue to utilise natural capital at a greater rate than it can be replaced by the biosphere, then the operation is unsustainable, and current profits will be a temporary aberration. As natural capital is depleted the Earth will be threatened by bankruptcy[7]. Unfortunately, natural capital has not been adequately valued[8], which has meant that there is an economic incentive to exploit it unsustainably.

The World Wide Fund for Nature has estimated[9] that the Earth's natural ecosystems have declined by about a third over the last 30 years while the ecological pressure of humanity on the Earth has increased by about 50 per cent over the same period. As the destruction of natural capital exceeds the ability of the biosphere to regenerate itself, considerable effort and change will be required to achieve a sustainable world. One estimate[10] suggests that materials consumption will need to be reduced by half worldwide and by forty times in developed countries.

A response to this environmental crisis has been incorporated into the concept of ecological sustainability development, which was introduced by the former Prime Minister of Norway, Gro Harlem Brundtland[11], when she headed the World Commission on Environment and Development. That Commission defined sustainable development would be: "development that meets the needs of the present without compromising the ability of future generations to meet their own needs".

The concept of sustainability requires fundamental changes to how business is done. It moves away from the simple idea of environmental protection, which involves avoiding doing direct harm. Instead it seeks to reduce the ecological footprint[12] of human activities. It has been estimated that the global ecological footprint increased by 50% between 1970 and 1997, a rise of about 1.5% per year[13]. This data supports the conclusion that natural capital is being unsustainably eaten up to support human activities

While there is no specific data on the ecological footprint of tall buildings, it has been estimated that building inputs consume 40% of raw materials, 36–46% energy and take up 20–26% of landfills[14]. Finally, the shape of our cities is influenced by commercial centres, where tall buildings predominate, and therefore their footprint also needs to include infrastructure such as the transport system which delivers people to their workplaces.

DEFINING THE SUSTAINABLE BUILDING

Architects have taken the lead in incorporating the concept of sustainable development in their principles, and in practice.

According to Robert Berkebile, founding chairman of the Committee on the Environment (COTE) of the American Institute of Architects', sustainable building design is an act of restoration and renewal, contributing to the social, economic and environmental vitality of the individual and of the community[15].

Translating such aspirations into design outcomes, William McDonough[16] formulated a set of principles for sustainability of the built environment now known as the *Hannover Principles*. They are:

1.　Insist on the rights of humanity and nature to coexist in a healthy, supportive, diverse and sustainable condition.

2.　Recognise interdependence. The elements of human design interact with and depend upon the natural world, with broad and diverse implications at every scale. Expand design considerations to recognise even distant effects.

3.   Respect relationships between spirit and matter. Consider all aspects of human settlement, including community, dwelling, industry and trade, in terms of existing and evolving connections between spiritual and material consciousness.

4.   Accept responsibility for the consequences of design decisions upon human well being, the viability of natural systems, and their right to coexist.

5.   Create safe objects of long-term value. Do not burden future generations with requirements for maintenance of vigilant administration of potential danger due to the careless creation of products, processes or standards.

6.   Eliminate the concept of waste. Evaluate and optimise the full life-cycle of products and processes, to approach the state of natural systems, in which there is no waste.

7.   Rely on natural energy flows. Human designs should, like the living world, derive their creative forces from perpetual solar income. Incorporate this energy efficiently and safely for responsible use.

8.   Understand the limitations of design. No human creation lasts forever and design does not solve all problems. Those who create and plan should practice humility in the face of nature. Treat nature as a model and mentor, not as an inconvenience to be evaded or controlled.

9.   Seek constant improvement by the sharing of knowledge. Encourage direct and open communication between colleagues, patrons, manufacturers and users to link long term sustainable considerations with ethical responsibility, and re-establish the integral relationship between natural processes and human activity.

To develop these principles, McDonough has teamed up with Michael Braungart. Together they have analysed[17] how materials can be cycled through the biological or a technological metabolic process, ideally without down cycling from higher grade to lower grade products. They argue that material flows must now become cradle-to-cradle.

In the literature there is a significant body of information on technologies to support these design principles[18], often demonstrating that extra construction costs are offset by reduced running costs[19].

PARADIGMS

The ability to identify paradigms[20] is important if existing practices that operate against achieving a sustainable building are to be challenged, and creative responses developed.

It is therefore useful to contrast some of the assumptions of various aspects of the creation and use of a conventional tall building with that of what we might expect of a sustainable building. Inevitably this is a coarse approximation, but the contrast between the two sets of expectations provides a useful starting point from which to challenge existing assumptions and paradigms.

The Issue	For a Conventional Building	A Sustainable Building
Limits	Physical constraints Cultural constraints (eg planning) "Height matters"	Ecological constraints Resource limitations "It's not height that matters, it's what you do with it"
Relationship to natural capital	Parasitic, using natural capital with a net deficit	Synergic, both using and contributing to natural capital with no net deficit
Design objectives	Economically viable investment Public statement – place of importance – visibility Meet basic standards of habitability	Sustainable life support system Interactions (eg transport, staff, materials, wastes) Involving occupants in sustainability mission
Location	Interactions, in which business-to-business and business-to-customer proximity are prime criteria	Efficiency of access (people, goods, support systems)
Shape	Statement Maximum useable floorspace	Harvesting resources (eg air, light, water, visual amenity)
Fabric	Economic constraints (capital cost) Conventional aesthetics	Provision of services (eg air, light, view) Aesthetics of a sustainable culture
Technology	Comply with standards, based on conventional practice	Outcomes focus based on best practice, such as employing innovative ways to maximise effectiveness of resource use

The Issue	For a Conventional Building	A Sustainable Building
Use	Fixed at the time of hand-over managed turn-key systems	Subject to ongoing environmental management, custom fitting, re-fitting and maintenance, Active environmental management
Economics	Focus on initial capital cost Rate of return on investment	Total life cycle evaluation
Management	Fragmented between developer, owner and lessees	Total asset management[21]
Personal Space	Generalised De-personalised	Recognition of workplace conditions as business-specific fundamental
Social	Ignore	Quality of life Use of time
Externalities	Compliance with regulations and codes, with no responsibility taken for outcome	Incorporated into planning and development phases, with focus of maximising outcomes
Effectiveness measures	Based on return on investment, usually capital component. Also includes the rentability of the building.	Measures the effectiveness of the building on its ability to cost effectively deliver staff productivity and reduce the ecological footprint

This list is not exhaustive, and other similar lists exist[22]. However it is presented to illustrate the mental changes that will need to happen for those currently producing conventional buildings if they are to produce the sustainable building that will be required in the future.

FRAMEWORK FOR THE SUSTAINABLE BUILDING

There is considerable difference between a sustainable building and a building with sustainable features. The two are often confused, with most projects being in the second category.

A sustainable building requires more than identifying solutions to specific problems, but changes to attitudes, paradigms, processes and systems to deliver the project.

If the process of identifying suitable technologies is not done within a systemic framework, it becomes piecemeal and may even become tokenistic[23] as specific solutions are lauded, while, overall, the sustainable performance of the building is poor.

The principles enunciated above provide a useful guide, though providing little information on how they can be translated into a building with truly sustainable outcomes.

In looking at the processes that currently are involved in delivering a tall building, it is little like being with Alice in Wonderland[24], as she chats to the Cheshire cat:

"Would you tell me, please, which way I ought to go from here?"

"That depends a good deal on where you want to get to," said the Cat.

"I don't much care where –", said Alice

"Then it doesn't matter which way you go," said the Cat.

"– so long as I get somewhere," Alice added as an explanation.

"Oh, you're sure to do that," said the Cat, "if you walk far enough".

Unless there is a common objective in a tall building, beyond being "tall", then it is not surprising that projects lack direction in comprehensively addressing the objective of sustainability.

Current systems are characterised by the following features:

- Rigid processes and procedures;

- Clear demarcation between professional "silos";

- Controlled (and limited) engagement between "silos";

- Fragmentation of problem solving;

- High level of automation, with heavy reliance on standards to define design features and services; and

- Disincentives for professionals to produce energy and resource efficient outcomes.

Institutional inefficiencies in various components of a project are described in more detail by Lovins[25] and Hawkins et al[26].

Few projects are defined in terms of sustainable outcomes, which is not surprising with the structure of the industry that is fragmented between developer, financier, designer, construction team, lessee, lessor and facilities manager, each with their own objectives. It will take some significant changes in the relationships between the various stakeholders in the "assembly line" that produces a tall building to achieve such a common goal, which is an essential pre-condition for delivering a sustainable tall building.

Once a goal based on sustainable development is in place then it is possible to develop more detailed objectives for a sustainable tall building and performance measures by which it can be assessed.

Existing systems and processes will need to change to bring together the various components of the project, so that financial and professional inputs are aligned with the sustainability goals of the project.

Translating the sustainability goals and objectives into design and construction specifications will require significant changes in the framework by which problems are solved.

There are a number of such frameworks available. The one that has gained the most widespread acceptance is The Natural Step (TNS), developed by Karl-Henrik Robèrt[27] as it is grounded in fundamental scientific principles[28]. They are summarised into four system conditions:

Nature is not subject to systematic increasing:

1   concentrations of substances extracted from the Earth's crust

2   concentration of substances produce

3   degradation by physical means

and, in that society

4   human needs are met worldwide

TNS has been effectively used in the construction of the University of Texas Health Science Centre, resulting in a reduction of energy usage by 30%, installation of a graywater system to reduce water consumption and ensuring that a high level of recycled materials were incorporated in the structure[29].

TNS not only identifies direct impacts of activities and materials used on a project, but considers supply chain issues such as whether timber is sourced from a sustainable harvesting operation. It is a structure to identify the nature of sustainable solutions rather than delivering applied solutions.

To complete the process it is useful to combine TNS with other analytic tools such as Cleaner Production[30], Industrial Ecology[31] and Dematerialization[32], where a range of case examples demonstrate the effective deployment of technologies that can be used to address the issues raised by TNS.

One of the advantages of aligning various stakeholders involved in a project is the lack of common language in which to communicate what it means to deliver a sustainable tall building. TNS provides such a common language.

CONCLUSION

Ken Yeang has incorporated the principles of sustainability in his "bioclimate" skyscrapers[33]. His buildings seek to create a place responsive to local climate and able to deliver a high level of comfort for occupants. In achieving these ends he makes innovative use of natural light and ventilation and of internal ecologies such as sky courts and vertical landscapes.

However such examples are few, although many buildings do incorporate elements sustainability, this is often done in a piecemeal manner.

There is a trend emerging where corporations are likely to take more interest in performance of their accommodation in terms of sustainability, to address commitments made in their environmental policy.

While the end user often has little impact on the supply chain that delivers a tall building for their use, this has occurred in other industries such as motor vehicle manufacturing[34]. Top-down decision making will not be replaced by bottom-up, but decision making at all levels and all along the supply chain is likely to be increasingly susceptible to community expectations on sustainable development.

While this paper has addressed some of the processes and drivers that could be used to deliver a sustainable tall building, it has not addressed the question of we will have tall buildings in the Third Millennium. While there is little value speculating about this question, what can be said with some certainty is that unless we learn to work within the ecological restraints of nature, then the tall building will have no future.

ACKNOWLEDGMENTS

The authors would like to thank Virginia Kneebone (VK+Assoc), Ros Magee (Spowers) and Rives Taylor (University of Texas – Houston) for there assistance in preparing this paper.

REFERENCES

1   Fernández-Armesto, F, *Millennium: A History of our Last Thousand Years*, Black Swan, 1996. See Figure 5.10 which shows the towers of San Gimignano dominating the skyline.

2   The link between Medieval Verona and a modern American city was made by Baz Luhrmann in his film *Romeo + Juliet* (Twentieth Century Fox, 1996). In his contemporary update of Shakespeare's classical story, the Montagues and Capulets are represented in the first scene as

aggressively competing companies, with their names emblazoned on the front of their corporate skyscrapers.

3   This is known as the sick building syndrome, and relates to the circulation of contaminants such as radon, formaldehyde and bacteria within the built environment. Most indoor air pollution comes from sources inside the building. For example, adhesives, upholstery, carpeting, copy machines, manufactured wood products, cleaning agents or biological agents via the air conditioning and ducting system.

4   Meadows D H, Meadows D L and Randers J *Beyond the Limits: Confronting Global Collapse, Envisioning a Sustainable Future*, Chelsea Green, Boston, 1992.

5   Natural capital has been defined by Herman E Daly in *Beyond Growth* (Beacon Press, Boston 1996) as being functions and services provided by the biosphere, such as ecosystem goods (such as food) and services (such as waste assimilation) which represent the benefits human populations derive, directly or indirectly. When natural capital is used, often at minimal charge, and combined with manufactured and human capital services it is integral to contributing to human welfare.

6   Costanza R, d'Arge R, de Groot R, Farber S, Grasso M, Hannon B, Limburg K, Naeem S, O'Neill R V, Paruelo J, Raskin R G, Sutton P and van den Belt M. "The Value of the World's Ecosystem Services and Natural Capital", *Nature*, Volume 387 no 6230, 1997.

7   Two examples of collapse of different parts of ecological systems can be illustrated by experience with fish stocks in different parts of the world. Falling stocks of cod in the North Sea, possibly linked to Greenhouse warming, resulted in quotas being instituted in 2000 (see Casey J "Prospects poor for North Sea cod", *Fishing News*, 8 October 1999). A similar collapse in cod occurred off Newfoundland in 1992, which was mainly due to over exploitation of the resource (see Haedrich R L and Hamilton L C "The fall and future of Newfoundland's cod fishery", *Society and Natural Resources*, 13:359–372, 2000). Both examples illustrate the link between economic and natural systems.

8   Costanza *et al*, op cit.

9   World Wide Fund for Nature *Living Planet Report 2000* see http://www.panda.org/livingplanet/lpr00/downloads/lpr_2000_full.pdf

10   Schmidt-Bleek F, "MIPS and the Ecological Safety Factor of 10 revisited", Unpublished paper, Wuppertal Institute, Germany 1993.

11   World Commission on Environment and Development, *Our Common Future*, The World Commission on Environment and Development, Oxford University Press, Oxford, p 43, 1987.

12   The ecological footprint is a convenient indicator of the amount of land and resources people impact by their production and consumption behaviours. In doing so the footprint demonstrates the connections between local behaviour and environmental quality across the earth. The footprint is a measure that relates any activity that processes energy and materials by relating it to an area or volume of land, sea and atmosphere that is necessary to support that activity. Usually expressed in hectares per person the ecological footprint is used as a comparative tool for assessing the sustainability of a given geographic area. For example, were the rest of the world to operate at the level of environmental impact of the US we would need three planets with the regenerative capacity of Earth. For more background on this concept see Wackernagel, Mathis and Rees, *Our Ecological Footprint: Reducing Human Impact on the Earth* (New Society Publishers, Gabriola Island, BC, Canada, 1996).

13   World Wide Fund for Nature, *Living Planet Report 2000* see http://www.panda.org/livingplanet/lpr00/downloads/lpr_2000_full.pdf

14   Yeang K, *The Green Skyscraper*, Prestel Munich 1999, p 9.

15   Zeiher L C, *The Ecology of Architecture*, Whitney Library of Design New York, p 31, NY, 1996.

16   William McDonough launched *The Hannover Principles* as a manifesto to EXPO World Fair in Hannover. See http://www.virginia.edu/%7Earch/pub/hannover_list.html, 1992.

17   McDonough W and Braungart, M, "The NEXT Industrial Revolution", *The Atlantic Monthly*, October 1998, www.theatlantic.com/issues/98oct/.

18   See *Sustainable Architectures Bibliography* at http://www.arionline.org/ari/info/programs/sustainable.htm.

19   Edwards B (editor), *Green Buildings Pay*, E&FN SPON, London. 1998.

20   Kuhn T S, *The Structure of Scientific Revolutions*, University of Chicago Press, Chicago, 1970.

21   Total asset management is defined as including all management initiation, processes, investment analysis, project management (including design, procurement and construction) and facilities management.

22   Van Der Ryn, S and Cowan S, *Ecological Design*, Island Press, Washington, 1996, pp 26–27.

23   A number of companies have been accused of "Greenwashing", where they publicise ephemeral achievements while continuing to conduct their business unsustainably, and sometimes causing significant harm to the

environment. For more details see Kenny Bruno and Jed Greer, book
Greenwash: The Reality Behind Corporate Environmentalism (Penang,
Malaysia: Third World Network. 1996)

24   Lewis Carrol, *Alice's Adventures in Wonderland*, Vintage Books Edition,
NY, pp 71–72, 1976.

25   Lovins A, "Institutional Inefficiency", *In Context*,
http://www.context.org/ICLIB/IC35/Lovins.htm.

26   Hawken P, Lovins A B and Lovins L H, *Natural Capitalism: The Next
Industrial Revolution*, Earthscan Publications Ltd, London 1999.

27   Robèrt K-L, "Educating the Nation: The Natural Step", *Context*, No 28,
Spring 1994.

28   The Natural Step is based on the principles of thermodynamics. It argues
that the biological and other elements of the biosphere that have evolved
together and over a long period of time have established an effective
working arrangement on which the biosystems are now dependent.

29   University of Texas, *Sustainability Report* 200.

30   Cleaner production is the continuous application of an integrated
preventive environmental strategy to processes, products and services to
increase efficiency and reduce risks to humans and the environment. It
includes the conservation of raw materials and energy, the reduction of
toxic raw materials, and reduction of the quantity and toxicity of all
emissions and wastes.

31   Industrial ecology is an interdisciplinary framework for designing and
operating industrial systems as living systems interdependent with natural
systems.

32   Dematerialization is the process by which lesser amounts of material are
used to make products or perform the same function as their predecessor.

33   Yeang K, *The Skyscraper Bioclimatically Considered: A Design Primer*,
Academy Group Ltd, London, 1996.

34   Through the vehicle of ISO 14000 certification, large motor vehicle
manufacturing corporations are requiring companies that supply them to
also be ISO 14000 certified. This requirement has seen large companies
requiring their suppliers to also be ISO 14000 certified.

Sustainable Cities

Patrick Troy

SUMMARY

Although we have not specified what we mean by sustainability and therefore what we mean by sustainable cities we have accumulated enough scientific evidence to suggest that the effect of human agency on local and global ecosystems resulting from the way we live is unsustainable.

This paper argues that the form and structure of the city contributes to the present generation of environmental stress and that we can significantly reduce that stress without sacrificing living standards by changing the form and structure of our cities.

INTRODUCTION

Australia's urban areas are the locations where the greater proportion of the nation's wealth is created and held and where the culture of the nation finds expression. In this most urbanised of nations, the cities and urban areas are also the locations where the great proportion of the population engage in a wide variety of economic, social and political pursuits.

Our cities are our most intensively shaped landscapes. They are the most heavily overlaid and inscribed by the pipes, wires, roads, tracks and cultures that reflect and represent our aspirations for the present and our images of the future. They are the locations of much pollution, are the generators of much of the greenhouse gas emission, the greater part of the waste stream and the demands for water, which have such a devastating effect on many of our catchments.

Over the past century we have witnessed increasing concern over a range of environmental issues. We have expressed that concern in a variety of ways and taken a series of initiatives to minimise what we used to call 'externalities'.

The development of water supply, sewerage and drainage systems, the regulations to control water pollution, the introduction of clean air regulations, the separation of residential development from industrial activities, the introduction of town planning all had a concern over environmental issues at their heart.

We now see that concern over the viability of the biosphere itself is rising and that rather than be seen as 'externalities' environmental issues are of central importance.

We have begun to recognise that the natural ecological systems cannot continue to function as they have done if we continue our present practices in the exploitation of natural resources. Although our understanding of the natural processes which occur, and how they might be affected, is incomplete there is enough evidence to accept that we have exploited or driven various species of flora and fauna to extinction and have compromised the lives of others, in the process reducing the diversity of the biosphere.

Our farming practices have led to desertification, and our clearing of forests has increased flooding, erosion and salination. The consumption of fossil fuels has led to a great increase in CO_2 in the atmosphere. Our activities have created major holes in the ozone layer with consequential threat to the stability of the biosphere. We suspect that forest clearance and the increase in CO_2 lead to climate change which, together with the holes in the ozone layer create major threats to life sustaining ecosystems. We are struggling to cope with mountains of waste which themselves become major threats to ecosystems.

Clearly, cities and rural regions exist in a symbiotic relationship so I do not intend here to imply some kind of city versus the country tension. That is simply an unproductive avenue of exploration but it is important to recognise that many of the stresses on our rural ecological systems originate in demands placed on them by the demands of urban populations for food, water and other natural resources. Attempts to establish some kind of city versus country dichotomy in the approach to resolution of environmental issues is currently used by the Commonwealth as a device to focus on the regions. The specific exclusion of cities from the initiative to combat salination is simply one illustration of this myopia. Unfortunately this dichotomy also a device used by many in the city to avoid accepting responsibility for the stresses the city places on the environment.

Nonetheless we have enough scientific evidence about the effect of human agency on local and global ecosystems to accept that the way we live is ecologically unsustainable.

SUSTAINABILITY

The word 'sustainability' however, means what speaker and listener want it to mean. On some estimates it is now differently defined in 150 pieces of legislation in Australia. Some argue that this is a strength, that there is value in the lack of precision. We might accept that there is great value in our kind of democracy in public processes of debate but this should not preclude striving for consensus on what we might mean by sustainability at any time.

In 1992 Dover and Handmer defined *Sustainability* as 'the ability of a natural, human or mixed system to withstand or adapt to, over an indefinite time scale, endogenous or exogenous changes perceived as threatening. *Sustainable development* is a pathway of deliberate endogenous change (improvement) that maintains or enhances this attribute to some degree, while answering the needs of the present population.'

They went on to make the point that 'sustainability is thus a long term (and probably fanciful) condition, and sustainable development the variable process of moving closer to that condition.'

These definitions lack precision which has often led to governments, including local government, appearing to create uncertainty, confusion and ambiguity among residents and developers. The lack of precision is not a strength in our legal system which inevitably seeks to establish clarity and to reduce ambiguity. If the planning system, including the process of public debate, does not provide definition or give meaning to the notion of ecologically sustainable development, the courts will.

While ecological sustainability is seen by some as problematic and by others as unattainable the reality is that Australia must adopt the strategy of a transition to sustainability by attempting to systematically reduce environmental stress. Nowhere is this strategy more important than in the cities given their central role as sources and locations of environmental stress.

It is important also because systems which import all their energy, water, food and raw materials and export their waste are, by definition, unsustainable. Modern cities are examples of such unsustainable systems. The challenge is to make them less unsustainable.

In pursuing the transition to sustainability, general economic settings and mechanisms, including the suite of pricing strategies and taxes, are important. However, planning tools and aids to decision making are needed which will complement them and allow the introduction of location and space issues into the consideration of policy options and development proposals.

One of our tasks, then, is to progressively invest the word and concept with a meaning which could be socially enacted with substantive outcomes. Another is to develop a planning system so that sustainability issues can be considered systematically and democratically in the planning and development of the cities and regions.

In developing such a planning system the first thing we must acknowledge is that there is no 'end point'. We must recognise that we are engaged in a process by which we set goals and targets which we strive to meet over a specified period knowing that the goals and targets will be continuously revised. That is, the notions of 'sustainability' and 'sustainable development' will be incrementally, progressively defined and we will periodically arrive at new consensual definitions.

URBAN PLANNING

For half a century we have employed land use planning as a way of pursuing social, environmental and economic goals in the development and operation of urban and regional areas, although economic goals have been given the greatest weight.

It has always been difficult to say with certainty that particular intensities or arrangements of uses would lead to specific outcomes. The identification of land uses were, at best, only ever crude approximations of the nature of activities, the connections between them and the 'externalities' associated with them.

That is, land use planning relied to a very large extent on the precautionary principle in pursuit of these goals.

Planners were for a time able to convey confidence that their prescriptions and recommendations about the uses to which specific pieces of land should be put, based on this precautionary principle, would produce the felicitous social, environmental and economic outcomes collectively sought.

A great deal of regulation was justified and built on this expression of trust – this belief in the efficacy of the decisions made by planners – and there is no doubt that it frequently produced congenial results. But there were also many instances where the lack of specificity or where the relationship between the activities proposed for a particular piece of land and the social, economic or environmental outcomes of those activities were vague or contested and which led to courts becoming involved in providing the precision or determining the relationship. A great deal of case law evolved to buttress this, which might have been a comfortable outcome for many lawyers but it thrust courts and judges into playing the role of planners – a role for which they were not necessarily well suited, nor did they welcome it.

By the 1980s the notion of scientific planning was unpopular. Urban planning which had never been seen as central in government decision making was even more marginalised. The public appeared to lose faith in politicians and instruments of government. The role of government was challenged and large areas of administration and service provision were deregulated or privatised. For a variety of reasons, not all of them due to the fallibility of planners, the land use planning practices followed were not able to cope with changing demands to accommodate growth or with the simultaneously increasing concern over environmental issues.

From the mid 1970s State and Federal governments were seized by a preoccupation with 'short-termism' a kind of virus which degraded the sensibility of politicians as they succumbed to the pressures and blandishments of entrepreneurs and their financiers. They were led to the belief that if they did not make the quick decision and respond instantly to the imperatives of the 'market', and especially to the international entrepreneurs and financiers, cities would 'miss out' and be left behind.

The current expression of concern over sustainability can, to some extent, be seen as a reaction to the limitations of narrowly conceived considerations of environmental impacts arising from pressure from entrepreneurs and their financiers for quick responses to their proposals. That is, the concern stems from anxiety that long term environmental consequences of developments are overlooked in favour of alleged short-term benefits.

The current expression of concern is also recognition of the global impact of many of the environmental stresses cities experience and generate. This recognition is tending to force governments to seek ways of ameliorating both the source and effects of the stresses.

How should we respond to this situation? How should the threads of the frayed town and regional planning system be pulled together to weave a stronger web which takes fuller account of environmental issues in the pursuit of social, environmental and economic goals?

PATH DEPENDENCY

To provide discipline or focus in this kind of dynamic planning we typically set a period which is 'realistic' – not too short to have no effect yet not so long as to be fanciful or regarded as so far in the future it will not affect behaviour or expectations. For the purpose of much of our planning we have tended to set horizons of twenty years.

While this period is arbitrary it is long enough to enable us to make significant changes to the infrastructure of our cities such as water supply, sewerage and drainage systems, road and rail networks, together with the rail rolling stock, and ferry and vehicle fleets, etc. It is long enough to make significant changes to other elements of the built environment, assuming current levels of building and construction activity. It is also long enough to make progress in achieving sustainability goals in the production and consumption of a range of services and in commercial and manufacturing processes.

The major disadvantage of such a horizon is that it is beyond the political cycle. This is particularly important in the approach to ecological sustainability. It has taken two centuries for some of the environmental issues to become critical and it is fanciful to imagine that we can quickly solve the problems. Salination, for example, has been with us for some time but it is only recently that its full magnitude has been accepted and the need for an imaginative large scale, continuing effort over a long period to reduce or eliminate it has been recognised. Even here, however, for a variety of reasons, the significance of salination and waterlogging in urban areas remains unrecognised.

Over the next twenty years the inherited form and structure of the cities will largely affect the provision of urban services in our cities, their provision will also, of course, be affected by the characteristics of the present investment in them.

This is not to say that planning for transition to sustainability is governed by the path dependency created by a city's past pattern of investment in fixed capital in buildings and structures – the physical fabric of the city – or in the fixed rail rolling stock, vehicle, or ferry fleets. It is simply that recognising the significance of the past helps identification of the difficulties which must be anticipated and planned for in proposing how urban services may change or may be changed. It also helps in the assessment of the environmental benefits that may be expected to flow from such changes.

Although the process of urban change is generally slow the process of change in some areas may be rapid (especially in areas where tall buildings are built). That is, the form of development in critical areas of the city may change very rapidly and certainly much faster than the capacity of the infrastructure that supports them can. This inevitably leads to stresses in the systems and tends to distort the patterns of investment in urban infrastructure.

Current land use planning approaches cannot provide an appropriate assessment of the nature or magnitude of the changes or whether they are more or less sustainable.

TRANSITION TO SUSTAINABILITY

A new approach is needed which integrates the concerns of the scientist/ecologist and the measures they can provide of the environmental effects of exploitation of resources with those of the urban planner who can facilitate the introduction of social, economic and aesthetic considerations into the expression of technocultural choices.

The components of an ecologically sustainable future for our cities can only be achieved by deliberate transitions from current practices to different ways of acting. One way of facilitating the transition would be to develop a different approach to the planning for and accommodation of the activities carried out in the city.

This implies a departure from the present approaches to land use planning, one which would allow planners to assess alternative development strategies for the physical fabric of the city, for investment options in urban services and in the structure and operation of manufacturing, warehousing and retailing. It also implies both increasing the awareness of people at all levels to the importance of ecological sustainability objectives and their inclusion in decision making designed to the achieve those objectives – this requires a philosophy of participation in decision making.

The word 'transition' suggests that we need to facilitate the expression of a range of interpretations of the existing situation in a city before we can begin to evaluate possible alternative directions for their growth and management. The word implies a state of flux, of development over time. It means a preparedness to accept that there are different problems which may emerge in making the transition from the present to the alternative futures to make them more sustainable and that it will be necessary to identify and adapt to the problems in different ways at different times.

Although we could draw up a long list of the objectives we might pursue in a transition to more sustainable cities the three most important aspects of sustainability in the city are:

1. energy consumption and its relationship to greenhouse gas production,
2. water consumption and its relationship to sewerage, drainage and water pollution and the effect of sequestering river flows to provide the urban water supply on the ecology of their catchments, and
3. solid waste production, its minimisation and its recycling.

Making cities less unsustainable in terms of consumption of energy from non-renewable sources and of water consumption would also have the effect of achieving other sustainability objectives such as reduction in air and water pollution and protection of biodiversity.

The significance of energy and water consumption is that they both affect, or have the potential to affect, the form and structure of the city whereas the connection between urban form and structure and other sustainability objectives is weaker.

The production of solid waste, its minimisation and its recycling is a major issue in Australian cities. While much of its production is a function of the consumerist nature of society the opportunities for minimisation of waste and its recycling are affected by issues of urban form. Opportunities for recycling may also be affected by city structure.

By 'form' I mean the nature or density of development. All major cities in Australia are essentially low density, especially in their residential areas, although recently city centres have been developed to high density.

By 'structure' I mean the spatial relationship of services and activities, that is, whether they are structured in linear relationships, are highly centralised or whether the city is structured as an interconnected set of nodes around which development is arranged. All the large cities in Australia – Adelaide, Brisbane, Melbourne, Perth and Sydney – are highly centralised and have been since their foundation. This is partly a function of the period of their settlement and growth.

The centralisation of the city raises a profound and, to some extent, unavoidable paradox. There are social and economic benefits to be derived from a degree of centralisation of activities and social investment in cities. However, pressure for centralisation produces demand for more people to be at the centre of economic and political power and influence. This demand in turn becomes a demand for tall buildings to provide accommodation for the commercial and governmental activities we pursue. It may also become a demand for tall buildings for residential accommodation

It is also clear that at some point in city growth alienation, anomie, segregation and diseconomies of scale and centralisation arise.

The structure of the city is the main source of its inefficiency. The greater the degree of centralisation of the city and therefore of its urban services, the greater the inefficiency.

Continued focus on the development of the CBD leads to continuously increasing demand for several urban services, including especially, transport. Increased centralisation forces travel through the centre even of those who do not have the centre as their destination. This in turn leads to the kind of congestion problems now experienced on road and rail networks in Australian cities.

We of course need and desire a degree of centralisation but the challenge is to find the degree which gives us the best trade-off in terms of the sustainability of the city.

A NEW PLANNING PARADIGM

The present planning and development process fails to give weight to consideration of the ecological sustainability of the city.

Governments and environmental groups have tended to focus on pursuit of global targets for reduction in consumption of renewable energy and therefore greenhouse gas production and on general targets for reduction in water consumption. They have tended to focus on simple market mechanisms including the transferability of pollution rights and to pursue carbon trading schemes and the like. There is no doubt that general targets and market mechanisms are important but to achieve the targets we must employ mechanisms which allow greater discrimination in the use of location specific measures. That is, we have to be able to direct the development where we want it to occur and of a form we want to minimise the degree of unsustainability.

We need a new approach which recognises at once that we must look at the way a city develops and is operated to identify potential ways to reduce energy and water consumption and waste production.

ENERGY CONSUMPTION

At present we are developing energy rating schemes for dwellings to encourage designers, builders and owners to become aware of the energy expended in their operation. This initiative is highly contested (Williamson 2000). At its best it might be a useful first step but the desired economies in energy consumption achieved this way can only be achieved by particular behaviour which may be inconsistent with the way households actually live in or use a dwelling.

Of greater concern is the fact that the energy rating system does not refer to the energy embodied in the dwelling. Nor does it take into account the context of the dwellings or relationship of buildings and structures to one another. That is, the energy rating system does not relate to the urban space created by the agglomeration of buildings and structures.

Embodied energy

Tucker, Salomsson and Macsporran in (1994) estimated that the embodied energy of the Australian national building stock at 22,500 petajoules was equivalent to about nine years of total energy consumption. About 40 percent of this embodied energy was estimated to be in the residential building stock. This estimate of embodied energy takes no account of the embodied energy destroyed by building demolitions nor does it estimate the embodied energy in the infrastructure.

Tucker and Treloar (1994) further estimate that 'CO_2 emitted over the years in stock production ... was estimated to be approximately 2200 MT (million tonnes)'. If the building stock is growing by 1 percent per year and the average life of buildings is 100 years that would mean 450PJ of energy was being used per year. (The average life of buildings in the stock is, and is likely to continue to be, significantly less than 100 years which means that this is a significant underestimate energy consumption in embodied energy.)

Suffice it for our purposes to note that Tucker and Treloar (1994) acknowledge that the 'construction sectors are one of the main contributors to energy consumption and CO_2 emissions nationally.'

Because taller buildings use more energy expensive materials we might expect that their embodied energy per unit area is higher that that for low rise buildings.

I am unaware of definitive studies of the embodied energy per unit area in tall residential buildings compared with tall office buildings, although conventional wisdom has it that, for all practical purposes, they are the same for buildings of equal height and that both are much higher than for conventional forms of dwellings.

We do not have a citywide breakdown of the embodied energy in either residential or the non-residential buildings so we cannot comment with any accuracy on the proportions of the embodied energy invested in different forms of development. That is, we do not know what proportion of the embodied energy of either residential or nonresidential buildings is in the form of tall buildings. It would be safe to assume, however, that the proportion is increasing because we now have many more tall buildings than we did, say, 20 years ago and the proportion of the residential stock now in the form of tall buildings has increased.

Current research into the embodied energy of detached, semi detached and attached two storey house types (Fay, Lamb and Holland 2001) indicates that there is virtually no saving in embodied energy between the three types of housing generally found in Australian cities. The research suggests however, that the traditional detached houses have a greater capacity to use materials with lower embodied energy. The authors conclude that the 'compressed suburbia' produced by the preoccupation with consolidation policies might lead to increased urban density but it does so at the expense of the pursuit of sustainability objectives.

Weatherboard houses embody one-sixth the energy of brick veneer houses and timber framed houses store carbon whereas steel framed houses permanently release carbon to the atmosphere. Yet in some of our cities we energetically construct yet more houses with steel frames. The full effect of the substitution of timber for more energy expensive materials depends to some extent on the source of the timber, that is, whether it comes from old growth forests or plantations and whether it has to be transported over long distances (Holland and Holland 1995).

There may be few opportunities, given present technology, for the production of tall buildings with lower levels of embodied energy. Whether we can reduce the embodied energy in tall buildings depends to some extent on whether we can devise new approaches to their heating and ventilation and to some extent on the substitution of low embodied energy materials for the high embodied energy materials currently used. The magnitude of such savings in embodied energy is an empirical question for which we have not begun the research.

The transition to higher density housing, including that in tall buildings has increased rapidly over the last decade. Almost half of all new dwellings are in the form of tall buildings in some of the larger cities. Although we cannot provide accurate figures we know that the energy embodied in the favoured newer higher density residential developments greatly increases the amount of carbon permanently released to the atmosphere. That is, the forms of residential development currently favoured by governments and their planning advisors are significant sources of environmental stress.

These twin issues of the operational energy and embodied energy of dwellings are, of course, only two aspect of the operational and embodied energy of buildings and structures in the city. The issues become of great significance when we contemplate the embodied and operational energy of tall buildings.

Currently we permit the development of tall buildings even when we know that the materials from which they are built, that their methods of construction and the waste material from their construction are expensive in energy terms. The embodied energy at all steps in the manufacture of the materials, their fabrication and the construction of the buildings permanently releases CO_2. In sum the greenhouse gas equivalence of the energy embodied in tall buildings constitutes a significant proportion of the annual release of CO_2.

The embodied energy in infrastructure is also a significant element of energy consumption. We do not have much empirical evidence of the stock of this energy investment although some research is indicating the significance of this aspect of our cities. Pullen (1999) reports on a study carried out in an Adelaide suburb measuring the embodied and operational energy costs of the conventional provision of water supply, sewerage and storm water services. His comparison of the conventional provision of a water supply with on site collection and storage indicates that in areas with regular rainfall individual tank storage has lower energy consumption depending on the size and type of tank. This suggests that it may be appropriate for some areas of the city to foster the site collection and storage of water rather that rely on the traditional approach to the provision of a water supply. This response could be appropriate in both new development areas and in areas in which the existing infrastructure needs to be renewed.

We have so far only limited evidence but similar approaches to the local management of sewage especially by using new biological treatment processes may also lead to lower energy consumption. One of the beneficial aspects of such an approach is that the stormwater runoff problems would also thereby be reduced.

Operational Energy

Most tall buildings rely on mechanical air conditioning and on the provision of lifts to make them habitable. That is their 'fixed' operational energy is also a significant source of greenhouse gas release. The energy used in the activities carried out in them is also from elaborately transformed sources, which are significant contributors to greenhouse gas production.

In some areas a high proportion of even low-rise buildings are mechanically air-conditioned. This may be due in large measure to the fact that modern developments allow few opportunities to plant trees and shrubs to naturally moderate local extreme climate variations. It is also due to inadequate attention being given to the design of buildings to minimise operational energy consumption.

In addition to the contribution of the 'fixed' operational energy of tall buildings to greenhouse gas production their heating and air conditioning may detrimentally affect the ecology of their local area. This might be reduced if the buildings were more energy efficient and more energy independent.

We similarly need more detailed information about the operational energy requirements of the city. In addition to the 'fixed' operational energy we need to develop our information about the transport energy demand of the city.

One area of inquiry into transport energy demand, which is advanced, is that relating to congestion. We express our concern over the contribution urban road congestion makes to greenhouse gas production – currently estimated to be 13 million tonnes – and we explore the use of economic mechanisms such as congestion pricing and parking charges etc. to reduce demand. Some proponents of such measures claim reductions of greenhouse gases of as much as 40% could follow from the introduction of location specific road user charges (BTE 2000).

These savings would only be achieved if behaviour was modified by pricing mechanisms, which in turn led to changes in development patterns. It would make more sense to employ such general mechanisms if they complemented policies to distribute development across the metropolitan area to reduce the centralisation which generated the congestion in the first place. This would have the accompanying beneficial effect of producing cities which were also more equitable in their access to employment and the range of cultural and recreational opportunities the city has to offer. Such a policy might lead to fewer tall buildings distributed across nodes in the metropolitan area.

Life-Cycle Energy Consumption

Current research into the life-cycle energy consumption of housing is revealing fascinating new results which gives hope that cities may be made less unsustainable. In a study of the life-cycle energy analysis of a 'green home' built in Victoria Fay, Treloar and Iyer-Raniga (2000) conclude that as operational energy becomes lower through a combination of efficiency improvements and life style changes the embodied energy becomes relatively more significant. They suggest that attention given to design flaws in new buildings and the substitution of low embodied energy materials for the high-embodied energy materials could reduce building embodied energy. They also suggest that renovation of existing buildings may also be a productive approach to the reduction of embodied and operational energy savings in the city.

Pullen (2000) found that the annual operational costs of 25 typical dwellings in Adelaide were 4 times that of the embodied energy of the houses, assuming a 50-year dwelling life. The estimate of embodied energy tends to be conservative which suggests that it is more significant that the ratio implies. This research does not separately identify the proportion of the operational energy which is consumed in heating and cooling. Heating and cooling is thought to account for about one quarter of the operational energy of the dwelling which means the embodied energy is approximately equal to the energy required to make the dwelling habitable.

WATER CONSUMPTION

The form of development affects the opportunities for reduction in water consumption and recycling and for moderating storm water runoff.

I do not have the space or time here to explore the situation in all Australian cities but use Sydney as an illustrative example where around 69% of current metered water use is for residential purposes.

Over the last decade a suite of demand management strategies has resulted in reductions of approximately 9% in the per capita daily consumption of water for residential purposes, including gardening which accounts for almost one quarter of consumption and equal to the amount used for showers.

Traditional forms of residential density permit a high level of water harvesting and storage. Coupled with modern recycling technology this form of development could lead to a significant reduction in the demand for water from the major storage and reticulation systems. That is, the consumption of metered water could be reduced if water was harvested from roofs etc and the water used in laundries, baths and showers was recycled. The demand would be reduced even more if the trend to changed gardening practices to make greater use of native plants and the use of mulching was increased and if the recycling of water was to a standard which permitted the recycled water to be used in showers etc.

High density residential development, on the other hand, creates fewer opportunities for water harvesting although recycling could be, and has been, incorporated into some developments making them more independent of city sewerage systems. Per capita consumption of water in high rise residential developments is lower because there usually is a smaller area of garden associated with the developments.

The traditional lower density residential development produces less stormwater runoff than high density development which means it creates less stress on the local environment.

Commercial undertakings in Sydney consume about 9% of metered water use. The amount used for these activities has remained relatively stable over the last two decades largely due to pricing strategies which encourage undertakings to cut consumption.

Industrial water use in Sydney accounts for approximately 12% of consumption. The lower consumption is largely due to greater efficiency and recycling encouraged by water pricing strategies.

The level of per capita consumption of water varies between the cities as does the opportunities for water harvesting and recycling. In seems clear, however, that Australian cities could be made significantly more sustainable in terms of their water consumption if they were to introduce water pricing strategies which encouraged residential harvesting and recycling of water. This would have the added benefit of reducing the flows to sewage treatment plants and, ultimately, to sewer outfalls which typically are to the ocean. This option is more viable in forms of development in which there is the space to harvest and store the water and to accommodate the small scale recycling plants needed to process the water. One of its benefits is that it does not require changes to the traditional form of Australian cities although it may require a commitment to reduce the degree of their centralisation.

A strategy of greater reliance on local water harvesting and recycling assumes greater significance as Australian cities face the twin problems of shortage of capital for infrastructure investment in new developments and of replacing the present obsolete or worn out water supply and sewerage systems. The paper by Pullen (1999) referred to above which examined the embodied energy costs of different ways of providing water supplies supports this contention.

A strategy of greater self reliance and therefore independence of housing may also be forced on cities as the demand for water reaches the limits of the water available from near city catchments.

WASTE PRODUCTION

The production of waste is more a function of attitude and life-style in a high-level consumerist society than it is of structural elements or the physical fabric of the city. It is important to note however that the reduction of waste in the building process in the re-use of 'waste' materials and in the recycling of material from demolition of buildings etc. has been significant.

The campaigns to get residents to reduce or separate out the flow of waste materials have been successful so that much waste which once was sent to land-fill sites is now used to produce compost or is recovered for use.

The reduction in waste in the form of sewage flows might also be achieved by changes in the behaviour of residents and changes in commercial and industrial processes. For example it may be necessary to place a high tax on the use of high phosphate detergents and cleaners etc. in favour of detergents and cleaners which do not create additional loads on sewage treatment processes. In residential areas where the sewage is relatively benign such an action would make it easier to operate local sewage treatment plants to produce effluents which could be re-used locally.

COMMENT

What does this exploration of the transition to sustainability tell us?

The first lesson to be drawn is that if we are to be serious about sustainability and expect others to take us seriously about our pursuit of this objective – especially when it bears on international obligations to pursue targets such as those to reduce greenhouse gas production – we need to develop the measures of energy and water consumption and waste production in our cities.

We need such measures to be able to show how well we are achieving the targets we have set or have agreed to. We need the measures to be able to convincingly make the case for the policies we must follow in pursuing substitution of high embodied energy materials for lower embodied energy materials. We need such measures to develop the performance based regulatory framework we aspire to.

We already have regularly updated spatially organised data related to a large number of the factors which bear on energy and water consumption. At present these data sets are separately collected and maintained according to different definitions of economic and social activity and expressed on different spatial bases. It wants for us to take the simple decision to coordinate the collection and presentation of the data in such a way that we can produce measures of the performance of our cities at different times. That is, it seems to be a simple problem of coordination between the agencies and corporations, most of them in public ownership or at the very least subject to a high degree of public regulation, to produce a data set which relates the consumption of energy and water to the areas in which the energy and water is consumed and the people who consume the energy and water.

The second lesson to be drawn is that while we cannot be sanguine about it we already have encouraging evidence from empirical research in Australian cities that we can achieve many of the sustainability objectives in reducing energy and water consumption without sacrificing standard of living and amenity of the cities.

The third lesson is that the indications are that with a better spatially based data set relating to energy and water consumption we no longer need to pursue urban development policies which are not soundly research based.

The fourth lesson is that to achieve progress in the pursuit of sustainability we must be prepared to critically explore some of the fundamental assumptions, often implicit, which underlie the approach we take to the form and structure of the city and of the way we meet demand for services such as energy and water supply. This might well force us to re-examine the institutional framework of decision-making – an aspect of the pursuit of sustainability which has not been explored in this paper but which deserves to be.

While most of the research referred to here explores issues related to urban form and may be seen by some as a contribution to debates about compact cities its underlying concern is about issues of structure. The point is that once we are able to assemble data about the energy and water consumption and waste production of the existing city structure we are able to consider alternative arrangements of activities. Once we recognise that we can dispose our activities in whatever arrangement best suits our pursuit of sustainability objectives we begin to accept that we need not be dependent on the provision of highly centralised services. This frees us to consider how we can structure our cities in the most congenial manner and to be able to measure the kinds of tradeoffs we must inevitably make in each city to achieve that goal.

Meanwhile, we can reasonably expect our planning authorities and agencies to provide leadership in adopting a strategy of transition to sustainability.

REFERENCES

Bureau of Transport Economics, 2000
URBAN CONGESTION – THE IMPLICATIONS FOR GREENHOUSE
GAS EMISSIONS. *Information Fact Sheet 16*, 4.

Dover, S. R. and Handmer, J. W., 1992
UNCERTAINTY, SUSTAINABILITY AND CHANGE. *Global
Environmental Change*, 2, 262–276.

Fay, R., Treloar, G. J. and Iyer-Raniga, U., 2000
LIFE-CYCLE ENERGY ANALYSIS OF BUILDINGS: A CASE
STUDY. *Building Research and Information*, 28(5), 139–148.

Fay, R., Lamb, R. and Holland, G., 2001
DOES COMPRESSING AUSTRALIAN SUBURBIA ACHIEVE
ECOLOGICALLY SUSTAINABLE DEVELOPMENT? *Journal of
Architectural and Planning Research* (forthcoming)

Holland, G. and Holland, I., 1991
APPROPRIATE DESIGN DECISION MAKING, IN TECHNOLOGY
AND DESIGN, *Proceedings of the 1991 Australian and New
Zealand Architectural Science Association*, Adelaide: 197.

Holland, G. and Holland, I., 1995
DIFFICULT DECISIONS ABOUT ORDINARY THINGS: BEING
ECOLOGICALLY RESPONSIBLE ABOUT TIMBER
FRAMING. *Australian Journal of Environmental Management*,
2(3): 157.

Pullen, S. F., 1999
CONSIDERATION OF ENVIRONMENTAL ISSUES WHEN
RENEWING FACILITIES AND INFRASTRUCTURE.
*8th International Conference on Durability of Building Materials
and Components*, Vancouver, June.

Pullen, S. F., 2000
ENERGY USED IN THE CONSTRUCTION AND OPERATION OF
HOUSES. *Architectural Science Review*, Vol. 43, No. 2. pp. 87–94.

Tucker, S. N., Salomsson, G. D. and Macsporran, C., 1994
ENERGY IMPLICATIONS OF BUILDING MATERIALS
RECYCLING, *Buildings and the Environment, Proceedings of the
First International Conference*, Building Research Establishment,
Watford, UK.

Tucker, S. N. and Treloar, G. J., 1994
    ENERGY EMBODIED IN CONSTRUCTION AND REFURBISHMENT
    OF BUILDINGS, BUILDINGS AND THE ENVIRONMENT,
    *Proceedings of the First CIB Task Group 8 International
    Conference*, Building Research Establishment, Watford, UK.

Williamson, T. J., 2000
    A REVIEW OF HOME ENERGY RATING IN AUSTRALIA:
    POLITICS, EVOLUTION AND EFFECTIVENESS. *ANZAScA
    conference*, Adelaide. December.

Cities in the Third Millennium

Josef Konvitz

INTRODUCTION

The city is the most complex form of social organization. Perhaps this explains its temporal endurance and global diffusion. It antedates the beginnings of the world's great monotheistic religions, and has outlived the empires of antiquity to which its origins can be traced. From the perspective of urban time, therefore, the beginning of the third millennium of our era does not appear to have any particular significance. Indeed, if we are to speak of cities in terms of millennia, we had better adjust our chronology. In Western terms, the first urban millennium probably began around 800 BC and ended with the collapse of Roman civilization in the third century AD. The second millennium began around 800 AD and ended around 1800. The third urban millennium started around 1800 at the conjuncture of three revolutions: the industrial revolution which harnessed technological innovation and scientific inquiry to more productive uses of energy and new uses of materials, the political revolution which enshrined individual rights and democratic process in law, and the demographic revolution which pushed back the average age of death and increased the size of the population. The third urban millennium is already two hundred years old.

From another perspective, however, the transition into the 21st century does represent a major turning point. Within a few years, more than half of the world's population will live in cities. This makes learning how to manage space better an urgent priority. Our ability to plan wisely for this future is grounded in some two hundred years of human experimentation and experience. The record has not been uniformly positive. Urban growth has often accompanied immiseration for millions of people; class tensions in industrial cities were a factor in the diffusion of two of the most sinister ideologies of the 20th century, communism and fascism, both of which discredited planning; and the environment is still not easily cleansed of the wastes and pollution which accompanied urban growth in the past. How can we manage space better?

There is a paradox about planning. Urban and regional planning is an exercise that commits resources for decades. Yet the circumstances and factors that will affect life in cities, even just 10 or 20 years from now, are almost impossible to predict. There is no alternative to planning in the context of uncertainty, but many of the policies and procedures that have shaped decision-

making in the 20th century are unsuitable for a more flexible and strategic
approach. To manage space in the 20th century, large-scale bureaucratic systems
regulated what can or cannot be done in different places.

This system has not been flexible enough, given the rate of economic,
social and technological change. A city zoned into single-purpose districts, with
uniform mono-functional buildings and land use patterns that risk becoming pre-
maturely obsolete, lacks a degree of latent adaptability or resiliency.
Nonetheless, control has remained the operative mechanism, the major contrast
between modern Western cities and cities in developing countries being the
degree to which development is controlled. Planning is at a critical state.
Deregulation would simply make the city a residual, the product of social and
economic forces. On the other hand, those who believe that the future of cities
should rather be guided by a vision of the kind of city we want to inhabit have
not yet solved the problem that different groups hold conflicting views about
what future they want.

To move forward, we need to ground urban policy and planning in a better
understanding of the nature of urban growth and change. The challenge lies in
adapting policies to the nature of cities as dynamic and complex social and economic systems, and to the scope for freedom and creativity that urban life gives
to people. In 1968, Edmund Bacon, the Philadelphia planner, wrote that the
failure of cities "is brought about by the failure of intellectuals to generate a
viable concept of a modern city and a modern region." Of course Bacon did not
mean that ideas alone matter; he meant that people would act on the city according to flawed ideas and assumptions if they could find nothing better. Implicit in
his statement is the assumption that people who make decisions about the future
of cities want to be better informed about critical variables and trends.

We must not underestimate the handicaps under which we labor. Because
our major challenge is to learn to manage space in ways that are better for
people, we need to understand some of the reasons why spatial thinking is so
difficult in our culture. Politics, science and economics have all contributed to
this state of affairs. Three hundred years of classical physics have propagated
the assumption that for the purposes of understanding the laws of matter and
motion, space is everywhere uniform. The system of modern statecraft which
emerged from the hands of Richelieu and Louis XIV at the same time as Newton
gave rise to the notion that the workings of the state and the application of laws
operate uniformly within its borders, or in other words, in disregard of territorial
differences. And finally, the concepts of modern economics which also date
from this very decisive era of 1630–1730, and are associated with the teachings
of Adam Smith about the wealth of nations a generation later, assume that
although the workings of the economy depend upon comparative advantages
which are distributed geographically, the laws of the economy apply everywhere
in the same way. The global system, from this point of view, is grasped through
an abstract form of reasoning; geographical knowledge therefore is descriptive
in value, but not analytic. As a result, the ability to think in spatial terms has
been devalued as a part of general culture. Yet spatial problems are more pressing than they have been for decades.

We may be at a decisive turning point, one of those periods when people question the goals and methods which have become normative because they realize that things will not improve unless there is change. A turning point of this order of magnitude, which represents a paradigm shift, may occur well within the space of a single generation. Much depends on where we think we are in this process. Assuming that this transition commenced in the early 1990s (the Rio Conference of the UN Commission on Sustainable Development in 1992, the Single Market Act in Europe in the same year, the rise of the internet, etc.), and is likely to last about twenty years altogether, then the years 2002–2008 may well be critical. In other words, we are only partway through this period of transition, but approaching its climax.

In what follows, I will divide my remarks into three parts. In the first, I will outline what some of the features of a new approach to the management of space might look like in due course. In the second part, I will ask the question where some of the problem areas may be. And in the third and final part, I will consider what modernism and postmodernism contribute to the politics of urban development.

PART ONE: HOW TO MANAGE SPACE BETTER

There are essentially three challenges confronting those responsible for the future spatial shape of the city: how to make progress toward sustainability, how to cope with the rapid rate of economic and technological change, and how to promote social integration and civic responsibility. What tools and concepts might the policymaker of 2020 have at his disposal to guide the development of cities and regions? What variables might he consider to be important, perhaps decisive, in deciding how, when, and where to intervene? What could be the criteria for success?

The answers to such questions call attention to an emerging policy field which at the OECD is called territorial development. This concerns the economic organization of space, based on the assumption that competition is increasingly a matter of how well each city, each region, can achieve its potential, based on their assets. (There is a paradox in the use of the term "territorial" in English, because the term in English has connotations of exclusive political control or even aggrandizement, whereas in our usage it refers to the spatial structure of the economy in terms of settlement patterns and networks that permit interaction and exchange. We no longer refer to the economic base of a city in terms of its hinterland, which is territorial in the old style, but in terms of its social and human capital). Territorial policies are not the simple addition of spatial planning, regional, urban and rural policies. They are a comprehensive set of actions which can be carried out by government to foster development in all its territories, to limit the disparities between them, and foster a more balanced distribution of population and economic activities so that remote rural regions are not abandoned, and major metropolitan areas are not overcrowded.

Territorial policies recognize that the changes associated with globalization and technological innovation, and with steps toward sustainable development, are concentrated at the sub-national level. Territorial capital refers to the stock of assets which form the basis for endogenous development in each city and region, as well as to the institutions, modes of decision-making and professional skills to make best use of those assets. The unique role of territorial policies is that they make the spatial dimension explicit. Now that monetary, fiscal, sectoral and structural policies are increasingly similar across the countries of the developed world, governments are recognizing that territorial development policies may be the principal means by which countries distinguish themselves.

At present, spatial, social and economic policies for territorial development face three sets of problems: they do not have a high political priority; there are considerable difficulties in achieving cross-sectoral integration even by applying a territorial perspective; and institutional changes such as decentralization and devolution sometimes transfer greater responsibilities to regional and local authorities, but without adequate resources. These constraints can be overcome. In twenty years, the field of territorial development policy will give policymakers some powerful new tools and concepts. They might look like this:

Territorialised Budgets

How much is spent on territorial development, public and private sources, is a mystery at present. Indeed, scarcely a mayor or provincial leader knows how much is being spent within his jurisdiction from all public sources. Cost-benefit analysis is therefore almost impossible. No minister knows whether he is spending too much, or not enough. Under-spending might be even more serious than over-spending, because a modest increase in expenditure might yield significant benefits if a threshold is reached, but achieve very little below that level. Today, budgets are sectoral, not spatial. National budgets are designed without regard for where the money will be spent. Cross-sectoral integration however takes place at the territorial level. But for the most part, regional and local authorities cannot combine and adjust the sectoral streams which reach them.

A territorial approach would organize the sectors according to the needs and opportunities of different cities and regions. In 2020, national ministries for health, labor, education, transport, environment, etc., will compose their budgets on the basis of sub-national, territorial units, so that all the sectoral streams coming into a region will be transparent and can be co-ordinated.

Territorial Impact Assessment

Many of the spatial consequences of macroeconomic, sectoral and structural adjustment policies are simply ignored by central governments. Just as environmental impact assessments are now standard practice in government and are often required of private sector investments in many sectors, territorial impact assessments will become normative. Territorial impacts which are the implicit and perhaps unintended consequence of other policies may be positive, or negative. And some sectoral policies arguably have a greater impact on territorial development than territorial development policy itself, such as highway construction or military research in the United States, or the Common Agricultural Policy in Europe. But in the absence of analysis, we do not know the answers to these questions. Many of the social, environmental and economic costs are internalized in the budgets of firms and households, reducing their productivity and wealth. These costs – or externalities – can be reduced if the territorial impacts of economic activity can be assessed better, ex ante, before, rather than ex post, or after the fact. Many policies for territorial development today address problems inherited from the past which are not remediated by economic growth alone. The opportunities for cross-sectoral integration will be enhanced considerably by 2020 by the widespread use of territorial impact assessments, which the technologies of digital mapping and global positioning should facilitate.

Reducing Disparities and Realizing the Potential for Growth

Disparities of income and opportunity within cities are most visible in the formation and spread of distressed urban areas, which now include between 5 and 10 per cent of the urban populations of OECD Member countries. The size of the urban population in Asia is expected to double, becoming twice as large as the entire population of the industrialized world today. The size of the middle class in Asia could be as large as the population of the European Union. We know however that poverty and underemployment will also exist.

Reducing disparities means not only implementing area-based strategies which often take 10 or 20 years to show dramatic results; it also means devising pro-active policies that reduce the social and environmental costs of the normal processes of structural adjustment and growth. Moreover, the gap between leading and lagging regions within countries may widen even as national economies continue to converge. In the past, redistributive polices raised income levels for many, but failed to improve the rate of growth.

By 2020, the practice of territorial development policy will give all regions the means to increase the stock of natural and man-made assets, and to grow faster by best use of endogenous assets, their territorial capital. The economic, social and spatial policies for territorial development will set a benchmark standard for policy integration and coherence.

International Markets for Urban Goods and Services

In the past, it was assumed that because buildings are assembled locally, construction is a non-tradable sector. This view was extended to cover a wide range of urban goods and services. But in recent years, globalization, including the opening of public procurement and public contracts to competition, has meant that firms are competing internationally for everything from the provision of street lights and parking meters, to the accounting systems for public water works, the design and construction of airports, and the installation of streetcar systems. Architectural and construction firms operate internationally. In the first half of the 1990s, the largest construction firms in Europe, the United States and Japan all expanded their international sales at a faster rate than their domestic markets. But many countries still do not record the income earned from this growing aspect of the global economy, which instead shows up as part of the trade on infrastructure or the service sector. This sector is critical to the provision of cities in Asia and Africa undergoing rapid expansion, and represents an unacknowledged resource for innovations when local, domestic markets are too small or specialized.

The traditional distinction has broken down between the goods and services for export, called "basic" in the literature because they pay for what a city must import, and the goods and services traded locally, called "non-basic". The number of manufacturing jobs, previously the mainstay of the basic category, is declining; the growth of services for export has been equally dramatic. The "non-basic" category now includes many things once considered to be non-tradables that are now part of world trade. Examples include, not only goods and services traded locally such as entertainment and fine dining which are important in competitiveness and help to attract business from outside, but also a wide range of products and services ranging from urban furniture, transit systems, management systems, design consultancies and architectural commissions which previously were highly localized and are now part of the service sector in global trade. Cities may exist to trade, but no longer is it possible to understand urban economies on the basis of a distinction between local production and consumption, and goods and services produced for or imported from elsewhere.

By 2020, countries will further liberalize their markets for urban goods and services by removing many of the remaining non-tariff barriers to trade in the form of local specifications or licensing rules. Countries will record the sales earned abroad by individuals and firms in the field of city-building and territorial development, and will measure the value added in city-building and management.

Sustainable Development Indicators

The OECD definition of an ecological city is not one which is already very clean, but rather is a city distinguished by the degree to which environmental considerations are incorporated into decision-making in public and private sectors alike. By this standard, a city with poor environmental quality has as much potential to become an ecological city as one which already has an environment of high quality. Achieving sustainability is a process that does not end. The environmental pressures associated with urban settlements and lifestyles, principally the generation of waste, air pollution associated with power and transport, water consumption and quality, excessive noise levels, and the contamination of land, are giving rise to better indicators, environment management systems, and the setting of targets. By 2020, only some 20 per cent of the built environment of existing cities in OECD Member countries will have been created or renovated through the regular process of growth, thus raising questions about how to adapt to new standards and norms the vast majority of buildings and districts that already exist. Notwithstanding the gains achieved through incremental changes in land use patterns, in the ecological efficiency of buildings, and in various technologies, new problems will have emerged, but urban observatories will give early warning so that corrective steps can be taken.

By 2020, the precautionary principle and the concepts of carrying capacity and of the ecological footprint will have been internalized into routine practice. Economic instruments will be in place which assess property, less for its market value, and more for its environmental efficiency and latent adaptability. Today the more a property is worth, the more it is taxed; tomorrow, the more a property is ecologically sustainable, the less it will be taxed.

More important will be new and powerful sets of indicators that can be used in everyday life. As things stand, indicators more often measure phenomena that can be tracked quantitatively, not necessarily those which are most important. Furthermore, indicators are usually in the hands of experts, and not accessible to the public; and the updating of trends is infrequent. Let me make a comparison with the weather, which is a highly erratic and unpredictable system the evolution of which is monitored with a wealth of data, verified against historical records and the theory of physical phenomena, calling on the largest of super computers. Yet the result is a series of written and graphic summaries, easily interpreted by people with no knowledge of the laws of thermodynamics or of the behavior of gasses under pressure, who can thereby decide whether to carry an umbrella or wear a lighter garment. We will someday have a system of indicators which will help us to adjust our daily behavior – energy and water consumption, and appropriate travel modes, and the like – as easily as we now consult the morning weather report.

Metropolitan Governance

Fragmentation of administrative jurisdictions within metropolitan areas results in a lack of correspondence between administrative and functional territories, and inhibits cross-sectoral policy integration. This situation, which tends to be defended by vested interests, results in a complex policy environment in which area-wide consensus is difficult to reach on important medium and long term goals such as environmental quality, economic development and competitiveness, social cohesion, equitable public finance, and the level and quality of public services across the urban region.

This situation is compounded by the strain on the financial/fiscal ability of local authorities in metropolitan areas. In many countries, decentralization has been used as an opportunity for upper levels of government to transfer responsibilities to the local level without introducing the corresponding, but politically difficult, financial and fiscal reforms. It is not surprising to discover that there is a lack of transparent, accountable decision making processes and of clear political leadership at the local level.

There is no single ideal model of governance for metropolitan regions. However, some common elements that underlie good governance in metropolitan areas have been identified. The Principles which have been adopted by the

OECD Territorial Development Policy Committee are non-binding and do not aim at detailed prescriptions for national or sub-national legislation. National/state governments have an important responsibility for shaping an effective framework that allows cities to function effectively and to respond to the expectations of citizens and other stakeholders (OECD, 2001). It is up to local government and the communities to decide how to apply these Principles in developing their own frameworks for metropolitan governance. (A full set of the guidelines and principles appears in the Annex.)

By 2020, institutions will be more accountable and effective, better able to achieve policy coherence. The administrative and jurisdictional boundaries of metropolitan areas will have been adjusted, perhaps several times, to conform better to their functional area. Principles of good governance will have been normalized, leading to voter participation at local levels comparable to or greater than that at national levels, thus giving new life to the old adage, that city air makes men free.

Education for Urban Living

That the economy is increasingly knowledge-based is no longer questioned. A recent OECD study on the learning city-region concludes that it is not individual learning *per se*, but the use of that learning in firms that matters for regional performance. Lifelong learning and the concept of the learning city-region are still abstractions. The problems of integrating more diverse urban groups, and of preparing people to enter or remain in the workforce, will however provoke major innovations at the local and regional levels. Meeting the needs of international students, both children and adults, will become a priority as cities compete for knowledge workers. Organizational learning, for firms and communities, will become mainstream. Given the rapid rate of change in technology and knowledge, people will learn better how to "unlearn" information and practices that are outdated or inefficient. The creation of the Land-Grant colleges and universities in the United States and of the so-called "red brick" or polytechnic universities in the United Kingdom shows that it is sometimes necessary to set up a new institutional infrastructure to cope with research and education.

By 2020, many new institutions will have been created to promote learning, taking the place of existing institutions that fail to adjust rapidly enough to the new environment.

R and D Programs for Cities

Developed nations compare how much they are spending on research and development. Science and technology are considered key factors in economic productivity and competition, and econometric studies are routinely undertaken to show the relationship between investment in research, and output. Urban issues scarcely matter to the defense establishment and other government agencies, and to industrial firms, which set national research agendas. We know more through historical research about how people acclimatized to cities 150 years ago than we know about the process of adjustment today. In a provocative essay written ten years ago, Ditha Brickwell pointed out that cities ought to have a direct influence on research policy, and she suggested two key global issues: how to cope with the problems of housing and feeding the poor, and how to respect and guarantee human rights. Problems of safety and density, to mention two, call attention to the need of cities for a social technology. Brickwell stipulated that the criteria for technological choices should favor intelligent systems which are adaptive to demands and which enforce demand-side management; technologies which allow implementation in steps and components, thereby making adjustment to changes of facts and attitudes continual; techniques which allow for the optimization of material flow with reduced environmental costs, and technologies which enable the participation of all sectors of the labor force, which offer jobs for training and better qualification, and which sustain arts, crafts and services.

By 2020, research budgets will have been adjusted to reflect the importance of urban issues, and new firms will have emerged to develop and bring to market a new range of urban services and technologies that improve everyday life and the agglomeration effects which are so important to business in the city.

Urban policy today is far removed from what a framework for cities should be. Heretofore, policies have been largely remedial. Policy has failed to keep pace with urban change and the growth of the global economy. Everything in the previous section is based on the assumption that policies for cities and regions can become pro-active and forward-looking. A territorial approach might be called utopian, but it is not impossible.

Territorial policy calls for a conspicuous if not decisive role for the state, with a high degree of public participation. The private sector, composed of individual owners, developers and corporations, will still control most of the funds invested in urban development. But it shares responsibility with elected officials and government for the framework of ends and means by which private capital creates places for living and working in cities. Privatization and regulatory reform have changed how government works, but they have not changed the nature of the problems facing government. To be realistic, unless there is a broad consensus which can extend over electoral terms and economic cycles, medium-term strategies, especially if they bring change to the status quo, can be difficult to sustain.

PART TWO: PROBLEM AREAS, OR THE GEOPOLITICS OF URBAN DEVELOPMENT

Between today and 2020, crises will occur. Can we be so confident that there will be no major war involving the great powers and the economies of the developed, industrialized world between now and 2020? Will popular support in favor of greater trade liberalization, immigration, and economic reform remain strong, or will a protectionist and nationalist movement, basically capturing the anti-urban ideology of community, emerge from the right in a marriage of convenience between agricultural and industrial producers, as happened in the 1890s? Will there be another Great Depression to depress investment, create unemployment, undermine security and confidence, and destabilize democracy? In this section of my paper, I want to call attention to some of the fault lines and problem areas that show how urban issues might intersect with other major sources of conflict. Sometimes crises serve to accelerate the introduction of changes when reforms are blocked. This could be favorable. But in a period of crisis, policymakers may never have a chance to implement a territorial development strategy. Cities could become dystopian, not merely victims of forces that they cannot control, but perhaps also agents of their own destruction.

External Crises

The highly interdependent nature of economies in a world in which 500 000 000 people live outside their country of birth (approximately 5% of humanity) means that cities are exposed to shocks of different kinds. A simple listing would include global climate change, natural disasters, energy crises and a great depression. Half of humanity lives on the coast of a sea, and this trend toward coastal urbanization is certain to continue, as there is no evidence that people intend to move to upland or mountainous areas if they have a choice. A large portion of humanity is at risk of flooding or severe water shortages. Meanwhile, an increase in the sea level could well expose many cities in both developed and developing countries to the fate of Venice. Efforts to adjust to climate change may not be undertaken gradually, but in response to overwhelming threats which test the solidarity of nations. Geological fault lines make it likely that Istanbul and perhaps Tokyo will suffer major earthquakes within 20 years, and these would be disasters on a scale that cannot be compared with any previous catastrophe. Finally, the next century could mark the end of the petroleum era, which would then have lasted about as long as the coal era which preceded it, about 250 years. And the systemic risk of a major economic depression, however unlikely, cannot be excluded.

Most worrisome is the evidence that governing structures are far from prepared to handle major crises. The lessons of past failures are difficult to apply pro-actively. Perhaps each generation has to pass through its own crucible. The point is that cities are exposed to major disasters of different kinds, some of which could occur concurrently, and lack the means to cope effectively. Meta-disasters by definition are a supreme test of the capacity of social and economic systems to recover or die.

Geopolitical Fault Lines

Many cities exist in regions which can be described as fault lines that are geopolitical, not geological in nature. Many of the largest cities and many rapidly urbanizing regions are in this category, which includes the East Asian archipelago extending from Korea to Hong Kong via the straights separating Shanghai from Taipei and Kaoshing; the North African littoral from Casablanca to Cairo, the US-Mexico border from San Diego-Tijuana to Juarez-El Paso, the Baltic triangle of Helsinki-St. Petersburg-Tallin, and the network linking Berlin, Warsaw, Prague and Vienna. These regions may be a positive factor in the growth of trade and interdependency, but tensions can also build up along borders between different economic zones. Cities in these regions are particularly exposed to changes driven by political decisions and trends. In developing countries, the biggest issue may be the benign attitude of government to accept rapid urbanization on terms which avoid necessary infrastructure investment and enforcement of building codes and planning norms.

Cities may be strategic assets as countries try to take advantage of change: this is the case, for example, of Berlin, Vienna and Warsaw, or of Seoul, Fukuoka, and Shanghai. The interests of cities however may conflict with the interests of nation-states, especially if strong regional disparities in nation-states exacerbate tensions that the nation-state cannot manage without trying to control cities, historically a futile exercise unless carried out by force. In such a circumstance, some cities, such as those in the East Asia archipelago linked by maritime networks, might find that they have more in common with each other than with their territorial hinterlands. The possible disintegration of the nation-state however would marginalise backward areas even further, thus accentuating the tensions and disparities which already are one of the principal drivers of international migration. The costs of nations pulling apart are fortunately based on but a few examples, but those show how expensive it can be. This does not necessarily dissuade people from pushing things too far.

War strengthens nation-states and circumscribes the autonomy of cities. The possibility of war cannot be excluded simply because for the past ten years following the end of the Cold War, international conflict has not erupted in most parts of the world where international tensions had often been high between 1945 and 1989. The expansion of NATO to the East in Europe is only one indication that the geopolitical system for regulating conflict is operating under new constraints which have not been tried in crisis.

Greater autonomy for cities is easier when the risks of conflict are low, as the experience of Western Europe within the European Union demonstrates. The relation between cities and states works in both directions. Historically, cities – or rather urban culture and institutions fostering freer communications, enforceable contracts and property rights, rule of law, and the resolution of conflict through negotiation — have acted as a check on the arbitrary power of the executive state. Cities without states can be defenseless; but states without cities can be autocratic.

Internal Fault Lines

We cannot ignore the internal fault lines in cities that may indicate fundamental problems in social and economic systems. Thus we find cities that have a disproportionate share of the wealthiest and poorest people, of the best maintained and most polluted landscapes, of the safest and most dangerous areas. The mental hospital and the university library, the public square where people mingle freely and the overcrowded prison – the twinning of opposites is a literary art practiced by all who have written about cities or featured cities in their creative fiction. Critics of cities often seize on the fact that cities are full of contradictions, as if this is some fatal flaw. This is an issue dear to polemicists and researchers alike: does globalization intensify social polarization in cities? If this is the case, then we indeed face an impossible choice: what if economic growth, which reduces poverty and improves the quality of life, also undermines the social conditions which provide the foundation for trade, innovation and investment?

We should be careful lest we jump to the conclusion that urban development itself is doomed by internal contradictions that by definition cannot be resolved.

Some problems endemic to cities, such as epidemics of cholera or other infectious diseases which were responsible for high urban mortality rates, as well as promiscuity and overcrowded housing, are largely a thing of the past, replaced perhaps by concerns about AIDS and fears of biological terrorism. But let us be honest with ourselves by admitting that our list of fears and concerns is more than a little subjective. The urban underclass and the informal economy are often exploited by the media. Mental health questions, for example, are largely ignored not only in the literature, but in the work of major organizations with a responsibility to cover the urban agenda. There is a risk that problem areas and groups will be contained through strategies of repression, isolation and surveillance, but at a price by restraining the freedom and invading the privacy of everyone. The result could be a decline in the strength of cities and in their problem-solving capacities.

What matters is how people react. Daniel Patrick Moynihan, when still Senator for New York, wrote a penetrating article called "Defining Deviancy Down" in which he argued that Americans have a tolerance or capacity to handle only a small number of social issues or problems at a time. When a new problem emerges which demands attention, others simply get assimilated into a category of things people learn to accept and live with, as if this is simply how things are. It would be worth comparing societies internationally from this perspective, because the range of sensitivities and the thresholds at which urban circumstances are perceived as problems certainly do vary from country to country.

The future may be characterized by a splintering urbanism, to use the title of a new book by Steve Graham and Simon Marvin. They may be right, but this outcome would deflect political attention and investment away from cities. Urban problems may be tackled if people are frightened enough of the consequences, but I submit that an optimistic and uplifting ideal has been responsible for more initiatives to transform and rebuild cities. The inspirational ideal is that of the cosmopolitan city, the city that is the home of mankind.

Having tried to identify some of the major risks facing cities, some with the potential for far-reaching effects on economic and social systems, it is worth recalling that the theoretical origins of strategic bombing theory had its roots in an analysis of the strengths and weaknesses of cities. In the 1910s–30s, the belief was widespread that urban society was so dependent on complex technological systems that it could not survive their destruction. Not only did it prove more difficult than expected to destroy these systems; urban people demonstrated that they were far more resilient and adaptive than urban social analysts had predicted.

The greatest disasters of the 20th century include the destruction of San Francisco and Tokyo by earthquakes, the shelling of Reims and Louvain, the bombing of Lorient and London, of Hamburg and Hiroshima, and the sieges of Warsaw, Leningrad, Stalingrad and Berlin. Although external threats cannot be ignored, the danger of urban mismanagement is no less real for being insidious and acting slowly, with consequences that are, ultimately, more long-lasting. The destruction of Rotterdam was made good decades ago, but the decline of Detroit festers.

PART THREE: RESOLVING THE CONTRADICTIONS

In the first part of my presentation, I outlined an admittedly optimistic scenario whereby those responsible for urban development would have available to them a range of policy tools and analytical concepts of great potency and effectiveness. This was followed by a discussion of some of the likely crises which might lead not only to disinvestment and a flight to security, but a breakdown in the very institutional and social fabric by which large urban economies function. The truth probably lies in the middle between these extremes. But even this middle ground will be difficult to achieve. Nonetheless, the goal remains to reconcile economic growth, social cohesion and environmental sustainability.

This is a conference about tall buildings and the urban habitat. There can be no future for tall buildings if large cities are not viable. In recent decades, two powerful trends have been acting upon the physical, social and cultural structures of cities, urban decentralization and urban deconcentration. These two trends are of course linked, but they are different. Urban decentralization has replaced dynamic, centripetal urban cores or centers with centrifugal, dispersed, polycentric patterns. Urban deconcentration has replaced the crowding of people at higher densities with the spread of people at lower densities over larger areas. They have made the challenge of making cities viable in the 21st century more difficult by exacerbating many negative social and environmental trends, complicating decision-making and the formation of strategies the public will support.

There is widespread concern about how large cities should be, what form they should take, and what will be there future. Very few countries have a balanced, polycentric urban system – Germany, and to some extent, Australia and Canada come to mind. In most countries, the dominance of a single primate city is characteristic. These cities tend to offer a range of opportunities and a quality of life superior to what can be found elsewhere, thus feeding on their own success up to the point that their size generates dis-economies and dis-incentives. At that point, governments often discourage expenditure to make these cities more attractive and functional, out of fear that they will grow even larger and absorb more resources and people, compromising the prospects for development elsewhere in the country. This conceptual formulation of urban dynamics ignores the critical role that some cities play in both national and international circuits of exchange and markets, a role that facilitates economic activity in the very regions the government would like to help. The loss of efficiency of Seoul, Mexico City, Tokyo, of Helsinki and Milan, of Bangkok and Tel Aviv, may not benefit other places but simply hurt the country as a whole. The world's mega-cities may not be consistent with an ideal pattern of sustainable development, but their collapse could be even worse than their growth.

The spatial structures and distribution of activities in these key cities are changing. Their cores, their historic central areas, are often weakening, to the benefit of newer edge-city districts. This is the case in Rio and Mexico, Washington and Johannesburg and London. There is a paradox to this pattern: the high end of the global economic is concentrated in a small network of cities; but cities which are increasingly integrated into the global economy, which is thought to be a source of strength, may be undergoing processes of social and spatial change internally threatening their competitiveness and sustainability. The social-spatial fracture is the more acute form that this takes, weakening the stakes that people feel in the future of the city as a whole, and strengthening their demand for security at the neighborhood level and their willingness to seg-regate.

The dynamics of social and economic relocation within an urban region, when new business and residential districts are being built, may help drive the construction sector locally, but can ultimately weaken social cohesion, and raise the public sector costs for infrastructure. The processes of de-concentration and decentralization within a single large metropolitan area are however largely ignored in macroeconomics. To a macroeconomist, the city remains something of a black box phenomenon: he knows that cities raise national output and pro-ductivity, but exactly what happens in cities, and the impact of spatial organ-ization on the outcomes, cannot be easily observed or analyzed. This gap in our understanding leads many to conclude that cities do not need a center. I want to explore the implications of this issue on the coalition and consensus-building needed to support urban development strategies.

The erosion of city centers has probably been greater in the United States in recent decades than elsewhere. Given the tendency to consider that the US today is an indication of trends which will be manifest elsewhere in the future, we should ask what the record of that experience indicates for the future of policy. The Fannie Mae Foundation (US) commissioned a survey that asked urban scholars to rank the key influences shaping the past and future American metropolis (Fishman, 2000). One hundred and fifty scholars responded, selecting the top ten influences for the future from a list of 50, in order of their importance. The ten items considered most likely to influence the American metropolis for the next 50 years are:

- Growing disparities of wealth.
- Suburban political majority.
- Aging of the baby boomers.

- Perpetual "underclass" in central cities and inner-ring suburbs.
- "Smart growth": environmental and planning initiatives to limit sprawl.
- The internet.
- Deterioration of the "first-ring" post-1945 suburbs
- Shrinking household size.
- Expanded superhighway system of "outer beltways" to serve new edge cities.
- Racial integration as part of the increasing diversity in cities and suburbs.

Most of the items on this list are negative, posing social, economic and governance challenges for which solutions are neither obvious nor easy, especially if the suburban political majority (2) is passive or hostile. Only one, the last item, is explicitly positive. This list shows that the preconditions for a political solution to urban problems may not exist in the US.

When the social basis for collective living has stretched too far and too thin, the economic and political basis for investment in urban centers is eroded. This condition became dramatic in the US, but there are also success stories. A recent American government survey of urban revitalization concluded that the "dynamic, committed leadership from elected officials and private sector and nonprofit partners" is critical to a successful transition to a 21st century economy. No surprise here. The largest cities in this study were Denver and Oakland, CA, the rest being cities of intermediate size such as Akron, Ohio, Fargo, North Dakota, and Wilmington, Delaware. Downtown redevelopment was first on the list of strategies for a successful economic revitalization. This is perhaps easier in cities that are not very large, especially when considering the prominence given to the social–spatial fracture in the Fannie Mae survey cited above. By contrast, a polycentric urban region can compensate for the erosion of the traditional urban center. Ironically, the largest urban areas, which are the places most conducive to the construction of tall buildings, and which have the greatest opportunity to improve sustainability through the efficiency of density, are also the urban areas which may have the greatest difficulty in reversing a pattern of decline in their centers.

The problem of forging a consensus on what needs to be done for cities can be illustrated further by exploring the differences between modernism and post–modernism which represent two very different views of urban society, and two different approaches to problem-solving.

Modernism had its greatest impact on cities during the period 1880–1960. Modernism was grounded in the assertion that there are principles and rules by which buildings and cities can be ordered. One can in fact talk of a tradition of modernity: a spirit of reform linked to an architectural and planning vocabulary suitable in a great variety of places and at many different scales, based on principles of reason and the criterion of meeting human needs. From this perspective, the Gothic revival of the mid-nineteenth century was just as much a phase of modernism as was the neo-Classical revival of the late-eighteenth.

Modernism recognized that the scale on which planners work is far greater than the scale which individuals inhabit and use on a daily basis. As a result, the techniques for giving form to urban space, to prepare them for development, have tended to shade the differences between people, to standardize around the average. This was above all typical in the Fordist era of mass production, when building and planning by rules and norms made possible the progressive expansion of the city while eliminating a range of environmentally unsound and unsanitary practices.

The lessons and achievements of the modernists are often forgotten now that technology provides many of the physical elements needed to make life comfortable. Modernism emphasized the need to improve environmental conditions and to give people access to light and space; it created public spaces appropriate to large urban crowds yet still often intimate enough for people to be alone; and above all, it asserted that people of different backgrounds must understand the city to make best use of it – hence the pursuit of a visual language designed to communicate clearly and meaningfully.

Postmodernism, by contrast, rejects the very idea that design can meet the needs of different people in a coherent manner, based on the argument that people are too diverse, and that any effort to develop a coherent style involves a relationship of power. It has been sensitive to issues of race, gender and ethnicity that found little place in modernist views of urban society.

Postmodernists hold that a holistic understanding of the city is impossible because cities by their very nature are marked by, and indeed generate, insurmountable social, economic and cultural divisions. From the postmodernist point of view, there are no objective standards against which a city and the conditions of life of the various groups living in it can be evaluated. Even the observer has no fixed point from which to take in the city and its experiences. Thus, postmodernist architecture violates classical norms about proportion and decoration as a way of proclaiming that there are no standards of beauty independent of a given time and place. Many postmodernists attack the shared public space and culture as an agent used by hegemonic groups to maintain their position and to dissimulate their power. Policymakers and politicians who are still trying to build coalitions and bridge differences do not see the city this way, however. They need a way to understand and act upon the city as a whole.

Postmodernism reflects the widely-shared perception that the future is uncertain and problematic; but the future has always been thus. Modernists may simply be more confident that the challenges of the future can be met. The problems of social fragmentation, environmental degradation, and structural economic change are real, but postmodernism is better at analysis than at providing the basis for policies that can address them. Perhaps we can indulge ourselves with postmodernist culture; we cannot afford postmodernist politics.

Ultimately, the problem of urban policy is a problem about how political advances can keep pace with economic change. Each of the three major periods of urban development since the Renaissance expanded political rights and economic opportunities, albeit through a process of change that was often highly conflictual. The late 17th and 18th centuries witnessed the creation of capital and commodity markets for the first major metropolitan centers of the Atlantic world, but also checks on arbitrary government and on the dominion of the military over cities, as well as the emergence of individual rights enshrined in law. Urban growth in the period 1880–1920 accompanied the introduction of modern telecommunications, infrastructures, electrification, mass production and retailing, as well as modern social welfare systems and universal suffrage. The economic opportunities of our era, combining globalization, environmental gains, and information and communications technology, are fairly clear to discern. But their implications for the exercise of democratic rights and for the protection of the rights of the individual are not so apparent.

Ever since the Renaissance, critics of the city have argued that because urban conditions are difficult to control and predict, cities – or at least very large cities – are a threat to economic, social and political stability. In the past, apocalyptic predictions of the collapse of cities have usually exaggerated their vulnerability to internal fractures and external shocks. The possibility of catastrophe cannot be excluded, but the history of cities shows that people can learn to resolve problems that are as difficult to anticipate as they are complex. Cities are not only the place where innovation occurs; the process of innovation can be applied to the solution of urban problems.

To summarize and conclude: a symbiosis is needed between the best of modernism and the best of postmodernism, not as styles, but as modes of reflection about cities. This means taking on modernism's concerns about efficiency on the one hand, and postmodernism's about identity on the other. These are two of the defining parameters of our time. But the one looks so objective, and the other so subjective, that a symbiosis looks impossible. They are in fact interdependent. The capitalist economy increases choices, giving people more control over their lives, more information, and more responsibility. Identity is about self-worth and dignity, the opportunity for individual mobility and self-development, the cohesion of communities, and freedom of expression. Both affirm the variety inherent in human experience; both explain why democratic systems are more tolerant, and cope better with problems, than autocratic and centrally-controlled ones. Social diversity is a positive factor in economic development; the modern economy supports a vibrant, complex society.

What has this to do with the conference then, tall buildings and the urban habitat? The principles of modernism operate at the macro scale which is appropriate to a large labor market, and problems of the environment, land, housing, transport and energy at the regional level. At this level, the location of investment is shaped, and major strategic issues resolved. Postmodernism is relevant to the micro scale at which people find meaning and understand everyday life in a small part of a city-region where issues of health, security and education are upper-most. At this level, families are formed, children are nurtured, culture and civil society flourish. If this symbiosis is found, then the viability of cities for investment in tall buildings will be coherent with the demand of people for cities as better places in which to live and work.

REFERENCES

Brickwell, Ditha, 1991
    RESEARCH AND TECHNOLOGY DEVELOPMENT FOR CITIES. In
    Ekistics, Nos. 350/351, pp. 324–29.

Fishman, Robert, 2000
    THE AMERICAN METROPOLIS AT CENTURY'S END: PAST AND
    FUTURE INFLUENCES. In *Housing Policy Debate*, 11, 1,
    pp. 199–213.

Graham, Steve and Marvin, Simon, 2001
    SPLINTERING URBANISM: NETWORKED INFRASTRUCTURES,
    TECHNOLOGICAL MOBILITIES, AND THE URBAN
    CONDITION, (London: Routledge).

Moynihan, Daniel Patrick, 1993
    DEFINING DEVIANCY DOWN. In *The American Scholar*, 17–30.
    OECD, 2001 CITIES FOR CITIZENS (Paris: OECD Publications).

ANNEX

Principles of Metropolitan Governance

Despite the affirmation that there is no one model of metropolitan governance, it is clear that (in addition to the broad principles which underlie any adequate system of democratic government – accountability, accessibility, representativeness, constitutionality, and protection of fundamental freedoms) a number of principles can also be applied in order to define the adequacy of systems of governance for urban regions in the 21st century. Classified in alphabetical order, these are:

Coherency: This principle states that governance must be intelligible to the electorate. A system based on a welter of agreements, complex formulae and compromised principles is inefficient. One of the main complaints is that nobody understands who does what. Confusion breeds indifference and apathy which in turn provide the ideal atmosphere for corruption and demagoguery.

Competitiveness: Governance must be associated with competitiveness. Urban regions are emerging as the main units in the world economy. As national barriers to trade fall, and as the factors of production become increasingly mobile, there should be an emphasis on investment in social and human development and in appropriate hard and soft infrastructures, rather than on reducing municipal taxes as a means of attracting investment.

Co-ordination: Given the administrative fragmentation of metropolitan regions, co-ordination among local authorities across jurisdictions, and between elected authorities and various regional boards or agencies with functional or sectoral responsibilities, must be a priority. This is especially important to assure a basis for strategic planning.

Equity: The institutional and financial arrangements in metropolitan regions must be designed to achieve as high a degree of equity and equal opportunity as possible among the composing municipalities and between the various social groups making up the population of the urban region. This will require greater pooling of resources.

Fiscal Probity: Any system must be created with the explicit recognition that the costs of governing most urban regions must be reflective of benefit received. Debt load and the tax rates are high in the urban centers of many OECD countries, and cannot be sustained in the face of strong international competition for investment. If cities are to meet the key social, environmental and economic challenges of our time, they must ensure careful resource stewardship.

Flexibility: Institutions must adapt as necessary to cope with rapid changes such as urban growth and swings in economic conditions due to globalization.

Holism: Any system must reflect the potential and needs of the entire urban region because this is the area that defines the economic and the environmental challenge. Each part of an urban region affects all others: this does not necessarily mean that all parts of the region require the same system of governance, but it does mean that all parts of the urban whole must be considered in the analysis.

Particularity: The principle of particularity states that, except where the case for standard policy is founded on human rights and immutable standards, policies and institutions of government must be crafted to fit the unique circumstances of various parts of the country. This principle is also important to more localized policy and institutions because it permits construction of unique solutions for various areas within the urban region.

Participation: Governance must fully take into account, and allow for, the participation of representatives of community groups, women, the elderly and the young, the business sector, social partners and all levels of government involved in the metropolitan area. New technologies and methods of communication can encourage and support more inter-active policy environments, bringing government closer to people.

Social, not sectoral: The objectives and institutional frameworks of metropolitan governance should meet the needs of people, which requires a cross-sectoral approach rather than one based on sectoral divisions which tends to perpetuate a bureaucratic, functionalist approach.

Subsidiarity: For the quality governance to be the best (and the costs least) services must be delivered by the most local level that has sufficient scale to reasonably deliver them. The principle rejects functional duplication and overlap. The principle of subsidiarity and the principle of holism together suggest a major decentralization of service delivery responsibility to local governments within a context of powerful policies and guidelines promulgated by senior governments. This would set limits on the principal means of governmental oversight as it is now practiced – duplicate review, multiple agency veto, second guessing and ad hoc revision as a means of control over local municipalities.

Sustainability: Economic, social and environmental objectives must be fully integrated and reconciled in the development policies of urban areas. This means adopting an "outcome" oriented approach which is holistic and integrates short, medium and long term considerations. In environmental terms, it means managing the metropolitan region in the context of the wider bio-region, the qualities and potential of which must be enhanced and preserved for future generations and as a contribution to a sustainable planet. In social terms it means ensuring that social cohesion is maintained and strengthened.

Josef Konvitz is Head of Division, Territorial Reviews and Sustainable Development, Organisation for Economic Co-operation and Development, Paris. This article is published on his own authority, and does not necessarily represent the views of the OECD.

New Towns in Hong Kong: Planning for the Next Generation

Pun Chung Chan

Hong Kong is perhaps the city in this planet that is subject to the most intense development pressures. I said this not because Hong Kong has the highest population growth rate – there could well be other cities having higher population growth rates than us. I said this because our development pressures come not only from very high population growth, at an average rate of about one million every decade over the last 40 years; but also from the rapidly expanding economic activities in the Pearl River Delta region much of which are handled in some ways in Hong Kong.

Our responses to meeting the development needs have to be largely framed within the span of about 1100 square kilometers of our own territory, a size compares very unfavourably with other big cities. Worst still, we can only develop on about 50% of our land mass as the remaining land is mainly Country Parks, which are 'no-go areas' for development.

How do planners in Hong Kong cope with these development challenges?

Many of you may be familiar with our high density developments, visually perhaps if not about the substance. Less people, however, know how planners in Hong Kong are planning the new towns. The purpose of this paper is, therefore, to give a brief introduction of the planning of our latest generation new towns which are now almost near completion.

OUR POPULATION GROWTH

I earlier mentioned that our population has been growing at the magnitude of about a million every decade. The growth had, in fact, accelerated in the nineties during which about 1.3 million people were added to the population. Today, we have a population of approximately 6.8 million. It is predicted that the population could increase to about 8.1 million in year 2011, and then to about 8.9 million in year 2016. (Annex 1)

Our Main Urban Areas, covering Hong Kong Island and Kowloon Peninsula, has a capacity for roughly 4 million population. There is a limit as to how much taller we can stack our developments; 'spreading out', in the form of new towns, is, therefore, not a matter of choice.

NEW TOWN PROGRAMMES

Our New Town Programmes can be said to have commenced in the early 60s when the Government decided to reclaim Gin Drinker's Bay, part of Tsuen Wan (later became one of the first generation New Towns) located some 10 km to the northwest of Kowloon Peninsula. Since then, a total of nine New Towns have been designated and are currently at various stages of development. To date, about three million people, that is about 44% of the territorial population, are living in the New Towns. It is estimated that by year 2011, about 4.1 million, or about 51% of the population at that time, will be living in these communities (Table 1).

The present situation regarding the development of these New Towns is summarized in the following table. The locations of the New Towns are shown on the plan in Fig. 1.

Table 1 New Towns in Hong Kong

New Town	Present Population	Ultimate Population
1st Generation New Towns commenced in early 1960s & 1970s		
Tsuen Wan	779,000	900,000
Tuen Mun	483,000	620,000
Sha Tin	624,000	760,000
2nd Generation New Towns commenced in late 1970s		
Yuen Long	136,000	210,000
Tai Po	290,000	320,000
Fanling/Sheung Shui	225,000	250,000
3rd Generation New Towns commenced in early 1980s & 1990s		
Tseung Kwan O	246,000	470,000
Tin Shui Wai	163,000	300,000
North Lantau	22,000	320,000
Total:	2,968,000	4,150,000

Source: Territory Development Department, Development Programmes, various years

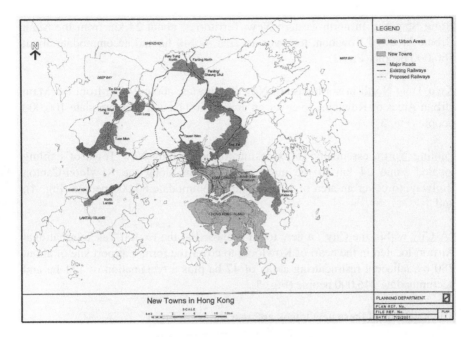

Figure 1 Locations of New Towns.

These New Towns have been successful to a large extent in providing a much improved living environment, which could not otherwise have been achieved in the Main Urban Areas. Development densities in these New Towns are, on average, considerably lower. Modern and efficient infrastructure including internal and external transport links are provided; and so are sufficient and modern community facilities such as schools, hospitals and community centers, to serve the needs of the residents. Above all, recreation spaces in the form of parks and playgrounds as well as amenity areas are provided at a much more generous rate (by Hong Kong standards) than the Main Urban Areas. A less successful element, though, is the provision of jobs, which has not been at the desirable level, resulting in a considerable amount of commuting to the Main Urban Areas for work.

THE NEXT GENERATION NEW TOWNS

When fully developed these New Towns, together with the Main Urban Areas, will provide accommodation for roughly 8.1 million people, which is still insufficient to meet the anticipated population level of about 8.9 million in year 2016. Creating development land and putting in place the necessary infrastructure on average takes about 12 to 15 years. That is to say, to cater for the population in 15 years' time, we must act now. This we have done, and are currently finalizing the planning of four new towns, namely:

<u>Hung Shui Kiu</u> in north-western New Territories, about 24 km from the Main Urban Areas of Kowloon, to cover an area of 440 ha and accommodate about 160,000 people (Fig. 2);

<u>Kwu Tung North</u> in north-eastern New Territories, about 24 km from the Main Urban Areas of Kowloon, to cover an area of 500 ha and accommodate 100,000 people (Fig. 3);

<u>Fanling North</u>, essentially an extension to the existing New Town of Fanling located some 24 km to the north of Kowloon along the Kowloon-Canton Railway, to cover an area of 192 ha and accommodate 80,000 people (Fig. 4); and

<u>"A City within the City'</u>, a new town proposed at the ex-Kai Tak International Airport located in the heart of Kowloon, to cover the former airport site of about 280 ha, adjacent restructuring areas of 47 ha plus a reclamation of 131 ha and accommodate 246,000 people (Fig. 5).

The locations of these new towns are also shown in Fig. 1.

Let me now briefly explain some important economic and social trends that have influenced the planning of these new towns, and the basic planning principles that we have adopted.

Economic Restructuring

Over the last twenty years, mainly as a result of Mainland China's 'Open Door' policy, Hong Kong has undergone fundamental changes in its economic structure, and has transformed from a manufacturing-based to a service-oriented economy. In 1980, the manufacturing sector employed some 890,000 people, or 50% of the total employment; today, manufacturing employment has drastically dropped to about 570,000, i.e. only 18% of the total employment (Annex 2). In 1980, the tertiary sector accounted for about 48% of the employment and contributed to about 68% of the GDP; today, 82% of our employment is from the tertiary sector, contributing some 85% of the GDP (Annex 3).

As a result of the economic transformation, the demand for industrial land in Hong Kong has diminished significantly, and one of the traditional functions of new towns in the provision of industrial land has also become superfluous.

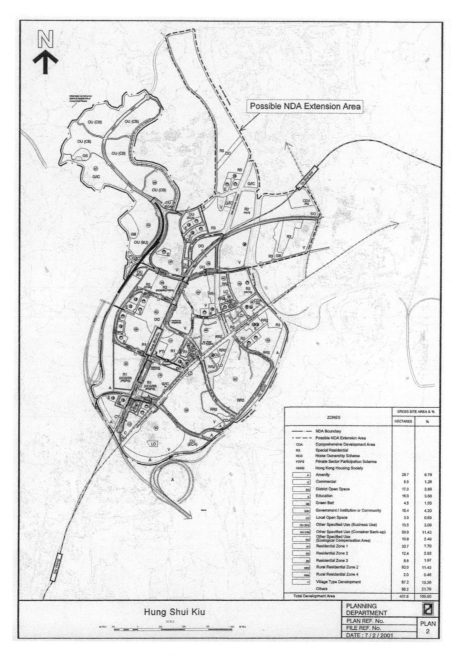

Figure 2 Hung Shui Kiu.

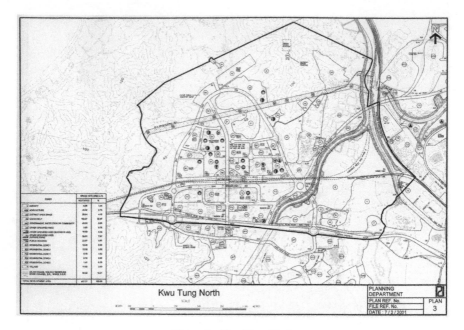

Figure 3   Kwu Tung North.

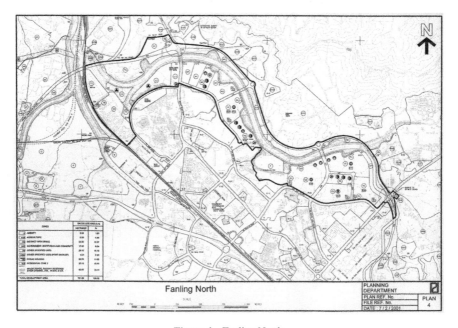

Figure 4   Fanling North.

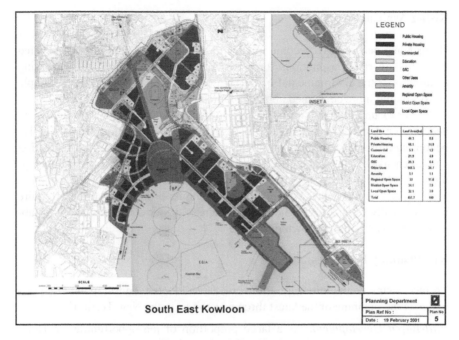

Figure 5 South East Kowloon.

Cross-Boundary Activities

On the other hand, as a result of the relocation of the labour intensive manufacturing activities from Hong Kong to Mainland China, particularly the Pearl River Delta region, our economic links with the Mainland have strengthened in many ways. For example, about three-quarters of our container throughput amounting to about 14 million twenty-foot equivalent units in year 2000 (ranks as the world's No. 1 busiest port) came from, or are destined to, the Pearl River Delta region. Social interactions with the Mainland have also intensified tremendously. For example, record-breaking numbers of people have been recorded every year traveling to the Mainland for leisure, visiting friends and relatives, and for business.

The increasing economic and social interactions between Hong Kong and the Mainland have given rise to a rapid increase in cross-boundary travelling. For example, in 1996 a total of about 65 million cross-boundary passenger trips were recorded at the boundary control points; in 1999, the number had increased to 99 million.

To capitalize on the development opportunities offered by increasing cross-boundary activities, the latest generation new towns proposed in the New Territories are all located in close proximity to existing and planned boundary crossings to Mainland China.

Rising Aspirations

Hong Kong people have become much more affluent and educated. Associated with this are rising aspirations for a better living environment and increasing awareness of the environmental quality. In tandem with the world trend, people are becoming more concerned about sustainable development.

For example, many are beginning to query the wisdom of building more and more extensive road networks to resolve traffic congestion. They are also becoming more concerned about the quality, not only the quantity, of recreation and green space. They have also demanded better urban design for our developments. Above all, they are also expecting more involvement in the planning process, and be more closely consulted on planning proposals. All these require us to critically rethink about the appropriateness of our conventional and established approaches to new town planning.

Key Planning Principles

To meet rising community aspirations, we have adopted the following key principles in the planning of the latest three new towns in the New Territories.

Rail-Based Development. As a large proportion of job opportunities in Hong Kong, in particular the high-end service sector jobs, will remain in the Main Urban Areas, it is expected that like their existing counterparts there will be a high degree of daily commuting from the New Towns to the Main Urban Areas. To reduce the reliance on roads, we have selected the locations and planned the three New Towns to cluster around and above the railway stations of a new railway line currently under construction (i.e. the West Rail).

The developments are designed to minimize the noise impact of the railways and maximize convenience in access to the stations for mass transport. In addition, railway station developments will cover a full range of retail facilities to provide convenient shopping.

Balanced Development. A mix of housing types and a full range of supporting commercial, community and recreational facilities are also planned for the New Towns. To take advantage of their strategic locations, proposals have also been made for business and hotel developments to capture the opportunities provided by cross-boundary activities. The intention is to create as balanced communities as possible, not just new towns for dormitory purpose.

Urban Design. Special attention has been given to the built forms to create distinct identities and characters. For example, to take advantage of the mass transport system, residential developments with plot ratios of between 5.0 to 6.5 are located in close proximity to the railway stations. A range of lower-densities, and lower-rise buildings, are placed on the edges of the new towns, nearer to the countryside. Special landmark buildings at key locations are also planned to create identity. Urban design master plans have also been prepared to provide comprehensive guidance to the design of building and streetscapes.

Environmentally Friendly Transport. These new towns are planned as compact and coherent urban communities. Majority of the population are proposed to be accommodated in developments located within a 500 meter walking distance from the railway stations, so as to encourage in-town walking and maximize the use of the railway. To encourage the use of railway transport, enhanced pedestrian facilities (e.g. travellators, landscaped walkways) are also planned to connect activity centres, residential blocks and the railway stations.

For developments located further away from the railway stations, the potential of using environmentally friendly modes of transport, such as electric buses or liquefied petroleum gas vehicles, are being explored. In addition, well-designed networks of cycle paths are planned to encourage cycling both as an alternative means of transport and a form of leisure activity. The intention is to reduce the environmental impacts, such as air and noise pollution, as far as possible.

Green Neighbourhoods and Attractive Pedestrian Networks. Green and leisurely designed neighbourhoods are very much a luxury in the congested Main Urban Areas. To provide a 'greener and softer' living environment, extensive district and local open spaces are planned as an integral part of the New Towns. Networks of landscaped pedestrian corridors are proposed to permeate the New Towns linking up these open spaces. It is intended that these people-oriented designs will create a leisurely environment and encourage social interactions.

Minimize Road Impacts. Where roads are necessary and unavoidable, they are planned to minimize their environmental impacts. In Fanling North, for example, the main road is aligned along the periphery of the New Town to discourage traffic penetrating into the residential neighbourhoods. Another example is the main road passing through the Hung Shui Kiu Town Centre, which is designed as a sunken road to reduce the visual and noise impacts. An added benefit of such a design is the creation of a car-free pedestrianized civic square for community enjoyment.

A key feature of the Kwu Tung North New Town is a Central Green Spine to link up the future town centre with a nearby lushly-vegetated knoll. This Green Spine comprises a bus-only lane, landscaped pedestrian way and amenity areas, designed in such a way as to segregate pedestrians from the bus traffic.

Conservation of Natural Heritage. A positive planning approach has been taken to ensure conservation sites are well preserved and enhanced. The Long Valley site in Kwu Tung North New Town is a good example. The ecological study undertaken as part of the planning for the New Town has identified the area as a very important freshwater habitat. Twenty-one different types of bird species warranting protection, including four globally threatened species, have been identified in the area.

To ensure that the new town development would not jeopardize this valuable ecological site, an area of about 40 ha is being considered for designation as a Nature Park, with private properties within the park to be all resumed and proper management measures introduced to place the area under full protection by the Government. Facilities would, however, be included at fringe locations for students and visitors to help raise awareness of the importance and value of conservation.

Enhancement of Cultural Heritage. Within the proposed New Towns and in their vicinities, there are many sites of historical value, such as traditional Chinese walled villages, study halls and ancestral halls. Some of them are already integral components of existing heritage trails. As part of the new town developments, access to these trails will be enhanced to facilitate both local and overseas visitors. Appropriate enhancement schemes for face-lifting of these heritage assets are also being devised.

'A City Within the City'

The relocation of the international airport from Kowloon to Chek Lap Kok in Lantau Island has opened up an invaluable development opportunity in the heart of Kowloon for building a 'new city' within the city. It has also opened up an important section of the central harbourfront for quality and creative developments. The key planning proposals are highlighted below.

Sustainable Transport. Like their counterparts in the New Territories, this new city is designed to encourage the use of public transport. The existing Mass Transit Railway Kwun Tong Line and the future East Kowloon Line will serve as the backbone of the public transport service. The use of these railways for inter-district transportation will be integrated with a local environmentally friendly transport system, possibly in the form of a light-rail or a trolley-bus system

Urban Design Aspect. In addition to the adoption of different building height profiles and urban design requirements, a special feature of the urban design plan is the incorporation of view corridors for Victoria Harbour and Kowloon Hills. In addition, some 5.4 km of new waterfront promenade is proposed for the enjoyment of local residents and visitors, and that a large tract of land, about 25 ha in size, is also earmarked for a regional park, to be named as the Metropolitan Park. These developments will be the future focus for tourists as well as local activities.

Environmental Aspect. The 'new city' development also offers opportunities for new environmental designs and facilities, such as depressed roads with wide landscaped decks to reduce environmental impacts, pedestrianized areas with spectacular view corridors, travellators, district cooling systems for buildings and automated refuse collection systems, as well as the utilization of solar energy for special projects.

Tourism and Leisure Developments. To promote tourism, part of the harbourfront, amounting to some 23 ha, is earmarked for leisure and tourism-related facilities to include such possible uses as a cruise terminal, hotels, a entertainment/retail/dining centre, an aviation museum, a IMAX theatre, a children's discovery centre, a monumental observation tower, etc.

Public Participation

I mentioned about the community wanting more and closer involvement in the planning process. In firming up the planning of the latest new town proposals, in addition to the statutory requirement we have also adopted a much more pro-active approach to consult the public. Soliciting public views through public forums, meetings with stakeholder groups, expert group discussions, etc. have been organized extensively. We hope a more open planning process, particularly at the early stage of planning, will not only promote mutual understanding and facilitate the building of community consensus; it will also help secure public ownership of the plans and minimize confrontation in the course of implementation.

CONCLUSION

In conclusion, I hope I have given you a brief and yet informative synopsis of how planners in Hong Kong have taken up the formidable task of planning and developing new communities to accommodate the fast growing population and transforming economy. I welcome everyone to visit our website (http://www.info.gov.hk/planning) where more detailed information on these new towns are available. Finally, thank you for your attention and patience.

Birmingham: Inventing a New Life for the 21st Century

Rod Duncan

During 1998–99 I was fortunate to participate in an international professional exchange to the United Kingdom, working in and researching innovation and best practice in urban design and regeneration. Across a nation now strongly focussed on addressing and managing change in its cities, perhaps the most dramatically successful example I observed was Birmingham, in particular the new sector of the inner city known as Brindleyplace. This may have been emphasised by negative impressions gained during a brief visit to the city 20 years earlier. The following account is based on detailed discussions with key officers involved in the process (particularly Geoff Wright), supported by wider research and recent updates.

There are valuable lessons to be drawn from the Birmingham experience, most obviously from identifying the factors influencing its successes, but also in how outcomes and processes might have been more effective. Many of these lessons will have applicability to cities throughout the world dealing with change.

Birmingham, with around a million residents, rivals some other contenders as England's second city. It has a rich history as a prosperous product of the Industrial Revolution that successfully made the transition into the twentieth century, maturing into a proud civic-minded community with a healthy manufacturing economy, large enough to be cosmopolitan, yet small enough to retain its own identity and ethos. However, turbulent economic change in recent decades has challenged the city's capacity to make the transition into the new millennium as smooth or comfortable.

Part of Birmingham's strength has been its diversity. Whilst dominant in some major industries (such as motor manufacturing and armaments – admitting to arming both sides in several conflicts) the city had a broad variety of industries, being known as 'the city of a thousand trades'. Some of these industries, such as metal-working and jewellery manufacture, are characterised by numerous small operations. These sustained a diversity of ethnic and religious subcultures within the city. It is one of the few places in Britain boasting of its 'multi-cultural' character in tourist literature. There is a culture of civic innovation, arts patronage and social benevolence, such as the Quaker Cadburys' initiatives in Bournville.

Birmingham suffered from the attentions of the Luftwaffe, but like many other cities much of its physical destruction was self-inflicted in the post-war period. Embracing modernism, by the 1960s central Birmingham was dominated by elevated motorways feeding duplicated ring roads with interminable round-abouts threading between – or through – concrete towers of offices and housing, whilst pedestrians were channelled into a disorientating network of vertigo-inducing aerial walkways or intimidating underpasses. Perhaps this reflected the city's readiness to embrace innovation, adopting what was 'best practice' for the era – and now a lesson in the risks of pursuing the latest fad. Perhaps it was structural: a large, prosperous municipality with high autonomy. Perhaps it was cultural: this was the city that spawned the Mini in a similar period.

My memories of briefly visiting Birmingham in 1977 have coalesced into a collage of this imagery, with the flicker of some grand Victorian civic sur-vivors stranded in the centre. Shades of 'Clockwork Orange'. Retrieving my photo archive, I found that, tellingly, I had taken just one shot during that visit. This confirms the accuracy of my impression: the Bull Ring shopping centre straddling the Queensway inner ring road ... with an ironic pebble-paved compass point vainly offering some sense of orientation.

Birmingham's diversity of industries, particularly manufacturing in its inner areas, moderated the impacts of de-industrialisation relative to Britain's more northerly industrial cities, often vulnerably narrowly based on big indus-tries like coal, steel and ship-building. However the economic changes during the 1980s, attributed to the policies of Margaret Thatcher, wrought devastatingly upon Birmingham. Employment in manufacturing in the inner ring of the city halved in a decade, falling from 60% to 30% of the workforce. This activated the 'stimulus of desperation' that can by-pass cities that either slowly decay or atrophy in the warmth of their comfort zone. Perhaps Birmingham got clever out of desperation.

The City of Birmingham could draw on a successful and lucrative edge in the new economy. It owns the National Exhibition Centre, a massive complex developed beyond the urban area in the 1970s on a motorway interchange adja-cent to a railway station and airport. This venue hosts most major exhibitions and trade fairs, conveniently positioned to serve the national and European catchments. As well as convincingly illustrating the growing value of the service economy, it provided the City with established expertise and a handy cash-flow.

Municipal leaders and managers recognised that the service economy was a lifeline, perhaps the sole lifeline, to the city's future. Substantial investment has been made in the International Convention Centre and National Indoor Arena, both located in the inner west of the city complementing a growing hos-pitality precinct. The City's reputation for patronage of the arts was reflected in an impressive new Symphony Hall, providing a home for its famed orchestra whilst doubling as the plenary venue for the ICC. Hosting the G8 Summit in 1999 confirmed the venue's global perspective.

The City also elevated planning from a peripheral activity to a critical obligation. It adopted a philosophy that 'Good environment is good business.' A search conference of high profile participants in 1987 (termed the Highbury Initiative, after its venue) set a course that had the resilience to weather political volatility, including change of Council control and a marginal national electorate. This initiative generated a City Centre Strategy and supportive structures and commitments.

In 1989–90 Francis Tibbalds undertook BUDS, the Birmingham Urban Design Strategy, that examined the form, function and spirit of central Birmingham. This report was an identification of opportunities, a prompt for self-examination, and insightful observations, rather than prescriptions. (He was invited to Melbourne to prepare a similar report at about this time.) This informed refinement of the strategy for the whole central area, policies and statements for each of its seven 'quarters', with subsequent elaboration for component sub-areas still continuing. This methodical hierarchy of increasingly specific and detailed strategies and action plans appears somewhat theoretical, but it is simple to follow and readily understood, linking the big picture to local relevance, and appearing to ensure focussed effort. The chain of linkage to the central objectives is also supported by sustained political commitment and resource allocation.

An early conclusion was that the tight inner ring road restricted the central core's capacities, whilst stifling flow-ons that could fuel adaptation and revitalisation in the encircling inner 'quarters'. The cry went out to 'break the concrete collar', lowering and downgrading vehicle routes, humanising the forms, providing at-grade and direct pedestrian links throughout the central area. The heavy engineering of the 1960s has now been largely re-engineered and a variety of destinations have become a comfortable and pleasant walk away, with more planned.

The most apparent changes to date are in the western and northern sectors of the inner city. The latter is the Jewellery Quarter, still housing large numbers of artisans and small businesses, and a trade and arts school in impressively restored buildings. Support extends to some simple but effective initiatives such as Council leasing a building then sub-letting small studio spaces to artisans and students, and nurturing a specialist tourist theme that complements jewellery production and prompts investment in restoration of an extensive but fragile stock of heritage buildings.

But it is the western axis from the city centre that has undergone the most dramatic transformation. An established pedestrian route from the civic focus of the impressively upgraded Victoria Square now extends via a broad walkway over the lowered ring road to Centennial Square. The Square has been paved in representation of an oriental carpet, celebrating one aspect of the city's diversity and providing a large resilient space is capable of accommodating major public events. The Square forms a forecourt to Symphony Hall, the International Convention Centre, a regional theatre and war memorial, with major pieces of public art reflecting the city's history and spirit.

The pedestrian axis follows through the atrium joining Symphony Hall and the ICC complex, emerging beside part of Birmingham's extensive canal network, long ignored and hidden behind factories. Across a feature bridge is the redevelopment known, after a pioneering canal engineer, as Brindleyplace.

BRINDLEYPLACE

When completed soon, this seven hectare site will house about 10,000 workers in about 100,000 m^2 of offices and a mixture of other service and hospitality uses. (To reflect on the change between the millennia, this will exceed the number of workers at Rover's Longbridge car plant – if that is still in production.)

This was previously a decaying low-rise manufacturing area, assembled in the 1980s by the Council and sold for an approved development to a company that failed in the 1990 crash. By then, Council had spent the proceeds (26 million pounds) on building the adjacent indoor arena.

When a parent company sought to retrieve something from the asset, it brought the modest expectations that a former (Tory) Planning Minister termed 'the benefits of bankruptcy'. Having already spent its money, Council and senior management was also relaxed in its expectations. The Council officer that re-negotiated the site approvals attributes the success to 'lots of unpressured thinking time' and a high level of autonomy to work at the professional level to articulate strategic objectives with a sympathetic developer. 'It was delivered because no-one believed it would be.'

Negotiations with Argent Properties centred around lifting the cap on office floor space and the capacity to pursue fresh 'planning gain' benefits that Council could have required for a new development proposal. The return negotiated was improved quality and certainty, with a lot more attention to detail than usual. This included upfront detailed master planning by John Chatwin with layout, circulation network, and enabling creation of self-contained lots that could be developed separately (providing flexibility for the developer).

A commitment to high standards of urban design and architecture was achieved, along with the early provision of substantial public infrastructure, including a feature pedestrian bridge, two major squares and rehabilitation of the canal-side. The first stage included establishment of active elements including waterside bars, restaurants and shops that cost substantially more to construct than their completed market value (a 'loss leader' in property parlance), but were seen as essential (and financially justifiable) to add vitality and profile to the wider area.

The City's detailed strategies, urban design analysis, and practical guidance for redevelopment equipped it to know what it wanted, and where, in great detail, and gave officers the confidence to demand it. The quality and clarity of this information also assisted the developer to recognise not just why Council wanted this, but how it complemented the private sector's interests. Having a developer with long-term commitment and deep pockets clearly helped, but it is now clear that going for top quality on this site will be a very lucrative investment.

The arrangements negotiated for Brindleyplace articulated the site with a fine grained network of routes, extending the strong pedestrian axis established from the central city across a feature bridge. Ground floors on main routes are generally active frontages with retail, hospitality and nightlife activities, and new public spaces of exceptional quality are required to be completed <u>before</u> the stage each services. Key existing buildings were incorporated, including the Oozells school, converted to a contemporary art gallery.

A wide variety of uses have been achieved along with the dominant office function: a community theatre, gymnasium, aquarium, gallery, variety of restaurants, retail (principally tourist-focussed), and a 900 vehicle car park, along with a Novotel developed separately. The central focus of the early stages is Brindleyplace Square, featuring understated landscaping and a temple to the modern religion: a small but bold coffee kiosk that won Piers Gough/CZWG a national architecture award.

There is not a residential element in the principal development area, but an adjacent triangle of land separated by a canal was sold early and has been developed as Symphony Court, a 140 unit town house enclave. Gated private access and the intervening canal limit its contribution to Brindleyplace, and constrict circulation linkages with destinations to the north. The value of residences within this development are reported to have increased to 250% of their sale price in the first few years, reflecting the extent of re-imaging of the locality.

Office buildings comprise about 100,000 square metres total floor space in ten separate structures from 4,000 m^2 to 12,000 m^2 in floor area, with flexibility to respond to individual tenants in units down to 500 m^2, and serviced suites for smaller users. The development company builds and retains ownership with a variety of predominantly national tenants. Each building is architect designed with a high standard of finish, distinctive from its neighbours, but of a consistent height and bulk. Curiously, Argent utilises two major construction companies, who each opt for different construction technology nation-wide: one builds only in reinforced concrete, whilst the other prefers steel frame structures, apparently with comparable economics.

The dominant scale of office elements is around 7 storeys, subsidiary elements generally two or three levels. Individual office buildings have gained positive design reviews and architectural and office industry awards, enhancing the rapid uptake of prime occupants.

In later components, the developer has become more adventurous toward mixed use in a single structure. One recent building includes flexible office space (including hot-desks and short leases), a corporate restaurant, five levels of serviced apartments and the capacity for office conversion to more residential or even retail if the demand is shown. The developer sees the benefits of mixed use making an area that is a destination in itself, along with an attractive work environment that helps retain or attract skilled employees in a competitive environment.

LESSONS AND OBSERVATIONS

Birmingham's experience demonstrates some lessons for others. Hopefully these will be pursued more deeply than trying to copy the success through simply mimicking the built forms. Having a sound, agreed strategic basis helped it target key opportunities, and to provide incremental infrastructure improvements that maximised these opportunities coming to fruition. Having a clear logical and publicised basis for negotiating quality, supported by Council's own investment, enhances prospects of developers responding positively, and treating their involvement as a positive alliance. None of this ensured that the essential spark would light, but it ensured that when an opportunity did come it generated the most positive outcomes and magnified the benefits.

The result has been spectacularly successful at several levels. A whole new urban locality has been inserted close to the centre of a major city without being monochromatic, monofunctional and quickly dated. It is also clearly a raging commercial success. (Hosting the British Council for Offices conference last year must rate as the top prize in the commercial world, with the hard-nosed financiers and property managers drooling over the complex.) In turn this private success has dramatically challenged Birmingham's image across Europe, and will doubtless stimulate increased inward investment.

There must have been moments since 1983 when questions were asked as to whether the efforts being put into strategic planning, and sensitive nurturing of the city's diversity and character was the way to go. Liberalising standards in a downward dutch auction is attractive to the short-term thinkers both near and far. Birmingham's nerve and backbone is now paying off, as lesser wills slide deeper into mediocrity.

On the downside, there are some shortfalls in the development. In the absence of a residential component (discounting the enclave across the canal) it falls short of the mixed use ideal. However it does contain a diversity of other uses, particularly at street level and along route frontages. It also lacks integration with the city's rail network for commuter access, with no lines close to this locality. However it does front the major commuter bus corridor in Broad Street. The site's large car park encourages commuting worker traffic that conflicts with sustainability objectives, (and, it would appear, national planning policy). There are plans for a light rail service, including a line to the north of the site, but timelines for this are unclear. Introduction of a new mode could further strain achievement of integration in the city's already fragmented transport system.

Pedestrian routes dissipate across the site from the strong city centre link to the east, avoiding servicing, commuter and client vehicular traffic that enters from the west and disperses, minimising conflicting paths like interleaved fingers of two hands. However, this would seem to terminate any future prospect of further extending west this successful pedestrian axis.

Brindleyplace also contains what is generally conceded as Norman Foster's worst building (apparently illustrating the pitfalls of design-built procurement), the Sea Life aquarium, an unconvincing scallop shell that aims to front a corner of the principal square expanding back to fill a dead corner site. Ironically, this aquatic experience does not embrace its extensive waterway frontage. It is unfortunate that rather than a feature, this is a fill-in between office buildings of greater architectural pretention. The offices are above average and generally achieve diversity with consistency that gives a comfortable variety. Whilst the anchor building (by Porphyrios Associates) linking the canal frontage to the new Square reflects a curious parentage in Neo-Venetian/Classical vein with Italianate campanile clocktower in red brick and reconstituted stone, it brings a landmark presence that is also comfortable at street level and unoffensive.

CONCLUSIONS

With the success of the Brindleyplace site, and equipped with its clear, practical strategies for the rest of its central areas (and refreshed confidence in their effective application), Birmingham appears to be well-placed to succeed in the transition to another century. This success has jolted aside the image of Birmingham as a manufacturing dinosaur, being the envy of many European cities in attracting the service economy to its heart.

Whilst perhaps good luck, good fortune, fortuitous circumstances and location may have helped, there has clearly been a lot of clever work to ensure that when the opportunity arose it was seized, and then maximised. Birmingham's thorough forward-planning and focus on quality in design and urban environments ensured that redevelopment of the Brindleyplace site was elevated from just another office complex to a landmark development that has repositioned the city for the new millennium.

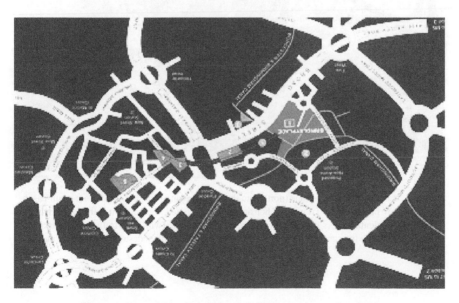

Figure 1 Central Birmingham (encircled by inner ring road at right).

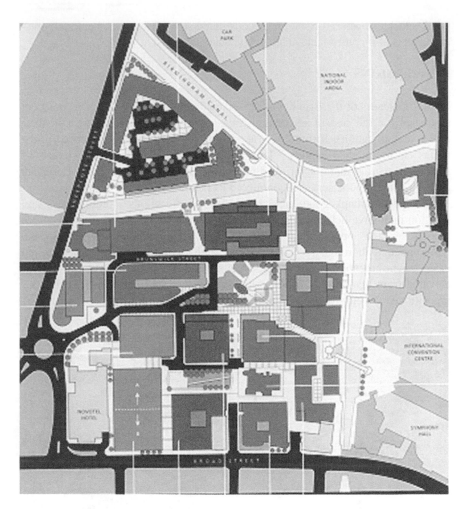

Figure 2   Layout of Brindleyplace (north is to the top), showing pedestrian axis from south-east to principal new public space. Triangle of housing to north is gated, separating it from the balance of the area. Vehicular access from the west disperses into the site (with most traffic to the parking structure beside the canal at the north), minimising conflict with pedestrian areas and routes.
[Argent Properties]

Figure 3 Artist's impression of completed Brindleyplace facing west. Note restaurant, café and tourist retail (marked B) on canal frontage and pedestrian routes to principal square (A) and smaller Oozells Square (S) fronting converted school (C). [Argent Properties]

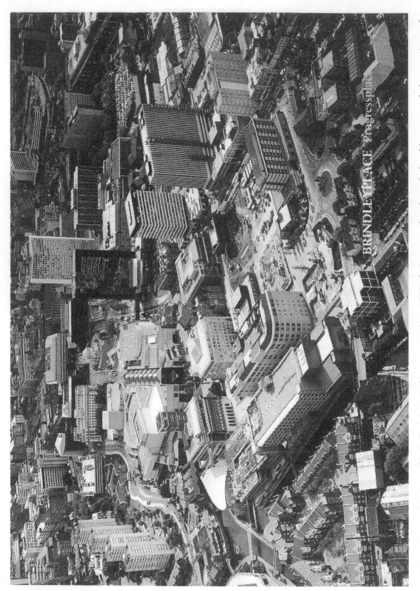

Figure 4 Aerial view of Brindleyplace during construction (1998) facing east, showing relationship to central city area (central middle and rear ground) and Broad Street arterial on right. [Argent Properties]

Figure 5   Brindleyplace Square, featuring award-winning coffee shop by Piers Gough/CZWG and No. 3 Brindleyplace office building (with clocktower) by Porphyrios Associates, with pedestrian link (shaded) to footbridge, International Convention Centre and city centre via Centennial Square.
[Argent Properties]

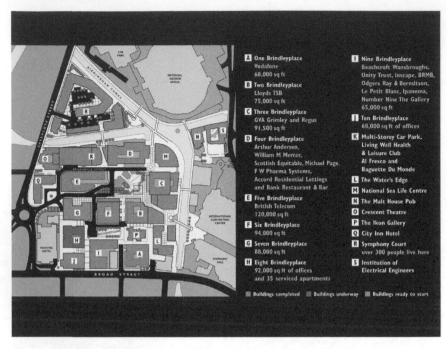

Figure 6   Site plan showing details of buildings, construction dates and architects.

Tall Buildings in Melbourne: Challenging Policy Frameworks

Lyndsay Neilson

INTRODUCTION

This paper provides a brief overview of the history of tall buildings in Melbourne, Australia, and discusses the policy context within which they have developed through various phases. Although the policy techniques have varied, I will argue that they share a common feature of being effectively reactionary in approach.

The need for a more specific design context approach is discussed. The antecedents to such an approach can be found in the 1985 Melbourne Strategy Plan and the 1991 Planning and Urban Design Guidelines; whereas the present Melbourne Planning Scheme has adopted a strict performance based approach that has tended to result in a reversion to a site by site reactionary approach.

Finally, I will discuss the characteristics of a possible design context lead approach and the precedents for this within the Southbank Strategy Plan (1999).

EARLY HISTORY

The 19th Century period of Melbourne's development was naturally characterised by low rise buildings, and the dominant character of the land form is evident in early paintings, where the two hills along Collins Street and the consequent valley along Swanston Street are clearly evident.

With the discovery of gold in the 1860's, the population and development accelerated, and by the 1880's Melbourne was booming – a period know as "Marvellous Melbourne". It is pertinent to note that at this time Melbourne was one of the most confident and advanced cities in the world, adopting new technologies such as railways and electric light within a year or two of their introduction internationally, and developing buildings of international standard in terms of both scale and technology. Had this trend continued into the 20th century, Melbourne might well have been one of the first to develop very tall buildings.

As it turns out, some of the confidence within the Melbourne society was lost in the depression of the 1890's; and although it remained quick to adopt new technologies such as the hydraulic lift, the city would not seek to rival New York or Chicago as birthplace for the skyscraper.

Through the 1950's a height limit of 40m was imposed, based on the effective reach of fire brigade ladders. The introduction of alternative strategies including automatic sprinkler and alarm systems had lead to the removal of similar restrictions in New York in the 1930's and in Sydney in 1957, but it would not be until 1968 that height limits for buildings in Melbourne would be lifted.

Even then, there was little policy discussion regarding the change, and it seems that the removal of height restrictions was largely in response to developer demand rather than an active policy interest in the tall building as a built form.

THE EARLY TALL BUILDINGS

With the removal of height restrictions from building regulations in 1968, a series of office developments well in excess of 40m commenced. Not surprisingly, the initial focus was to locate them where land values were highest and where views and outlook were available. For Melbourne, this was Collins Street, the location of some of the grandest of the Victorian era buildings. The towers were generally achieved by accumulating several sites and removing existing buildings.

The collection of buildings along Collins Street is evident in this image from around 1980.

By the 1970's, Victoria had introduced extensive heritage protection that identified and protected nominated buildings, however many lesser buildings that were lost, while not necessarily of high heritage significance, did provide a context for the significant buildings. A second grouping of buildings breaking the 40m level occurred along St Kilda Road, a broad tree lined boulevard that was flanked by stately mansions with generous garden setbacks. Planning controls protected the garden setbacks, so that the towers which are generally of lower height than those in the central business district also have garden settings.

Until recently there has been little pressure for tall buildings outside the CBD and St Kilda Road, as Melbourne continued to expand outwards with low density development. However since around 1985 there has been an increasing move towards consolidation and a focus on attributes such as Port Phillip Bay and the Docklands and Yarra river areas. It is becoming clear that a more effective policy framework is needed to guide the location and design of tall development generally, and that the experience within Melbourne's CBD is now relevant to many places.

TOWARDS A PRO-ACTIVE POLICY

The first urban design policies aimed at guiding the location of tall buildings in Melbourne appeared in the Melbourne Strategy Plan (1973). This proposed that tall buildings in Melbourne should be located to reinforce the natural geography, and identified two locations where height might be located, termed the "East End" and "West End". Several reasons or objectives are given for this policy, the most prominent being to reinforce the two main hills of Melbourne and to maintain views in and out of the city by restricting height around the perimeter of the CBD.

Development through the 1980's followed this principle and it was reinforced in planning guidelines including the 1985 Strategy Plan and the 1991 Planning and Urban Design Guidelines. A notable exception was the construction of the two slab towers on the river side of Flinders Street for the Gas and Fuel Corporation. The public dislike of these towers lead to controls to protect

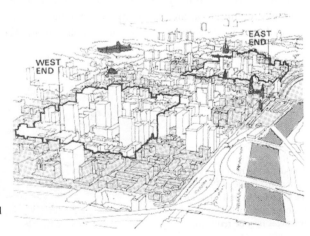

Illustration from the Melbourne Strategy Plan 1973, arguably the first formal encouragement for tall buildings in Australia.

Illustration from early 1980s
clearly shows the "valley"
reinforced by development
on the two hills.

overshadowing of the river, and the eventual demolition of the towers in 1997 to
make way for the Federation Square project, now under construction.

THE 1985 STRATEGIC PLAN

The City of Melbourne Strategic Plan (1985) began to provide much more detailed
policy to attempt to integrate tall buildings into the city's built fabric. While the
notion of the two hills was retained, two additional concepts were introduced.
Firstly, high rise development was encouraged around the underground rail loop,
leading to the Melbourne Central Project which stands as an exception to the
"valley" effect; and the recognition of the Southbank Precinct that had been identified in the State strategy "Framework for the Future" (1984) as an area for
increased commercial and residential development as a partner to the CBD. In a
sense, these two concepts were in competition with the East-West End concept.

This diagram from the 1985
Melbourne Strategy Plan indicates
the 40m height limit area, and also
indicates areas where active
frontages may be required.

This image of 101 Collins Street shows the tower set back with a podium to address the street.

The 1985 strategy also proposed height limits of up to 40m for the "retail core" of the city, where the traditional fabric was most intact. This was done in order to maintain a variety of smaller tenancy opportunities and to reduce the upward pressure on rents, both of which were seen to reduce the viability of retail in the CBD. This period coincided with the growth of regional shopping centres in the middle suburbs.

The strategy also introduced requirements for active frontages on key pedestrian streets, and the concept of podium and tower in order to provide a reasonable context to older fabric, and retain a "human scale" for the street environment.

The strategy plan also set guidelines for setbacks; areas where the building line should be maintained to the street, and the podium form to provide a workable relationship between historic built form and the newly introduced towers.

The Central City Urban Design Guidelines (1991) built on this strategy and introduced a number of provisions to guide the design of tall buildings.

These included guidelines about the tops of buildings; further guidelines about tower spacing; and provision of active street edges to protect the street environments and overcome the negative effects of the "tower-in-park" model.

Introduced controls for protection of views and vistas; and encouraged the creation of new public spaces and arcade links to add to Melbourne's already rich structure of pedestrian arcades and lanes.

The 1991 Guidelines were based on a plot ratio model adapted from the New York planning guidelines of the late 1960's. This provided base level plot

This recent image from the Southbank area illustrated the earlier mid rise housing and two recent residential towers utilising podium treatments to moderate their scale at the street level.

ratios with bonuses, often to a factor of two that could be achieved by meeting the design objectives such as arcades and active frontages.

The "new format" Melbourne Planning Scheme introduced three years ago removed plot ratio controls and has moved to a fully performance-based format where design objectives are guided by policy alone. There are no bonuses now.

This has effectively provided greater encouragement for market-led solutions and with height limits now treated as part of policy rather than a control. In the absence of broader design policies, the result tends to be a site by site negotiation which presents a challenge for managing the built form overall.

The Development of the Southbank precinct and later Melbourne's Docklands encouraged taller buildings and more intensive development outside the city grid. Attempts were made to extend the grid into Docklands but with limited success.

A progressive easing of policy in recent years also allowed taller buildings along the Bay foreshore (for example Beacon Cove, Port Melbourne).

Southbank, Docklands, St Kilda Road, and Beacon Cove have all provided opportunities for the dispersal of taller buildings and 'breaking' of the original policy frameworks that located the tallest buildings on the two hills of Melbourne.

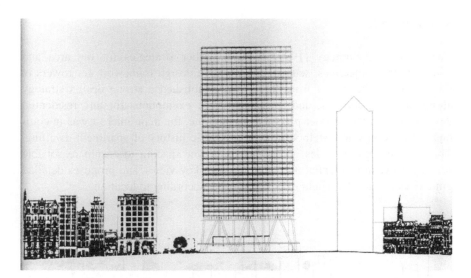

A recent proposal for a new tower over a former open space in Collins Street illustrates the challenge of inserting large tower forms into a more traditional scale environment.

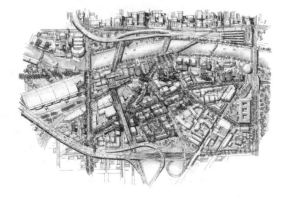

SOUTHBANK

The Southbank Strategy (1999) reviews earlier strategies for the area and amongst other objectives, sets out to provide a design framework for towers in this precinct. The policy from the 1980s established a strong design strategy along the river corridor, and provided height exemptions for any residential development in the broader precinct. At the time, the main interest was in commercial buildings and Melbourne had very little history of apartment dwelling. The new strategy provides guidelines for tower spacing to ensure reasonable privacy between residential uses and maintain key views, and proposes detailed guidelines to vary the height of developments according to location.

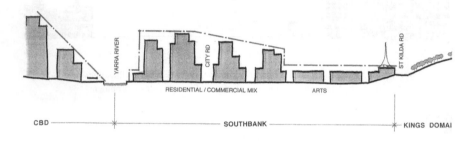

PROPOSED SOUTHBANK BUILT FORM PROFILE
(FOLLOWING SOUTHBANK BOULEVARD ALIGNMENT)

Proposed built form profile from the Southbank Structure Plan 1999.

Guidelines also established design attributes for the pedestrian environment, and nominated landmark sites where taller development was expected, such as the development site that has now become the Eureka Development. Never-the-less, the Eureka proposal exceeded the predicted heights, and at 300m plus, will be the tallest building in Melbourne and it is claimed it will be the tallest residential tower in the world.

The Eureka Development located on a designated "landmark" site, will become the tallest building in Melbourne. (now under construction).

DOCKLANDS

Among the proposals for the renewal of the Docklands precinct a proposal for the world's tallest building has been developed. This is still a potential project, although competing against a number of other bids to gain access to the site.

Proposed Grollo Tower in Docklands, Melbourne.

A tower of this scale presents real challenges not just in its wider impacts of shading and wind. In fact, the design panel assessing the proposal spent most of its time negotiating the conditions at the base of the tower, in order to achieve an urban environment that could provide the pedestrian friendly link between the CBD and the new waterfront.

CONCLUSION

Proposals of this kind represent a major challenge to the resolution of urban design principles for Melbourne. It also demands that we resolve our attitudes to the concept of taller buildings, as we enter the century of sustainability.

The policy history in Melbourne suggests we have only partly been successful in reacting to the tower building on a site by site basis. It could be argued that this is because the model that most contemporary tall buildings have sprung from remains the "tower in park" vision of the early modernists. Devices such as podia and required active frontages can moderate this, but the sheer size of the footprint of a contemporary tower makes it very difficult to integrate with traditional urban structures.

A preferable direction would be to establish an urban design vision for a precinct in terms of the desirable heights; massing; distribution of activities and open space; linkages and access; environmental sensitivities (eg places that should not be shaded) and so on; which would lead to more definite controls. The 1985 Melbourne Strategy Plan and subsequent Planning and Urban Design Guidelines (1991) provide a model for how detailed controls might work; whilst the Southbank Strategy (1999) provides a model for the development of a vision that takes into account both market aspirations; community policy objectives and the need to control environmental impacts.

REFERENCES

Department of Infrastructure, 1999
    SOUTHBANK STRUCTURE PLAN 1999, Department of Infrastructure
        in conjunction with Melbourne City Council, Melbourne.

City of Melbourne, 1991
    CENTRAL CITY PLANNING AND DESIGN GUIDELINES, City of
        Melbourne, July.

City of Melbourne, 1987
    GRIDS AND GREENERY, THE CHARACTER OF INNER
        MELBOURNE, City of Melbourne.

City of Melbourne, 1985
    STRATEGY PLAN 1985, VOLUME 1: THE PLAN, City Strategic
        Planning Division, July.

City of Melbourne, 1974
    CITY OF MELBOURNE STRATEGY PLAN, Ennis & Jarrett Pty Ltd,
        Highett.

Department of Planning and Environment, 1984
VICTORIA. CENTRAL MELBOURNE: FRAMEWORK FOR THE
FUTURE. LAND USE AND DEVELOPMENT STRATEGY, F D
Atkinson Government Printer, Melbourne, December.

Docklands Authority, 1996
THE HUB OF VICTORIA, A HISTORY OF MELBOURNE
DOCKLANDS, Mercedes Waratah Press, Australia.

Interplan Pty Ltd., 1974
CITY OF MELBOURNE STRATEGY PLAN, URBAN DESIGN AND
TRANSPORT IDEAS, Noric Pty Ltd, Melbourne.

Victorian Department of Planning, 1982
MELBOURNE CENTRAL CITY DEVELOPMENT MANUAL,
VOLUME 2, F D Atkinson, Government Printer, Melbourne.

A Third Millennium City Prototype – Melbourne Docklands

John Tabart

INTRODUCTION

This Congress is timely – and very relevant. It provides an excellent opportunity to discuss and debate one of the greatest challenges for the Third Millennium – cities of the world, and the urban habitat.

We regularly refer to increasing globalisation, and the enormous steps forward in technology, communications, bio-science and finance. These are the high profile activities. But it is the quality of life in the cities in which an increasing majority of the world's people will live that will be paramount to local populations – shaping the physical form, politics, viability and sustainability of human endeavour. The subject is that important.

Within the theme of 'New Cities', we introduce Melbourne Docklands – Victoria's new waterfront, a 220–hectare waterfront precinct at the centre of Australia's second largest city. It is very unusual to find such a large re-development opportunity so accessible and so close to the centre of a major world city.

For decades, the Docklands area was an under-utilised port and rail yards, immediately adjacent to the CBD. Some of the nostalgic images are still in evidence and continue as an important part of the heritage of the area.

But the vision for Melbourne Docklands, and the practical planning, are very much focused on the future city concept – creating Victoria's New Waterfront for the Third Millennium. Already more than $Aust 1 billion of development has been completed, is in construction or is committed to start this year, a further $Aust 1.5 billion is contracted with major developers, and the estimated total investment over the next 15 years is more than $Aust 6 billion.

By 2015, it is estimated that Melbourne Docklands will attract 20 million visitors annually as the venue for up to 200 land and water events – all focused on the 40 hectares of water that makes Docklands a special part of Melbourne. Local, interstate and international tourism will be an increasing part of the New Economy – and the water, arts, and events focus of Docklands will provide a major boost to employment in Victoria.

In 15 years, Melbourne Docklands is expected to become home to 15,000 residents, with both high rise apartments looking over the water and the central city and low rise apartments, with a water address, that will become part of a vibrant public realm.

Melbourne Docklands is expected to be workplace for 20,000 people in 15 years time. The nature of work places, globally, is changing rapidly. The high tech nature of the Docklands area, particularly our commitment to broadband communications links to the world, and across the whole area, will make it a very attractive area for many Century 21 work skills and enterprises.

Melbourne Docklands provides a prototype for cities of the Third Millennium. There are three key features of our approach:

- A focus on the projected and real needs of the people who will visit, live and work in the Docklands, and the community they will evolve.
- A robust partnership between the public and the private sectors to deliver the built and activity form.
- Substantial new public assets and attractions resulting from the private/public partnering

The partnership between the public and the private sectors will see the private sector providing most of the finance and the high tech development expertise, while the Government takes the planning and management initiatives to achieve a balanced outcome for the people of Melbourne and Victoria.

Substantial new public assets resulting from the private/public partnering will include seven kilometres of continuous public 30-metre wide waterfront, and a commitment to design excellence in all developments. This includes recognition of the rich history of the area, and integration of significant structures to fulfil our statutory heritage obligations.

The focus on the needs of the people who will visit, live and work in the Docklands, and the community will include: recognising the Aboriginal history of the area, building on the European settlement patterns begun in 1835, celebrating the multi-cultural nature of Melbourne, and anticipating the changing lifestyles of the children and youth who will benefit from the success of the endeavours of our generation.

These features of our approach are not just theoretical principles. They will provide practical outcomes that enhance private sector investment interest, support the values of the community, and achieve both an image and a reality that will attract local people and visitors alike.

WHAT IS MELBOURNE DOCKLANDS?

The key assets of Melbourne Docklands are threefold: its location – adjacent to the CBD and at the hub of the main waterways; seven kilometres of public waterfront, and 40 hectares of water surface; and excellent accessibility – at the centre of the metropolitan train, tram and road networks, and with key links being developed for pedestrians and cyclists.

Melbourne Docklands has made a strong start into its implementation phase. Colonial Stadium has been completed and operating for a year, with more than 1.8 million people attending events so far with 2 to 3 million expected this year. More than 1,100 apartments have already been sold in the last 15 months for a value of over $500m, and their construction is underway. More than $120 million of major infrastructure has been completed, funded by the private developers and coordinated by Docklands Authority; and the area has hosted many thousands of people at a wide range of events.

The results to date, and the projected outcomes, indicate a dynamic, new mixed-use urban habitat.

There are numerous examples of waterfront redevelopments around the world. We have studied them, learning both the successful and the less successful lessons, and applying them to the particular circumstances that make Melbourne Docklands a unique place. The extensive public promenades, and the integration of private, public and civic activities are key examples.

Making sure that the necessary provision for road traffic is achieved without compromising the pedestrian character of the public realm at the waterfront has led us to place the major new road away from the waterfront, adjacent to the railway. To connect Docklands to the central city and Spencer Street Station, new pedestrian friendly bridges and subway links have been constructed.

Achieving both financial viability and quality urban design are twin key goals of the development of Melbourne Docklands. We have produced Integration and Design Guidelines to assist developers and their designers to achieve high quality results, and we have the benefit of an Integration, Design, Amenity and Integration Review Panel – comprising international, national and local specialists – to advise us on these matters.

A THIRD MILLENNIUM CITY PROTOTYPE

Despite its proximity to the CBD, the 220 hectares of Melbourne Docklands has been something of a hidden area – set aside for important port and rail uses, but obscured from view, inaccessible to the general public, and to private investment. The port and rail complex continues to be Australia's largest, but has moved downstream with modernisation of operation techniques and to deeper waters suitable for growing size of cargo carriers.

Melbourne Docklands is planned to be a vibrant new part of inner Melbourne, and estimated to include 7,000 apartments. This is a relatively small proportion of the total projected resident increase for the whole of inner

Melbourne which is projected to see appropriately 90,000 dwellings in the next 20 years.

Also projected for Docklands for the next 15 years are: 450,000 m^2 of high-tech commercial space; 77,000 m^2 of retail space; more than 2,000 hotel rooms; and 100,000 m^2 of entertainment space. The area will be well serviced by car parking and 500 marina berths, as well as facilities for ferries, water taxis and charter boats.

THE PRECINCT

There are seven multi-use precincts in Melbourne Docklands, providing the dynamic mix of residential, commercial, retail and leisure activities. Based on current projects and the bids, the estimated proportions of development represent a strong mix of complementary land uses.

Colonial Stadium – located on the edge of the Central Business District, in the heart of the Docklands, has been purpose designed as the jewel in Melbourne's entertainment and sporting landscape. The secret to Colonial Stadium's success lies in its unique flexibility and world class facilities – which together create the best entertainment Stadium and corporate facilities in Australia

The Channel 7 Broadcast Centre is a $40 million development that will be the home of Channel Seven Melbourne and Australia's first purpose-built digital broadcast complex. The Melbourne Broadcast Centre will be a unique addition to the city landscape, with a design which aims to bring people closer to the magic of television production. A café looks into the news broadcast studio. The complex itself will be a showcase for Channel 7's operations.

One thousand, one hundred (1,100) apartments already sold and are now under construction in the Mirvac Yarra's Edge and the MAB NewQuay development. Further residential developments are included in bids for precincts yet to be decided. Eventually it is expected that about 50% of the total development will be of residential accommodation for a wide range of people – singles, couples and small families attracted to the inner city and water-based lifestyle.

NewQuay residential waterfront projects will integrate with new places to shop, socialise and sail, new places to work, dine and relax. NewQuay will have a strong artistic flavour with a contemporary Australian art gallery on Victoria Harbour, along with urban art, sculpture and water features throughout.

Mirvac's Yarra's Edge development is planned to have 2,000 apartments on its 14.5 hectares site – it will be one of Australia's most outstanding residential developments. It will comprise high rise towers, with a 30 metre wide public waterfront and three parks, and low rise apartments, all sunny north facing across the water toward the central city and the other Docklands precincts.

Digital Harbour@Comtechport is an exciting $300 million development that will focus on the growth of new technology businesses, research and industries in a working campus that has few precedents world-wide. Global changes in IT technology and practice are rapid and radical. This strategic initiative will attract major corporations, research and educational institutions, small/medium enterprises, and technology-based start-up businesses to co-operate in a very dynamic way, and play a role in the global IT environment. The interaction of these organisations will be a major step towards establishing Comtechport as a centre of excellence in the commercialisation of innovation, gaining international recognition for Docklands and, in turn, Melbourne as a 'High Tech World City'.

The remaining precincts (Victoria Harbour, Batman's Hill, North West Precinct, and Northern and Southern Stadium precincts) present a potential investment value of a further $3 billion over the next 10–15 years, and quality developers bidding for the waterfront and CBD edge precincts include some of Australia's and Asia's most credible organisations.

SPECIAL DETAILS

Achieving the broad vision for Melbourne Docklands is the main focus for our work, but it will only be successful if the details give it special qualities. They include the following aspects of the development.

Providing residents, the work force and visitors alike with a safe and attractive environment is fundamental. The recently launched Community Development Plan recognises that supporting the evolving strength of the social capital of Docklands – the people and the community of which they are a part – is a vital element in achieving a successful development overall.

Major infrastructure works have already been established, prior to demand – to encourage investment in the area. Most of it has been funded by the private sector. Items include improved public transport, new infrastructure for water transport, new roads and bridges, and high tech communications links.

Docklands Authority is committed to environmental improvement projects, a very necessary part of transforming an historic industrial area into a Century 21 living environment. They include the $Aust 45 million remediation of an old gas works site, and decontamination of soil and rebuilding of wharf structures by the private developers as part of their precinct developments. The Authority has also replaced the antiquated stormwater system, and installed litter traps to improve water quality in Victoria Harbour. Catching fish is once again a viable pastime in the area.

Docklands is the most accessible location in the whole of Victoria, at the hub of Victoria's arterial road and rail system, and excellent links to the CBD, tram, pedestrian and cycle routes, and the Spencer Street Railway Station have already been put in place.

When fully developed, more than 27 hectares of public waterfront, promenades, open space and public parks, including the 2.7 hectare Docklands Park which is part of the Harbour Esplanade spine – the major civic focus for Docklands.

Creation of Melbourne Docklands as an 'iPort' is a cornerstone of the development high tech focus. The immediate provision for high bandwidth services to all access seekers in the region is a major step in making Docklands a city for the Third Millennium. The base telecommunications infrastructure, including a high capacity data spine, is already installed.

IN SUMMARY

The key elements of the Docklands Vision are:

A place for all people – over 2 million people visited Docklands last year, a ten-fold increase on the 200,000 people who visited in 1999. When fully developed Docklands is expected to attract over 20 million visitors per year.

To be integrated with Melbourne – to complement rather than compete with other parts of central Melbourne and adjacent areas, with governance by the community for the benefit of all Victorians.

High quality design and public amenity – with development not only being commercially successful but combining high quality design functionality, sustainability, and aesthetics. Buildings, landscape, generous public open space and urban art will be designed together, to ensure Docklands is a destination with a unique character and sense of place which meets the variety of community needs.

A water, arts and events destination – where Victoria Harbour is the 'Blue Park' of Docklands, a significant international boating and activity node, and the best integrated boating, arts and events destination in Victoria – for Victorians and visitors alike.

Docklands will add an exciting new dimension to life in Melbourne, as well as providing a major economic boost for the State. Docklands is forecast to provide on average 3,000 construction jobs per year for the life of the project, and an estimated 3,000 new permanent jobs by 2015.

The people of Victoria have adopted Docklands as part of their new image. Community support for the development, in both the metropolitan area and the country regions, continues to be very high.

But the interest is spreading further, with interstate and overseas investors joining in. A wide range of people, young and old, are now voting with their feet, coming to live in or visit Melbourne Docklands – a city for the Third Millennium.

Docklands is now well and truly under way. We couldn't stop it now if we wanted to. It will accelerate and slow down with market demand. Its offerings of public attractions, waterfronts, proximity to the city centre, public and private transport access, a preferred location for high capacity bandwidth organisations and a promise of a new exciting living and working place make it a focus of Melbourne's progress for the next 10 years.

Is There an Afterlife for the New Asian City?

Stuart Brogan

From the Sudirman Central Business District in Jakarta to Pudong in Shanghai, the master planners of the world are conceiving new city developments for the emerging economies of Asia. These cities are grand in vision and architectural design, spacious in plan with abundant green space, modern masterpieces replacing the old Asian slums, creating, according to the Shanghai Municipal Housing Development Bureau "a grand symphony". Do these grand plans translate into good urban habitat, places for people to live, work and recreate, do those that hold the future of the Asian cities need to change their approach to the development of their cities?

As a regular visitor to Asia, my experiences and the processes that formed my opinions influence my views. However, as an urban planner by training and interest, my work is around me, I cannot help but observe, and I hope that some of my observations will be of interest and thought provoking.

The following is a response to what I see happening in planning from the perspective of a designer of new communities and urban centres in varying environments, both in Australia and Asia. What are some important issues in the design of our future cities, our urban habitats, what should be some of the prime objectives of the designers? What is the future of these new city centres, what do we want them to be, how are we going to make them that way?

The concept of urban habit is a broad area, from the mega cities of the world, down to the small country towns in central Australia.

NEW CITY CENTRES

As we are here at a conference for Tall Buildings and Urban Habitat I would like to look at the New City Centres in Asia, the places where the Architects of today scramble to fulfil their dreams of designing not only the Tallest Building in the world, but the tallest in all four categories. Three of the four tallest Buildings in the world are in Asia, with Petronas in Malaysia holding position one and two. If you log onto the World's Tallest Buildings web site, www.worldstallest.com, some of the leaders in the race to take the crown from Petronas are:

- Kowloon MTR Tower, Hong Kong, SAR at a height of 574 metres;
- Shanghai World Financial Centre, Shanghai, PRC at a height of 460 metres, which is or has been redesigned to put it even higher, and;
- the Taipei Financial Centre at a height of 508 metres, although our office in Taipei tells me that this has been lowered because of flight path restrictions.

I understand Chicago wants to fight back with the SOM designed project known as 7 South Dearborn around 600 metres, and of course Melbourne has the Grollo Tower. The three buildings mentioned in Asia are all proposed in new development areas within existing cities, as opposed to being built within the existing urban fabric. This paper however is not concerned with the heights of the buildings, but more about what is happening at the other end of the building, at the base, where the building meets the ground, where we as people enter the buildings, move from building to building; where the buildings sit in the urban habitat. The part of the building that often seems to be forgotten when viewing these buildings as illustrated by the cover for the handout that is supplied to visitors when visiting the observation deck of the SOM designed Jin Mao Tower. A magnificent building that I was told cost US$1 billion.

Cities or urban habitats have been around for a long time. Mesopotamia is said to be the birthplace of cities, over 3,000 years BC. Others say the cities of Erech and Ur that developed on the Euphraties in about 3,000 BC were the first examples. Since then there has been rises and falls in the profession of city planning, but it has been 5,000 years, so one would think we would be starting to get it right by now. In many cases we have, but for some reason, when we climb onto those aeroplanes, leaving our pedestrian orientated cities in the West, cities like Melbourne, our concept of what is a city centre changes. We create new urban habitats that would have been rejected, and are currently being rejected in the west on many fronts.

It would be very obvious to those of you visiting Melbourne, some for the first time, and if you have had a chance to walk around the city centre, that there is something different about the Melbourne CBD and the new city centres being developed in Asia. The Melbourne CBD is not unlike other city centres in Australia and some existing centres in Asia. Walking through the streets in older parts of Shanghai, with the Plane trees arching overhead, shops and residents fronting the street, reminded me of Melbourne, especially as it was cold and grey (Fig. 1). It is nice to revisit Melbourne at a warmer time of year than my previous trips.

Figure 1   Shanghai.

Beijing, is another very pleasant place to walk about in, despite the cold when I was last there (Fig. 2).

Figure 2   Beijing.

Brisbane as the photo illustrates is one example of an Australian city centre that has a good Urban Habitat, the city I spend my time in when back in Australia (Fig. 3).

Figure 3   Brisbane.

This leads me into what I would like to discuss today. The new city centres of Pudong, Shanghai, Sudirman CBD, Jakarta, and Hsin-I District, Taipei to name a few I have been able to visit are not like Melbourne, or any CBD in Australia.

What is different about these new CBD areas and their existing and proposed Urban Habitats, and what future they may have. I would also like to offer some solutions that may give them an afterlife, because in my opinion, as places for human habitat, in the sense that we often talk about, that is habitat being a place where humans or any species are in harmony with their environment and each other, these city centres are already dead.

Before you look at the program and say, wasn't this a talk on Resources and Sustainable Development, well in a way it is, but I will avoid stating what others such as Dr Peter Newman know more about. These new city centres do not make good use of their resources nor are they sustainable. Good well planned cities make the best use of their resources and provide a good Urban Habitat, something I am sure all here would agree on, but the concept of what is good planning will cause us to differ.

My first experience with the concept of developing a totally new CBD within the urban fabric was when I worked for 8 months in what is known as Jakarta's New CBD, or the Sidirman CBD, at the southern end of Jalan Sudirman. This experience was just a taste of what I was to find elsewhere in Asia in the following years. Unfortunately I do not have any photos of the area, as at the time I could not see anything worth photographing. The following images give an indication of the vision of the designers.

In the book "Cox Architects", which is part of the Master Architect Series the designer's summary for Sudirman CBD states that:

"The plan is based on an internal ring-road system accessing all precincts, with predominantly commercial buildings inside the residential buildings outside the ring. The ring will provide for vehicle movement on ground and sub-ground levels with a monorail loop elevated to first level. A secondary layer of space links provides continuos pedestrian links at grade which avoid conflict with the vehicular system."

It sounds like the designer has good intentions, but apart from the fact the monorail does not exist and probably will not for a long time, the pedestrian space links that are discussed are not designed into the existing buildings. The one way 4 lane road system is in plan, and I guess at a time, better roads use a big priority.

An important point is that this was an opportunity lost, in a city that desperately needed some focal point to, as stated by the designers, "contrast the previous commercial sprawl across Jakarta". If designed as a true Central Business District, that is a prime destination within the city for those wanting to do business, there should be no need for monorails or space links. Buildings should be designed to front the street, not the "porte cochere", where vehicles jostle for space to pick up and drop off their passengers, people queue for taxis and pedestrians generally struggle to get through all this, to get out to the main road where they can find a bus.

For lunch we would walk out the building down the access drive way, along the narrow footpath in the heat or the rain, and usually both, across the fast moving one way roads, jump a couple of garden beds, across another road, up the driveway of another building, just for some variety in our diet.

Looked good on plan, nice perspective, nice Architecture but not much different to anywhere else in Jakarta's commercial sprawl. Individual buildings on busy roads, and in this case fast roads that are used as rat runs; the only difference is that the the kampongs in between have been removed. I agree, at that time the designers would have had a set brief, The Mass Transit System was not proposed and the solution was innovative for Jakarta. But planning is always evolving, Krismon, the Asian Economic Crisis came along and things have changed.

Thanks to Krismon, I now spend some time in Shanghai, staying in Pudong, because that is the part of Shanghai clients think you want to stay in, as it shows that the Chinese are modern just like us. I would like to talk further on this new city development, as the Chinese Economy is still booming and construction in Pudong is continuing while Sudirman is an hold.

It was only just over ten years ago that the Chinese Government announced its intention to develop the area that is now known as Pudong across the Huang Pu River from the Bund, the historical business district developed by the Europeans in the 1920s. Along with the decision to move forward with the development, the Shanghai Government decided to consult foreign consultants, and ultimately choses firms, one being local, as part of an International consultation rather than an international competition.

The brief proposed 4 million square metres of development that included all the uses that would normally be found in a city centre of this magnitude. This was to be developed in an area that had little infrastructure and connections with the existing city. The International consultants prepare proposals with the knowledge that the Chinese had already devised their own Master Plan. Despite a number of good plans and ideas, the Chinese went ahead with their plan anyhow.

On April 18, last year Pudong celebrated 10 years in existence, and the spending of US$ 33 billion on fixed assets. The local officials are very proud at what has been created, and have often been quoted saying "only ten years ago, the whole area was only green fields and orchards". Which leads me to ask why they build all the old looking stuff in the foreground, and why isn't Pudong claimed to be the largest urban redevelopment project in the world (Figs. 4–6).

Significant amounts of that money has gone into the development of major infrastructure, including a new airport that will serve greater Shanghai, deep water port, road tunnels and bridges. The downtown area of Pudong has expanded from 17 sq. km to 100 sq. km. With an average annual growth rate of over 20 percent, the economic zone will by the year of 2010 be extended to 200 sq. km. and its population will grow from one million to more than two million. (Monday, April 17, 2000, *People's Daily*)

A large amount of the money has also gone into the construction of tall buildings (Fig. 7).

Some has gone into the Urban Habitat (Fig. 8).

Figure 4 Pudong skyline.

Figure 5 Pudong skyline.

Figure 6 Pudong.

Some of the money was spent in well designed civic spaces, and a lot has gone into roads (Fig. 9).

The designer of Century Boulevard put forward a comparison recently at a conference in Shanghai. He compared the road with the Champs Élysées in Paris, but I saw a comparison with the new M1 from Brisbane to Gold Coast (Fig. 10).

So why did cities new city centres develop, what were the causes behind the resultant designs. As can be seen, a lot of good ideas were put forward for Sidirman CBD and Pudong, but those were either rejected or did not eventuate.

It is unlikely that we will put a single factor forward as the causation of the development of these urban areas, and it is important to remember that as we develop our cities, that they are not the result of a single force, but <u>dynamic</u> in their development and organisation. Our concepts of city design are also evolving, so what was seen as good design will be looked at completely differently today.

Perhaps, especially in the case of Pudong, the designers had Le Corbusier's utopia, the Radiant City in mind for, after all, along with the "City Industrille", he did have socialistic objectives. Concepts he put forward such as; the neighbourhood is the high rise, the high rise a method of increasing density, there was plenty of open space to be overlooked, traffic was segregated, the city centre was a multi-level traffic interchange, pedestrian seemingly eliminated or at least limited, buildings were autonomous in the landscape.

Figure 7   Tall buildings in Pudong.

Figure 8 Sidewalks.

Figure 9 New streets.

Figure 10   Highway from Brisbane to Gold Coast.

Sixty-story towers with lower densities were on the fringe. The city was a grid, something Le Corbusier saw as the ultimate expression of man's superiority over nature. The photo shows the centre of Pudong as viewed from the Jin Mao tower, a large traffic interchange, over the entrance to the tunnel under the Huang Pu River to old Shanghai.

Le Corbusier's model city had no nature, was neutral, it was on a flat site with no physical features and no climate. Pudong is flat, and now we can create climate within the buildings.

The goals that Le Corbusier set were to;

1.   De-congest the centre of our cities;
2.   Augment their densities;
3.   Increase the means of getting about;
4.   Increase the parks and gardens,

These goals sound familiar today, especially in the cities of China. Yet despite these relevant goals, Lewis Munford called Le Corbusier "Anti City", and I would tend to agree, his city lacked what I feel is the most important element of a city, a sense of community, a sense of plan.

So is Pudong "Anti City"? I will leave that up to you to decide on your next visit to China as I would like to move from the negative side of my presentation and be more positive.

What I would like to talk about now is where are we going from here, and how could we build the cities of the future? I guess you now realise that I do not like the trend that is occurring in the planning of new city centres, and are wondering if, in my roll as Planning Manager for the Jiangbei New City Centre we are doing it any different. The answer is yes and no in some cases, but we are trying to change the trend.

Chongqing is the world's largest metropolitan region by population, which totals over 31,000,000 residents and is the fourth and largest metropolitan area to come under the direct control of Beijing. Last year the central government nominated Chongqing as the capital of Western China as part of China's western development policy. The CBD is the first stage of the redevelopment of the northern sector of the city, an area known as Jiangbei, which will eventually cover over 70 square kilometres. It is touted in China as the largest urban redevelopment project in the world and will occur over the next 40 years. It is definitely bigger than any brown field projects we might find in Australia.

The original plan for Jiangbei was prepared for the city as part of a competition, and had to satisfy certain design criteria. Some of the objectives of the plan were not dissimilar to those of Le Cobusier discussed earlier. The ultimate objective of any team entering a competition was to develop the best plan, to win the competition, and be awarded ongoing involvement. In order to do this, the criteria set by the local committee must be satisfied, no reviewing of the development of the plan during the competition phase. The planning criteria is developed locally and the design team goes away and works in isolation to some extent.

The resultant plans are therefore constrained in the development of other ideas, and ultimately must look good as a plan. The fact that Jingbei is a very hilly site was an important consideration for our team, and we were honest in depicting the constraints of the site. One criticism of the plan was the treatment of the interface with the river, but on further inspection of other plans it was noted that the issue of slope was not addressed. Other issues such as lots of green on the plan impressed the judges.

Following the appointment of our group to firm up the plan and take it through the statutory planning and on, we had an opportunity to talk to the various government departments and discuss their requirements and desires. We also had the opportunity to raise various issues with them, and introduce our own ideas and make modifications to the plan that affected the way it looked on the plan, but improved the way it sat on the site, and worked in reality.

This brings me to my first point, that is that plans developed for competitions do not necessarily make for good plans. They do not necessarily address all the issues, but those relevant to the competition. Most Government commissions in China these days must go through a competition phase.

The first major change since the competition, and the one we feel very strongly about, was to introduce a rail system into the site. This was not a requirement of the competition, but it had tentatively been suggested on the plan. Chongqing currently has a light rail project under construction that was planned prior to the competition. The plan shows the location of the new city centre along the proposed routes.

No routes currently take into account the new city centre. They are now mulling over the concept of linking this into the network in Chongqing and will get back to us. It's a tough one.

Access by road is restricted, without an efficient public transport system on dedicated routes the development is doomed. It will not work efficiently, as its location on a peninsular means there is limited access. Through traffic has been diverted under the site, or passes by on the existing road system. Unlike Pudong, where the traffic is brought through the city centre prior to entering the tunnel to go on to Shanghai, traffic entering Jiangbei will be there for a purpose.

My next point is that these new city centres must exclude traffic that is not there for a purpose relating to the function of the city. If we can do this, we can begin to reduce the size of the roads within the city centre, resulting in more room for activities and the sought after green space. Combined with more efficient transport systems such as rail systems, there is a better use of resources creating a more sustainable city.

The stations in Pudong are not integrated with the development as Figures 11 and 12 show, taken from above a station.

Figure 11   Pudong station.

Figure 12   Pudong station.

Having looked at the past and present trends, we need to form the basis of where we want to be and identify the problems we want to solve. As can be seen from this paper, the problems of our urban environments cross cultures and time. Le Corbusier's goals as stated earlier were included in the conditions for the competition for a new city centre for Chongqing. His solutions differed from ours.

We must continually be flexible, allow for changing ideas, so that in the future the cities can evolve. The planning for Jiangbei is already evolving, as can be seen in the design of the Civic Square. The competition plan had a very formal square, a very grand square to satisfy the requirements set. We now have a multi level, multi function square, much less formal, but suitable for that role.

Although we can mould and adjust the function of the square, the planning system now in place does not provide the same flexibility for the commercial and residential areas.

The planning of these cities needs an overhaul of the current statutory planning systems, as ours here in Australia have, and are continuing to do so. This also goes for engineering standards. In the case of Jiangbei it was a stated objective that international standards be introduced, but it is a constant battle to vary the Chinese national standards in any way.

Planning must be flexible and evolving. What was a good idea ten years ago does not necessarily suite today's conditions.

When planning our urban environments we should consider what type of Urban Habitat we want, and more importantly, what do the users of a particular place desire. This phase is the development of a vision for our urban habitat. Of course, one of the reason teams of foreign experts are called in to design these new city centres is that the local city leaders are not really clear on how they are to achieve their goals, those goals may not be clear.

As an Australian, and having grown up here, but then having lived in an Indonesian city for nearly five years, I would be a bit unique in what I like and don't like about cities, and I do not feel I am in a position to give you the solution.

The most important issue in my opinion is to go beyond the physical design of the city, the way it looks, what style its buildings are, and put more effort into finding out what makes the community tick, the dynamics of the population that will live there.

Look at the problems of the city that are being addressed by groups such as the New Urbanists, but go beyond their grids and their rows of Retro-housing. Le Corbusier was correct in the goals he set, but they lacked totality, and he saw physical design as the means to the solution. Perhaps through physical design the New Urbanists feel they can determine how people live, work and play, the design is a means to the desired end.

More importantly though, the designers, should look at those aspects that have been raised by the New Urbanism movement and others, and look at what type of community is desirable. One of the big issues within the New Urbanists movement is the reduction of vehicles in the urban habitat, creating walkable urban environments. These objectives seem relevant in our city centres in Australia, but seem to be ignored in Asia.

Why are the new cities of Asia being designed around vehicle movements, and not around the movement of people.

So how do we redevelop the new cities in China or else where in the world for that matter? How do we put in place the vision for these Central Business districts that is more in accordance with the creation of a good Urban Habitat?

Firstly, we do not ignore the physical design of our communities. It is a critical part of the process, and forms the basis, the foundations, for the development of the urban environments. Do we need traffic engineers on our team, or perhaps a whole new profession that is called people movement engineers.

The grid for example is efficient, and has many of the aspects of a city layout that are desirable, and that is why it has been around for some time and will continue to be. It has good permeability and allows for ease of pedestrian movement, except when to cross from one monumental building to the next in Pudong we take 55 steps from kerb to kerb.

This question needs be reviewed at a forum such as this, as we have a gathering of people that is rare and unique, we have the Planners of the cities and the Architects of the buildings within these cities, all with different answers and solutions.

We need not be afraid to embrace new ideas, and run them through our planning models. If they pass the test of analysis; give it a go, if it fails, at least we will be held in good company such as Frank Lloyd Wright and Le Corbusier.

The creation of a model for Asia, should not be based on the western ideas, but should not ignore them. We should not judge Asian cities against ours, by our values, but against their own needs, their own vision for their future communities.

The most important factor is to go beyond the roads, and look at the communities in which people live, work and play, those that we want to leave for our children. Communities have been around for a long time, a lot longer than the buildings in which they reside, or the roads on which they travel.

So finally, let us start planning our cities for people, not cars or buildings. Stop talking about vehicle movements in a city and talk about the movement of people, be that walking, by some form of public transport or in private cars. Design our cities around the people in the buildings, not the buildings themselves. Jin Mao Tower is a magnificent building when you stand back and look at it, but I would not want to work in it, it would be the same as my experience in Sudirman in Jakarta. Pudong, Sidirman are basically multi-story office parks.

Plan for dynamic evolving cities that are adaptable to change, are flexible in their planning.

My photos of Pudong were taken on a Sunday, and it was void of people, but when I moved across the river to Nanjing Road it was still Sunday, and the place was teeming with people. The place was alive, the MRT full, Shanghai was using the existing resources to the fullest.

I hope to return to Pudong one day when the 55 metre wide pavements are gone, replaced by dedicated light rail or tram corridors in the place of pavement, resulting in smaller roads with green space that is useful. The base of the buildings will come out to the street, with uses that add life to the streets. A city for people, not just cars.

We are doing it now in Australia and elsewhere, revitalising, bringing back to life degraded parts our cities, perhaps it is time to do the same thing to some of the new cities, give them an afterlife. At the best, let's do better more responsible planning for those cities we are currently planning.

Figure 13   Brisbane.

The Role of Accessibility for the Success of City Centres

Rolf Monheim

World-wide discussions about the success of city centres are focused on their accessibility. Generally, it is understood first of all as the accessibility for driving and parking cars. There prevails the attitude that this demand should be met without any restrictions. Therefore, there is a trend to continuously increase the infrastructure for car traffic. Regularly reoccurring shortages are not understood as a sign of a wrong concept but, to the contrary, a directive to continue the expansion. The common sentence "no parking no business" is misleading because it diverts the attention from the fact that first of all the excellence of a place determines the decision where to drive to. If a city centre is run down, even abundant parking will be unable to revitalise it, as can be seen in the USA.

In Germany, in contrast, city centres were able to defend their prime role. This is due to the combination of various public and private measures. The city administration, on one hand, improved accessibility by public transport, because it realised that this was the only means to bring large numbers of visitors to a densely built up area; it also improved attractiveness by a good design of the public open spaces now free from car traffic. The landowners and retailers, on the other hand, made the best use of this chance by investing in a better business and improving city marketing. Nowadays, in Germany even new shopping complexes mainly choose integrated city centre locations!

1   THE ROLE OF ACCESSIBILITY

The discussion of the role of accessibility generally is focused on the access from outside to the city centre, including parking. It underestimates the role of "internal accessibility", i.e. the walk from the place of arrival to the various activity locations and then back to the place of departure (mostly a public transport stop or off-street car park) (Fig. 1). The main reason for this is the orientation towards the "one-stop-shopping" of a suburban mall and the neglecting of the fact that most city centre visitors, at least in Europe, go to several destinations within the centre and like walking to them.

The discussion of traffic infrastructure furthermore neglects the fact that it does not necessarily determine travel decisions as such but by the way that it is perceived. As a consequence, the marketing of accessibility is important. Unfortunately, for a long time complaints of retailers on poor car-accessibility have resulted in a negative perception among the citizens. However, surveys show that visitors rate accessibility much better than retailers believe.

External accessibility	**Internal accessibility**
Trip from the starting point to the city centre – Access roads – Parking facilities – Public transport (net, stations and service) – Access for pedestrians and cyclists	Trips after the arrival in the city centre to all activity locations until leaving the city centre – Pedestrian street – Traffic calmed / shared street (without / with separation of vehicles and pedestrians) – Boulevard – Area with speed reduction

Figure 1 Measures for the access to the city centre.

Due to the limited availability of space for the construction of additional road- and parking infrastructure in densely built up areas a distinction has to be made between necessary and less necessary car use. The latter refers especially to commuters. This is because of the long parking duration and the problems of the rush-hour. In Germany, therefore off-street car parks for new office buildings in a city centre with good accessibility by public transport are permitted mostly only in a limited number. Instead of this, employees may get tickets for public transport at a reduced fee; this is also less expensive for the employer. The parking needs of visitors now are often served better by guidance-systems that show the availability of parking, as well as by other means of parking management. This requires a cooperation of all parking providers.

Traffic calming and pedestrianization often result in a reduction of on-street parking sites. This regularly causes complaints by local shop owners. A survey by the Association of Large and Medium-sized Retailers, however,

Source: Surveys by Bundesarbeitsgemeinschaft der Mittel- und Großbetriebe des Einzelhandels (BAG) among ca. 900 retailers in 300 pedestrian precincts and traffic calmed areas in 1987

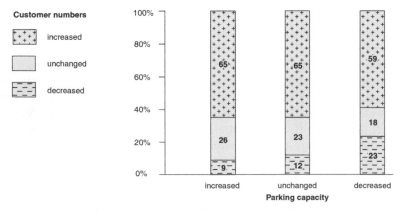

Figure 2 Change in customer numbers after change in nearby parking capacity due to traffic calming measures.

showed that even out of those retailers that had suffered some reduction in nearby parking capacity 59% noticed an increase in customer numbers against only 23% noticing a decrease (Fig. 2). Increases would occur even more often if car-parking were managed more adequately, i.e. if the parking duration would be limited and more effectively controlled (see Monheim, ed., 1989).

The "internal accessibility" can be improved first of all by pedestrian precincts. Within the last 30 years, in Germany they have expanded enormously up to more than thousand. In the beginnings, they were introduced only in streets with the highest concentration of large shops and pedestrian volumes. It was feared that otherwise people would have to walk too long distances. Later, the great popularity of pedestrian precincts encouraged planners and politicians to expand them step by step, in many cases up to networks of 4–9 km streetlength.

2 NUREMBERG AS AN EXAMPLE FOR GOOD URBAN DESIGN

Nuremberg, the centre of the Middle-Franconian conurbation, is a good example (490,000 and 1.2 million residents respectively; for details see Monheim 1996(b), 1997(a) and 2000). The city centre had been almost totally destroyed during World War II and was reconstructed after controversial debate in accordance to the historical pattern. Only a few streets were widened for car traffic. In 1966 for the first time car traffic was banned from a shopping street (Fig. 3). It was a great success. Therefore six years later the city councillors decided to expand the pedestrian precinct to include large parts of the shopping district. Finally, the city centre was divided into five traffic precincts where cars could not drive directly from one to the other. This reduced traffic volumes within the city without endangering accessibility. In 1996, a new conservative city government reopened some small connecting streets; however, car drivers meanwhile had learned that they do not need to drive around so much within the centre and use these connecting streets quite rarely.

What were the results? Traffic engineers had warned that the closure of roads with heavy traffic crossing the city centre would provoke chaos. But only 20–30% of the previous volume of traffic showed up on the remaining streets. The parking facilities were left nearly unchanged, but they were better organised in favour of visitors and residents, discouraging commuters.

Retailers had warned that the city centre would be cut off from its potential visitors. But their numbers increased clearly; many were attracted from large distances, especially on Saturday. This resulted from several factors. A new subway and parking management improved accessibility from outside and 9 km pedestrian streets improved the internal accessibility. Many new shops opened, the most important ones KARSTADT with 24,100 m^2 and a shopping centre with 12,200 m^2 (several more are under construction). A good design of public open spaces as well as changing lifestyles encouraged the expansion of street cafés and restaurants. A multiplex-cinema with nearly 5,000 seats and, in addition, restaurants for 1,200 guests added to the entertainment function. A large number of museums, theatres and other cultural institutions strengthened the identity of the city centre. They became connected by a "Culture Mile" and a "Historical Mile". Their route was advertised through large billboards. A city

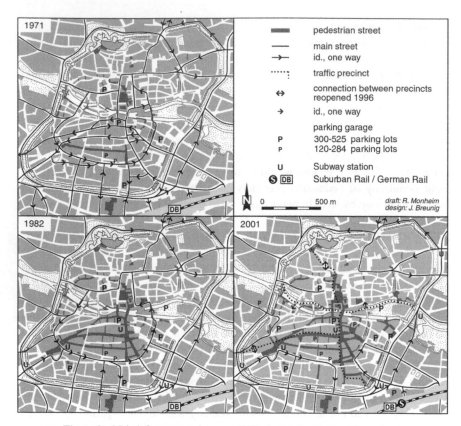

Figure 3   Main infrastructure for accessibility in the city centre of Nuremberg.

management was introduced in public-private-partnership, supporting the attractivity of the city centre by a great variety of events. This broad mix of shopping, recreation and culture has led to an organically developed urban entertainment centre. This may be seen as a model contrasting with the new urban entertainment centres in the USA as planned and managed by one single developer.

The example of Nuremberg demonstrates the necessity to subordinate car traffic according to the needs and capacities of the city centre and to put great emphasis on good public transport. This makes it possible to strengthen the internal accessibility by a sophisticated system of pedestrian streets which connect the various activity areas.

3   ACTIVITIES AND PREFERENCES OF CITY CENTRE VISITORS

Successful urban design depends very much on its suitability for the needs of the city users. Surveys demonstrate how they behave and what they prefer. Each of the five cities presented in this paper has its own character which modifies the general trend (for more details see Monheim, 1998 and 2000, Monheim *et al.* 1998).

Lübeck (217,000 residents, close to the Baltic Sea Coast)) has a large old city located on an island and declared as "World Cultural Heritage". In 1989 it started a stepwise traffic calming process, limiting for residents access to and parking in large parts of the city centre. Several other streets are accessible for cars only at certain hours, whereas only a few streets give unlimited access. Pedestrian streets, however, have developed to a very small length of 1.6 km. Traffic calming was accompanied by strong complaints of retailers. This was based on their believe that the city centre could not be reached any more by car. However, this was not the case. These controversies provoked a negative marketing effect for the city centre. In addition, deficiencies in the urban design and the small size of the pedestrian precinct caused discontent. In Bremen (540,000 residents, located at the North-Sea) there are also some deficiencies in urban design. Regensburg, in contrast, the smallest among the five cities (142,000 residents, located in Bavaria), is called the "Medieval Wonder". It has maintained its historical character; many small retail shops and restaurants make the best use of it. Munich, the capital of Bavaria, is considered the most attractive German city (1.2 million residents in the city and 2 million more in the conurbation area). The well designed city centre and pedestrian precinct support this image. Munich together with Frankfurt has the highest pedestrian volumes in Germany with about 15,000 in the peak hour during the week and 20,000–27,000 on a Saturday (London: 8,100 and 12,900, New York 7,000 and 4,600, Sydney: 6,400 and 12,000 respectively).

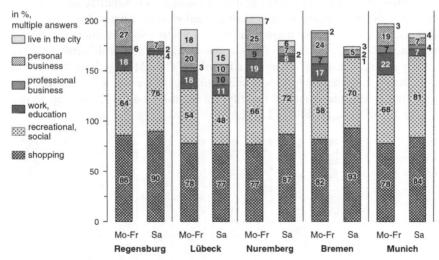

Source: Pedestrian surveys by Section for Applied Urban Geography, University of Bayreuth, and Omniphon (Leipzig), 1996 - 2000 (9 am -7 pm)

Figure 4 Activities of city centre visitors.

Visitors of German city centres on average have about two different purposes for their visit (Fig. 4). Most of them combine shopping and leisure, the latter nearly equalling the share of the former. Looking more in detail at leisure activities we see that visitors often combine several ones, especially strolling and eating (Fig. 5). They are most frequent at Munich and least frequent at Lübeck, with 164 and 65 different leisure activities respectively per hundred persons surveyed.

In addition, many visitors in modern societies consider shopping itself a leisure activity.

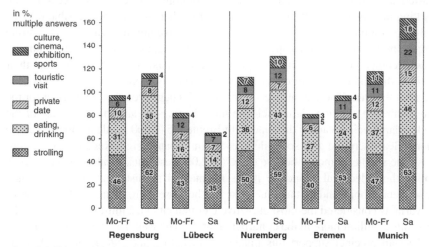

Source: Pedestrian surveys by Section for Applied Urban Geography, University of Bayreuth, and Omniphon (Leipzig), 1996 - 2000 (9 am -7 pm)

Figure 5   Leisure activities of city centre visitors.

Most shoppers visit a large number of shops – on Saturday about half of them five or more. Many do not have a fixed plan of what to buy. Therefore spontaneous purchases are frequent as well as visiting shops without purchasing anything. The walk from one shop to the next, together with strolling as an activity in itself, result in a long duration of the visit. Most visitors stay two or more hours, only a few less than one hour. Again we see a marked contrast according to the attractivity of the city; on Saturday only 11% stay up to one hour in Munich against 49% in Lübeck.

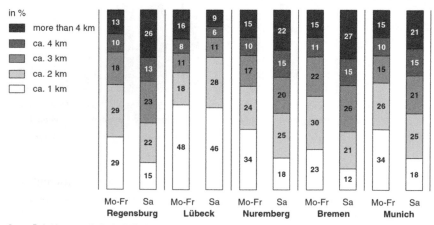

Source: Pedestrian surveys by Section for Applied Urban Geography, University of Bayreuth, and Omniphon (Leipzig), 1996 - 2000 (9 am -7 pm)

Figure 6   Length of the walk within the city centre.

The distances walked within the city centre are quite considerable (Figure 6). Only a few visitors estimate to walk up to one km and the majority more than two km. Only in Lübeck, due to its small pedestrian precinct, the longer distances are quite rare (15% against 36–42% more than 3 km).

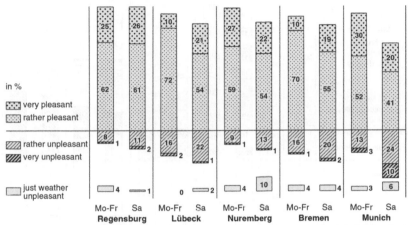

Source: Pedestrian surveys by Section for Applied Urban Geography, University of Bayreuth, and Omniphon (Leipzig), 1996 - 2000 (9 am -7 pm)

Figure 7   Opinions of the walking experience within the city centre.

A large majority of visitors feel good or even very good about their walk, only very few feel bad. Sometimes they complain only about bad weather (Figure 7). Longer distances are not linked to negative feelings, sometimes even the contrary occurs: those walking a long distance feel better. This, on the other hand, is logical because otherwise they would not have walked so far. Discontent is caused mainly by the hustle and bustle resulting from too high numbers of pedestrians, especially on Saturday. The major risk for large German city centres and their pedestrian precincts nowadays is not a lack in accessibility but an overcrowding, especially during peak hours. This is also the case because shop rents explode so that many old established retailers cannot compete any more with national or international chain stores. Some fear that, as a result, the local identity of the city centre will disappear.

At the end of our interviews the visitors were asked what they liked most and what they did not like at all. The percentage of those not liking anything in particular shows a clear trend with Lübeck and Bremen performing weakest and Regensburg the best (Figure 8). In all cities except Munich the historical townscape is mentioned most frequently (40–56%). Shopping is much less important. People and flair have about the same importance. They are connected with the pedestrian precinct as a stage where the visitors can see and be seen.

The percentage of those not particularly disliking anything is quite high (33–50%), which is a good sign for the city centres (Figure 9). At Bremen and Munich most criticism refers to the overcrowding, especially on Saturdays (21–33%). The Lübeck visitors complain especially about dirt and, a bit less, about accessibility and parking. The latter is the result of a misled publicity.

Notwithstanding the extensive traffic calming, at Lübeck, Munich and particularly at Regensburg there are still complaints about traffic.

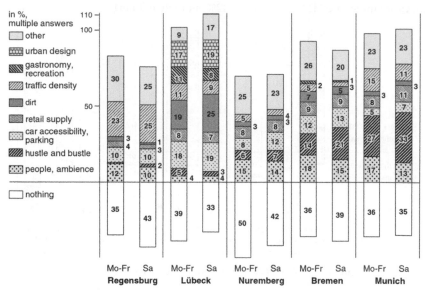

Source: Pedestrian surveys by Section for Applied Urban Geography, University of Bayreuth, and Omniphon (Leipzig), 1996 - 2000 (9 am -7 pm)

Figure 8   Particularly appealing aspects of the city centre.

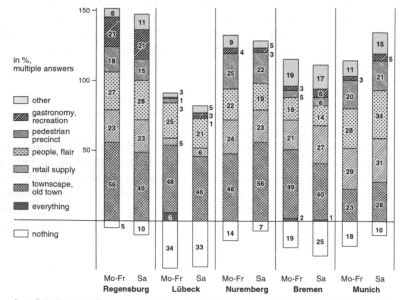

Source: Pedestrian surveys by Section for Applied Urban Geography, University of Bayreuth, and Omniphon (Leipzig), 1996 - 2000 (9 am -7 pm)

Figure 9   Particularly unpleasant aspects of the city centre.

A special question whether city centre visitors would prefer a better car accessibility or less car traffic in the city centre, in all cities shows a clear majority for traffic calming. Retailers in the same centres, on the other hand, vote with great majority for a better car access. These different perceptions can be found also when both groups score car accessibility: visitors, especially those coming by car, give much better marks than retailers. These differences in perception can explain a great deal of the criticism expressed by retailers against planning and politics for the accessibility of city centres. In the opinion of citizens, there is a great desire for traffic calming and no other planning measure has a stronger support than that of pedestrian streets.

4 CONCLUSION

As a conclusion one can state that

1. the importance of the external accessibility to the city centre is overestimated and the potential of an adequate parking management policy is underestimated;
2. the importance of the internal accessibility, i.e. the conditions for walking within the city centre, is underestimated, including the benefits of a good urban design (which, however, is considered very important in every planned shopping complex!);
3. the importance of an attractive mix of functions, including leisure and entertainment, corresponding to modern lifestyles, is underestimated as well as the marketing and management of the city centre (again, they are considered important in every planned shopping complex and urban entertainment centre).

Planners should not try just to copy shopping complexes and urban entertainment centres. They should, however, learn some principles to strengthen the competitiveness of traditional city centres. This is best done by working out their identity and uniqueness, whereas there is no chance to successfully compete in respect to car accessibility. To the contrary, only an excellent accessibility by public transport enables the density and mixture of functions required for an attractive urban core to occur. This principle is also applicable to cities outside Europe. The recent development of Melbourne is a good example.

REFERENCES

Bahrenberg, G., Mevenkamp, N. and Monheim, R., 1998
   NUTZUNG UND BEWERTUNG VON STADTZENTREN UND
      NEBENZENTREN IN BREMEN, (Bayreuth: Arbeitsmaterialien zur
      Raumordnung und Raumplanung, H. 180).

Baier, R. and Schäfer, H., 1998
    INNENSTADTVERKEHR UND EINZELHANDEL. HINWEISE ZUR
    BERÜCKSICHTIGUNG DES EINZELHANDELS BEI DER
    ENTWICKLUNG VON INTEGRIERTEN
    INNENSTADTVERKEHRSKONZEPTEN, (Bremerhaven: Berichte
    der Bundesanstalt für Straßenwesen, Verkehrstechnik, H. 52).

FGSV, Forschungsgesellschaft für Straßen- und Verkehrswesen, 1993
    „AUTOARME INNENSTÄDTE" – EINE KOMMENTIERTE
    BEISPIELSAMMLUNG, (Cologne: FGSV-Arbeitspapier, Nr. 30).

Frehn, M., 1998
    WENN DER EINKAUF ZUM ERLEBNIS WIRD. DIE VERKEHRLICHEN
    UND RAUMSTRUKTURELLEN AUSWIRKUNGEN DES
    ERLEBNISEINKAUFS IN SHOPPING-MALLS UND
    INNENSTÄDTEN, (Wuppertal: Wuppertal Papers, 80).

MASSKS, Ministerium für Arbeit, Soziales und Stadtentwicklung, Kultur und
    Sport des Landes Nordrhein-Westfalen, 1999
    STADTPLANUNG ALS DEAL? URBAN ENTERTAINMENT CENTER
    UND PRIVATE STADTPLANUNG – BEISPIELE AUS DEN USA
    UND NORDRHEIN-WESTFALEN, (Düsseldorf: MASSKS 1322).

Monheim, R., 1980
    FUSSGÄNGERBEREICHE UND FUSSGÄNGERVERKEHR IN
    STADTZENTREN IN DER BUNDESREPUBLIK DEUTSCHLAND,
    (Bonn: Dümmlers Verlag = Bonner Geographische Abhandlungen,
    H. 64).

Monheim, R. (ed.), 1989
    PARKRAUMMANAGEMENT UND PARKRAUMMARKETING IN
    STADTZENTREN, (Bayreuth: Arbeitsmaterialien zur Raumordnung
    und Raumplanung, H. 75).

Monheim, R., 1992
    THE IMPORTANCE OF ACCESSIBILITY FOR DOWNTOWN
    RETAIL AND ITS PERCEPTION BY RETAILERS AND
    CUSTOMERS. In *The attraction of retail locations*, IGU-
    Symposium, Vol. I, edited by Heinritz, G., (Kallmünz/Regensburg:
    Verlag M. Laßleben), pp. 19–46.

Monheim, R., 1996 a
PARKING MANAGEMENT AND PEDESTRIANISATION AS
STRATEGIES FOR SUCCESSFUL CITY CENTRES. In
Sustainable transport in central and eastern European cities, edited
by OECD/ECMT (Paris), pp. 53–143.

Monheim, R., 1996 b
DIE STADTZENTREN VON MELBOURNE, MÜNCHEN UND
NÜRNBERG IM SPIEGEL IHRER BESUCHER. In *Stadt und
Wirtschaftsraum*, edited by Steinecke, A. (Berlin: Berliner
Geographische Studien 44), pp. 47–65.

Monheim, R. (ed.), 1997 a
*,AUTOFREIE' INNENSTÄDTE – GEFAHR ODER CHANCE FÜR DEN
HANDEL?* Teil A/B, (Bayreuth: Arbeitsmaterialien zur
Raumordnung und Raumplanung, H. 134).

Monheim, R., 1997 b
THE EVOLUTION FROM PEDESTRIAN AREAS TO, CAR-FREE'
CITY CENTRES IN GERMANY. In *The Greening of Urban
Transport*, edited by Tolley, R., (Chichester: Wiley), pp. 253–266.

Monheim, R., 1998
METHODOLOGICAL ASPECTS OF SURVEYING THE VOLUME,
STRUCTURE, ACTIVITIES AND PERCEPTIONS OF CITY
CENTRE VISITORS, *GeoJournal*, **45** (4), pp. 273–287.

Monheim, R., 2000
FUSSGÄNGERBEREICHE IN DEUTSCHEN INNENSTÄDTEN.
ENTWICKLUNGEN UND KONZEPTE ZWISCHEN
INTERESSEN, LEITBILDERN UND LEBENSSTILEN,
Geographische Rundschau, **52** (7–8), pp. 40–46.

Monheim, R., Holzwarth, M. and Bachleitner, M., 1998
*STRUKTUR UND VERHALTEN DER BESUCHER DER MÜNCHNER
CITY, UNTER BESONDERER BERÜCKSICHTIGUNG DER
AUSWIRKUNGEN DER NEUEN LADENÖFFNUNGSZEITEN*,
(Bayreuth: Arbeitsmaterialien zur Raumordnung und Raumplanung,
H. 177).

Yencken, D., 1994
CENTRAL MELBOURNE SURVEY 1991 AND 1993. (Melbourne: Faculty
of Architecture and Planning. The University of Melbourne).

DEVELOPMENT AND MANAGEMENT

The Cyber Cities of Malaysia: Realising the Vision

Azman bin Haji Awang

INTRODUCTION

At the threshold of the New Millennium, we face a critical challenge. And Malaysia is no exception. The challenge is to ensure that Malaysia remains competitive in the 21st Century; a challenge that is essential if we are to maintain the rapid economic growth, political stability and racial harmony we have long enjoyed. Compounding this challenge are two significant trends in the global economy: increasing globalization and its competitiveness; and a revolution that is driven by rapid changes in the information and communications technologies (ICTs). It is the convergence of these technologies that has transformed and is transforming our societies and communities. This transformation has long been anticipated, but there is clear evidence now to show that the process of change is exceptionally exhilarating. Based on this backdrop, Malaysia like many other countries is attempting to shift forward towards a knowledge-based economy or k-economy – an economy which is directly based on the production, distribution and use of knowledge and information[1]. For most countries, this would require significant micro- and macro-economic reforms, suggesting that there is a considerable gap between the reality of what can be achieved, compared to political rhetoric. But for Malaysia, it has taken an innovative step forward to create one of the most wired regions in this part of the world by setting up the Multi-media Super Corridor (in short MSC or a Smart region)[2].

On February 1, 2001 Putrajaya the newly created administrative capital city of Malaysia and a component of the Multimedia Super Corridor (MSC) was officially declared a Federal Territory[3] thus ushering Malaysia into the digital era and the new information economy. It is testimony of the Malaysian government's commitment to the convergence of various information–based, telecommunication, broadcasting and mass media telecommunication technologies. In the words of Mahathir Mohamad, Malaysia's fourth Prime Minister, the architect of modern Malaysia:

"Malaysia must develop into an information-rich society through coordinated development and effective use of information technology (IT). The Malaysians must go through a process of acculturation in order to participate meaningfully in an environment characterised by the centrality of knowledge in all spheres of activities. Opportunities for individuals and collective development will be generated through the provision of equitable access to information resources. The proper utilisation of knowledge will contribute to the creation of an intellectual and economic edge made possible by a supportive information infrastructure and an enlightened program of research and development" (The New Straits Times, 1995)

Thus this paper is an attempt at *raison d'etre* – the justification for the Malaysian government's initiatives in formulating a cyber plan in which the first cyber cities – the Putrajaya and Cyberjaya in Malaysia are being built. It will highlight the concept and the components of the MSC plan. In the process, it will identify the challenges and issues that have to be addressed by the government and the private sector and how the two parties are being orchestrated for the common goal. It is the government's commitment to position Malaysia through multimedia and information technologies into the 21st century's digital age and to enhance its global competitiveness.

THE IMPACT OF INFORMATION AND COMMUNICATION TECHNOLOGIES (ICT)

What will the future be with Information and Communications Technologies (ICTs)? Will the functions of future cities be the same as those of today's? Or will ICTs alter the functions of cities? These are some pertinent questions lingering in our minds.

Many have argued that information economy is replacing the industrial economy in the new millennium. The global marketplace and its entire information infrastructure (info-structure) are in many ways shifting from the manufacture and distribution of physical goods to processing and exchange of value-added information. The ability to generate information and to add value to information create the opportunity for this information to be sold as a commodity and delivered through a combination of advanced digital networks that largely override geographical boundaries. We see this as a drive to decentralise cities into satellite towns and theme clusters and electronic communities. This challenges the traditional way of doing business in cities and between spatial organisation.

The ICTs will enrich our lives by increasing time for leisure. It will generate more opportunities and choices in leisure pursuits, kinship, social interactions, work and civic spheres of each person's life. There will be less need to travel for business or government transactions. Almost all interactions with various government agencies will be possible through the national info-structure. Shopping will be done at home, paid for electronically, and the items delivered at the doorstep. Such cash-less transactions, like tele-shopping and tele-banking, are made more efficient. As a consequence people have more quality time and this will eventually lead to greater satisfaction of our levels of needs such as culture, spirituality, voluntarism and selflessness.

Many city-wide functions will be provided through the computers. The implications of this will be on the local government as the provider of public services. It is projected that its role will be greatly changed or diminished. Thus citizens' perception of public places like city centres, neighbourhoods, schools, houses of worship as loci of social interaction will change. This will have an effect on the socio-political behaviour of the urban population and thus will further change the way people are governed. Many studies in the US, the Netherlands and Japan have supported this scenario. The question is what will draw communities together, to socialise to make those personal bonds? These are some of the possible scenarios of what would be the impact of ICTs on citizens' behaviour and activities.

THE MALAYSIAN POLITICAL LEADERSHIP

Clearly a country which desires to be better prepared for the 21st Century will pay a price for this transition. It will need to re-skill its human resource, retool them, question old ways of doing things, challenge the vested interest or perhaps induce reinventing the government (Azman, 1995). But this assumes long term vision on the part of political leadership with a certain amount of political risks and the necessity of enjoying popular support from the citizenry. In Malaysia this vision is encapsulated in what is known as "Vision 2020"[4] that is in the words of the Prime Minister, Mahathir;

"In the computer age that we are living in, the Malaysian society must be information-rich. It can be no accident that there is today no wealthy developed country that is information-poor and no information-rich country that is poor and undeveloped. The second leg of our economic objective should be to secure the establishment of a competitive economy … it must mean among other things an economy driven by brain power, skill and diligence in possession of wealth of information with the knowledge of what to do and how to do it." (Mahathir, 1991).

The MSC is envisaged as an advanced technological milieu and contributing to the realisation of Vision 2020, the Prime Minister's existing long-term goal of making Malaysia into a fully developed country by the year 2020. The MSC represents a continuation of prior aims and objectives of the national economic and industrial plan, though somewhat on a different technological platform. Two somewhat paradoxical strategies might be identified, namely integration and independence (Bunnel, 2001). The development of the most wired region in Asia is designed to be globally integrated with the world's economy facilitated by the advanced ICTs infrastructure. The imperative strategy to be technologically independent has been the hallmark of the Malaysian policy thrusts since the 1970s. The acquisition of knowledge and new technology is thus believed as a means of preventing domination by the technologically and economically powerful nations. Thus ICTs are seen to be an imperative of technological and economic upgrading and thus circumventing foreign domination.

WHY THE CYBER APPROACH?

Why then does a newly industrialising country like Malaysia opt for a cyber plan? What are the compelling reasons to account for adoption of this untrodden path? Are there risks? These are some of the issues to be resolved by the government's think tank. Since then, a task force was set up in the Prime Minister's Department to prepare a plan for Malaysia for the 21st Century. Among the issues discussed were mechanisms to achieving the Vision 2020, taking into consideration Malaysia's structural weaknesses and compelling strengths.

There are a number of options open to Malaysia. A pre-requisite, however is for Malaysia to sustain its economic Gross Domestic Product (GDP) of at least 7 percent throughout to year 2020, and its economic base to move from manufacturing to service and knowledge-based industries. It has been determined that based on the current industrial approach, Malaysia would have a potential of achieving USD 5,000 per capita by 2020. However to achieve the Vision 2020 agenda Malaysia needs to have a per capita income of say USD 10,000. On the basis of the findings, the only plausible approach to achieving the Vision 2020 is through the information technology approach, assuming other factors remaining the same that is the GDP growth of 7 percent, political stability and unforeseen natural disaster *(Table 1)*. However the contagion of the

Table 1   Justification for Malaysia: The Vision 2020

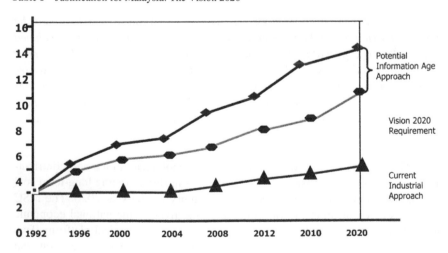

Source: EPU, IMF; World Bank, McKinsey 1992[5]

Asian Monetary Crisis starting from 1997, right through 1998 and 1999 had rendered Malaysia gravest concerns as there were doubts that the Vision 2020 and the developed nation status are achievable within the stipulated time frame. Fortunately, the adoption of the much-criticised capital control mechanism by Malaysia has proven to be very effective in containing the contagion, much to the chagrin of the critics.

The Malaysian economy was setbacked by the lackadaisical performance of the 1998 GDP (real) negative growth rate of −7.4 percent. The rebound of 1999 and 2000 of 5.8 percent and 8.5 percent (estimate) respectively, however, has provided an all round optimism that a new trajectory provided by the ICTs will determine that the Vision 2020 is achievable.

The National Information Technology Council (NITC) of the Federal Government is manned by members drawn from the government, the corporate sector, IT companies and academia. With the Prime Minister as chairman and the Malaysian Institute of Microelectronic System (MIMOS) as secretariat, they will determine the information technology agenda for the whole country. The national information technology agenda has two major components. First, the MSC with its seven flagship applications; and second, the demonstrator applications which focus on three key components of a developed nation or civil society; people development, info-structure development and application and the linkages between them.

Though the Task Force managed to identify perceived obstacles to Malaysia's success, in such factors as availability and quality of telecommunications, shortage of skills, Bumiputra policies,[4] restricted access to foreign expertise, intellectual property etc., Malaysia nonetheless enjoys compelling strengths. Factors such as multi-cultural links with the biggest Asian markets (China, India, Indonesia), English as a widely spoken language, cost advantages compared to the "tigers" of the region, no entrenched interests, laws which are easily changeable, stable political climate and highly committed top political elite have rendered ICTs a forgone conclusion. Having strategised all these, the government moved swiftly by setting up the National Information Technology Council (NITC) responsible for preparing the blueprint for the national IT plan or the National Information Technology Agenda (NITA). The delineation of a proposed intelligent region of the world, the Multimedia Super Corridor (MSC) and the setting up of the Multimedia Development Corporation (MDC) responsible for the development of the MSC with the corporate sector are the manifestations of the Government's seriousness and commitment. The appointment of members of the International Advisory Panel (IAP) consisting of the ICTs' giants such as chairpersons of Sun Micro-system, IBM, ACER, NTT etc. to advise on the MSC gives added credibility at the world stage. Areas outside the MSC are assigned to the respective states for implementation but their programs must fall in line with the NITA agenda and guidelines set by the Federal Government.

THE MULTIMEDIA SUPER CORRIDOR (THE "SMART" REGION)

The MSC, covering an area of 750 km^2, the size of New York City and even bigger than the island of Singapore is part of Malaysia's strategy to shift to high-technology knowledge-intensive industries with a special focus on information and communications technologies industries. The MSC is a conscious creation of a conducive environment of a world class standard for multimedia firms and effort to catalyse local and foreign IT firms to become globally competitive. The MSC has three objectives namely;

- the MSC will help Malaysia achieve Vision 2020 goals by catalysing productivity led-growth;
- the MSC will help leapfrog Malaysia to leadership in the information age by promoting smart partnership between foreign and local firms, and
- the MSC will build global bridges between Malaysia and other intelligent cities for mutual enhancement. The MSC is to accelerate the development of information-age technologies that will fuel the growth needed to achieve Malaysia's national goal of a developed, industrialised and information-rich nation status by 2020.

In August 1996, the Malaysian Prime Minister, Mahathir Mohamad announced that a region of 15 km by 50 km squares stretching southwards from Kuala Lumpur city in the north and the Kuala Lumpur International Airport (KLIA) in the south would be developed as the Multimedia Super Corridor (MSC). This "Smart Region" contains two new cyber cities which are under construction: Putrajaya, the new administrative hub of the Federal Government and Cyberjaya, an "intelligent" city (MDC, 1997) for foreign and local multimedia companies. In Malaysia, MSC is seen as a vehicle to laser-head into what is called as the – k-economy, informational economy and society. It is also an effort to provide a global connectivity as a means to national success in the new technological era (Bunnel, 2001). The MSC is not intended merely as an industrial park, but rather as what has been termed elsewhere as a "technopole" (Castells and Halls, 1994). The former is seen merely as a place of investment, while the latter is directly involved with research and development and the creation of new technology at Silicon Valley in the US.

The MSC combines the state-of-the-art urban planning and design with information and communications technologies (ICTs) in attracting the foreign multimedia companies to participate in the multimedia utopia (MDC, 1996) and thus incubates an ICTs culture and innovations that Malaysians will initially participate and eventually contribute. The MSC as a viable and attractive node in the high-tech economy (Graham and Marvin, 1999) and the ambience of the ICTs innovation will "catalyse" the development of a highly competitive cluster of Malaysian IT and would multimedia companies that eventually become world class (Zainuddin, 1997). Thus the MSC is envisaged to be a planned ICTs region for the long-term competitive enterprises of Malaysians which are globally connected where "an environment of collaboration, creativity and risk sharing are fostered" (MDC, 1996). Besides, the MSC will be a global community living on the leading-edge of the information society supported by world-class facilities and sensitive to conserving and protecting the natural environment, predominantly low density area with nature-ways and green-ways of landscaped parks and conservation reserves.

The MSC is planned to world class standards, incorporating the notions of the aesthetics and functionality. To that end special teams were sent abroad to study some of the world's best practices from Irvine and Florida in the US to Sophia Antipolis in France, Perth and Gold Coast in Australia, Japan and Europe. The highlights of the MSC's physical environment include:

The Kuala Lumpur City Centre (KLCC), the world's tallest twin towers at the heart of the capital – Kuala Lumpur city with sophisticated and intelligent features. Sited at the northern gateway of the MSC, the KLCC offers unparalleled commercial, recreational, entertainment and retail facilities in planned parkland.

The Kuala Lumpur International Airport (KLIA) at Sepang to the south, a recently opened state-of-the-art airport complex and integrated logistics hub, commissioned in 1998 to 21st Century requirements. Designed to cater to an expected traffic of between 25–50 million passengers per year, the KLIA is equipped with 80 gates and two 400-metre long parallel runways, six check-in islands with a maximum provision for 216 counters, and excellent road and rail links to Kuala Lumpur city. In addition to serving as a regional logistic hub, KLIA will form the centre of Malaysia's emerging aerospace industry.

Putrajaya the new seat of government and administration is a garden city catering for 250,000 people. Termed as Malaysia's first intelligent Garden city, Putrajaya will become a vital development catalyst due to the role it will assume as a model city and marking a new chapter in the history of modern city planning in Malaysia and the world. It sits on a magnificent 4581 hectare spread. Its Master Plan is designed along an axial tangent which runs from the north-east to south-east taking full advantage of the natural surroundings. The design concept of Putrajaya is the harmony between man and his creator, man and man and man and nature. The "intelligence" part of Putrajaya is to create a comprehensive and integrated electronic community where products and services are accessible anytime, anywhere and by anyone. Its undulating terrain treats visitors and residents to commanding vistas of the environment. About 40 percent of Putrajaya is natural. Lush greenery, botanical gardens are spread across the landscape enhanced by large bodies of water and wetlands. It is designed as paperless government in a bold experiment at *e*-government. Putrajaya also houses the office of the Prime Minister of Malaysia. It will offer efficient, on-line services to citizens and businesses, streamline the government's internal machinery, and increase productivity.

Cyberjaya, another garden city of 7000 hectares, the first MSC-designated cyber city to have a population of 240,000 people when fully completed. As the nucleus of the MSC, Cyberjaya is envisaged to be a multimedia haven, a first choice site for innovative companies. Strategically located within the MSC, Cyberjaya will offer a full package of incentives and facilities – including high speed fibre optics 3–10 gigabytes, a balanced development of enterprise, residential, commercial and public precincts, world class homes, restaurants and shopping facilities and large parkland. Cyberjaya also aspires to be a near "zero-emission city" through strict zoning policies and environmental guidelines. Freedom of ownership guidelines will allow MSC-status companies to own 100 percent of the land and buildings in designated areas.

The highlights of Cyberjaya are:

- A world-class urban development, an attractive low density development revolving around green axis and reserves and excellent infrastructure.
- A human-friendly urban environment that comprises balanced development of enterprise, commercial and residential precincts with plentiful recreation areas and public facilities, including facilities for the physically handicapped and vulnerable group.
- An eco-friendly sustainable environment. Cyberjaya encourages solar and waste power generation and rain water utilisation. The city prohibits use of poisonous and hazardous materials and emphasise harmony with the existing topography and ecology through careful planning of roads and buildings.

The Multimedia University, a world first, started its operations in Cyberjaya in May 1999 with an initial intake of 3000 students. The University provides multimedia specific programs and caters to the skill requirements of companies located in MSC and Malaysia. It aims to be a world-class university in the promotion, acquisition, generation and application of knowledge in areas related to multimedia.

Telecommunications and communications, supporting the MSC is a high-capacity, digital telecommunications infrastructure designed to the highest international standards in capacity, reliability and pricing. This information network is part of an integrated logistics hub enabling rapid distribution of products along modern land, air and sea links. Key network features that will link MSC to regional and global centres include:

- A fibre-optic backbone with an unprecedented 2.5–10 gigabits per second capacity;
- Open standards, high speed switching and multiple protocols including ATM;
- Performance guarantees including installation of telephone services within 24 hours; ATM circuits within 5 days and a 99.9 percent service availability;

- High capacity links to international centres to ensure information, products and services free flow;
- Competitive pricing including flat-rate low pricing for basic network services compared with other regional centres and an open entry policy for value-added network services;

Amenities and Facilities, the state of the art eco-friendly commercial and enterprise estates, residential and housing suburbs, international schools and other academic and leisure and entertainment amenities.

Modern Transportation and Highway Systems incorporating rapid train links from the cyber cities to the Kuala Lumpur Metropolitan city. The systems include the KL-Seremban Highway, the North-South Central Expressway Link on the west of Cyberjaya, the South Klang Valley Expressway, the Damansara – Puchong Expressway and a highway providing direct link between Kuala Lumpur and the KLIA. Efficient public transport systems delivering efficient commuter rail and LRT services will also be available within the cyber cities.

MULTIMEDIA DEVELOPMENT AND ENVIRONMENT

To spearhead the development of the MSC and to give shape to its environment, seven primary areas of multimedia applications (flagship applications) have been identified. These applications contain a challenging opportunity unprecedented before for local and international multimedia companies to participate in the real world applications in the MSC. If successfully applied they can be replicated elsewhere in the world. The MSC flagship applications are divided into two separate categories: a) the Multimedia Development flagship applications offering concrete business opportunities to facilitate the MSC's development; and b) the Multimedia Environment flagship applications providing optimal environment that supports multimedia companies entering the MSC.

The Multimedia Development applications projects have long term objectives that reach beyond MSC's borders. These applications are e-government, multipurpose card, smart schools and tele-medicine. Each application will provide companies with excellent opportunities to collaborate with the Malaysian government in creating and implementing innovative multimedia solutions in a unique environment unprecedented elsewhere in terms of environmental location. Thus, these projects will use MSC as a global test bed for multimedia and IT development.

An Opportunity to Re-invent the Government (e-Government)

With e-Government, the government is to reinvent the way the government operates. e-Government will improve both how the government operates internally as well as how it delivers services to the citizens. It seeks to improve the convenience, accessibility and quality of interactions with citizens and businesses, simultaneously, it will improve information flows and processes within government to improve speed and quality of policy development, co-ordination and enforcement and hence good governance. The vision of e-government focuses on effective and efficient delivery of services from the government to the citizens, enabling government to become more responsive to the needs of its citizens.

The objectives of the e-government are two pronged. One is to reinvent the government through connectivity and the other is to catalyse the MSC. The e-government is to redefine the relationships of government to citizens, to business and to itself.

Tool for the Information Age (Multipurpose Card)

This application seeks to develop a single and common platform for a multipurpose card that will enable the government and private application providers to implement smart card solutions without duplication of effort and investment. The MPC will be a plastic card embedded with a chip or micro-processor that has the capability to perform a wide range of functions, including data processing, storage, and file management. Eight applications have been selected for inclusion. They are the National ID, Driving Licence, Immigration, Health Card and Electronic Cash and Other Financial Transactions.

Education for a Smart Society (Smart Schools)

The MSC's smart school initiative responds to the need of Malaysians to make the critical transition from an industrial economy to a *k*-economy. This requires for a technologically literate, thinking workforce which is able to perform in a global environment and use ICTs tools to improve productivity. An integrated set of strategies will be employed to focus on thinking ability, vertical integration of students' progress, teachers as facilitators and learning to become self-directed.

A New Paradigm in Health Provision (Tele-medicine)

Tele-health is a process of providing individuals greater access and increased knowledge in healthcare. It empowers the individual to manage one's own personal healthcare and integrates information to allow the smooth flow of services and products throughout the healthcare system. In other words, the tele-health initiative aims to keep people in the "wellness" paradigm. By taking advantage of the existing multimedia and information technology and also developing new technological solutions, this application will ensure Malaysians enjoy a high quality of health care. It would facilitate Malaysia's becoming a global hub for tele-health services. There are already four pilot applications namely mass customised personal health information and education, continuing medical education, tele-consultation and lifetime health plan.

For each of these applications, teams from MDC – the lead agency and the private sector would develop Concept Request for Proposal that describe the requirements of identified pilot applications and give consortia of private sector companies the flexibility required to innovate and deliver the best solutions.

The Multimedia Environment applications will provide both Malaysian and international companies with the opportunity to operate in an environment of close co-operation with leaders in the multimedia industry, research and academic institutions, and customers, in one of the world's most attractive business regions. These applications will also allow companies to build centres of excellence for their R&D activities, create hubs to efficiently deliver value-added services to companies throughout the region, and innovate entire businesses by taking full advantage of the MSC's unique environment and infrastructure. The Multimedia Environment applications projects are: R&D Cluster, World-wide Manufacturing Webs and Borderless Marketing.

Next Generation Multimedia Technologies (R&D Cluster)

Technological development and advances in multimedia and information technologies have been made possible through substantial investments in research and development. It flourishes in an environment where the necessary infrastructure is in place, where creative and risk-taking activities are promoted and shared, and where experts find living conditions most attractive (MDC, 1998). The MSC guarantees such an environment and Malaysia strives through this unique environment to promote the development of the next generation of multimedia technologies by forging collaborative R&D efforts among leading edge corporations, public research institutions and universities. Research-driven companies are encouraged to take advantage of the MSC's strengths, including government commitment and support, the MSC comprehensive package, conducive research environment, and the growing business opportunities in Malaysia and the region. Other R&D advantages in MSC are Malaysia's location and its cultural diversity which can easily relate to China, India –two of the world's fastest growing economies and Indonesia, growing business opportunities and dedicated facilities. These include the Multimedia University, other universities within the region, the Technology Park Malaysia and the Malaysian Institute of Microelectronics Systems (MIMOS). MIMOS' objective is to develop an indigenous capacity in multimedia technologies.

Building Best Practices in High-tech Operation (The World-wide Manufacturing Web)

This application aims to provide a conducive environment for high value-added manufacturing activities using multimedia technology. The WMW brings together Malaysia's unique attributes as well as those of the MSC to enable companies to support their regional operations using multimedia technology. To foster companies to build links between operation centres around a wide range of support services such as R&D, design, engineering, manufacturing control, procurement, logistics and distribution support. The combination of reliable, state-of-the-art technology and infrastructure, combined with a central Asia-Pacific location, all contribute to making MSC an ideal place for regional research and development. Foreign MSC-status companies will enjoy freedom from strict immigration laws and procedures, the 10-point Bill of Guarantees – a comprehensive and realistic framework of Cyber-laws and intellectual property laws, already committed by the government. The laws are: the Digital Signature Act 1997, the Copyright (Amendment) Act 1997, the Computer Crimes Act 1997, the Tele-medicine Act 1997 and the Communications and Multimedia Act 1998.

New Frontiers in Commerce (Border-less Marketing)

The borderless marketing is premised that multimedia technology can be used by businesses to move efficiently and effectively to serve customers across different time zones. The traditional barriers of time, space and form will be eliminated as a result. Tele marketing companies will find the MSC an opportunity to centralise their call centre operations and to tap the vast potential of the Asia-Pacific region robust economic growth.

REALISING THE VISION

No other Malaysian project but the MSC, has attracted so much world attention. Attention has been expressed in the form of scepticisms, criticisms, commiseration or plain foolhardy diatribe. And some expressed plain fascination that a third world country would dare to attempt such a technological feat. We are unsure whether the controversy relates to the nature and extent of the MSC project or the man behind the MSC or both. We ourselves are uncertain. But what we are certain is that the MSC project is very innovative, unprecedented, futuristic and of a scale quite grandiose. From the professionals to the politicians, the ICTs savvies to the transnationals, investors and Marxian economists have penned their opinions on MSC in journals, newspapers and books. Such is the nature of the MSC! Now, what are the concerns of all these critics and sceptics? Are these concerns valid? What are the strengths of the MSC or the line of defence that the optimists and the pro-activists employed? These are interesting issues that need to be addressed. We will, therefore, try to make an objective assessment of MSC in the context of its likely scenario to achieve what it is designed to achieve. At the same time we should be able to make critical observation of the government initiatives and strategies to transform MSC into an ICTs' habitat for innovation and entrepreneurship (Miller, 1998) of the world.

William Miller, Herbert Professor at Stanford University, the leading researcher who made an objective analysis of the strengths and weaknesses of the various "silicon valleys" or smart regions of the world, made several observations. He uses 11 criteria to be applied across the globe and comes up with a competitiveness analysis. The eleven criteria are knowledge intensity, quality of workforce, mobile workforce, rewards risk taking, open business environment, community collaboration, developed venture capital, university interaction, quality of life, government involvement and indigenous companies (Miller, 1998). Based on these criteria and scaling applied, we find that topping the scale is Silicon Valley of the US. The MSC is way below the list, in the same league as Gifu of Japan. Singapore One is better only in respect of university interaction score (*Table 2: Regional Summary*).

Table 2 Regional Summary

	1	2	3	4	5	6	7	8	9	10	11
Silicon Valley	H	H	H	H	H	H	H	H	H	L	H
North Carolina	H	H	M	L	L	M	L	M	H	H	L
Austin, Texas	H	H	M	M	M	H	M	M	H	L	M
Singapore (One)	H	H	L	L	L	M	L	H	H	H	L
Taiwan (Tsinchiu)	H	H	M	H	M	L	M	M	M	M	H
Malaysia (MSC)	H	L	L	L	M	L	L	M	H	H	L
Gifu, Japan	H	L	L	L	M	M	L	L	H	H	L
Sophia-Antipolis	H	H	M	L	L	M	L	L	H	H	L
Australia MFP	H	H	M	M	L	L	L	M	H	H	L

H – high   L – low   M – Medium

1.	Knowledge Intensity	7.	Developed Venture Capital
2.	Quality of Work Force	8.	University Interaction
3.	Mobile Work Force	9.	Quality of Life
4.	Rewards Risk Taking	10.	Government Involvement
5.	Open Business Environment	11.	Indigenous Companies
6.	Community Collaboration		

Source: adapted from Miller (1998)

The strengths of the MSC vis-à-vis the Silicon Valley as benchmark are in the areas of knowledge intensity and government involvement. To spearhead the development of ICTs and to give shape to its environment seven flagship applications of knowledge intensity, of which four constitute multimedia development applications and three are multimedia environment applications have been identified. A dedicated university namely the Multimedia University is specially built in Cyberjaya to produce ICTs' professionals required for the MSC and Malaysia. This is over and above several universities located within the region.

The government's involvement in MSC is viewed in a positive light. In terms of its role, its commitments in the physical planning and development, providing hard and "soft" infrastructures, introducing financial and fiscal incentives, creating conducive business environment with the necessary legal instruments such as cyber laws and intellectual property laws, besides the ten-point Bill of Guarantees. The Government's initiative in inducting ICTs' global players into the International Advisory Panel Committee is a smart strategy and has injected high level of credibility to the MSC. The visionary attributes of the current Malaysian Prime Minister Mahathir Mohamad cannot be underestimated.

The MSC can boast a high quality of life made available to the residents. Putrajaya and Cyberjaya are developed as a model intelligent city with a unique green-field environment within which one can work, live and play. There are sites for offices, enterprises, shopping facilities and recreational facilities. There is a wide choice of hillside mansions, lakefront houses and condominiums to suit varying family needs as well as commercial precinct comprising of shopping facilities, first-class resort hotels, convention centres, food outlets and service apartments to accommodate business professionals and activities. The whole development is supported by an advanced telecommunications infrastructure and world-class facilities including Multimedia University, a tele-medicine hospital, an international school, a large public park and natural wetlands.

However what is interesting and perhaps a dominant attribute of MSC is the "newness" and the "dedicatedness" to ICTs compared to the rest. The rate of development and the quality of project execution are impressive when we con-sider the ground-breaking ceremony was done in August 1996. In five years time, the Prime Minister has already moved his office and official residence in Putrajaya. Being new MSC enjoys certain advantages such the flexibility of physical design, technology and learning from other people weaknesses.

Cities and city regions are commonly understood as the key players in the information economy (Castells, 1989, 1996). City regions have become motors of the global economy especially on the account that spatially concentrated human relationships which are essential for tacit learning, innovation and cre-ativity (Storper, 1997, Crevoisier, 1999, Hall, 2000). Cities or city regions are thus understood as to be more flexible in adapting to the changing conditions of market technology, and culture in the so called information age (Castells and Hall, 1994). The fact that MSC lies within the Kuala Lumpur Metropolitan Region with Kuala Lumpur city as the hub has provided powerful synergy and critical mass to the mutual development of both entities.

The most telling factor is the creation of a habitat for the high-tech innov-ative community. The MSC has taken the right step in terms of knowledge intensity which are translated into various flagship applications. As what is hap-pening in the Silicon Valley, knowledge intensity applications have generated new quality jobs. Knowledge quality applications require the presence of high quality work force in which the universities and other training institutions must provide the kind of skills required. In order to create the ambience of an open business environment, including international linkages, business climate should reward risks taking and does not punish failure. It must be a positive sum game (Miller 1998). Another important factor in spawning the habitat is the presence of the community dynamics of collaboration between business, government, independent sectors such as universities, foundations, councils etc. The other factor which MSC is trailing behind other smart regions is the venture capital industry that understand high-tech. Venture capital industry in MSC is still weak. The other locational factor of the MSC which can be taken advantage of is the presence and the accessibility of the Multimedia Universities and other universities within the Kuala Lumpur Metropolitan Region. The physical prox-imity of these universities must be translated into effective interaction with industry to generate co-evolution of ideas. It is these interlocking institutions which sustain the social relations which are the basis of continuous learning, innovation and therefore growth (Riain, 1997).

In implementing the MSC or cybercity initiatives we are faced with several intriguing and unprecedented challenges and issues. These include:

• *Systems Integration*

In the top-down approach of MSC initiatives implementation the Federal Government takes centre stage and provides the impetus to the development of various components of the MSC, especially the flagships. The states' initiatives are later rationalised through the National Information Technology Council (NITC) whose members comprise state representatives from the respective states. In a 2 or 3-tier government at local, state and federal levels, the highest level of coordination is necessary in terms of timing, project implementation etc., and they must be done collaboratively.

The issue of standardisation of systems and protocols must be addressed to ensure effective implementation and application. This issue will be further exacerbated in addressing standardisation between countries.

• *Cultural Integrity*

Quite often we tend to resort to engineering or IT solutions to solve IT problems. We tend to presume that people are receptive to new ideas and better ways of doing things and that the learning curve is always short. But in practice we find that people are not receptive to change or fear change. People do not want to be de-skilled and to lose their jobs. The older they are the less receptive they become. But we do find that people in the private sector are more receptive and adaptable and hence adoptive and the resistance is lower compared to the public sector, which is often less creative and innovative.

As the world is intertwined and wired, we find that access to information and knowledge is made easier and faster. As a consequence there will be more interface and cultural assimilation or enculturation between and among cultures of the world. Undoubtedly there are the good and the bad elements. Cultural purity, in the future would be a thing of the past. Positive cultural borrowing and adaptation are not undesirable if they are done cautiously and creatively. The MSC with its Bill of Guarantees has brought to our attention some of the attendant problems.

• *Cyberlaws and Digital Crimes*

MSC with its Bill of Guarantees and cyberlaws are meant to ascertain that the test-bed companies and other knowledge-based companies are given the right environment to compete. Their intellectual property rights are protected. The scope of cyberlaws on digital contract, digital signature and their admissibility as legal evidence in court are still being debated. Cyberpayment, multimedia intellectual property, cyber crimes are still unenforceable between countries. Other aspects of societal cyberlaws such as privacy protection, consumer rights protection, equity and access, cyberfraud are still being discussed.

The Malaysian government commits that it would provide world class infrastructure, unrestricted employment of knowledge workers, freedom of ownership, competitive financial incentives, freedom of sourcing of capital globally etc. But one of the moot points is to ensure no censorship of the Internet. This has created some concerns in certain quarters.

- *Concept of Government and Governance*

Governance is the way in which people are managed. In a democracy, governance is ideally by the people and for the people through a process of authority levels of administration, empowered through legislation to make things happen. In good governance, innovations and creativity make way for the best things to happen. Governance normally encompasses a large range of concerns, including effectiveness of institutional arrangements, decision-making process, policy formulation, implementation capacity, information flows and nature of relationship between the ruler and the ruled. IT makes good governance possible. It is participatory, transparent and accountable and encourages the creation of a civil society. This is something which is constructive and must be set in motion.

- *Technology*

We need technology to enhance our lives. But technology is only a means to an end. The advancement of information and communication technologies has altered positively the way we do things – more effectively and efficiently. But where is the limit? There will be a time when technology becomes king and we are subservient to it. Information and communication technologies (ICTs) should not take away our fundamental rights to liberty, freedom and privacy. This is an ethical issue to be addressed by all.

I am sure there are other pertinent issues and challenges that we ought to address and be prepared to face them. Since we are delving into a new territory a great deal of research has to be done especially the impact of ICTs on human behaviour and social interactions.

CONCLUDING REMARKS

The paper has put forward a critical review of the rationale and justification of Malaysia embarking on the ICTs path and strategising itself to achieve the Vision 2020 through the ICTs' mode. It argues that with the government commitment, Malaysia's impressive economic growth and its ability to rebound from the economic crisis of 1998–1999 and the tremendous effort on the human resource development, MSC will have every chance to succeed.

The optimism surrounding the MSC's potential is boosted by the successes of Malaysia in attracting global investment and promoting international business and industry. The advantages of Malaysia as a regional hub are many.

Among other comparative advantages for Malaysia are its strategic location, stable economic growth, on the average of GDP 8 percent, political stability, quality of life, its multicultural and multiethnic environment, cost advantage and its development track record.

NOTES

1   k-economy is now the buzz word the Malaysian Government has officially adopted to instil the new economy which is based on knowledge.
2   the MSC or Smart Region of Malaysia has been used interchangeably to refer to the 15 km × 15 km area incorporating the Kuala Lumpur Twin Tower in the north and the Kuala Lumpur International Airport in the south.
3   Putrajaya becomes the third Federal Territory after Kuala Lumpur and Labuan. Under the Malaysian 3-tier system of government, the Federal Government acquired Putrajaya by paying certain compensation to the State Government of Selangor, the original land owner of MSC.
4   Vision 2020 is the Government of Malaysia development's philosophy which was launched by the Prime Minister, Mahathir in 1991 charting the future course of Malaysia and the formula to go about attaining the objectives of developing Malaysia into an industrialised, developed, democratic country based on civil society.
5   McKinsey Consultants were employed by the Malaysian Government in mid 1990's to advise on the development strategy for Malaysia.

REFERENCES

Azman, A., 1998
    THE NATIONAL CYBERPLAN – THE MALAYSIAN EXPERIENCE.
    A paper presented at the Conference in Bremen, Germany,
    Weltweite Erfahrungen und Visionen zu Regionalen Elektronischen
    Gemeinschaften. June 22–26.

Bunnel, T., 2001
    CITIES FOR NATIONS? EXAMINING THE CITY-NATION-STATE:
    RELATION IN INFORMATION AGE MALAYSIA, Department
    of Geography, NUS, Singapore.

Castells, M., 1989
    THE INFORMATIONAL CITY: INFORMATION TECHNOLOGY,
    ECONOMIC RESTRUCTURING AND THE URBAN REGIONAL
    PROCESS, Blackwell, Oxford.

Castells, M. et al., 1994
    TECHNOPOLES OF THE WORLD: THE MAKING OF THE
    TWENTIETH CENTURY INDUSTRIAL COMPLEXS, Routledge,
    London.

Castells, M., 1996
THE INFORMATION AGE: ECONOMY, SOCIETY AND CULTURE,
VOLUME 1, THE RISE OF THE NETWORK SOCIETY,
Blackwell, Oxford.

Crevoisier, O., 1999
INNOVATION AND THE CITY. In Malecki and Oinas (eds.), *Making Connections: Technological Learning and Regional Economic Change*, Ashgate Publishing, Aldershot.

Graham S. & Marvin S., 1999
PLANNING CYBERCITIES? INTEGRATING
TELECOMMUNICATIONS INTO URBAN PLANNING, *Town Planning Review 70: 89–114*.

Hall, P., 2000
CREATIVE CITIES AND ECONOMIC DEVELOPMENT, *Urban Studies* 37, 639–649.

Mahathir, M., 1991
MALAYSIA: THE WAY FORWARD, Malaysia Business Council, Kuala Lumpur, Malaysia.

Mahathir, M., 1991
MALAYSIA: MOHAMAD ON THE MULTIMEDIA SUPER
CORRIDOR, Pelanduk Publications, Subang Jaya, Selangor, Malaysia.

Multimedia Development Corporation, 1996
INVESTING IN MALAYSIA'S MSC: POLICIES, INCENTIVES AND FACILITIES.

Multimedia Development Corporation, 1997
CYBERJAYA; THE MODEL INTELLIGENT CITY IN THE MAKING.

Multimedia Development Corporation, 1998
BUILDING THE MALAYSIAN MULTIMEDIA SUPER CORRIDOR
FOR WORLD CLASS COMPANIES.

Miller, W., 1998
POLICIES AND CONDITIONS FOR AN ENTREPRENEURIAL
ECONOMY, a paper presented at the Conference in Bremen,
Germany Weltweite Erfahrungen und Visionen zu Regionalen
Elektronischen Gemeinschaften, June 22–26.

Riain, S., 1997
   COMPETITION AND CHANGE, Harwood Academic Publishers,
       Amsterdam.

Storper, M., 1997
   THE REGIONAL WORLD: TERRITORIAL DEVELOPMENT IN A
       GLOBAL ECONOMY, New York: Guilford.

The New Straits Times, 1995
   THE NEW STRAITS TIMES, Malaysia.

Zainuddin, M., 1997
   PLANNING OF CYBERJAYA: THE MULTIMEDIA CITY FOR
       MALAYSIA. Paper presented at South & South Mayor Conference,
       Putra World Trade Centre Kuala Lumpur, Malaysia, July 3–4.

Cities Spanning the Millennia: Cairo/Alexandria

Policies for Directing Urban Growth in Greater Cairo Region: A Critical Analysis

Abdalla Abdelaziz Attia

ABSTRACT

Throughout history, Cairo and Alexandria have attracted migrants from Egypt's rural and less developed communities searching for better jobs, better income, and better livelihoods. This, however, had increased the two cities' problems. Indeed, congestion, pollution, squatter settlements and slums, breakdown in urban services are just few implications. For example, the population of Greater Cairo reached about 15 million inhabitants in year 2000, 30 percent of which live in slums. In addition, the urbanization phenomena had been decreasing agricultural land and thus risking the country's food supply. Accordingly, the last 3 decades had witnessed the preparation of many planning studies to cope with the population and urbanization problems. The main objective of the national development strategy had been to direct the urban growth of Egypt's mega cities away from the agricultural land in the Nile Valley and Delta and towards the desert areas.

Hence, the focus of this paper is Cairo. The paper attempts to critically analyse the urban development plans and policies that aimed to solve Cairo's underlying problems particularly the policies for directing the city's urbanization towards the desert areas away from the scarce agricultural lands. The paper will analyse the different studies for Cairo's development that had been carried out from the early 1970s. The aim of this paper, therefore, is to draw the lessons that could be learnt from the experience of the last 3 decades. Accordingly, the paper reviews Cairo's urban development policies for dealing with problems of unemployment, migration and the increase of inhabitants in the city, the expansion of the built area of the Greater Cairo Region, and the decrease of agricultural land. The paper will also analyze the policy for directing urban growth that was formulated in 1997.

The paper calls for the re-assessment of the national urban development policy in order to redistribute the new towns and new urban communities throughout the whole country so as to release the pressure on Greater Cairo Region from the growing urban concentration. It also highlights the need for an even distribution of public and private investment throughout the whole country to ensure equity among all Egyptian social groups. It argues that urban development policies should be for the public interest rather than for personal interests that serve short-term political and economic benefits. Finally, it points out the importance of capitalizing on the information technology to assist in the policy formulation, implementation and monitoring processes in order to manage and control high population densities, congestion and environmental problems and achieve sustainable development.

Abbreviations

GOPP	General Organization for Physical Planning
GCR	Greater Cairo Region
CAPMAS	Central Authority for Public Mobilization and statistics
Ha	Hectares ($=10,000$ square meters).

1   CAIRO'S URBAN DEVELOPMENT PLANS IN THE PERIOD BETWEEN 1970 AND YEAR 2000: AN INTRODUCTION

The idea of directing urban growth towards the desert areas away from the densely populated Greater Cairo Region (GCR) was raised in the early 1970s. The idea was to formulate urban development plans based on the concept of creating satellite towns in the desert areas, east and west of the built area of Greater Cairo to direct urbanization away from agricultural lands.

The implementation of the urban development plans that aimed at directing urban growth towards the desert has been taking place since 1970 through successive master plans that proposed establishing new cities, new towns, new settlements, and new urban communities. However, it was not until the late 1980s when the implementation of the development plan was realized due to the delays in implementing the infrastructure and the roads' networks.

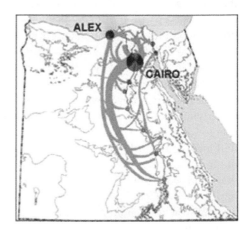

Successive alterations to the master plan were carried out in order to promote a quicker urban growth towards the desert. The

Figure 1   Egypt Repulsion Movement to Major Urban Centers, Cairo & Alex. (GOPP 1992).

development plan was formulated in 1970; its first amendment was authorized in 1983. In 1991 another revision to the master plan took place. Later, in 1994, the plan was amended, and finally, in 1997 the development plan was approved and authorized. The following sections will analyze the successive development plans for the GCR.

2 EVOLUTION OF THE MASTER PLAN FOR GREATER CAIRO REGION (GCR)

2.1 Greater Cairo Master Plan (1970)

In the early 1960s, the rate of increase of population in Cairo was 4.15% (CAPMAS 1966). This was due to the push factors from the rural and less developed areas in Upper Egypt and the Delta to Cairo and Alexandria Cities being the two largest urban centers of Egypt.

This had necessitated that a high planning committee was to be formed in 1965 to plan for the development of Greater Cairo Region (GCR) to cope with the problem of population explosion. The committee determined the boundary of the GCR, which included the following areas:

a) All inhabited areas in Cairo Governorate and the adjacent inhabited areas of Giza and Kalyoubia Governorates.
b) The economical or social activity areas that service the above mentioned inhabited areas.

Figure 2   Greater Cairo Region (GOPP 1982).

c) The areas on which the urban extensions would physically extend during the following 50 years in order to direct and control the extension of the built area away from the agricultural lands and archaeological areas.

Accordingly the GCR total area was estimated to be 685,000 feddans, (287,700 hectares) where the primary land uses were distributed as follows:

Uses	Area	%
Desert areas	168840 ha	58.7
Agricultural areas	87780 ha	30.5
Built areas	26040 ha	9.05
Water areas	5460 ha	1.75
TOTAL	**288120 ha**	**100**

The central city area of Cairo represented 6% of the total built area where the governmental, commercial, and financial activities were concentrated. The built areas of Cairo City center included the following uses:

Uses	Area	%
1) Housing areas	8983.8 ha	34.5
2) City Center	1562.4 ha	6.0
3) Natural areas around GCR	651.0 ha	2.5
4) Secondary Services	3906.0 ha	15.0
5) Industrial areas	3515.4 ha	13.5
6) Cemeteries	651.0 ha	2.5
7) Military areas	6770.4 ha	26.0
TOTAL	**25200 ha**	**100**

2.1.1   The analysis of Cairo's population growth

The GCR planning committee estimated that the population of GCR would amount to 14.8 million by 1990 assuming the decline of the rate of increase of immigrants to Cairo (2.9%), which represented 70% of the rate of increase of GCR population, and also assuming that the natural rate of increase was constant (1.25%) which represented 30%.

It is necessary to point out that in the period between 1947 and 1960 the rate of increase of population in the city central district amounted to 2.6%, while the rate decreased by (–0.7%) between 1960 and 1976; and by (–16%) between 1976 and 1986; and by (–1.5%) between 1988 and 1996.

Despite the decrease in the population of Cairo's central district (as shown in the following table), the population of the Greater Cairo Region (GCR) has been growing and urbanization has been increasing replacing the scarce agricultural lands (as Figure 3 illustrates)

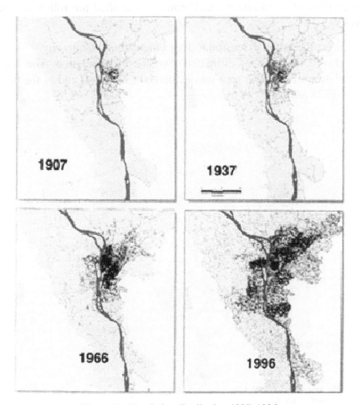

Figure 3   Population distribution 1907–1996.

Period	Increase or decrease of population in the city center	Average yearly increase or decrease of population	Percentage of increase or decrease
1947–1960	+443 637	+34 126	+2.6
1960–1976	−207 783	−12 948	−0.7
1976–1986	−245 107	−24 511	−1.6
1986–1996	−198 979	−19 898	−1.5

2.1.2   Objectives of the 1970 master plan

The master plan of 1970 (shown in Figure 4) identified the following planning objectives:

a)   To direct the urban extension to four cities around Cairo city.
b)   To absorb the 42% of estimated increase of population. The density of population of the built area was estimated (325 p/ha) and of the new cities to be (238 p/ha).

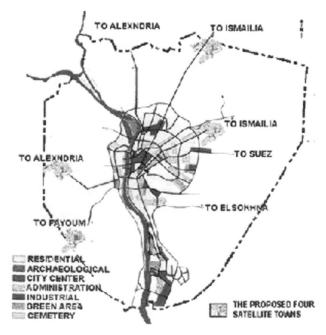

Figure 4   1970 GCR Development Plan (GOPP 1982).

c)   Protecting the agricultural lands from being replaced by built areas.
d)   Establishing a green belt around the built area of the GCR to limit its growth.
e)   Creating a rapid transit facility (Metro) around GCR to facilitate the population movement and to connect the existing communities by the new communities.

2.1.3   Planning policies during the seventies

2.1.3.1. In 1973 the General Organization for Physical Planning (GOPP) was established to be responsible for all physical plans in Egypt. Its prime objective was meeting the national goal of directing the urban extensions away from the Nile Valley and Delta.

2.1.3.2. In 1977 the Presidential Decree No 475 sub-divided Egypt into 8 economic regions (later became 7 regions in 1983 by merging Alexandria and Matrouh regions) as shown in Figure 5.

2.1.3.3 In 1979 a new organization was founded (General Organization for New Communities) to become responsible for planning, implementing and managing new cities and new settlements that were being built outside the boundaries of the existing urban agglomerations.

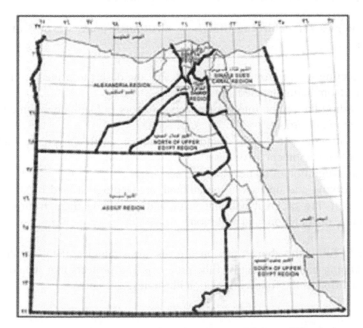

Figure 5   The Seven Economic Regions (Source: Ministry of Planning).

2.1.3.4 As a result of the social and economic changes in the Egyptian society in the late 1970s, especially in Cairo, the nature of investment of the public and private sectors in urban development projects had experienced significant changes. These changes had strong implications on implementing the 1970 master plan for the GCR. Accordingly, the changes had also necessitated that a revision and an alteration to the master plan take place.

2.2   Greater Cairo regional plan 1982–2000

A planning unit within the GOPP was founded in 1981, to be responsible for reviewing and updating the planning strategy for the regional urban development.

Accordingly, it established a long-term regional urban plan for Greater Cairo Region (GCR), which was approved and authorized in 1983 by a Presidential Decree.

The regional plan estimated that the GCR population in year 2000 to become 16 million inhabitants[1]. The planning goals of 1982 plan could be summarized in two main points:

1.   Promoting economic growth.
2.   Upgrading and enhancing the living environment.

The strategies for achieving the above-mentioned goals were identified as follows:

*   Directing urban growth of GCR towards the desert to protect agricultural land from being invaded and decreased.
*   Dividing the region into homogeneous sectors to facilitate urban development management.
*   Encouraging the private sector to participate with the public sector in urban development activities.
*   The integration of all development sectors.
*   Preserving and upgrading the cultural, archaeological, and historical heritage, being of great national value and protecting them from disintegration, degradation, and pollution.

In order to fulfill the above-mentioned goals and strategies, the 1982 development planning had also identified the programs and projects that would enable implementation as the following section illustrates.

[1] In year 2000 the population of GCR was 14.9 million, distributed as 6.8 million in Cairo Governorate, 4.8 million in Giza Governate, and 3.3 million in Qalubia Governorate.

2.2.1 Strategic recommendation of the 1982 master plan

2.2.1.1. The construction of a ring road around the existing GCR built area in order to achieve the following objectives:

- Avoid through traffic in the existing built area.
- Controlling the increase of the built area within a defined framework.
- Directing urban growth towards the desert.

2.2.1.2. The creation of new urban communities in the desert areas outside the ring road.
2.2.1.3. The division of the region into homogenous sectors, each sector with its homogeneous social, urban, and economical characteristics. The region, therefore, was subdivided into 16 homogeneous sectors (shown in Figure 6). Each sector included 1–2 million inhabitants living in a homogeneous social and urban pattern. Each sector would offer sufficient job opportunities for its inhabitants, as well as sufficient basic services.

The usage of existing road networks that link the GCR to the desert for directing the urban extensions towards new development axes in the desert areas outside the ring road.

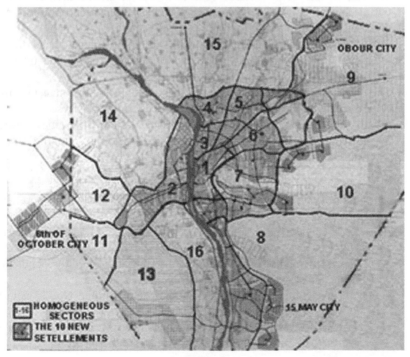

Figure 6

2.2.2   *Implementing the 1982 development plan*

According to the GCR urban development plan (1982–2000) the built area of
Cairo City was to be limited to 20,580 ha while the population was not to exceed
6.3 million inhabitants (i.e. a population density of 304 p/ha). Such built area
was meant to include new settlements to be located east of the ring road and also
include 15th of May City (located to the south of Cairo City) around which 2
new settlements were planned to be implemented. In addition, a green belt was
planned to be implemented to the east of the existing built area in order to limit
its growth towards the 4 new settlements which were planned to be located on
the east side of the ring road (GOPP 1982).

   The growth of Cairo's built area amounted to 23,010 ha in the period
between 1981 and 1986 (i.e. the built area exceeded what was expected by an
additional 2,100 ha). Moreover, the population of Cairo city reached 6,068,000
inhabitants in 1986 (a population density of 264 p/ha) (GOPP 2000, CAPMAS
1996). Accordingly, the implications of the 1982 long term urban development
plan for GCR, in general, was positive particularly in areas such as:

*   The decentralization of service activities to be in sub-centers within the 16
    homogeneous sectors.
*   The rate of loss of agricultural land in favour of urbanization decreased
    during the 1990s to 25% of the rate of decrease that occurred during the
    previous 3 decades.
*   The development axes and roads networks were implemented according to
    the regional urban planning of 1982. Indeed, the ring road and its intersec-
    tions are almost completed.
*   The metro as a rapid transit system was also established to connect Cairo's
    central area with the south, the north, north-east, and western sectors.
*   Finally, the urban growth was also directed to the new cities located
    north, east and west side of the ring road and also to the south of Cairo as
    follows:

    *   Obour City and new settlement No. 10: were to absorb the popu-
        lation from, and serve, sectors No. 15 and No. 8 north and north east
        of Cairo.
    *   Badr City: were to absorb the population from, and serve, sector
        No. 8
    *   New settlements Nos 1, 2, 3, 4, and 5: were to absorb the population
        from, and serve the eastern sector No. 10.
    *   6th of October City and new settlement No. 6 and No. 7: were to
        absorb the population from, and serve the western sectors.
    *   15th of May City and new settlements No. 8 and No 9: were to
        absorb the population from and serve the southern sectors.

It was recognized that several political, administrative, and economic
factors had constrained the implementation of many of the 1982 planning objec-

tives. However, it was necessary to capitalize on the positive outcomes of policy and to dissipate the negative results.

2.3 GCR development plan 1986–2000

The General Organization of Physical Planning (GOPP) had to revise and amend the 1982 regional urban development plan because urbanization continued to decrease agricultural lands and the rate of increase of the GCR's population continued to rise.

Indeed, the average annual growth of Cairo City's built area alone in the period between 1982 and 1986 exceeded 832 ha per year (compared to the annual average growth rate which was 382 ha per year between 1970 and 1981). In fact, the total built area in the Greater Cairo Region (Cairo, Giza and Kalyoubia) exceeded 4158 ha. This had necessitated the revision of the 1982 master plan

The main objective of the updated development plan for the GCR (of 1986) was to limit the growth of urbanization to be within the ring road. The new settlements located on the east side of the ring road became 5 settlements instead of 4. A green belt was recommended in order to limit and direct the city growth eastwards.

However, the 1986 planning failed to achieve its objectives, which necessitated another revision and amendment to the master plan.

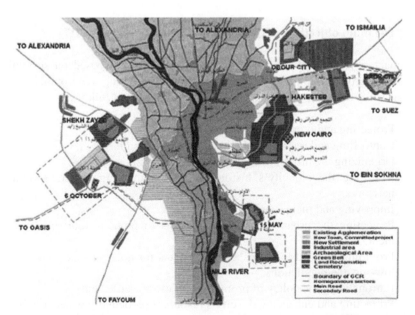

Figure 7 The GCR Urban Development Planning of 1991 (GOPP 1992).

2.4 The 1991 development planning for the GCR

Due to the failure of achieving the goals of the 1986 GCR plan, the GOPP decided to re-update the plan in 1991. The new updated plan had to deal with the miss-use of land by prohibiting the growth of the built area towards the east and north-east in order to protect agricultural land. It was also recognized that the green belt had to be established to limit the fast growing urbanization.

With respect to the population growth, it was noticed that the annual average growth rate decreased during the period between 1986 and 1991 from the annual average growth rate during the period between 1981 and 1986. The average annual increase of population was 208,000 inhabitants per year (between 1986 and 1991) compared to 95,454 inhabitants (between 1981 and 1986), while the population growth of Cairo City reached 104,000 inhabitants in 1991. Accordingly the overall population density decreased from 264 p/ha to 219 p/ha (CAPMAS 1996, GOPP 2000).

However, the urbanization phenomena continued to pose a threat to the country's scarce resources especially agricultural land. The urbanized area of Cairo City exceeded 5,166 ha during the period between 1986 and 1991. The annual average urban growth was 1,033 ha per year, which exceeded the annual average growth between 1981 and 1986, (which was 842 ha per year). Also, Cairo City Center continued to push the population away towards other parts of the city. Therefore, this had necessitated another revision of the GCR's development planning.

2.5 The main objectives of GCR development plan (1997–2017)

The main objectives of the 1997 development plan were based on sustaining economic growth, upgrading, and improving the living environment. The strategies for realizing these goals could be summarized as follows:

- Protecting and preserving agricultural lands.
- Controlling and restricting the unplanned expansion.
- Organizing the urban structure and providing public services.
- Restoring the old city's historical areas. Improving the transportation networks.
- Improving and increasing the infrastructure.
- Protecting the urban heritage as a means to promote tourism development.
- Providing planned locations and substitutes for housing especially for the low-income population.
- Rationalizing the policy of promoting industrial settlements.
- Protecting and enhancing the city's resources particularly water supply.
- Controlling and regulating the sources of pollution and disturbance.

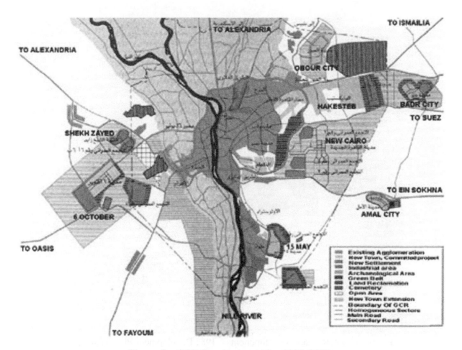

Figure 8 GCR Development Plan 1997–2017.

The following analysis indicates the components of the development plan of 1982–2000 compared with the components of the development plan of 1997–2017.

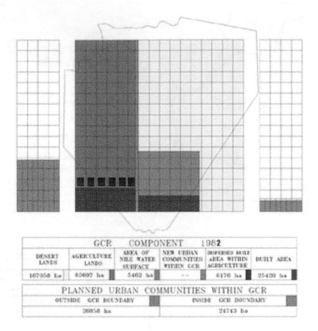

Figure 9   GCR Land use components – 1982 Development Plan Analysis.

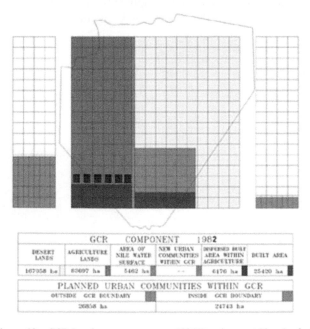

Figure 10   GCR Land use components – 1997 Development Plan Analysis.

The development plan of 1997 had included strategies and recommendations that attempted to improve the quality of life and protect the environment in the region through the following actions:

- Creating a peripheral way around Greater Cairo region.
- Building new urban communities on the desert lands adjacent to the urban agglomeration.
- Applying the homogeneous sector system as means for managing and controlling urban development. Such homogeneous sectors (16 sector) were determined in the original 1982 development plan as mentioned before.
- Determining new axes for the growth of the GCR's urban development.

The remainder of this section attempts to analyze and evaluate the recommendations and scenarios of the 1997 development plan and the new scenarios and decisions that emerged prior the 1997 plan particularly in a new revision of the plan in year 2000. The focus will be on the fundamental actions that constituted the main elements of the 'scenarios' particularly the establishment of a peripheral way around the Greater Cairo Region and the establishment of New Cairo City to strengthen the east-western development axis.

2.6 The action of creating a peripheral way around greater Cairo region

The first ring road, which was determined in the 1982 development planning, gave opportunity after its realization to open the desert areas to the urban extension. Such ring road was planned as a peripheral way around the GCR built area to limit its extension and to decrease traffic.

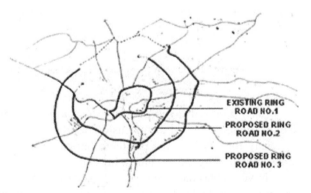

Figure 11 The Three Ring Roads.

Another ring road was proposed as a peripheral way around GCR at 30 km far from Cairo's city center to limit the new urban extensions east and west of the first ring road that was already executed, and at the same time, to directly connect the new urban communities without passing through the existing Greater Cairo's built area.

A third ring road was also proposed in year 2000 (50 km far from the city center) in order to facilitate the national-scale freight transportation and to avoid crossing the central urban area – forming with the former 2 ring roads a concentrically urban extension scheme, <u>which would be a very risky action because of the expected sprawl shaped extensions in all directions.</u>

2.7 Proposal to strengthening the east-west axis

Since the first ring road had succeeded in linking the desert areas with the urban extensions, it was proposed to strengthen the east-west axis by two north-south tangential corridors along the existing ring road. Such corridors would be linked by north-south links without building an additional ring road.

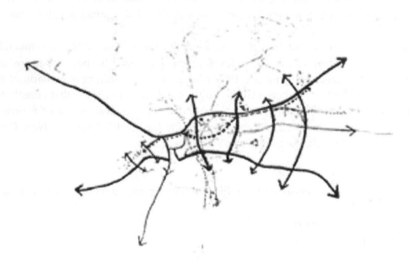

Figure 12   East-west axes with a regional mass transit system downtown.

Such proposal involves establishing a mass transit system (metro) at the regional scale and avoid the extension of urbanization within the protected areas especially the northern and southern agricultural lands and the archaeological zones in (including the pyramids) in Southwest of Giza Governorate.

In year 2000 the ring road at a distance of 50 km from the city center was approved so as to facilitate the national freight transportation. <u>There are risks involved as a consequence of this action</u>:

- A new boundary for the GCR would be established;
- The urbanized area within the new ring road would increase; and consequently
- Agricultural land that lies within the new boundary would decrease.

This calls for a new regulatory framework to enforce the protection of agricultural lands and prevent additional urbanization within the new proposed boundary of the GCR.

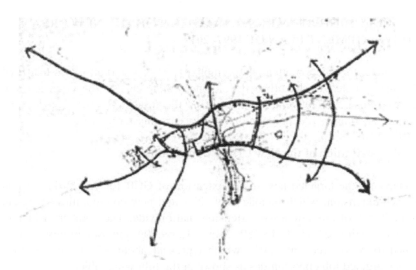

Figure 13 East-west axes with a regional mass transit system from the outskirts.

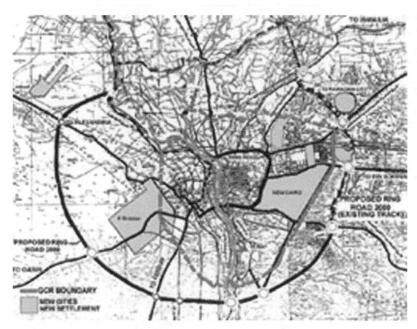

Figure 14 (lower) Proposed ring road number 3.

3   CONCLUDING REMARKS: EVALUTION OF THE NEW URBAN DEVELOPMENT PLAN OF 1997–2017

To conclude, this section attempts to emphasize five important issues that need to be incorporated in the urban development planning for the Greater Cairo Region to ensure effectiveness and to avoid the loss of the time, money and effort that accompany the preparation of successive master plans:

3.1   Redistribution of the New Urban Communities

By analyzing the targeted new urban extension of GCR for year 2017 we find that such extension, which include new cities and new communities, represent around 49.5% of the new urban extensions nationwide. The number of inhabitants was estimated at 8,113,250, while 1,269,009 job opportunities were planned to be offered until 2017 which represent about 42.40% of the total national targeted job opportunities as shown in the following table:

Economic Regions		New Extensions	%	Job Opportunity	%	Inhabitants	%
1	Cairo	740.00 km²	49. 5	1,269,000	42.4	8,113,250	50.9
2	Alexandria	168.50 km²	11.3	271,361	9.0	1,504,000	9.4
3	Suez canal	215.00 km²	14.4	492,750	16.5	2,152,000	13.5
4	Delta	64. 64 km²	4.3	179,000	6.0	770,000	4.8
5	North upper Egypt	128.82 km²	8.6	362,000	12.1	1,485,500	9.3
6	South upper Egypt	122.51 km²	8.2	262,000	8.8	1,232,000	7.7
7	Assiut	55.80 km²	3.7	157,000	5.2	690,000	4.4
	Total	1495.27 km²	100	2,993,111	100	15,946,750	100

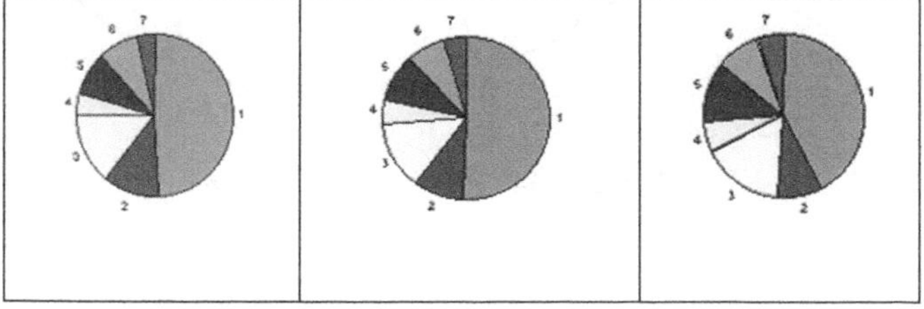

Figure 14   Policies for New Urban Communities with respect to Job opportunities and Population in the 1997–2017 Development Planning.

This means that the new urban extension will continue to be more and more centralized within Cairo region. Therefore it is of utmost importance to reformulate the national policies in order to redistribute the new urban communities, the job opportunities, and the inhabitants throughout the 7 economic regions, while special attention should be made for directing urban extensions to the southern regions. Also it is necessary to decrease the rural-to-urban migration and manage the factors that could realize the decrease of migration from the countryside to Cairo, (which ranges between 100,000 and 150,000 inhabitants every year). For example, government administration should be decentralized and allocated outside Cairo, and higher education institutions should be allocated in the new cities to attract people to settle there.

3.2 Effect of New Information Technology Revolution on Decentralization Policies

The revolution of the new information technology should be invested for directing decentralization policies in order to relieve Cairo as a mega city from high population densities, transportation, traffic and environmental problems.

However, Cairo as a mega city will be a major future component of the globalization epoch. Therefore, major activities and facilities of international magnitude should be considered in the future development plans.

3.3 Community Development Policies and Comprehensive Development Plans

Community development policies should guide and direct new urban extensions and new urban renewal policies in order to achieve integrated national development plans on the local, regional and national levels. Therefore, national public and private investment should be carefully directed to realize the welfare of the different Egyptian social groups all over the country and not to be only concentrated in major cities such as Cairo and Alexandria.

The rate of change of the city dynamics, city problems, urban fabric, and people's needs is faster than the time taken to formulate and implement policies that are prepared to cope with the problems in the cities. Indeed, the speed of change challenges the traditional method with which urban planning had been conducted. Planning, therefore, should be for sustainable development so as to cope with the fast changes in the urban environment, which has been fuelled by the fast change of technologies and of the revolution of sciences in all fields, which has its direct effect on humanity and the living environment.

3.4 Rehabilitation Development Plans as a Major Constituent of the National Urban Development Planning

Rehabilitation development plans should be carefully studied as important components of the national urban context. National development policies should be directed to solve the problem of informal housing in Cairo and Alexandria and in other cities in the Nile Delta and in Upper Egypt, especially because about 40% of Egypt's population live in informal housing (CAPMAS 1996).

Also it is of great importance to continue the national efforts, which started in the last few decades, which aimed at solving the problem of informal housing within the existing urban areas.

Finally, public and private investors should participate in implementing development projects in valuable locations, which are exploited in present day by squatter settlements and polluting activities. Low and medium income groups living in such valuable locations should be offered new houses of suitable rents. Mortgaging systems should be adopted as a tool for providing housing owner-ship's for these groups.

3.5   Abiding By the National Goals

A final remark that this paper attempts to point out is that the national goals are clear, rational and purposeful. It is, therefore, necessary to abide by the objec-tives of the national development strategy in all development plans, and pro-jects. Indeed, policies for directing urban growth of the GCR should be formulated to fulfill the three primary national goals, which are:

- Protecting agricultural land;
- Reducing population density, and
- Raising the standard of living for the population.

The above goals could be disaggregated into more specific objectives. The objectives are environmental, social, economical and developmental objectives.

The main environmental objectives are achieving the maximum use of natural potentialities, achieving ecological balance, protecting national reserves, and protecting the environment from water and air pollution.

The fundamental social objectives are working to raise the social, economical and cultural standard of living of the inhabitants, decreasing migration to the major cities and attractimg inhabitants to move to the new development areas.

The main economical objectives are protecting agricultural lands, encouraging horizontal extension of reclaimed lands, exploiting natural resources, encourag-ing private sector to invest in marketing Egyptian products, and decrease unem-ployment.

The main developmental objectives are erecting new urban communities in the desert, strengthening the connection axes between the existing urban areas and the new urban communities, determining priority urban development areas, facing the problem of squatter zones, emphasizing the relationship between the five year plans and the urban development plans, managing and controlling urban growth, and achieving the maximum return from archaeological, histor-ical, and touristic potentialities.

REFERENCES

Abouzeid, H., 2000
　　URBAN DEVELOPMENT IN THE NEW COMMUNITIES: AN
　　　　ECONOMICAL POINT OF VIEW, Proceedings of Al-Azhar
　　　　Engineering Sixth International Conference, September 1–4.

Ali, A., 2000
　　THE ROLE OF THE ECONOMIC BASE IN THE URBAN
　　　　DEVELOPMENT OF NEW CITIES, Proceedings of Al-Azhar
　　　　Engineering Sixth International Conference, September 1–4.

CAPMAS, 1966
　　CENTRAL AUTHORITY FOR PUBLIC MOBILIZATION AND
　　　　STATISTICS.

CAPMAS, 1986
　　CENTRAL AUTHORITY FOR PUBLIC MOBILIZATION AND
　　　　STATISTICS.

CAPMAS, 1996
　　CENTRAL AUTHORITY FOR PUBLIC MOBILIZATION AND
　　　　STATISTICS.

Executive Organization for Cairo's Development Projects, 2000
　　THE RING ROAD AROUND GREATER CAIRO, Ministry of Housing,
　　　　Infrastructure and New Communities.

GOPP, 1982
　　RING ROAD ACTION PLAN, Report Presented to the Ministry of
　　　　Housing and New Communities, General Organization for Physical
　　　　Planning.

GOPP, 1992
　　THE RING ROAD AROUND THE GREATER CAIRO REGION,
　　　　Ministry of Housing and New Communities Press.

GOPP, 1998
　　RECONSTRUCTION AND DEVELOPMENT MAP FOR EGYPT TILL
　　　　YEAR 2017, Ministry of Housing, Utilities and New Communities.

GOPP, 2000
　　GREATER CAIRO ATLAS, Center for Greater Cairo Region, Ministry of
　　　　Housing, Utilities and Urban Communities.

Kamal, M., 2000
        INSTITUTIONALIZING INVESTMENT IN HOUSING PROJECTS IN
                EGYPT, Proceedings of Al-Azhar Engineering Sixth International
                Conference, September 1–4.

Khairy, B., 2000
        SQUATTERING IN THE NEW COMMUNITIES, Proceedings of Al-
                Azhar Engineering Sixth International Conference, September 1–4.

Sakr, H., 1997
        THE RING ROAD: AN EVALUATION, Report Presented to the Ministry
                of New Communities, General Organization for Physical Planning.

Soliman, M. and Sharaf El Din, I., 2000
        MANAGING AND CONTROLLING CAIRO'S URBAN GROWTH,
                Proceedings of Al-Azhar Engineering Sixth International
                Conference, September 1–4.

DEVELOPMENT AND MANAGEMENT

The Creative City: Harnessing the Imagination of Urban Decision Makers

Charles Landry

The world is changing dramatically in ways that amount to a paradigm shift affecting the role of cities and how we run them. So we need to re-think and re-assess the purpose and objectives of urban policy for the 21st century. The idea that the future will resemble the past is long gone. Our capacity to cope is stretched when so many key ideas and ways of doing things are changing at once. We need new skills well beyond new technological literacy, including new approaches to thinking about the city.

It is no wonder the world is confusing for the urban manager. The anchors of our lives are transmuting before our eyes, affecting how we plan our cities: How we *create wealth* is more to do with manipulating data into knowledge and added value and less with manipulating material goods. Competition has moved from immobile, physical resources like coal or gold towards the new gold of brainpower and imagination transacted through cyberspace. What is the role of the city in a cyberworld? It highlights the city's role as a centre for logistics, trade and ideas exchange, whose quality of life requires an ambience that is both buzzy and encourages the unexpected yet is safe and predictable. *Governing and governance* is recognized as being less effective when run by hierarchies. Adapting has revolutionised government and civil society institutions as they have moved to flatter structures, devolved responsibilities, partnership structures, networks or even virtual organisations. Pre-prescribed *lifecycles* are breaking up as society moves away from the education–work–retirement continuum towards lifelong learning and portfolio lifestyles. Learning environments in this context go well beyond the school and will include, for example, any cultural institutions from the museum to art gallery affecting in turn their programming and what a local authority can legitimately fund. With global markets, the 24-hour city is already with us transforming our notion of *time*. Instantaneous communications in real-time across time zones brings in its wake changed working patterns, yet most cities are still locked into a 9 to 5 rhythm. Cyberspace and virtuality in turn shatters our concept of *place*. Good quality places, which are distinctive and with which people can identify, become even more important for once sedentary people who are now hyper-mobile travelling between multiple real and virtual locations. Our expectations too of what urban spaces can look and feel like is affected by the convergence of word, image and sound and the new forms of interaction they make possible.

The above affects our concept of *thinking* itself. The 'modernist' idea that everything progresses in a regular linear progression has broken down with an increased awareness of 'holism', with its belief in the intrinsic connectedness of everything. This reshapes fundamentally our idea of *development* away from the image of the city as a machine towards that of a living organism. This in turn shifts our focus away from a concentration on physical infrastructure towards urban dynamics, overall well-being and health of people and their lived experience of the city. So the idea of sustainability broadens out too beyond the environmental to be understood in economic, social and cultural terms too. As a result we need to broaden our understanding of *measurement and evaluation* beyond the limitations of the simple financial calculus .

The world of city planning and management has not as yet 'stood back' and fully assessed the implications of these new conditions. But the new will envelope the old. The emerging post-industrial system lives side by side with the old, the pre-industrial and industrial. Much will stay the same—people will still take buses and cars to work and houses will still look like houses—but the inner logic of the knowledge-based economic system will increasingly frame industrial society. Old patterns of behaviour will overlap and collide with the new needs which emphasize fluidity, portability of skills and adaptability.

Yet why have some cities been able to surf the waves of change and make them work for them, whereas others seem to be passive victims of change, simply allowing it to happen to them? Cities like Barcelona, Melbourne, Seattle, Vancouver, Helsinki, Bangalore, Ahmedabad, Curitiba, Rotterdam, Dublin, the cluster along the Emscher River in the Ruhr in Germany or around Zürich, Karlsruhe, Strasbourg spring to mind.

Successful cities seem to have some things in common—visionary individuals, creative organizations and a political culture sharing a clarity of purpose. They seem to follow a determined, not a deterministic path. Their planning is strategically principled based on an ethos of how they want to run their city, yet they are tactically flexible and anticipatory. There is open-mindedness and a willingness to take risks and say 'yes' rather than 'no'; a clear focus on the long term; a capacity to work with local distinctiveness where standardized solutions give way to unique, locally diverse responses often finding a strength in apparent weakness. Key is change in mindset and organizational culture, ever more necessary as we cannot solve 21st century problems with 19th century mindsets: where central power is more devolved; where institutions seek less to control and more to influence and instead of working in isolation partnership is preferred; where leadership is widespread permeating public, private and voluntary sectors and seen as much about enabling participation and encouraging commitment to place as about taking decisions; where a low risk/high blame culture is supplanted by a higher risk/lower blame one; and finally where there is an awareness of the interdependence between economic, social, cultural and environmental factors and recognition of how inadequate narrow functional specializations are in solving urban problems. In sum a culture that allows us all to think creatively, plan creatively and act creatively.

But the urban manager has a thankless task. Pushed and pulled in many directions by interconnected problems over which they have no control as they seek to match individual desires with political, social and budgeting priorities. For example everyone wants a car, but without reinventing what a car is, the problem of pollution can only increase—not to mention ambient pollution like light, noise, traffic jams and parking problems. Developers' demands can result in skyscrapers blocking out light or distorting historic townscapes. Effluents souring the air or poisoning water supply—the environmental list could go on.... The extra stresses are largely borne by the worst off who cannot afford to live in pleasant places. Such frustrations with poverty and unemployment which breed hopelessness, unfulfilled expectations, and boredom can change whole areas into ghettos with self-reinforcing cycles of deprivation. Meanwhile the rich can create their own ghettos to protect themselves from the perceived or real threat of the poor, which is even evident in large British cities like London, Manchester or Newcastle.

Yet the urban utopia already exists in the dispersed experience of global best practice. Taking an eagle eye view of cities around the world it is astonishing how many ordinary people show leadership qualities to make the extraordinary possible when given the chance. The network of local exchange trading schemes (LETS) around Britain where people barter skills and services is an innovative way of creating employment; South Shore Bank in Chicago fosters business by profitably providing banking services to the deprived; Tilburg's or Porto Alegre's citizens budgeting process has made urban governance more efficient as well as more accountable. Helsinki's digital city applies technology imaginatively so you can find your way around a city's services on the net; Berlin's 'rent a granny' scheme unleashes forgotten undervalued skills of the elderly. There is inspiring architecture that speaks to a city's soul and identity, as the 'township Bilbao' the Gugasthebe Arts Centre in Langa in Cape Town jointly conceived by local architect Helen Smuts and the local community; there are clever energy saving devices such as the eco-house in Leicester or using George and Harriet a pair of hawks as pigeon predators in Woking town centre so saving costs in clearing faeces. There is public transport that is a joy to use as in Strasbourg's tram system. There are retail environments that merge entertainment and learning as in Amsterdam's network in intimate lanes around the Grachten or its imaginative solution to lack of space, too much water and too many cars—floating carparks; and public spaces that encourage urban buzz and celebrations that capture the unusual, the uplifting and the creative as around Rome's Piazza Navona, which is creative by staying as it always was. Widely dispersed they are hard to see or learn from. We see instead the city as a place of fear, crime, pollution and degradation.

We know how to transform the buildings, it is the easy bit. We have tried and tested formula for reinventing physical infrastructure to new purposes but we have little experience in changing our mindset. A defunct power station turns into a national museum; the under-used St. Paul's church in Walsall into a shopping and entertainment's centre, with worship squeezed into the rear. Emblematic transformations capturing the mood of the post-industrial age, where increasingly the city seeks to feed inspiration and aspirations. Factories and industrial warehouses have been swept away or re-used as housing, offices, high-tech studios as the physical urban fabric is wrenched into place for new functions.

The primary task now is to create the conditions for innovative problem solving in Britain. This requires a new approach to urban planning which helps open out an ideasbank of possibilities from which innovations will emerge. In passing through transition mistakes are inevitable. This is the cost of living through a paradigm shift on the scale of the industrial revolution. Solutions depend on experiments that can be managed by organizations open to new ideas.

For cities to find their new niches to survive in the new economy consciousness of culture will be key. Culture is the values, way of life and talents of those living in a city expressed in different ways. Culture is what shows that a place is unique and distinctive representing assets and resources to be exploited replacing coal and steel. Culture is the soil and raw material which by the creative method can be exploited and provide the momentum for development. Even cultural heritage is a resource that is reinvented daily whether this be a refurbished building or an adaptation of an old skill for modern times: today's classic was yesterday's innovation. A cultural approach to urban strategy involves looking at each area like economic development, social affairs or housing culturally.

In health we can ask: Are there indigenous health practices we can build on that might foster preventative care? In social affairs: Are there mutual aid traditions that can be adapted to provide support structures for drug users or lonely elderly people; or alternatively to kickstart the setting up of the increasingly popular local exchange trading schemes (LETS) where people barter skills and services? In terms of job creation we could undertake an audit of older craft skills in a city and assess how they can be attuned to the needs of the present. We could participate in the enthusiasms of the unemployed youth and see whether economically viable businesses can be created from their pastimes. To attract tourists we could scan history and traditions and seek to rediscover local cuisine or craft potential that could help brand the city. One could invent celebrations or congresses that chime well with a city's aspirations for the future, yet build from the soil of the past. Educational institutes can be looked at afresh to assess whether circumstances are providing triggers for action. Derry in Northern Ireland, for example, used the fact that it was a centre of the 'troubles' to create the world renowned Centre for Conflict Resolution. Looked at so every facet of culture from history or contemporary events; a quirky circumstance or how a city has dealt with its topography can be seen as a resource to be turned into an opportunity; a tradition of running an arts festival can be metamorphosed by using the core skill of organizing to become an all year congress industry as in Adelaide; a style of music or sound can be branded by a city thus transform a city image or develop a cultural industry. The approach thus does not look at policy sectorally, such as policies for the development of housing, theatre or transport. Its purpose is to see how the pool of cultural resources identified can contribute to the integrated development of a locality. The key issue is not so much their specialism, but their core competences—the capacity to think creatively, open-mindedly, laterally across disciplines, entrepreneurially and to be managerially and organizationally competent.

We need to develop a new idea of urbanism, the discipline which helps understand the dynamics, resources and potential of the city in a richer way. This means looking at the city from different perspectives to gain more penetrative insights. By seeing the city through diverse eyes potential and hidden possibilities from business ideas to improving the mundane are revealed. Traditionally urbanism has been dominated by planners, architects and urban designers thus focusing too exclusively on physical solutions. The new urbanism should draw on the insights of cultural geography; urban economics and social affairs; urban planning; history and anthropology; design, aesthetics and architecture; ecology and cultural studies as well as knowledge of power configurations. Each discipline should be seen as an equal partner, the synergies can be immense.

Planning is a generic concept that applies to any activity with an intention to achieve an objective methodically. Yet in the urban context it is nearly exclusively associated with land-use and development control important as this function is. It causes severe problems. It implies planning falls under the jurisdiction of the land-use specialist putting them towards the top of the hierarchical tree, so downgrading the power of people based departments from business start ups to community development specialists. There is a danger that focusing on land use planning oversimplifies issues and looks at problems through one lens. Finally the tight association of planning with land-use is the source of the major criticism that planners do not have a sufficiently broad base of knowledge. If land-use planners are to maintain a pre-dominant role it is essential they understand more about culture, history or social dynamics as well as acquire new skills concerned with understanding networks and communications dynamics. This greater focus on soft infrastructure highlights concern with issues of ambience, feeling and behaviour. It means land-use planners should be equal members of a team.

To identify new niches a city can develop requires planners to be more creative. Civic creativity will be key. The 'civic' and the 'public' have come in for a battering over the last decades. A string of negatives are associated with them: bureaucratic, red tape, hierarchical, inefficient, social welfarist, lacking in vision, machine like spring off the tongue. They are linked to underachievement, lack of strategic focus and failure. Yet for most of the last century they stood for self-development, social improvement and modernisation. Can these purposes be renewed within a 21st century framework 'Civic creativity' is imaginative problem solving applied to public good objectives. The aim is to generate a continual flow of innovative solutions to problems which have an impact on the public realm. 'Civic creativity' is the capacity for public officials and others oriented to public good aims to effectively and instrumentally apply their imaginative faculties to achieving 'higher value within a framework of social and political values'.

This might need new people in charge and certainly new training. What is the philosophy, purpose and future role of urban planners and what should their education be. The 'team' that plans cities, both administrative insiders and the strategic private sector or community groups outside, should combined represent a diversity of roles: facilitators, visionaries, leaders, public servants, investors, advocates and technical specialists, with a wide range of intellectual resources to match. Planning should be a vision driven process and deliberative, rather than narrowly technical. Planning can be effective in a market economy if it is underpinned by a principled ethos with regulation at a minimum and a forceful and ingenious incentives structure focused on taxation and other fiscal control measures. Planning needs more power and less power. More to create imaginative incentives to get the private sector and other actors to contribute and less of powers that stifle creative action. The urban strategy making suggested is thus both much broader in scope than the classic town planning idea involving a wider range of expertise, indicators and criteria for what it is trying to achieve and also more strategic in that it is more conscious of eco-system implications, regional impacts and creative needs.

So what of the next wave of creativity? The conditions for future urban success will be determined just as *much by software solutions as by hardware solutions* to generate wealth and social cohesion. How we manage and organize the complex logistics of urban life; how we govern and lead cities; how we create atmosphere, feeling and a milieu where ideas exchange and chance encounter can emerge requires many imaginative solutions rather than technological fixes. We need to develop projects that '*add value and values simultaneously*' as the emerging economy requires an ethical value base to guide action. For example, projects such as 'Dorset Reclaim furniture recycling' in Poole that creates employment, training and a sense of community as well as adding value to the seemingly useless. We should see urban potential through different eyes. By '*valuing varied visions*' we can bring together the creative contributions coming from the young or the old; from women or men; ethnic majorities and minorities; the well-connected or the excluded and to harness each form of cleverness and imagination to the overall goals of the city. Finally we should '*recombine the old and new imaginatively*'. Easiest and most visibly impressive are the interventions in the built fabric, such as transforming Bankside into the Tate Modern, inserting the Pei pyramid into the Louvre, placing an intricate wire structure on top of the Tapies museum in Barcelona or the conversion of a munitions factory in Karlsruhe into a massive Centre for Media Technology. History can be mined for the future in myriad other ways. It is an asset and erasing memory for its own sake a waste. Yet the past is left too exclusively to antiquarians, nostalgics and historians. The presence of the past in the present gives a sense of weight and significance to any urban endeavour and the constructive clash and recombination of past and future can reap unusual rewards. The art of blending new and old provides immense scope for action: From re-capturing a city's period of intellectual prominence thus giving a sense of lineage to cope with the knowledge economy to re-visiting big urban questions such as what does the good life in the city feel and look like.

Based on *The Creative City: A toolkit for Urban Innovators* by Charles Landry. Published by Earthscan, available from EARTHSCAN FREEPOST, 120 Pentonville Road, London.

DEVELOPMENT AND MANAGEMENT

Emerging Trends in Retail and Entertainment Developments

Ro Shroff

The Retail and Entertainment industry has been going through a noticeable evolution over the last decade. From Melbourne to Madrid to Manhattan there is continual experimentation of newer models of developments, obfuscation and overlapping of several segments of the industry and births of new and exciting attitudes and trends. Retail and entertainment based developments have historically acted as catalysts and precursors for economic growth. As the world gets more urbanized and cities continue to be the experimental grounds for changing lifestyles, it is an appropriate juncture to step back, evaluate and prognosticate as to what these developments mean and portend. Trends generally tend to have different life spans, ranging from generations to short term fads. The retail and entertainment segments of the commercial development industry incorporate fairly long gestation periods prior to actual building. As such, trends tend to lag behind the social and technological changes that precipitate them. It is also equally possible, that the notions articulated below may no longer be valid or applicable a year from today, given the fast changing dynamics of new generations of shoppers.

SHIFTING PARADIGMS

Although the term Retail/Entertainment conjures up a singular entity, it is in fact an amalgam of several different components. They range from the much revered, imitated and maligned American model of the Shopping Mall to the Urban Entertainment Center, from Hypermarkets to upscale Department Stores and from themed environments to branding. The industry "mantra" today veers beyond just the physical environment and is succinctly articulated in the recent book "The Experience Economy" as "events that engage individuals in a personal and memorable way". Today, the sequence, ceremony and choreography of arrival and the mystery and anticipation of the journey to the destination are as important to consumers as choice, value and convenience. Similarly, service, scripting and synergism are being perceived of on par with merchandising and leasing strategies. "Story telling" has become a very important facet in the design and development of these facilities and the concepts of "bundling", "branding", "co-retailing", "precincting", "placemaking" and even trademarked names like "shoppertainment" are becoming the precursors to physical planning and architectural design. The emphasis is as much on communication, pulse and emotion as it is on physical form.

ALL THE WORLD'S A MALL

The Shopping Mall still remains an industry icon throughout the world. Pundits and prognosticators have, for a while, been citing the death of the shopping mall and the De-Malling movement as the next trends in retailing. The fundamental economic formula of the shopping center and its basic premise as a centralized repository of fashion, recreation and entertainment within a safe and controlled environment has not changed since its inception. However, the building type as we know it, continues to evolve into more dynamic and fluid versions. Starting with Country Club Plaza in Kansas City in 1912 to the first enclosed shopping mall at Northland Center near Detroit in 1954 to the seminal Horton Plaza (The Jerde Partnership) in San Diego in 1985, this American "invention" continues to evolve into several hybrid models.

Two recent projects articulate the trends and attitudes that are being currently explored or reaching maturation. They are Flatiron Crossing (Figures 1–3) outside Denver designed by Callison Architecture, and Bluewater, near London, designed by Eric Kuhne Associates. Flatiron Crossing, a 1.5 million square foot regional shopping center completed in 2000, defines an emerging and popular trend within retail developments of combining climate-controlled components of the conventional model with a reorientation to the exterior environment. The project incorporates a "village" offshoot from the central node populated with boutiques, restaurants and theaters reinforcing the "outdoor" ambiance and health conscious lifestyle of Colorado. The two storey mall itself exudes a resort like character with voluminous spaces appointed by natural materials and imbued with hospitality oriented amenities. The most significant feature of the design is the substantial transparency, up till now unknown for this building type, as well as a seamless connection to the exterior through large glazed walls and operable "garage" doors, thus minimizing the anonymity of the hermetically sealed environment. Overall, the project pioneers new territories in the layout and design and has evolved into a veritable trendsetter in the U.S.

An interest in urban living, a health conscious and relatively prosperous population and a need for community has encouraged a lot of retailers and shoppers to mandate exterior orientations in new developments. It does not hurt that this strategy defines better opportunities for tenant identity, contiguous parking and lower common area maintenance charges. Whether this evolution is ultimately a rebellion against the contained anonymity of shopping malls or a reflection of a more urbanized society, the shopping center prototype continues to be redefined and repositioned. Some projects like the recent Mall of Georgia (TVS Associates) in Atlanta have experimented with this hybrid model on a much larger scale with a veritable "Main Street" at its front door. How this format eventually evolves in terms of economics and mix of components, remains to be seen.

Figure 1 Flatiron Crossing, Broomfield, CO.

Figure 2 Flatiron Crossing.

Figure 3   Flatiron Crossing at night.

Bluewater, completed in 1999, on the other hand exemplifies the quintessence and maturation of the American model, ironically in England. The two million square foot center within a triangular configuration encapsulates all the perceived and expected attributes of the building type in an elegant and contemporary European vocabulary resplendent with voluminous spaces, themed courts, "branded" mall concourses as well as three intermediate entertainment nodes. These include an outdoor component called "Thames Walk", a "Winter Garden" and a "Water Circus", each articulated in varying architectural idioms and materials ranging from a historically accurate recreation of a Kent village to a lagoon and finally to a high tech crystalline conservatory. Sited within an existing chalk quarry, the project incorporates three upscale department stores and a collection of exclusive retailers and each of the three concourses are provided with ecologically sensitive skylight systems. Overall, although the project does not break any new grounds, it is accomplished with vigor, finesse and exuberance and has achieved somewhat of a cult status within the industry as the epitome of present day retail and entertainment.

INSIDE – OUT

Many semi urban or urban models of retail developments within the U.S., Europe and Asia are experimenting with the preference toward more eclectic styles and less isolated contexts. North Bridge (Anthony Belluschi/OWPP) in Chicago and Denver Pavilions (ELS/Elbasani and Logan) in downtown Denver are two recent projects that explore the "introversion" and "extroversion" of the retail and entertainment environment within vibrant urban patterns. The former is a three level enclosed Galleria and a primary component of a six block urban redevelopment off Chicago's famous Michigan Avenue while the latter is a two level open urban specialty retail center on four downtown blocks. Both projects celebrate the urban feel and format but differ in the way they engage the city fabric. North Bridge connects a hotel, an entertainment center and a major Department store with a curved climate controlled glazed link reminiscent of European arcades and provides a multi block link bridging over a highly traversed avenue. Denver Pavilions meanwhile, engages the urban contexts head on and defines a spatial district with predominantly lifestyle, food and entertainment tenants identified with bold forms, signage and graphics within a highly visible open air environment.

METROPOLIS

Cities like Berlin have on the other hand encouraged a truly urban experience by facilitating a critical mass of mixed-use developments with varied retail and entertainment components within a few contiguous blocks. Developments like the Sony Center (Murphy Jahn) and the multi block Debis (Richard Rogers Partnership) project around Potsdamer Platz have breathed new life into this previously no-man's land. The re-developments of Times Square and 42nd Street in New York and the Hollywood and Highland (Altoon + Porter) project in Los

Angeles have or will soon completely change the perception and commercial viability of once seedy and dangerous parts of these cities. Today upscale retailers, refurbished theaters and Class A office towers and hotels coexist in harmony and encourage further redevelopment within surrounding districts.

Development of dense retail and entertainment related facilities adjoining tall buildings within urban environments is also going through a stage of rejuvenation in spite of regional economies and international market downturns. The Time Warner Center (SOM) in New York, KLCC Galleria (Cesar Pelli Associates) in KL, Taipei Financial Center (C. Y. Lee/FRCH) in Taipei and the Grand Gateway (Figures 4 and 5) in Shanghai are recent examples of assimilating multi level retail and entertainment uses within their bases. Although not strictly a new trend, the concept of consolidating the components of "live, work and play" within the urban context continues to make economic as well as programmatic and urban design sense. Simultaneously, projects like Star City in KL and Core Pacific Plaza (The Jerde Partnership) in Taipei define the continuing trend of the mega mall within a dense urban context.

Figure 4   The Grand Gateway at Xu Hui – Shanghai, China.

Figure 5   The Grand Gateway at Xu Hui – Shanghai, China.

SPORTS, PORTS AND RESORTS

A concurrent phenomenon is the development of large scaled Urban Retail and Entertainment Districts in conjunction with sports stadiums and convention centers. Cities like Los Angeles and Cincinnati as well as Seattle and Seoul, all with planned or built sports stadiums and arenas have taken on the development of these urban entertainment districts interspersed with cultural facilities, museums and residential components. Appropriate merchandising strategies are prerequisites for these concepts to work and invariably they focus in on food and leisure related merchandise with some lifestyle retailing thrown in. Like airports, retail and entertainment venues within these contexts have the distinct advantage of a captive audience of thousands. Sponsorships and co-branding and naming rights simultaneously generate significant revenue streams.

New International Airports meanwhile, have integrated veritable shopping centers within the confines of the terminals, complete with gourmet restaurants, fitness centers and movie theaters commanding inflated rents and delivering high sales per square foot. Entire mini cities including shopping and entertainment centers, hotels and convention facilities are planned outside or over transportation hubs as exemplified by projects like The Grand Gateway in Shanghai, Kowloon and Hong Kong Stations in Hong Kong and Eurolille in Lille, France. The incorporation of major retail and entertainment components within or along with major resorts is also becoming more popular. They exude a relaxed mood of tourist destinations and incorporate a mix of exposition and theme park related facilities. Several such "Retail and Entertainment Resorts" are planned in the U.S. and Asia. Examples include Ala Moana Center (Figure 6) in Honolulu, the highest grossing shopping center in the U.S., Pointe Anaheim, a two million square foot hotel and retail resort outside of Disneyland and Emirates Experience, a major retail, hospitality and theme park based development in Dubai, all designed by Callison Architecture.

LUXURIES, LIFESTYLES AND VALUE

The consolidation of the luxury goods market and the proliferation of Value Retailing are trends at opposite ends of the spectrum, both having an impact on the Retail scene. Conglomerates like LVMH and Pinault Printemps Redoute are in the process of acquiring or having acquired major luxury brands from Gucci to Prada, Fendi, Donna Karan, DFS and most recently, retailing arrangements with the world's premiere diamond company De Beers and even venerable auction houses like Sotheby's. The luxury goods market is thereby leveraged in favor of a contained group of conglomerates and continues to generate higher profit margins while imparting a substantial cachet and prestige to the developments they are in. Meanwhile hip European fashion retailers like H & M and Zara as well as formally exclusive and high brow department stores like Harrod's and Harvey Nichols are expanding in places as far apart as the U.S. and Saudi Arabia.

The Department store industry itself is going through financial turmoil across the globe, primarily due to competition from discount retailers and ending up with mergers and consolidations. Although they continue to remain the mainstays of traditional shopping centers their leveraging ability is getting substantially compromised and a new breed of "anchorless" shopping centers are cropping up in several markets. International Department store chains like Sogo, Daiei and Seibu in Japan, Marks and Spencer in England and David Jones in Australia are experiencing similar problems in terms of growth and severe competition from more nimble retailers. Meanwhile, "Lifestyle" retailers catering to an urbanized population are also becoming the mainstays of these specialty shopping centers. Tenants like Crate and Barrel, Pottery Barn and Restoration Hardware in the U.S. and Conrans in London as well as large format books, music and sports stores continue to demand strategic locations and generate substantial foot traffic while paying less than average rents.

The growth of "Value Retailing" on the other hand is exemplified by the big box Hypermarkets all over the Americas, Europe and Asia. Large format stores like Wal-Mart, Costco and Carrefour along with large circuitous complexes like the various Mills projects in the U.S. catering to discount goods, are now ubiquitous from the heartland of the U.S. to the far corners of Asia. They offer a range of goods, groceries and services at leveraged and discounted prices

Figure 6   Ala Moana Center – Honolulu, Hawaii.

but in large freestanding formats that are a bane to issues of sustainability and continue to chagrin urban planners and city fathers. However, recent trends, especially in Europe suggest the incorporation of these stores within traditional shopping center formats, often replacing conventional anchor stores. Interestingly, one also finds a newer trend in the assimilation of both upscale and value retail within the same development, thus catering to a large spectrum of buying power and income groups simultaneously. The stratification of these categories of retailing is hence becoming very nebulous as shopping habits of the wealthy and those less prosperous tend to seemingly overlap or cross over.

CLICKS AND BRICKS

Much has been written about the impact of Internet based retailing on traditional shopping habits. In one sense, E commerce has been in existence for a long period of time starting from the days of catalogue sales. According to industry watch dogs the sheer scale of traditional retailing and entertainment is so huge that the fifteen billion dollars spent on non automobile related "E-tailing" in the U.S. last year constitutes less than one percent of the amount spent on "bricks and mortar" retailing. Within the next five years however, this amount is anticipated to increase to five percent of the expected $2.5 trillion retail industry in the United States. Interestingly, with NASDAQ still in turmoil and the sudden demise of several Dot.com companies, it's the traditional retailers like Barnes and Noble and Toys R Us, who are having the largest success stories with E commerce, the latter actually collaborating with Amazon.com, the world's most significant Internet based retailer. Some of the largest mall owners and REITS in the U.S. like Simon and General Growth are developing their own web shopping networks and collaborating with entertainment conglomerates like Time Warner to compete with the electronic retailing.

IF YOU BUILD IT, WILL THEY COME?

The cineplex industry is currently going through a major corrective stage in the U.S. after a period of phenomenal growth. Major U.S. theater operating chains like AMC, Lowes and Regal Cinemas, suddenly halted major expansion plans in mid 2000 after Wall Street clobbered their stock primarily due to a glut of obsolete and non performing assets. As technology makes the movie going experience more enthralling than ever, several older theaters are no longer able to compete with digital sound systems, stadium seating and gourmet concessions of newer multi-screen theaters. The optimistic prognosis is that once the corrective medicine is administered, the cinema industry will once again flourish to cope with Hollywood, Bollywood and Hong Kong, each churning out more output than ever before. Realistically it may take some time to achieve the industry position theaters enjoyed in the last decade. Meanwhile, large format theaters like 3D IMAX as well as video and electronic media oriented entertainment providers like Gameworks and DisneyQuest try to fill in the gaps wherever possible.

LIVING TO EAT

The dining experience has become an integral part of most Retail, Entertainment and Leisure driven developments and in some cases exceed 50% of the leasable areas. Beyond the quintessential food court, the industry is more focused on upscale dining opportunities, themed restaurants (despite the demise of Planet Hollywood) and "Food Precincts", a clustering of different types of restaurants and F and B outlets. Although the food court will continue to be a programmatic component, recent trends point more towards a leisurely experience and Al Fresco dining rather than the pursuit of fast food. The refurbishing of inner city districts like Clarke and Boat Quay in Singapore, Lan Kwai Fong in Hong Kong, 3rd Street Promenade in Santa Monica near Los Angeles and Cockle Bay Wharf (Eric Kuhne Associates) in Sydney suggest a long term trend of clustering dining and upscale food and beverage tenants within entertainment driven projects.

BRAND NEW

The creation of a "brand" today goes beyond products, conglomerates and themes. The "branding" of an experience takes as much precedence as that of a product. Coca-Cola, Disney and Nike are well-known product brands but today entire developments and designs are "branded" to appeal to targeted patronage. In a sense, "Branding" reflects the notions of familiarity, name recognition, advertising and repetition. Establishing Bluewater as a recognizable brand within the retail industry was achieved by communicating a striking series of four images: red roses, blue water, green grass and brown beans. The branding strategy was designed to arouse the senses and bring associations that were intended to convey the quintessence of Bluewater. Another part of the strategy was to work closely with the merchandising concept in developing unique and innovative strategies like the "male crèche", a cluster of targeted stores catering to male fashions and sports in close proximity to "female" stores.

The Irvine Spectrum (RTKL) and The Block at Orange (CommArts/ DaIQ) are two "entertainment" driven retail projects in Southern California also branding their designs and name recognition. The former takes a cue from Las Vegas and alludes to exotic marketplaces of Morocco and the Alhambra and the latter encapsulates the Hollywood culture through bold signature graphics and icons. Resort projects like Sun City in South Africa and Atlantis in Bermuda (WATG), developed by billionaire Sol Kerzhner define their own branded identity, whereas developments such as Universal City Walk (The Jerde Partnership) in Los Angeles extract and expand their brand from the name recognition of the movie studio. Similar examples abound in Asia, where a successful project and identity is expanded into several other cities.

Another impact of branding within the retail and entertainment industry is one of "star" architects and designers obfuscating the boundaries of product, art, sculpture and theater. Architects like Frank Gehry, Rem Koolhas, Philippe Starck, David Rockwell and Michael Graves are infusing Retail design with exuberance, vitality and dramatic flavor and essentially "branding" the products and environments they design. From the new prototypes of upscale fashion stores for Prada to flatware and utensils at Target, the trend is one of imparting the industry with flair, drama and theatricality. Disparate building types from casinos to restaurants, museums, airports and zoos, are enjoying a dose of idiosyncrasy and quirkiness. The recent Experience Music Project (Frank O Gehry) in Seattle personifies the ultimate confluence of art, sculpture and entertainment while the Sony Metreon Center (Simon Martin-Vegue Finkelstein & Morris) in San Francisco, the Rose Planetarium (The Polshek Partnership) in New York and the Millennium Dome (Richard Rogers Partnership) in London redefine the perception of conventional entertainment with a "branded" identity. When museum directors talk about the "guest experience" and retailers refer to "immersion" and "narratives and content", the era of branding has arrived.

LEARNING FROM LAS VEGAS

Perhaps no other city exemplifies the "Branded" Experience more than Las Vegas. Once the Mecca for Gaming, the city has transmuted into a vacation destination for families, complete with relatively affordable accommodation, an abundance of fine dining and the highest concentration of retail and entertainment venues in the world. Between the newly opened resorts over the past three years, the city boasts of 125,000 hotel rooms and more than five million square feet of retail and entertainment related spaces within a three mile stretch. Many trends experimented within the confines of this city are having repercussions globally. From the recently opened resorts and entertainment destinations like Bellagio, Desert Passage, Venetian and Paris, to the upcoming expansion of Fashion Show, Las Vegas offers an array of some of the world's most exclusive retailers and entertainment providers. In addition, there are journeys through exotic destinations, $30 million Dancing Fountains, Siberian Tigers, volcanoes and sinking ships and a 75% scale replica of the Eiffel Tower. Although themed environments have become commonplace everywhere, the Vegas model takes the phenomenon to a pinnacle and may ultimately tend to totally displace architecture with theatrics and "authentic" artifice. "Learning from Las Vegas" the well-known 1972 book and thesis by Robert Venturi explored the ubiquity of bold signs and "decorated sheds" as icons personifying Vegas. It suggested that architects be more receptive to the tastes and values of "common" people and be modest in their erections of "heroic, self-aggrandizing monuments". Las Vegas today has not only come full circle from this observation but also gone on to legitimize quirkiness, opulence and extravagance.

Interestingly, although Las Vegas is not the ultimate paradigm of "taste" in this industry, the city has been the recent recipient of venues for The Guggenheim Museum and Russia's venerable Hermitage. It may not be surprising if MOMA, the Louvre and the Tate may some day have their own exhibits in this city where the blurring of art, culture and excess is a flourishing phenomenon. Ironically and unfortunately though, Las Vegas is also being "commodified " into an entity much in demand with clients from third world countries bankrolling the recreation of their own proud histories and cultures within a "Vegas" garb.

ARCHITECTS, AUTHORS AND ARTISTES

The Design profession is continuously adapting to and redefining the changing expectations and nuances of the retail and entertainment industries. Whether it is by the written word or tome or the perception and appreciation of the built environment by the lay person, architects and designers are constantly at work to explore new thoughts and technologies. Retail design has traditionally been the bane of the architectural profession and justifiably so over the last few decades, wherein the design of shopping centers and outlet stores deserved little recognition. Today, however, with the blurring of retail and entertainment design with art, sculpture, advertising, industrial design, fashion and Hollywood, the design industry is achieving higher echelons of recognition. Design attitudes now range from the Disney inspired theatricality to utmost minimalism especially in the design of upscale fashion stores from London, Milan and New York.

Whether all this is manifested by internationally recognized design firms like The Jerde Partnership, Callison Architecture, RTKL (ID8), CommArts and the Rockwell Group in the U.S., BDP in London and The Buchan Group in Australia, or by minimalist designers like John Pawson, retail design expertise has been gaining substantial legitimacy worldwide and is in great demand. At the same time, architects like Rem Koolhaas, MVRDV or Philippe Starck continue to design, write and experiment with newer ideologies and pursuits. The peripatetic Rem Koolhaas is single-handedly involved in three concurrent ventures dealing with the shopping experience, including the upcoming book "The Harvard Guide to Shopping", the exhibit and publication "Mutations" and the design of high profile stores like Prada. Fashion and Architecture for once, seem to be more intertwined than ever before.

IS THERE A THERE, THERE?

What does this all mean? The emerging or maturing trends as articulated above are meant simply to familiarize us as to the disparate types and functions presently existing in this segment of commercial development. Some trends take on more permanence and most fads whither away. Ultimately, development trends are a direct function of the global economy and fast changing technologies. "Bigger, louder and quicker" is the general maxim of this industry and although the U.S. tends to be the experimental grounds of newer paradigms, they tend to mature and metamorphose in Europe and Asia. People across the planet will always need to shop, dine and be entertained and this need for consumption and human interaction will continue to be the generator of spaces, buildings and cities everywhere. Their venues, forms, and experiential attributes will however keep evolving, as the distinct types of functions tend to obfuscate and blend. Retail and Entertainment oriented developments will clearly continue to be the catalyst and glue for economic growth within future urban environments. Whether this is manifested by the re-creation of exotic destinations, story telling, branding or by the emphasis of service, choice and convenience, the industry will ultimately benefit and thrive, giving additional impetus for creating even more invigorating environments.

DEVELOPMENT AND MANAGEMENT

Existing Urban Conditions

Irene Wiese-v. Ofen

1 MAIN ASPECTS OF THE TODAY SITUATION

a) Changing demography

Urbanization throughout the world is growing rapidly. At the HABITAT II – Conference in Istanbul speakers of the World Bank reported, that in 2015 27 cities all over the world would have passed the 10 million mark of inhabitants. In Latin America and the Caribbean 89%–91% of the population are living in cities, in North America and Europe 77%, whereas in Asia and Africa where only 38% of the population are living in cities, this will go up to the year 2020 to 50%, USA and Europe then will have about 80% (that means a moderate growth comparing with Asia and Africa). That shows the urban growth and the large agglomerations are situated in the developing world. We see two different unbalances: rapid population growth in developing cities, aging in many developed cities, the most rapidly aging countries, such as Japan, Germany and Italy will exceed 40% of their population of riper old age – Cote d'Ivoire has with 50% of persons under 15 years – the youngest population.

This short review on the demography shows the extreme differences in the needs of population looking on infrastructure, health-systems, natural resources as land, water, air and energy and open space. Following these simple conclusions of different population ages housing and transport production as well as health and social care must be different.

The report on the Urban21–Congress in Berlin in June 2000 mentioned therefore different characteristics of cities and their urban conditions:

– The city of hypergrowth
– The city in demographic transition
– The mature developed city.

That seems to me a good characterization to categorize developments

b) Environmental problems

But there are some more impacts:

– By the year 2025 experts anticipate, that two-thirds of the world population will live in water stressed conditions,

–   The degradation of soil has already affected 20%, pollution is not reduced, especially the global emission of carbon dioxide (CO_2) and methane in the earth's atmosphere are believed to be enhancing the natural greenhouse effect, potentially leading to changes in global climate patterns,

–   Two thirds of people living in cities are suffering on noise and insufficient supply with fresh/drinking water and access to use,

–   More than half of the world's population lives within 60 km of the shoreline; population and infrastructure already damaged one-third of those coastlines. Possible climate change may have an impact on these situations as well.

–   All these statistics are taken from the UNEP 1999 report. We must be aware to what extent global facts exercise an influence on urban conditions.

c) Segregation/ethnic mixture/minorities

Following the Global Report on Human Settlements of the UNCHS the scale of international migration has increased to more than 100 million people living outside their own country with once more the same number of people thought to be refugees or asylum seekers. Many of the main migration flow have consequences of the globalizing economy. One of the consequences for instance is the fact, that more and more migrants are thought to be foreign workers, with a remittance of $71 billion (1991) back to their countries, so that this is a considerable item in international trade. Main reasons are:

–   The removal of the iron curtain, that has brought a large migration to Germany and Israel. This has direct consequence to housing questions in German cities as well as in Tel Aviv and Haifa in Israel especially on programs for integration and social infrastructure,

–   A large-scale recruitment of foreign workers was going on in the oil-rich Middle East,

–   An increasing number of refugees in Africa associated to the different wars and civil strife,

–   The changing nature of migration in Asia between labor importing countries (e.g. like Japan or Singapore) and predominating labor exporters (like China, India, Pakistan or Indonesia),

–   A great number of asylum seeking people stream to North America, Europe and Australia, trying to find a place for survival in the growing agglomerations of these continents.

–   The impact of the migration on urbanization is a very strong one. Ethnic mixture and minority problems induce often cities to focus on special needs, or to neglect them. Sometimes these ethnic mixtures are an explosive mixture.

d) Social exclusion

Therefore some keywords for the ongoing structural changes in the cities world-wide are:

– Increasing international competitive conditions, cities try to create a special profile and give priority to win the competition instead of developing comprehensive strategies to avoid social exclusion,
– Otherwise we see an increase of employees qualification requirements,
– New production and management concepts, and reduction of labor by rationalization.

This leads overall to the changing of labor markets, disintegration by unemployment, decrease of equity and by the way to social exclusion followed by political exclusion. Regarding to the fact, that 1.3 billion of people all over the world earn less than 1$ per day, social exclusion and the risk of a civil war between "the haves" and "the have-nots" is one of the main challenges for the urbanized world, because people are looking for an inclusive city and integrative labor market and their hope and dream is to find this in the cities. Millions of urban poor, of people staying in informal settlements and working in the informal economy hope for a sustainable human development that empowers people more than marginalizing them.

Perhaps education may aid most effectively against social exclusion. Obviously the existing urban conditions do not provide enough chances for people's choices and capabilities – the question is whether it will be possible in the ongoing process of urbanization to develop and to implement a sustainable offer of capacity building.

The insecurity in life conditions, corruption on all levels of the societies and the permanent changing of political responsibilities make the most existing urban conditions unstable. Cities are the focal points of today's social and ecological problems. On the other hand cities are the motor of development; they merge the different ethnic population to identify themselves as inhabitants, cities are the place where people meet and new thoughts arise. Cities are a monster and at the same time are a fascinating artificial cultural product of the highest level and the great hope of mankind to master the future challenges. Although countries in Africa or India and to a certain extent China have still the majority of their population in the rural area and persist on a rural development as a necessary accompanying strategy to urban development and concerning naturally food, water and land as the basic preconditions for our survival, in the cities the future will be decided. The city in my opinion will be never stable in the long term, but ever changing in its shape, falling and rising up again from the physical as well as spiritual ruins to new glory and new delights, but again only for a certain time as every process of change contains a beginning and an end leading to a new, different beginning, but this is more a philosophical view on urban conditions, and perhaps it may help a little bit to conquer the future.

2 PRINCIPAL TRENDS

Looking at the existing conditions and overview of the main points as done in the first section, urbanization will continue even in the developed countries, in the so called "mature cities". Because their people are able to finance more personal comfort and in the trend of satisfying individual preferences combined with a high level of mobility the urban sprawl will continue.

At the same time a special trend of disurbanisation is to be seen in the USA and Europe; the center, that often loses its outstanding importance of being the place of centralized offers of shopping, culture and services of all kinds and the directly connected quarters under the influence of the downwards tendencies in centers gives the initiative for several programs of revitalization. Especially the European cities struggle for a new version of spatial development following the tradition of the European "city of short distances", characterized by polyvalence, centrality, density and functional mix.

The program of the commission of the European Community "Europe 2000+" will enhance economic and technical co-operation within the European communities. Urban pilot projects are supported by financial alimentation, such as:

c)   Actions in relation to disposal of waste, air pollution, noise and other environmental issues to meliorate living conditions in the cities,
d)   Ideas how to integrate unemployed, especially young people into the mainstream of economic life in the cities, the decrease of unemployment in the cities and to avoid social segregation,
e)   Conservation strategies to revitalize the historic centers and protect the architectural heritage as the focal point of identity of the urban population,
f)   Transport and energy planning schemes to help the communities in developing and managing urban transport systems to avoid the still increasing trend of mobility by individual car instead of public transport.

By these programs "Urban" and "Interreg" the European Community tries to stop the trends of more space (with the effect of reducing the open landscape), more cars (with the effect of more noise and more carbon dioxide), shopping centers outside the traditional towns (and the effect of losing the traditional heart of the city) and at least through all these elements to stop a disintegration of the society.

The structural changes in heavy industries, especially in European mining, but also in the steel-industry has given back many "brownfields" to the cities. This opens great chances for the rehabilitation of wide parts of the cities, which had been put on heavy pressure in former times. Mostly these parts are centrally situated with good access to infrastructure and in the neighborhood of the settlements of the former workers. These examples just like Sheffield or Birmingham in UK or the Ruhr- and Saar-district in Germany, like the Bockinage in Belgium or l'Alsace et Lorraine in France – to call some European areas – but also the Golden Triangle in Pittsburgh or Detroit in USA, as the former largest automobile producing cities, will have to accept the industrial change as one of the most important trends: but a great chance in altering the city. Even in China we can see the first steps to an urban policy against the former huge state companies, which occupied central spaces in the cities. By turning the planned economy to a more liberal one and by global influences in production expires and logistics not only in industrial areas they change or lose their use but as well former military sites and transportation and railway terrain as well as docks and harbors. Barcelona in Spain e.g. has built up a new skyline at the coast by protecting the historical heart of the city. Urban renewal affects the city's overall tradition in cohesion with new demands. On a territorial level urban renewal, which aims to make better use of former built up areas, contributes to a reduction of using "greenfields" for urban purposes and thus helps to implement sustainability in urban conditions.

Urban reconstruction on brownfields is an ongoing trend to be foreseen all over the world, not only in the developed world. Cycles of evolution and decline are quite normal throughout the centuries, but we must care how declines may be organized; not by destroying, nor by renovation in any case, not by social degradation, nor by alimentation and investing in structures of yesterday – in long terms it is a balance of coming and passing just like a look to nature shows us this phenomenon in short terms <u>during</u> every year. Looking more to the advice, that nature may give us, one of the instructions, we should give more attention to is the supply and disposal of materials as well as to questions of abolition of buildings after the use has ended or the possibilities of reuse.

Building ecologically does not only mean better thermal insulation, percolation for rainwater, heat recovery and district heating systems for reducing energy, water and other resources, but to look at the waves of changing that bring today the decisions for tall buildings and high density as the only solution for business demands – and what will be the demands of tomorrow? How quickly those high raised "landmarks" can be changed into other using wishes in the overall accelerating globalising world? And what about the tradition, the abilities and the dreams of the people: that means their wishes for detached houses in a perfect environment? The trend all over the world is obviously the inability to solve this antagonism between economic constraints, ecological limitations, human rights to adequate housing and our dreams of a better world – but we must try it. And in many plans of the world, with different cultures and historical backgrounds, technical knowledge and economic possibilities the solutions will differ despite global influences. The challenge is to assess these impacts and to foster developments that are flexible and sensible for differentiation

Let me mention one further trend that seems to have an impact on our cities. That is:

a) E-commerce is thought to be booming. This applies to trade between companies (business to business) as well as the business with end-user (business to consumer). E-commerce can only assert itself on a growth sector if linked to corresponding logistics. Logistics is a determining factor regarding quality, service and maintenance of targeted costs for long-term success of online-business.
    The impact on cities is obviously:

- Large distribution centers with highly efficient logistic systems, which need to be established for storage, picking and distribution,
- Flows of goods on roads and railways, air cargo and a rapid increase of the number of returns,
- Changing of the inner cities shopping structure, the retailer of the future has to be a professional consultant.

b) Shopping for routine goods by large retail centers outside the inner cities and by e-commerce, shopping as entertainment in the specialized, heritage-protected inner cities or excellent designed artificially paradises of shopping centers. These trends will change the shopping structure: more exhibition character and more specialized professional consultative qualities of retailers will be looked for; the customer gets information through the Internet, emotional physical feeling in exhibition – shopping – entertainment – centers, delivering the bought goods by special logistic systems.

At the moment it is not yet quite clear, whether these tendencies will continue and how we react. But we have much more need of technical infrastructure and assistance in education and empowerment, which are beyond possible trends, concerning the facts, that:

- 850 million people are illiterate,
- 1 billion live without shelter,
- 2.7 billion outside of sanitation,
- 1.3 billion without water,
- 2 billion have no access to electricity.

And last but not least we have many cities of natural risks: hurricanes and other storms, earthquake, overflowing etc. demanding consequences for the construction of tall buildings as to be seen at many catastrophes in the past years. And think of the challenge to develop overall more ecologically designed buildings.

3   THE CHALLENGE FOR FUTURE URBAN CONDITIONS

There are many challenges for future urban conditions. Please allow me to simplify the complexity of this subject to 5 theses:

a) Sustainability

The Brundtland Commission 1987 laid down the following definition:

> "Mankind has the ability to establish sustainable development, to guarantee that it meets the needs of the present without endangering the ability of future generations to meet their needs."

In 1992 the UN Rio-Conference on Environment and Development established the Agenda 21 and 177 states committed to several actions to attain the goals and mission of the Brundtland commission. In nearly ten years no definition was spread so worldwide and changed especially in the consciousness of the civil society so much. But it is not enough. To reach sustainability is an ongoing responsibility. We have to fulfill the obligation to work on the triad of economical, ecological and social balance. Each of our decisions as politician, as manager or as member of any city based organization (CBO) or non-governmental organization (NGO) should be the result of a weighing process having in mind this triad.

b) Increased efficiency

We have to increase efficiency. This is meant not only in the question of energy concepts – sun and wind as renewal sources – but as a question to all of us and our way of life. Energy saving installations in housing, quality of design and over all the maintenance of the building stock, facility management and the consumption standards of our life are some possible answers. Energy saving is not only a question of the developed countries, water consumption in wide parts of China is four times more than in European countries per person, or the lack of organization in maintenance of the now privatised prefabricated building stock in the former communist countries are some examples all over the world that consumption standards are to be taken into account in any country and any culture. Increased efficiency is not only a question of research and better technology that has to be stimulated, but as well a question to political and civil socicties acting very often contrary to the dictates of conscience.

c) Responsibility

Worldwide cities have been developed following the Charter of Athens. In 1933 the CIAM (Congrès Internationaux d'Architecture Moderne) drew up a charter with 95 articles. The main article: "The concept of the towns is delimited by functions: Living, working, recreation and transport" grew up to the Magna Charta of urban planning.

We nowadays have the responsibility to change the results of monofunctional segregation into the inclusive city and to remember, that another article in the Charter of Athens is e.g. much more important. Article 2 states: "In addition to economic, social and political values, the psychologically and physiologically based values connected with human beings, bring considerations of an individual and collective nature into the debate. Life can unfold only to the extent that these two conflicting principles governing human personality, the individual and the collective, are compatible". We all together are responsible for the spirit of the city.

d) Gender equality and family cycle

Following a report of the Center for Human Rights in Geneva 1996 there was outlined that "following all information available it is abundantly clear that women across the world continue to suffer from discrimination in the attainment of all aspects of the right of housing:

– Land security and inheritance of rights to land and property,

– Access to credit facilities,
– Access to information essential for participation in housing activities and in contributing to the improvement of living environment,
– Availability of essential housing services and resources including potable water, sanitation, fuel and fodder and access to appropriate housing projects, upgrading schemes and resettlement areas".

In most countries of the world women have neither a right to the home in which they were born nor to the home they live in after marriage. The UNCHS therefore following the HABITAT-Agenda started the global campaign for secure tenure. Access to land is the precondition for solving housing problems. South Africa e.g. started several housing programs to develop within the urbanizing areas of Cape Town, Durban and Johannesburg low cost housing on little plots and infrastructure (water and sanitation) offered to each of the houses sold to families under a special income.

Housing is therefore one of the crucial challenges. It is necessary to meet the needs of the inhabitants of our towns and targets of more social and gender justice in distribution and chance must guide looking at urban conditions planning for future conditions. Efforts must be made to recognize and compensate more the disadvantages and risks of no or lower income and more vulnerable groups of the population (homeless, single parent families, foreigners and refugees, older people, unemployed poor). Furthermore socially compatible settlements must permit chances for self-determined action. In a pluralist society individual and equal gender life planning as well as self-organization and self-determination on the neighborhood must be possible for the family cycle with different cultural background and different scopes for maneuver.

e) Planning approaches

One essential element of urban planning is the duty for weighing the public and private interests. Mandatory weighing is a key element of urban land use planning. Effective land administration is one of the preconditions. The Bathurst Declaration of the International Federation of Surveyors 1999 explains, "changing humankind – land relationship and current global and local drivers such as sustainable development, urbanization, globalization, economic reform and the information revolution, demand land information responses and are forcing a new land administration paradigm". The declaration continues, that a property market, that is another vital element if urban planning is thought to govern urbanization processes, has several key requirements:

"– An appropriate legal framework aimed at minimizing risk and uncertainty over issues of ownership and use,
– Registration of interests in land and spatial land use planning".

The other element is valuation of rights in land as driving force in the functioning of a real estate market.

We must be aware that the international call for good urban governance includes knowledge and expertise in these fields of planning and surveying. Parallel to local efforts there must be created a legal framework. Within legal security the process of planning has to include and co-ordinate all activities of local authorities, private groups, housing companies, financing and other groups of stockholders, users, tenants and owners of private properties. An holistic approach may help to solve as well complex social problems, that are not sufficient enough solved by only social policies, as well as environmental concerns and economic pressure. Building and land using plans are not able to meet the dangerous developments of degradation and insufficient housing besides glamorous business areas just by zoning and property regulations. Common efforts are asked for. Architecture and planning have to reflect not only technical and commercial issues but also social consciousness. The role of architects and planners is a changing one: to work in consensus with other disciplines in an integrated approach to optimize the whole (not maximize their own discipline).

f) Recommendations

As we have seen, cities are the motors of production, innovation and social chance as well as social change. The urban job market follows changing urban conditions; macroeconomics factors cause urbanization. We must develop strategies to work in a resource efficient way in that directions, that resource not only are nature and environment but manpower, expertise, individual responsibility and emotional involvement of citizens. We must look at economic dimensions of new developments in our cities but as well as maintenance of housing, buildings and infrastructure. The latter is one of the preconditions of welfare and comfort in a city and the hinterland. The other is a democratic structure, which means elected bodies, with self-responsibility and certain independence even on the local level, public control of decisions and participation of citizens and groups of public concerns in planning processes. We as part of the societies of our countries can help to be a pressure group that accesses to land should not only be a part of the constitution but also come to realization. For economic growth and equal opportunities secure tenure and security in landownership as well as access to land especially for women may help in solving the problems of informal settlements. We have more opportunities to achieve this goal step by step than we think to have. Concerning the informal job market round table strategies that combine individual initiatives, educationally orientated project work and selective co-operation of different institutions and sponsors together may bring people in more secure self-responsibility.

Examples of South-America e.g. show, if the land is bought by funds, (supported e.g. by developed countries, World Bank etc.) and given to the people in the informal settlements, and they have to pay a small contribution for revolving the fund and giving them the consciousness to be independent and not alimented without rights, then they organize everything that is necessary like infrastructure, safety, education, market and logistic necessities. More people than we think want to be independent especially women. National security, big transport systems, legal framework and reliability of elected bodies and their decisions are the precondition of private initiative. If we take this as the vision of urban conditions, then we have to organize the frame of cities by the state and to give people reliable conditions to organize their living conditions for their own.

The exchange of ideas at conferences – like this 6. World Congress or other professional or non-governmental meetings about "best practices" – may help to learn from one another and to feel endorsed by common efforts.

Urban conditions – at present and in future – tell us what the cities always have been and will continue to be in the future: big, dense, noisy, impenetrable, lively, stimulating, eccentric, poor, rich, criminal, protective, damaging, beautiful and so much more – a fascinating challenge, an utopia?

I often have thought that the ideal of the "long-term securing of development" after the Rio conference has something Utopian about it when applied to the effects on the city, something comparable to the ideal cities of past centuries, which we link with the hope that if at some point we achieve this ideal situation, everything will be fine!

The task of securing long-term development consists not only of the ecological aspects of soil, water, air and energy – the elements by which human beings live; it also consists of securing opportunities for the development of future generations in an environment of care, education, health and sufficient nourishment, an environment that will allow them to develop their faculties freely.

The ideal situation is that these two themes complement one another; many people would say that they not only have to complement one another but that one is a precondition for the other. However it cannot be overlooked that ecological demands and anthropocentric priorities are in conflict. That this conflict may be insoluble makes it even more difficult for us to make balanced plans to protect and secure both basic requirements for development. Merely Utopian, then?

Pursuing Utopian demands, however, is a special challenge – it does not release us from the task of trying to attain them. For instance, it is indisputable that we have the responsibility to do everything in our power to save resources for our descendants and to continue to cherish and make the most of the social and cultural capacities we have developed. But I consider the hope of reaching a condition of perfection and stability only if we act exactly as our knowledge and all the sectoral studies recommend – seems to me – to be unrealistic.

Our desires will continue to lead us to attempt to secure our opportunities for life long-term – and such attempts should, without doubt, be supported with all our power and with the concentrated application of our knowledge. But when ever we come up against something that does not work, nor with which we cannot identify, where we remain on the outside or we have to accept losses and do not achieve our planned goals, we should remember that it is part of the human activity to make mistakes, and often enough good intentions lead to disadvantages, which we did not or could not identify in advance. We are looking for criteria of quality and restruggling to define what long-term quality assurance might involve: competition versus co-operation? Market versus planning? Equal opportunities versus power of the strength? In many cases we feel assured in quality – if we use the term "identity". Cultural identity requires a long period of growth and links with the past, and yet thrives just as much by taking on ever new influences, that is, by integration. This requires investment and care. However, the question of cultural identity in our cities is also essentially one of the preserving cultural varieties while at the same time respecting differences. This means above all that urban spaces must be kept open – democracy came into being in the open space of the Greek Agora. In the closed house of the Roman curia, it set into the mould of an authoritarian empire with a sole ruler.

Open space, allowed to enter for everybody is the beginning of a city and the essentials, the heart. We should remember to include in our calculations that quality of life may depend on totally different criteria for each individual! But public space offers a special quality and chances for equal opportunities; they should be safeguarded and not changed in private malls of perfect artificial worlds.

If we look at the existing urban conditions in which we live, for me, the most important conclusion from all kinds of international examples, is that when we recognize mistakes we must correct ourselves and yet trust ourselves to act anew on the basis of our new-found knowledge. At the same time we must remain patient and guard against those who, under the cloak of ideology, promise us quick solutions (for these are usually one-dimensional), that we must learn to accept that, even if at our next attempt the city seems as fragile as ever, because human beings are so fragile, our strength is that we know this, and therefore we can live with it. Then we can face anew, day by day, the challenge, which the city as a way of life poses for us and existing urban conditions are a promise for the future if we combine our efforts to master the ongoing change.

PLANNING AND ARCHITECTURE

Co-existing with the Past City – Building in Historical Places

Jong Soung Kimm, FAIA, KIA

Co-existing with the past requires the wisdom of diverse constituents of a given society. It might take the form, at the least desirable end of the scale, of outright replication and grafting of historical building types; giving new lease on life through preservation and adaptive re-use to historical structures; or building new structures that would genuinely enrich the lives of citizenry in a given culture through creative and judicious combination of analogy, metaphor and reference to the collective memory of that culture.

I begin my short presentation by looking at some preservation and adaptive re-use efforts of individual buildings in Chicago, as it is in Chicago that I first obtained my education as an architect. The Reliance Building of 1890–95 designed by Charles Atwood for D. H. Burnham Company was recently converted to a hotel, appropriately named the Burnham Hotel. An even older structure, The Rookery Building of 1886 designed also by D. H. Burnham Company went through a long period of neglect before being renovated to the former glory a few years ago. The main stair hall and lobby which Frank Lloyd Wright is said to have designed has been restored to its original splendor. The Flat Iron District in New York has numerous structures of 1890's vintage as do downtown cores of San Francisco, Los Angeles, Boston, etc. For examples of much older edifices actively in use, however, we need to look at the European or Asian cities. Among the numerous cases of successful conversion of historic buildings to new use in Europe, I would like to mention Castel Vecchio in Verona which Carlo Scarpa so ingeniously transformed into a museum. The National Museum of Roman Art in Merida, Spain designed by Rafael Moneo in 1980–85, is a happy instance of a new structure sitting literally over the Roman archaeological site in the old "Augustus Emmeritas".

Next, I would turn to new structures of public use which have been successfully incorporated into the urban fabric of historical cities. When the railroad was introduced in the second half of the 19th century in major cities in Europe, there emerged an opportunity for daring new structural inventions by architects and engineers to make their appearance. Breathtaking, glass-roofed shed houses over the platforms had to take a back-seat, however, to the generally eclectic designs of the station houses, as the general sentiment was not ready to accommodate starkly "modern" structures right in the heart of historic core in respective cities. What a remarkable transformation in public attitude these last hundred years have wrought! The JR (Japan Railroad) Kyoto Station designed by Hiroshi Hara Architects is a case in point. The entire grid-iron historic district of the twelve hundred years old Kyoto is a UNESCO World Cultural Heritage registered site. We assume that it was not without controversy that the design for the new station was announced in 1991, but since its completion in 1997, the JR Kyoto seems to have caught the public's imagination, and has become a favorite hang-out of citizens of Kyoto.

The museum as a building type is less than two hundred years old. Collections of sculpture from the antiquity used to be displayed in palaces, and gradually, spaces for displaying paintings and sculptures became an indispensable clement of palaces and royal residences. When the city fortification was razed and the Ringstrasse was created in Vienna in the 1880's, Gottfried Semper designed the symmetrical ensemble of the Art History Museum and the Natural History Museum to be an integral part of the Vienna cityscape. As museums are very often built in the historic core of old cities, museum design presents a special challenge to architects. To illustrate this point, I would like to discuss the City Museum of Seoul which we began to design in the mid-1980's, but was completed at the end of 1998. It is awaiting the opening exhibition in the spring of 2002. The City Museum is intended for exhibiting the history of the city and its art collection. It houses about 20,000M2 of galleries, seminar rooms, an auditorium, storage, and curatorial and administrative offices. The Museum is located on the site of Kyung Hee Kung, one of the Yi Dynasty palaces which was totally razed during the Japanese colonial rule. One programmatic requirement was the preservation of a 30-meter square area of archaeological excavation. The structure combines the spirit of today's technological age with a respect for and understanding of historical palace architecture. The placement of the archaeological excavation gave rise to the building's U-shape, with its clear axis flanked by galleries. The galleries are laid out in enfilade. Each is 12 meters wide by 14.4, 21.6 or 28.8 meters long, and is punctured by a series of skylights. The exterior rust-colored aluminum cladding recalls the painted hue of wooden palace structure, while the lead-coated copper roof echoes the grey palace roof tiles.

Through history, there have been cycles of change in public attitude toward building in historical context. Each generation of master builders strove to contribute buildings which in their eyes were better, newer, and more innovative than their predecessors' output. Witness the western façade of Chartres cathedral: the master masons who added the north tower had no qualms in building a design of a quite different sensibility. In the century which saw the emergence of modern architecture, after an initial resistance, the prevailing attitude in many societies for a long time was that new is better than the old. Today architects have generally become more sensitive to preservation issues, and the orderly rejuvenation of the urban fabric has become an area of great interest.

Some years ago, my firm was engaged by the Seoul city government to conduct an analysis of Insadong district in Seoul and make an urban design proposal. Insadong district is situated right in the heart of the historic core, to the southeast of Kyungbokkung Palace and to the west of Chongmyo, the royal ancestral shrine. Around the time of Japanese colonization in 1910, Insadong area was characterized by a significant cluster of traditional, clay roof-tiled houses, some large residences for former high court officials, a palace and a generous ground for a prince. The district itself, and neighboring areas contain many, significant historic sites and structures of 75 to 150 year vintage. Destinations of far greater significance, Changdukkung Palace and Chongmyo, both of which have been designated as monuments of UNESCO World Cultural Heritage are located a kilometer away. Insadong area, known to English speaking visitors as "Mary's alley" is a shopping destination for anyone looking for Korean chests; folk painting, mounted either as scrolls or folding screen.; calligraphy supplies and mulberry papers; folk dance masks and mother-of-pearl inlaid lacquerware. Not to be left out is a wide variety of restaurants and tea houses which serve up anything from authentic vegetarian Buddhist temple dishes to folksy traditional home cooking. I believe that food and eating habits of any culture are the most resilient and "die-hard" carriers of tradition, and the eating establishments in and around Insadong area deserve to be recognized and encouraged to play an important role. The traditional buildings making up a good portion of the Insadong district have been converted to restaurants and small shops; three to four-story high new constructions have sprung up housing art galleries, antique shops and calligraphy supply places. At the same time, some sections have been built up into mid-rise commercial structures. Inevitably, however, dilapidation has also crept in.

We set the objectives of our study thus: to retain the traditional ambience of the Insadong quarter; to create "Museum" – like street; to make Insadong cultural and tourism destination. First, we dealt with the obvious planning interventions such as traffic and pedestrian paths: to exclude through traffic from the diagonal street; to connect firelane network; to connect street parks and historical sites and to propose gateways at north and south entrances to Insadong. Also, we recommended that new parks be created by removing a few dilapidated buildings for use by the general public for folk performances. 15 years have passed since our study, and I am pleased to see that some of the recommendations have been implemented by the city government. Our vision is to preserve the contextual quality of the Insadong quarter, while at the same time making it more commercially viable and vibrant. It needs the collective will of the residents, merchants as well as the local and national authority to make it a reality. It could take the form of a public sector funding for structure(s) to improve the sense of place within Insadong. It could be tax incentives or an outright grant for maintenance, repair or conversion of single-storey traditional buildings.

PLANNING AND ARCHITECTURE

High Rise-High Density

Richard Keating

1   POST DIGITAL LIVE/WORK

Twenty years ago during the extraordinary building boom of the 80's, we saw a plethora of object buildings in the U.S. exalting the glories of demographically monochromatic corporate imagery. Celebrating power and wealth, these buildings presented a rich variety of materials and opulence in hopes of gaining a perception of permanence and a domination of success in each economic community in which they operated. Today almost all of the names of these buildings have changed creating havoc amongst architects who maintain slides and photography of their past work but also reflecting the enormous evolution of business in the past decade through mergers and acquisitions.

Figure 1.1   *Time* magazine celebrates architects.

There is perhaps one building that best represents this period and one company, AT&T. In New York the company created a Chippendale highboy antique image associating with a sort of Blue Chip / Old Boy / White Yankee / Laura Ashley / first wives / country club power base. They moved out of the building in the next decade.

In Chicago they chose to be the lead tenant with what could be described as the utmost in opulent buildings that maximized the variety of glitz and marble. This building makes Henry VIII's Hampton Court look like a brick dump and would even stretch the marble collection of the Vatican to excess. Not only did AT&T move out but so did many others embarrassed by the opulence over the next 5 years including the building developer.

Figure 1.2   A T & T Building and Sears Tower.

The 80's was a period of duality wherein the 80's style buildings of excess were adopted overseas by every tin pot petroleum dictator like Petronas Towers in Malaysia, the Golden Triangle of Jakarta and the *nouveau* economics of Pudong. Here in the U.S., hunker-down sensibility reigned with Hewlett-Packard, BMC Software, British Petroleum and design build projects looking towards exit strategy and commonality. The out growth of this period is fascinating.

Figure 1.3 Petronas Towers.

During the 90's, of course, the American business evolved completely as did much of our social values. Think dot.com plus affirmative action.

Today, in Los Angeles, for instance, the cost of rent in a Venice warehouse exceeds the 72nd floor of an I.M. Pei high rise downtown.

What does this mean? Is it a lasting trend? How can we envision the future based on what we can observe?

2   NEW DOWNTOWN

Will we fully adapt telecommuting? No, we have too much to gain from one another on a daily basis. Both the serendipitous encounters and the energy gained by working with others have sustaining value-supporting collaboration over anomie. We have also invested enormously in infrastructure that supports the urban core of our cities.

What is very real is that our demographics will outstrip the courage of our political leadership who are all too influenced by nimby-ism and term limits to actually look beyond today's re-election to the urban fabric needs of the very near future.

Our freeways will exceed capacity and many already do – think the 405 in Los Angeles. Our bridges are full and the built transit systems are at capacity. We have transit systems being built that don't go to airports until 20 years later. Could this be a taxi union influence or simple lack of political vision about connectivity?

We have cities that have evolved as clear linear development (ideal for transit) that for the fear of class intrusion would not allow transit – think Beverly Hills and Wilshire Blvd. We have airports at capacity with subsequently grown surrounding communities now limiting growth – think LAX, or Orange County.

We are finally learning that sprawl is the worst offender of draining our resources whether economic or energy. Building cheaply on the urban fringe is a quick return for the purveyors of the greed at the expense of open land and time for living rather than commuting.

Our only solution is infill in the existing urban form. The question is whether this will continue to take the form of high-rise office towers in the center and single-family houses in the surrounding areas.

Let's look at the occupants and what they will be doing in the future.

The dot.com revolution may not sustain itself in the same intensity we saw throughout the early part of the year 2000, but when you clearly look at dress-codes-lifestyle and the 24/7 work mentality, enough has changed since the norm of the 80's American high rise to substantially question the work place of the future.

Another aspect for consideration is that to the corporations of today the value of employee sustainability has surpassed the value of corporate imagery to the community.

What do these employees respond to and how are they different from 20 years ago? Think skateboards vs. the NFL. Or why has the NFL hired a woman from MTV to rework with their image?

One of the hidden aspects of the office-building boom of the 80's was the impact of women in the workplace. We essentially employed the other half of our population in one decade and almost entirely in the service sector. Think office buildings. Say what you will about tax credits to developers, when you couple this with the baby boom cohort-survival group going to work for banks, law firms, and accounting firms this created the 80's building boom.

Today many of these women are partners and are sustaining families at home as well. Today the office work population is a lot less white and draws from numerous parts of the city neighborhoods that surround our urban cores.

3 THE NEW TOWN IN TOWN/THE NEW SUBURBAN CORE

Because of this, there has been a movement towards urbanism both in the center cities as well as in certain of the peripheral suburban communities. In California we have seen a whole second tier of cities arise reflecting all the benefits and some of the problems of their pre-cursor cities. San Jose is now the second largest city in California. Walnut Creek, Glendale and Burbank are complete cities that have emerged from their prior existence as suburbs.

Then there is Yerba Buena Gardens in San Francisco as well as Mission Bay in San Francisco and Playa Vista in Los Angeles. This development represents the combination for desire for urbanity and available inner city land. Mission Bay was, of course, Southern Pacific Rail yards in its prior life and just like the Illinois Central property south of the Chicago Loop has outgrown its first function just in time for expansion of the urban form to the north. Yerba Buena Gardens was 30 years in the making and took the stimulus of the strategic re-location of the Convention Center as well as various cultural institutions to kick start its growth. Of course it didn't hurt to have $50-$70 sq.ft. rents in the adjacent built-out downtown as well as a desire by many young employees to live downtown. The central landscaped garden gives focus to the cultural district but housing and restaurants are left mostly to rehab and remodeled surrounding buildings.

Los Angeles, on the other hand has an enormous opportunity to create a truly new town in its downtown. Not unlike the former rail yards described above, there is available relatively inexpensive land immediately adjacent to the existing urban core.

Figure 3.1 Yerba Buena Gardens, San Francisco.

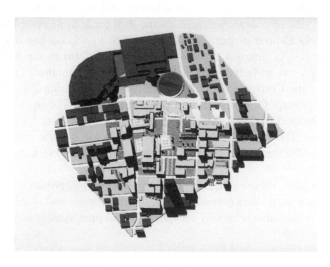

Figure 3.2 Los Angeles new town.

In its past it has been largely surface parking for the nearby high rises. At one edge, both a convention center as well as a sports and events arena has been recently built. There are two major freeways directly adjacent serving both north and south as well as east and west. There is an existing transit stop on the blue line which leads from downtown to Long Beach.

With this entire existing infrastructure, as well as the pre-existing 80's style downtown, the New Town in Town right next door can be the natural location for the increasing demand for alternative work style buildings.

Designed as a live-work environment and made cohesive by a clearly defined public open space, existing buildings can be rehabilitated as well as new construction can take place to create a large floor plate – low rise group of buildings. The function would be a balance of working spaces as well as retail and housing. While these types of uses will seek available and affordable land in locations that make sense, there is a moment in time in which the character of the neighborhood and a sense of place can be defined by the design of a major park and the spirit of the architecture. For Los Angeles that time is upon us. The public perception of commonality is the space created by the edges of private development. Following this is the continuity and integrity of architecture.

The problem is the gap between those of us with vision and our political leadership with the power. With politicians that sway according to short-term considerations and groups or individuals with special interests, it is often impossible to carry out a desire for major urban ideas. In a democratic society like ours, it takes a benevolent dictator, often operating at the edge of the system like Richard M. Daley of Chicago, or all too often a major catastrophe like the San Francisco earthquake of '06 or the Chicago Fire to give impetus to large scale urban initiatives. Interestingly the career of Dan Burnham and his oft quoted "make no little plans for they do not stir the soles of men", was literally made by these two events.

We desperately need to find a middle ground that can bring larger urban visions to reality without losing our democracy. Since the void spaces like Central Park, Champs Elysee and Tienamen Square provide the clarity and impetus for adjacent development, just as natural edges like Bays, Lakefronts, and Mountain Ranges have done in other cities, it would seem the easiest path for democratic societies to concentrate its focus on the opportunities of the public void. This armature will create its own character of adjacency in time. We cannot have the rule of the individual influence groups trumping the larger urban visions or needs if we are ever going to affect air quality, transportation and urban imagery in meaningful ways. In Los Angeles, our new city charter provides for neighborhood councils to give voice to local issues yet had the potential to leave the larger needs and concerns to a more regionally responsible governing body.

4   THE NEW OFFICE PROTOTYPE

For the past 30 years the modern office building has been idealized as a central core, rectilinear floor plate of 20–25,000 square feet with a core to window-wall dimension of 40–45 feet. This dimension has evolved as an optimization of perimeter offices plus assistant space and circulation, or more recently, office cubicles packed side by side. However, when you think about what activities this supports or the underlying genetic code, it is really about office workers talking on the phone, and passing papers to one another. Of course all of this activity can now be done on the back of a wristwatch, so are we destined for everybody in a home or remote office interconnecting with one another by laptop? No, and for the very reasons of our society, the joy and benefit of learning from one another. I believe that the new genetic code underlying our office environment is one of collaboration and meeting in serendipitous ways to advance a field of ideas. So what is the right setting for this type of activity? One has only to look at the recent trends in the market place. When a warehouse in Venice costs more to rent than a premier downtown high-rise, and when the dot.com revolution coupled with relaxed dress codes can generate exceptional new business that drive our economy within the time span of a few years in nondescript suburban buildings surrounded by surface parking, then one has to seriously depart from the AT&T model discussed earlier.

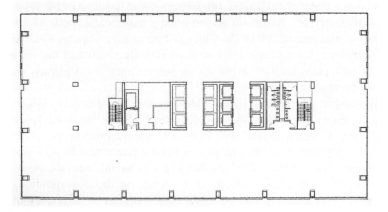

Figure 4.1   Floor plan.

If, however, we build warehouse after warehouse all surrounded by surface parking lots, it creates a miserable urban environment. Silicon Valley business in just such environments are now seeking alternatives that include some form of urbanity to afford their employees more that just a work environment. So can we create buildings with greater density that can accommodate this life style and still make sense from both a real estate and good city building point of view?

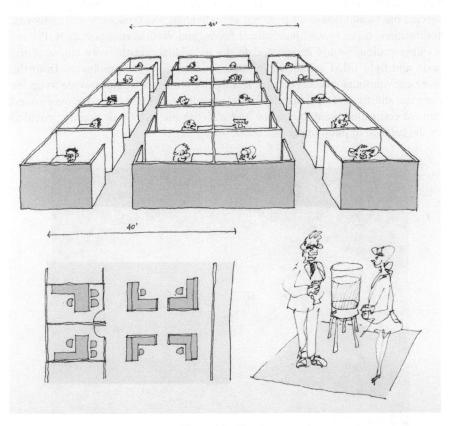

Figure 4.2   Sketch.

The following three prototypes indicate some possibilities.

Collaborative Space/Service Space – This building in Figure 4.3 provides large open loft like spaces with very tall ceilings that float within a garden environment. Not unlike the character of a warehouse, these spaces provide an optimum setting for collaboration and teamwork. Connected by bridges to one another and also to a service building, these spaces are bathed in sunlight. They can accommodate ever-changing furniture arrangements and should have all furniture and equipment to be on rollers to do so. The finishes can range from the bare concrete and exposed ductwork ceilings to a more finished system. The service bar would house all personal or individual workspaces, meeting rooms, auditoriums, toilet rooms, mechanical rooms and vertical transportation. Private docking stations would accommodate the individual, yet the very nature of the open and light filled collaborative spaces would draw the employees from the more conventional space. The park setting in Figure 4.4 would provide room for exercise, dining and day care facilities as well as generally provide year round climate controlled break out space and assist in the indoor air quality provided by the balance of plant material.

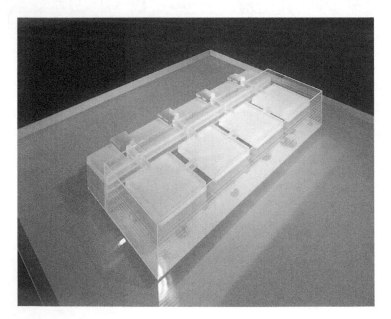

Figure 4.3   Collaborative space/service space.

Hybrid – This concept can also be a hybrid in combination with a traditional high-rise office tower and affords unique opportunities to deal with the necessary large garage aspect of buildings that are built outside of the urban core and away from the support of public transit. In this case a certain set of functions can be traditionally housed and the growing needs for larger floors to accommodate trading rooms, call centers and general collaboration spaces of high density can be in an acceptable and pleasant environment rather than endless windowless sameness.

Figure 4.4 Park setting.

Urban Version – A more urban version of this concept is to group the "warehouse" floors into quadrants leaving one quadrant empty for the park and separating each block of space by light filled linear atriums that, in turn, accommodate circulation on linear bridges. In this manner each block of space is surrounded on all four sides by light. Two walls are normal exterior walls and the other two are translucent glass panels that borrow light form the adjacent atrium. The park quadrant can also accommodate the vertical transportation and toilet rooms while the exit stairs at the end of the circulation can be organized as major structural elements and mechanical rooms.

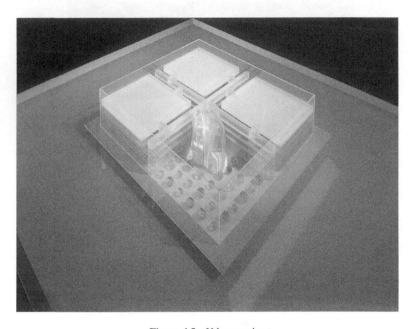

Figure 4.5   Urban version.

In each of these building types there is the opportunity to provide the ultimate in communication connectivity during the remaining time we may still need hard wire or cable connections. Undoubtedly there will be significant development in wire-less communication of data and images in the near future that may obviate the raised floor and the ceiling plenum as we know it today.

5   FASHION STYLE AND ARCHITECTURE

Neal Gabler's book "Life, the movie" points out the continued blurring of reality and imagery in which we are presently living. We should be mindful that so much of our architectural reality is now as affected by Ralph Lauren, Martha Stewart and Ian Schrager as any one else in our profession. No architect or designer knows this aspect of our society better than Philippe Starck who can translate an immediate sense of hipness to his every work for Schrager. While that may be dead-on for the long needed overhaul of the hotel imagery, we have also seen Jon Jerde redefine retail architecture and Frank Gehry personalizing our cultural institutions in free form pop images. The issue at hand is one of permanence and or temporary illusion. The icons of the power elite office building that was represented by AT&T are now the employee friendly suburban atrium buildings. It may be appropriate for retail to have a short shelf life resulting in architecture that is relatively disposable, and it may be fine for the super rich like Paul Allen to indulge an any form / any materials type building like the Experience Music Project, but does the computer really allow for blob architecture to make sense for all aspects of our built environment? How many architects are busy designing buildings with the magazine in front of them open to the Bilbao page only to find later that the pragmatism of budget constraints means that the only materials that are affordable are stucco and EIFS (Dry-vit) unless serious formal compromises are made? What are the appropriate means to contemplate the individual contributions of architecture that make a finer urban experience?

Undoubtedly, we will be influenced once again in our planning and architecture by yet another unforeseen set of events or forces beyond our current vision that will lead us to the future, but I submit that the true indicators of the future are the inevitable issues of demographics and the balance of need. Only in this way will urban sprawl be contained and the richness of densification be balanced by public vision and stylistic amenities.

Figure 5.1   Calvin Klein advertisement.

Figure 5.2   Gehry Model.

Figure 5.3 Earthquake damage to infrastructure.

The Future of High-Rise Buildings

George Schipporeit

For anyone interested in the relationship of urban issues and tall buildings, there is no greater challenge than this 6th World Congress theme of "Cities and the Third Millennium." The last few decades leading up to this threshold of time have demonstrated an unprecedented acceleration of transformation and change. Because there has been no long-term learning curve, the marketplace has become the judge of success and cities have been expected to just absorb the growth.

Yet, the future of high-rise buildings when viewed from their origins, and albeit brief history, do hold lessons to be learned and assessments that can be made. However, it should also be noted that from the famous Home Insurance Building, by William Le Baron Jenny, in 1885, through the beginning of the Great Depression, and from post-World War II through the year 2000, there are generally two 50-year increments, or a mere 100 years of development. Well intentioned perceptions must be expected to compensate for the luxury of an extended period of architectural history. Even with these limitations it is still necessary to look back before projecting forward.

It seems so natural now to understand how Chicago was destined to become the culture that produced the first tall buildings and the powerful verbal image skyscraper. The reconstruction of the city after the Great Fire was well under way, there was a vibrant creative architectural community that had been drawn to this construction cycle and there was a surge of commercial growth that motivated the investment in increased rentable floor area. But even more important was the resource of technology. The skeleton structural frame of the Home Insurance Building benefited from the first steel available from the adjacent mills, electric elevators had replaced the original steam lifts by E. G. Otis and, to compensate for the low-bearing capacity soils, new foundation systems were engineered from steel and concrete.

These three technologies, and the practical considerations of investment, became the drivers of an architecture known as the Chicago School. It was based on the form developing from a functional plan. Each building came down to the property line to include a street edge of retail. If the site were large, the narrow office configuration required for natural ventilation formed interior courts with the vertical plumbing, stairs and elevators given this inner orientation. But what was most evident was a fierce determination to develop an ornamentation that broke away from historical tradition to find its own appropriate architectural expression.

The vitality of this period continued into the turn of the century and also produced buildings in other surrounding cities, including Sullivan's Wainwright Building, St. Louis, in 1890, and Guaranty Building, Buffalo, New York, in 1895. But perhaps the most significant architecture of this time is the Reliance Building in Chicago, by Burnham & Root, in1895, with the profound lightness of its glass and terracotta skin drawing in as much natural light as possible with the undulating bay windows.

Figure 1   Reliance Building, Chicago: Burnham and Company 1894
(Courtesy: Chicago Historical Society).

Then with the construction of Burnham's twentytwo story Flat Iron Building, New York, in 1903, the momentum shifted east. By shear numbers, this domination continued during the balance of this first phase of high-rise construction. However, the East Coast architects were more heavily influenced by the European Beaux-Arts traditions and their tall buildings now expressed an historical eclecticism. And, as the engineering confidence in steel developed, the investor motivation of rentable area was reinforced with the identity of an individual building's domination of height. Zoning now became the only limitation.

Figure 2   Empire State Building, New York Shreve Lamb and Harmon, 1931
(Courtesy: Empire State Building).

Some of the crowning achievements of a mature architecture were designed during the final building boom of the late 20's and early 30's. Chicago and New York, along with many other U.S. cities, were unified with the rich expression of Art Deco and the search to create a modern style. New simplified vertical forms and massing replaced the dependence on historical ornamentation. In one dynamic shift, an exciting collaboration of architects, painters and sculptors enriched the exterior detailing, street edge scale storefronts and majestic lobbies with the luxury of a beautiful, artistic craftsmanship of stone and metal.

In Chicago, two exceptional tower examples were the Palmolive Building, in 1929, and the Chicago Board of Trade in 1930, both by Holabird and Root. However, it was New York that seemed to express this spirit best with three projects which have that unique quality of being both the best of an ideal and the end of a period.

The Chrysler Building, by William van Alen, in 1930, literally topped off the ultimate expression of Art Deco. It ended up being number two in the race for height at 1046 feet. Number one for over 40 years at 1250 feet, and still the best architecture of a tower on a base, is the Empire State Building, by Shreve, Lamb & Harmon. Opening on May 1, 1931, a little over one year after the start of construction, it set a standard yet unequaled for logical construction and contractor coordination.

The end of the era produced this century's finest example of an urban vision. Rockefeller Center, by Raymond Hood, collaborating with an architectural team that included Andrew Reinhard and Wallace K. Harrison, clustered fourteen buildings around the central axis of an arcade and sunken plaza, a concourse of retail shops and a double height lobby that connects Fifth Avenue with the Sixth Avenue subway. The RCA Building and Radio City Music Hall opened in 1932, with the construction of the remaining buildings struggling through the 1930's.

Figure 3 Rockefeller Center, New York Hood with Reinhard and Harrison, 1932–1939
(Courtesy: Rockefeller Center).

Even though economics were a constant concern, the plaza areas and lower rooftops were landscaped as elegant formal gardens with the skating rink added in 1936. Sculpture and extensive murals embellish the outdoor and indoor spaces, high speed elevators and one of the first applications of central air-conditioning established the quality level of Rockefeller Center in the marketplace. But it is the lofty formal massing and the intimacy of the outdoor spaces that still gives anyone coming off Fifth Avenue an unforgettable urban experience.

By the late 1940's, the post-war economy of the U.S. had made the transition from wartime industry to the expanded production required to meet the pent-up civilian demands. The last of the office space that had remained vacant since the Depression was rented and a building boom began with a construction volume unprecedented in history. New York was the first city to recover and shake off the gloom of office building investment. Most of these initial buildings were generic with wedding cake profiles filling out the maximum allowable zoning. While the exteriors generally reflected that which had been done twenty years before, there were major new contributions of technology. The market had changed to larger floor areas that were made possible by fluorescent lighting that was cooler and more efficient, central air-conditioning to provide comfort levels for both interior and exterior zones and faster automatic elevators.

During the intervening years, the architectural climate had also changed. Historical styles no longer seemed appropriate. There was a post-war optimism at all levels of society and the issue was not looking to the past but, once again, searching for the new. Modernism was an idea whose time had come and the architectural leadership was about to shift back to Chicago.

The creative forms and embellishment of Art Deco in the 1920's had grown out of an earlier arts and craft movement. With an almost parallel development, Modernism had its origins in the Bauhaus and the belief that technology was the spirit of the time. When Mies van der Rohe came to Chicago in 1938, to become director of the architectural program at what is now Illinois Institute of Technology, he was recognized as its leading practitioner. This stature was based on a few very visionary drawings, his role at the Bauhaus, furniture design and three modern low-rise structures. Chicago would provide the first opportunity to do high-rise buildings.

It is possible to review the dynamics of change through the development of the residential and office prototypes in Mies' office. Promontory Apartments, Chicago, 1949, was a 22 story reinforced concrete structure representing pre-war construction technology. The floor was a pan forming system with a plaster suspended ceiling, partitions were plaster on masonry units and the exterior was the exposed concrete skeleton with brick spandrels and infill. 860-880 Lake Shore Drive, Chicago, 1951, was a 26 story steel structure with a steel and glass skin that became the icon of Mies' influence. Sited along the shore of Lake Michigan and surrounded by vacant land with surface parking, the glass lobbies were recessed and the buildings were expressed as free-standing towers.

Commonwealth Promenade and 900 Lake Shore Drive, Chicago 1956, were two buildings that made the transition to flat slab construction and at 28 stories were the world's tallest reinforced concrete structures. The underside of the concrete slab was now the ceiling and the partition system was a combination drywall with a single coat of wet plaster. Air-conditioning had not been included in 860 and 880 because the lenders believed the additional costs were not financially feasible. These two projects now became the first residential buildings to have central air-conditioning. The exterior skin was a custom curtain wall of aluminum extrusions fabricated into a grid and glazed with tinted glass. The cycle of new construction technology was completed in 1958 when drywall systems completely replaced all wet plaster.

Figure 4   860–880 North Lake Shore Drive, Chicago Mies van der Rohe, 1951
(Courtesy: Chicago Historical Society).

All of the previous studies and issues of proportion were brought together with the opportunity to do the Seagram Building, 1958. The entire building with its special hardware, light fixtures, perimeter air diffusers, partition systems and all of the other new products were designed to convey a unified feeling of quality. This was to be Mies' first and only office building in New York. Understated with its curtain wall of extruded bronze and tinted glass, the tall slender 38 story massing set back from Park Avenue became the ultimate tower on a plaza.

Figure 5   Seagram Building, New York Mies van der Rohe, 1958
(Courtesy: J. A. Seagram).

This image, along with the growing momentum of the U.S. economy, positioned corporate identity towers as the dominant architecture for the next twenty years. Where zoning would permit, every major city was growing its own skyline of tall buildings. Along the way, the developer began to be the client and the corporate name was the marketing lure for a large block of office space. During the 1980's, this architecture truly became international. First in

Europe for more scaled down high-rise banking and office market requirements, and then for the balance of the century, in Asia and the Pacific rim countries to establish their status. For the developed cities, the high-rise buildings solidified their economic strength in the region. For the developing countries, the high-rise buildings attempted to establish credibility to attract foreign investment.

Looking back now over this very brief summary that represents 100 years of history, it is easy to give an unqualified answer, "Yes, there is a future for high-rise buildings." However, this is not a clarion call to Utopia.

The most glaring error is that high-rise apartment blocks must not warehouse low income subsidized families. While they may have been built with at least the good intentions of providing new, clean housing and the illusion of urban renewal, their failure has finally been accepted. Through the years the social programs increased the isolation, dependency and poverty. In spite of the politicians who have feared losing control of their constituency, virtually all of these remaining apartment buildings in Chicago will be demolished in the next few years. Buildings do not solve social problems.

There are also the issues of urban fabric and infrastructure. During the first 50 years, all of the office buildings were built out to the property line and maintained a sidewalk edge of retail shops. Even Rockefeller Center lined its inward arcade and sunken plaza with shopping, and the Empire State Building sets the example of retaining the street edge urban scale with a tower on a base. New assessments must be made.

Every city might welcome a dignified tower on a plaza. Open spaces also have an urban value. What did not work were the zoning changes that provided bonuses for ground floor set backs. When implemented, this completely destroyed the life-giving street edges of the city. Moreover, it also affected the architecture. It seemed that most high-rise buildings were being designed with presentation drawings and constructed as stand-alone objects expressing more of the architect's pastiche rather than their role in the urban setting. What began as a positive search for diversity has the danger of becoming an end unto itself. It is a fundamental dichotomy that high-rise buildings have achieved a confident sophistication of architecture and technology, but there is still much to learn about how to enrich the city with their development. One lesson is to look back again to the 1920's.

Returning from a trip to New York, Frank Lloyd Wright did a series of drawings titled, "Skyscraper Regulation," in 1926. His basic concept was to accept the grid and plan several city blocks at one time. Parking was to be below grade or above the second floor within the complex. The street level was maintained for services and automobile access with the streets widened for landscaped medians. The second level was pedestrian with sidewalks that bridged over the streets. Horizontal surfaces were landscaped gardens and terraces with some connecting pedestrian walkways also spanning the streets. All of the services including parking, retail, commercial and restaurants, were planned within the horizontal infrastructure of the base with the towers given their own open orientation into the natural setting of the roof level parks. The beauty and the logic of these theoretical drawings were Wright's last attempt to make sense out of the city before his dictum that it should spread out.

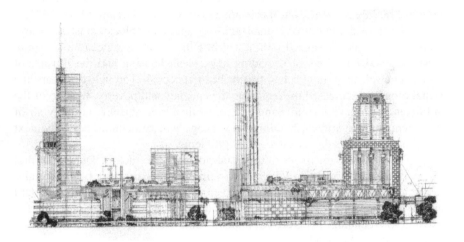

Figure 6   Skyscraper Regulation, New York Frank Lloyd Wright, 1926
(Courtesy: Copyright 2000 The Frank Lloyd Wright Foundation).

Some 40 years later, the graphics had changed to master plans that could address larger segments of a city. Incorporating the new terminology of urban design, these new images could represent streets with vistas and landscaped edges framing an attractive mix of high-rise buildings. However, when examined more closely, they were usually just identifying parcels of land for separate development without any provisions for integrating the buildings together with a cohesive infrastructure. It is a significant commentary on the present urban reality to note that even with the timeless message of Rockefeller Center, cities have continued to be built with stand-alone icons.

One significant exception is Embarcadero Center, in San Francisco, by John Portman, 1976-1981. This 8.5 acre site has a wedge-shaped atrium hotel on the east end near the Bay and extends west five city blocks with three additional blocks added in a second phase. All parking is below grade with the street level and two additional pedestrian circulation levels forming a three-story base for the towers. An existing east-west street through the center of the project was closed to vehicular traffic to form a three-level pedestrian spine. Bridging over streets and connecting the Golden Gateway housing on the west with Justin Herman Plaza at the Bay end, it becomes a wide landscaped multilevel access to the retail shops, restaurants and building lobbies. These three levels of public spaces are really people places with generous seating, trees, fountains and art work. If the provisions for financing to build out a complete complex at one time can be achieved, the potential of including similar quality amenities only enhance the success and secure the investment of the venture. The lesson here is not just assembling several square blocks for an urban development, the architecture must also produce a sense of place.

Figure 7　Embarcadero Center, East View, San Francisco, John Portman, 1976–1981 (Courtesy: John Portman and Associates).

All of these issues became the premise for a joint research project at the College of Architecture, Illinois Institute of Technology (IIT), that was supported by the resources of, and funded by, Hyundai Engineering and Construction (HDEC), of Seoul, Korea. Previous publications from the Council on Tall Buildings and Urban Habitat, including most recently Chapter 2 of the 1995 Architecture of Tall Buildings, have documented IIT's historic and current role in the development of high-rise buildings. In general, it has been the tradition that students at the Graduate and Doctorate level use the design studio as a research process to combine faculty from the academy with professionals from practice to form a unique study environment. This same format was used to assemble a research team of several graduate students, key professional consultants, along with full-time representatives of architecture, structural and mechanical engineering from HDEC that were all coordinated by IIT faculty.

The initial proposal was the research and design services to consolidate the globalization of communication and electronic industries with a 340,000 M² (3,650,000 sq ft) ultra-tall office building with a 600 room hotel in a planned area of Seoul, to be named The Hankang City Project. Including a Business Center and an Electronics Exhibition and market, the total area was programmed at 490,000 sq M (5,250,000 sq ft) with approximately 4,600 cars. During October 1996, four different architectural options, including their structural systems, for a 502 M (1,650 ft) tower with 134 floors on a four-level base, were presented to ownership. The selection process for the final tower to be designed in more detail during a Phase II of research also revealed some major problems. It became obvious that it would be virtually impossible to build a project this large in a relatively undeveloped location in Seoul without the required supporting infrastructure. The decision was made to position the ultra-tall tower within the context of a completely planned Hankang City.

To resolve the growth and economic needs of Seoul as a world-class city, the City Planning Committee had designated five strategic major centers within the metropolitan area that had the potential for major urban development. This proposed site of approximately 685,000 M² included a U. S. military compound in the center, railroad property, and a low density residential district with few social constraints. The priority of urban issues were identified as: vehicular and pedestrian traffic flow; multilevel streets; large horizontal base building to include the various functions and the orientation of the tower to open and closed spaces.

Since 1993, the Seoul City Planning Committee has been preparing policies for development schedules, criteria for planning, financial control and developer selection. The primary goal of the Hankang City Program was to bring together an appropriate mix of investment real estate to assure financial feasibility:

Hankang City Transportation Center	180,000 M²
Exhibition and Convention Center	220,000 M²
Convention Hotels	400,000 M²
Ultra-Tall Tower Complex	490,000 M²
Four Tower Office Complex	500,000 M²
Retail Shopping Center	300,000 M²
Market Rate Residential	740,000 M²
Public Service Facilities	60,000 M²

2,890,000 M²
(31,000,000 sq ft)
Not including parking and service areas

The challenge was to unify these functions with a design concept that would be cutting edge for improving the living and working environment with a new urban quality of life. Most cities have a central office core with residential high-rise and low-rise options distributed within the metropolitan area. This separation of office and residential forces the major shift of a commuting population every working day. For urban areas with accelerated growth, an already inadequate transportation infrastructure becomes intolerable. By bringing all of these functions together, Hankang City has the opportunity to demonstrate a new form of urban architecture and construction to accommodate urban growth for the 21st Century.

The most immediate planning concern before any design level decisions were made was the flow of all traffic to and from the site. These considerations had to include the total mix of varied traffic requirements, including office, retail, shopping center, hotel, convention, electronic market and residential, along with access to the new Hankang City Transportation Center. There is a six lane east-west access road at the south edge of the site, along with another four lane access road at mid-point, that both connect with the Riverside Expressway on the north edge of the Han River. The elevation of these roads are +6 M with long ramps that bring all traffic down to grade. A north-south high speed expressway next to the east edge of the site is planned at −6 M with ramps bringing all traffic up to grade with an interchange for Hankang City and another for the Transportation Center. Internal circulation has been planned to minimize any traffic light constraints.

As was previously stated, one of the most critical issues of international urban development is that the transportation infrastructure invariably lags behind the exponential growth. Hankang City brings together one of the most sophisticated intermodel transportation systems ever consolidated into a single Transportation Center. Within the classical scale of a transportation hub space of approximately 1000 M² (10,760 sq ft) with a 38 M (125 ft) high vaulted roof of glass are the interfaces of a new high speed train station, a railroad station for

surface trains along with express rail to the international airport, an existing
subway system, commuter bus terminal and expressway vehicular traffic. Three
levels of perimeter shops, restaurants and travel services surround the space to
provide all appropriate travel amenities, along with direct connections to a 7,500
car parking garage. This hub, in turn, is linked to Hankang City with a loop
transit system that has two shuttle trains with a continuous cycle. One train
travels clockwise from the station and the other counterclockwise for optimum
access to terminals adjacent to all buildings and functions.

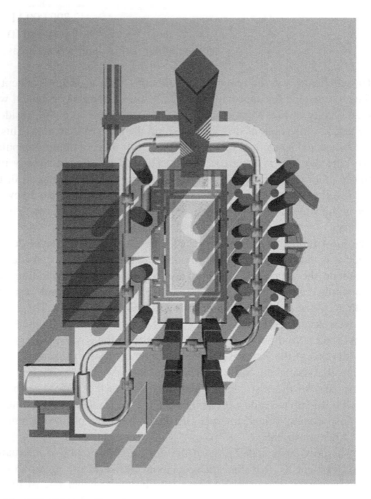

Figure 8   Hankang City, Plan View, Seoul, Korea IIT/Hyundai Joint Research, 1997
(Courtesy: Robert Krawczyk).

Two basic decisions established the design concept for Hankang City. The first was to simply raise the plaza level to a +12 M (40 ft) elevation. A single basement level permits automobile and pedestrian circulation under the grade level streets with typically four levels of parking above grade at 3 M floor to floor. This above grade structure for parking is not only more economical to build, it also becomes available for alternate functions during the evening and weekends i.e., office parking is also available for the Shopping Center and the Exhibition and Convention Center.

The other planning concept decision was to build the 100,000 M^2 (1,076,000 sq ft) exhibition building over the eight lane expressway to the east edge of the site. This recaptured site area becomes a Central Park with burmed landscaping, lakes and walking paths. The actual park is 134 M wide by 300 M long (10 acres), but when the landscaped open space is combined to include other landscaped areas, the open space is 215 M wide by 475 M long, or over 100,000 M^2 (25 acres). This Central Park landscaping and the scale of the open space balance the high density with nature.

The ultra-tall tower is located on the south end of Central Park and is complemented by a four office building complex on the north end. On both the east and west edges of this Plaza Level open space, the sidewalk is enriched with the scale of trees and the diversity of shops, restaurants and gathering places for outdoor dining and people-watching at both the sidewalk level and the outdoor Promenade at the Concourse Level. Also at this level is a continuous two story interior pedestrian arcade with a vaulted skylight of glass that encloses the shuttle trains.

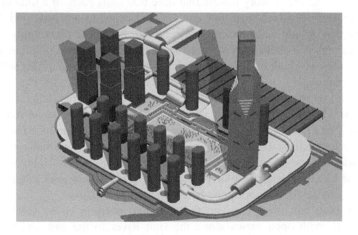

Figure 9   Hankang City, Southwest View, Seoul, Korea IIT/Hyundai Joint Research, 1997
(Courtesy: Robert Krawczyk).

To conveniently serve the needs of the Convention Center, two hotels have an exterior orientation to Central Park along with direct access to the Convention Arcade. Each has two 155 M, 50 story towers with 1,000 rooms each, or a total of 4,000 rooms. Meeting rooms and food services that are normally required to serve a convention facility are provided by the hotels. Basement unloading facilities are shared with the Convention Center. The large 95,000 M^2 exhibition space is divisible into smaller units, each with its own entrance from the +18 M Concourse Level. There is also a lower level interstitial space for services and smaller exhibitions at the +12 M Plaza Level.

Most of the 300,000 M^2 of retail is located in the Shopping Center on the west side of Central Park. The sidewalk edge and Concourse Level Promenade have an exterior exposure for a diversity of shops and restaurants. Two levels of shopping also line both sides of a continuous interior Arcade at the +18 M Concourse Level and the +24 M Shuttle Level. The west edge terraces down to grade level to link with the existing Seoul Electronics Market.

The Shopping Center Arcade becomes the Concourse Level main street and central circulation spine for pedestrian access to all of the residential towers. This city sidewalk edge has the shops, restaurants, cinemas, entertainment and support services, bringing the people places, color and sounds of a city together into a unique urban environment. At the axis with the residential buildings, there is the identity of a larger spatial intersection with smaller arcades leading to the secure lobbies of both buildings. The +12 M Plaza Level has lobby access for taxi and visitor parking. A 27 M diameter opening through the Shopping Center brings natural light down to this level. Six 122 M, 45 story residential towers have 500 units each, or a total of 3,000 units. On the west side of the Arcade there are six 152 M, 55 story towers that have 600 units each, or a total of 3,600 units. The roof over the Shopping Center is landscaped as a private park for the residents that would also include recreational facilities, community center, daycare centers and elementary schools. Hankang City makes it possible to have the convenience of both living and working within the diversity of this self-sufficient community.

At the north end of Central Park, four 245 M, 60 story office towers of 125,000 M^2 (1,350,000 sq ft) each are grouped around a common covered courtyard. The Plaza Level has the drop-off areas and the garage entrance for the cars and driver limousines. There is a single major shuttle train terminal with radiating arcade access through support retail to the individual lobbies at the Concourse Level. It would also be possible for the two towers on Central Park to each have small 200 room boutique hotels on the upper floors.

The ultra-tall tower is given the dominant position at the south end of Central Park with open views along the Han River to the east and west and across the river to the city views of Seoul. Observation levels and restaurants are located on the four top floors. The next twenty-two floors have the hotel sky lobby and 600 guest rooms with an additional seven floors of support services below the sky lobby. There are two floors of mechanical with the balance of the floors divided up into three office zones with mechanical floors between each zone. As each of the corners of the tower are dropped off to change the form to a rotated square, the stepped enclosure becomes a twelve story atrium planned as corporate community space. The lowest level would be landscaped as a terrace and gathering place with food service available throughout the day. The remaining floors could have private dining rooms, conference rooms and executive offices opening out to the space.

Figure 10   Hankang City, North View, Seoul, Korea IIT/Hyundai Joint Research, 1997
(Courtesy: Orlando Cabanban).

The horizontal structure is a very critical element of the project. It must accommodate a population of approximately 20,000 people, including the functions of the Business Center with its meeting rooms and auditorium, racquet club, the hotel banquet rooms and back-up services, the transportation node and support shopping center, lobby areas for the hotel and office portions of the tower and the circulation for the observation decks and sky restaurants. But perhaps its most important architectural significance is to provide a strong visual platform as a base for the ultra-tall tower.

Figure 11   Hankang City, Northwest View, Seoul, Korea IIT/Hyundai Joint Research, 1997
(Courtesy: Orlando Cabanban).

The final presentation included a complete engineering analysis to confirm feasibility. Both lateral and gravity loads were resolved with a structural system that combines an exterior structural steel frame tube and an interior structural concrete core-wall which drops off below the upper third of the building. At the transition zones, the lateral loads are transferred through the sloped faces and there are belt trusses at all mechanical floors to reduce the rotational reactions. Large diameter concrete caissons would carry the loads to bedrock which is estimated to be –28 M below grade.

Based on computer simulations and economic analysis, a hybrid combined cycle cogeneration thermal energy storage plant was designed on a modular basis not only for the ultra-tall project but is also expanded to produce all of the energy requirements for Hankang City. In general, heat recovery through heat pipes and thermal-wheel for all ventilation air is used throughout the project.

Optimum energy efficiency is achieved by balancing all of the different residential, hotel, retail and office mechanical requirements with the ability to transfer and store energy. In addition to the electricity produced from cogeneration and the supply system of chilled and hot water, the Hankang Utilities also includes bringing incoming domestic water into storage tanks to be processed through a water purification plant. All sanitary waste goes through a treatment plant to produce availability of gray water with the overflow entering the city sewers. And, all of Hankang City would have optical fiber access with Internet always online and the most advanced telecommunication systems available, including the co-location storage of Internet carriers.

High-rise buildings will always have the logic of being the most energy efficient and practical way to resolve high density urban requirements. This project demonstrates a cohesive mix of functions that are interfaced with a horizontal infrastructure concept that also brings life to the city. What had not been thoroughly researched was the ultimate height of a high-rise building.

When the ultra-tall tower was designed, the height was intentionally programmed to exceed the Sears Tower's 443 M (1454 ft) by approximately 200 ft. Since then, there has been another determined escalation to achieve that temporal distinction, "World's Tallest Building." One of the earliest was Norman Foster's proposed 840 M (2,755 ft) Millennium Tower, in Tokyo. All of the architects and engineers involved with these boundaries of height have confidence in the potential of today's technology. However, very significant planning issues for both the vertical tower and the horizontal base must always be resolved. As a postscript to Hankang City, there has been an ongoing study for a Half-Mile High tower to compare the feasibility of this major change of scale with the more conventional high-rise buildings. The architectural objective has been to not just reach for height but rather to design a viable vertical city with all required services and amenities. The assumption was made to establish the gross area at approximately one-half the size of Hankang City, or a total of a 15 million sq ft tower on a base of parking and retail. When the magnitude of these traffic and circulation patterns were analyzed in more detail, the criteria replicated the services that had been planned for the ultra-tall tower site. With the assumption that additional basement floors of parking could be added, the Half-Mile High was also positioned at the south end of Central Park.

The base of the tower is 360 ft × 360 ft with a slightly tapering vertical profile. There are 90 ft × 90 ft structural tubes at each corner connected together with large Virendeel trusses on a vertical spaced module of squares. At two locations, the infill floors between these trusses are eliminated to allow a complete flow of air through the tower. The vertical sequence of functions has a large atrium above the Business Center and Exhibition lower floors that serves as the lobby and restaurant level for 2,000 convention hotel rooms. The next vertical function is five million sq ft of office space with the upper floors of the tower planned with approximately 3,600 units of market rate and luxury residential and a luxury hotel. The top four floors would also have observation decks and sky restaurants.

There are several conceptual issues that determine the feasibility. If a major portion of the residents can both live and work in the tower, the circulation requirements at the base of the building are significantly reduced. There are similar efficiencies if many of the residential amenities of retail shopping, restaurants, cinemas and entertainment are also included in the tower. It would then be possible for residents to literally live in the tower for several days at a time and, when they did leave, it would not be during peak periods. A vertical infrastructure is required to move these large segments of the building population and distribute building services.

The scale of the vertical express elevator cabs would be similar to the shuttle train cars and would move on a continuous cycle to transportation nodes in the tower. These three story high sky lobbies would be large landscaped atrium spaces with appropriate support amenities that would also sort out the passengers to conventional elevators to their selected floor. At the residential levels, these landscaped atriums could also include day care centers, elementary school or community services. The final phase of this study will include an economic analysis to compare the Half-Mile High with other Hankang City functions.

All of this research must still relate back to the 6th World Congress theme. Making projections into the Third Millennium is a daunting endeavor. With our most recent experiences, even the 21st Century may be difficult to comprehend. At the beginning of the 20th Century, the U.S. had an agricultural economy, but industry was growing. By 1980, the industrial economy dominated and the number of people employed in agriculture was insignificant. From roughly that point in history, computer technology and the Internet began to take over. By 2005, it is estimated that the percentage of workers employed in industry will be less than 20%. What is making this happen is software. Instead of physical capital, the economy is now based on ideas.

A company's survival is directly related to the potential of its creativity and innovation. It must not only attract the best and the brightest, it must also hold them with a corporate culture that builds a sense of community. These same market forces are beginning to reshape the city. In Chicago, the suburban office vacancy rates are increasing and downtown construction has started again. This talented workforce, along with a growing population that is rejecting suburbia want to live in the city. The resulting residential construction is generating high-rise buildings supporting the density of the core, along with more modest developments on all the surrounding infill. This same infusion of a vitality of life back into cities is occurring in almost every industrialized American and international city. A similar transformation is also taking place in this next wave of the global economy. Even developing countries are compressing the time frame by marketing favorable government policies, the required urban infrastructure and their resource of an educated workforce.

While technology will continue as the delivery system, humanism will define the future. Making our cities more livable is now the idea whose time has come.

Hankang City Research Team

Graduate Students		Hyundai Engineering and Construction	IIT Faculty
Samir Abdelmawla	Hyeong-Ill Kim	Dr. D.W. Lee	Leonard Bihler
Amel Aboulla	Jin Hoon Lee	Han-Soo Kim	Mahjoub Elnimeiri
Alex Baumgarten	Steven Park	Sang-Min Kim	Robert Krawczyk
Fang Chen	Dharmentra Patel	Man-Kun Lee	George Schipporeit
Yongsun Choi	Hunseock Shin	Jae-Chul Lee	David Sharpe
Kamon Jirapong	Xu Sun	Do-Kyun Lim	John Urbikas
Kyu Hyung Kang	Firuzon Yasamie		

PLANNING AND ARCHITECTURE

Social/Community Impacts and Implications

P. Jerome

For cities in many parts of the world, the dominant issues for the 21st century arise from continued urbanisation and the difficulties of providing basic infrastructure services and mass housing. However, for cities in the developed world – whilst still concerned with sustainable provision of basic services – a new dynamic is emerging. It is a dynamic derived from long term demographic trends, greater divisions between rich and poor and sometimes ill-defined 'lifestyle' aspirations.

Using Melbourne as an example, this paper explores various aspects of this emerging dynamic and possible implications for planning and development of the urban habitat. It draws on a number of publications from the Research Branch of the Department of Infrastructure.

DEMOGRAPHIC TRENDS

Fertility

In common with many parts of the developed world the fertility rate in Victoria has declined over recent decades. In fact it has been below replacement level since the mid-1970s although the impact of this has been masked by high levels of immigration. From a rate of 2.95 in 1971 it had declined to 1.94 by 1981. It has continued to decline and by 1997 had dropped to 1.69, well below the nominal replacement rate.

The long term effect of this decrease in fertility rates is that the rate of growth of the population of Melbourne and Victoria will decline over the next twenty years.

There are some interesting geographical patterns in fertility rates. Areas with the lowest rates are heavily concentrated in the inner suburbs of Melbourne while suburbs with the highest fertility rates are generally on or close to the fringe growth areas. For example, the fertility rates for Cranbourne and Wyndham – two outer fringe growth areas – were 1.8 and 1.7 respectively in 1996. By contrast the fertility rates for Prahran and St. Kilda – two inner metropolitan suburbs – were 1.0 and 0.8 respectively.

Significant drops in fertiltity rates have also been observed in those inner suburbs which have undergone 'gentrification' associated with an influx of white collar professionals into previously working class neighbourhoods. One such area – Brunswick – has seen its fertility rate drop from 1.7 to 1.2 in the period from 1981 to 1996.

Age structure

We are an aging population. This is due largely to the natural aging of the post-war baby boom generation coupled with the decline in fertility rates previously noted. In a nutshell, the percentage of people aged 60 and over in Melbourne is forecast to rise from 16% in 1996 to 23% in 2021. By contrast, the percentage of people in the age range 0–17 years is expected to fall from 24% to 19% in the same period.

In absolute terms these figures translate to a net **decrease** of 48,500 in the 0–17 age range but a significant net **increase** of 381,447 in the 60 and over age range during the next twenty years. This last figure is startling. It represents a 74% increase over the 1996 population of 60 years and older. For planners it signifies a massive proportional increase in demand for the sorts of health services, recreation and leisure activities and accessibility requirements associated with these older age groups.

However, as with fertility rates, the geographical distribution of aging is not uniform. There are many suburbs on the fringe of Melbourne which continue to attract young families and have a high proportion of children. Wyndham, for example, on the fringe of the metropolitan area in a designated growth corridor, has an age distribution heavily biased towards the family age groups.

This type of age structure contrasts strongly with that found in Melbourne's older and more established suburbs. Monash, for example, is what we call a middle ring suburb. It was settled predominantly in the 1960s and at that time had an age structure very similar to that of Wyndham today. However, in the last twenty years Monash has experienced a marked loss in population within the 0–20 year age bracket and a significant increase in the number of people 75 years old or more. This pattern typifies the 'life cycle' of Melbourne suburbs. They witness an initial influx of young families, a period of growth as additional children are born, and a gradual decline and aging of the population as children grow up and leave home. Whilst recent evidence suggests that children are staying at home longer this has not resulted in any change to this underlying pattern of suburban life cycles.

By contrast, some of the older inner suburbs are being rediscovered by increasing numbers of younger people typically those on high incomes and without children. Port Phillip, adjoining the CBD of Melbourne, has gained population in the child-bearing age groups between 20 and 50 but has not seen any commensurate increase in children. In fact the number of children in Port Phillip has declined during the same period.

Household size

Changes in household formation and the aging of the population are two of the primary factors causing the steady fall in average household sizes in Melbourne from 3.5 in 1966 to 2.7 in 1996. It is expected to be down to 2.3 by 2021.

Traditionally in Melbourne larger households have been found in the outer suburbs. Conversely smaller households are mainly found in the inner city and surrounding suburbs with the fringe exhibiting very low numbers of small households. Overall the pattern shows a decline in average household size with increasing proximity to the inner city.

But decreasing household size is also a product of the life cycle of suburbs and other social factors such as divorce and the aging population with the effect that all suburbs are experiencing the same phenomenon. In Wyndham for example, the figure is expected to drop from 3.11 to 2.56 over the next twenty years; Monash will see a reduction from 2.77 to 2.54 over this period and Port Phillip will fall from 1.91 to 1.79. In Port Phillip, over 70% of households already comprise one or two persons.

The most obvious impact of this trend is in housing demand. It has been estimated that if average household size had not changed in the past 30 years, the current population of Melbourne could have been accommodated in 250,000 fewer dwellings – or the equivalent number of dwellings built in Melbourne over the past twelve years.

But in planning ahead, the converse is also true. By 2021 Melbourne will need 420,000 extra dwellings. About 270,000 of these will house the forecast population increase; the other 150,000 will be needed solely to compensate for the anticipated fall in household size.

Mobility

In each five year inter-censal period from 1981 to 1996 more than 50% of people in metropolitan Melbourne have indicated that they are not living in the same place as they were at the previous census. This suggests a massive amount of mobility. It must be tempered with the obvious movements associated with younger people leaving home, new households being established and elderly people either dying or moving into retirement accommodation. But the figures also indicate that up to 10% of people in all age groups move homes in any given year.

Traditional patterns of mobility in the metropolitan area have focussed on student accommodation around universities and a strong outward trend as new young households are formed. Outward mobility has been strongly sectoral. That is, people in the east moved out further east; people in the north moved further north etc. Whilst this remains the dominant pattern there are now some distinct counter trends which appear to be supply driven. New housing developments with emphasis on amenity and landscaping have attracted markets across from the east to the west; inner suburban developments have pulled people from the outward thrust to an inward movement. Many of these movements are said to be associated with life-cycle or life-style aspirations. Nevertheless, the underlying pattern of mobility remains strongly outward and radial.

URBAN DEVELOPMENT IN MELBOURNE – THE HISTORICAL PATTERN

Against this background of demographic trends, Melbourne has developed in a remarkably consistent pattern which demonstrated, until very recently, the triumph of institutions and regulations over community aspirations.

Unlike many cities, Melbourne's development took place after the industrial revolution. It did not develop the intricate patterns of older European cities. Rather, its development was shaped predominantly by railways and roads allowing the broad spread of the urban area from an early stage. With the exception of the 1890s depression it has been a city of ongoing and outward growth for more than 150 years.

The role of State government policy and planning strategy has been significant. For much of the last century the planning authority was the Melbourne and Metropolitan Board of Works (MMBW). Its planning policies directed the rapid growth of Melbourne from the 1960s into distinct corridors of development. As the centralised agency for the provision of Melbourne's water and sewerage infrastructure, as well as being the planning authority, the MMBW was able to directly influence the direction of Melbourne's growth.

Melbourne's growth in the post-war era was fuelled by high immigration from abroad, high fertility rates, rising levels of household wealth and easy finance for home and car ownership. The vast majority of population and urban growth occurred in the newly developing fringe areas in contrast to the stagnant or declining population levels in the inner suburbs.

Government incentives for home purchase and support from financial institutions allowed most Melburnians to purchase their own homes. The physical expression of urban development was controlled through the Uniform Building Regulations which were in operation from 1947 up until the early 1980s. The impact of these regulations was – as their name implies – uniformity, and the post-war urban development of Melbourne proceeded in a number of rigidly defined ways. Until very recently over 90% of Melbourne's post-war housing stock comprised single detached two and three bedroom houses. Typical blocks of land for these houses were between 500 and 900 square metres in area with standard frontages, depths and building set-backs. Even taking into account the pre-war inner city development of much smaller blocks and much larger blocks in the semi-rural fringe areas, almost 70% of all residential blocks in the metropolitan area still fall within this size of 500–900 square metres.

An attempt was made in the 1960s to provide for medium density development by a set of prescriptive rules governing height, set-backs and car-parking requirements. For several years this gave rise to a proliferation of small apartment blocks in some inner suburbs until popular concern about their perceived ugliness and social impacts led to the demise of these regulations. This was the only period on the post-war era when the population of two inner suburbs heavily developed in this form did not decline. These apartment blocks, locally known as 'six packs', now provide an important source of relatively affordable accommodation.

However, the popular reaction against this form of development effectively stifled medium and high density development in Melbourne during the 1970s and 1980s resulting in Melbourne having a much lower density of urban development than most other capital cities in Australia and the world. Restrictions on conversion of commercial buildings for residential use also precluded residential development in city areas prior to the 1990s.

Clearly there have been many other factors at work influencing Melbourne's development. But the monolithic control of the MMBW and the regulatory rigidity of the Uniform Building Regulations supported by the lending policies of financial institutions played a big part in establishing a framework of conservatism and a low density city increasingly dependent upon the motor vehicle as suburbs spread away from the limited alignments of public transport routes. It has also, until the last decade, been a self-fulfilling prophecy. The development industry has been very successful in persuading itself and the community that, because this is what they have, this is what they want. The status quo has prevailed.

It is now apparent that the demographic trends outlined earlier in this paper have given rise to a pent up demand for diversity in housing types and locations which the market has, until very recently, failed to deliver.

The impact on travel patterns of this low density sprawling form of urban development has also been predictable. Overall Melbourne is very car dependent. About 75% of all trips are made by car and every day there are over 500,000 car trips made shorter than two kilometres in length.

Car travel and car ownership continues to rise. Since 1996 there has been a 1.9% increase in car travel per year and there are now 0.79 cars for everyone over 18 years of age. This is despite the fact that 11% of households don't have a car at all and nearly 33% of people don't have a drivers licence.

If this pattern of car dependency continues, the total distance driven by Melburnians over the next twenty years will increase by 30%, significantly outstripping the increase in population.

Furthermore, the nature of trips has become more diverse as dual income households, in particular, pursue different destinations and often combine shopping, work, school and recreational activities in cross-suburb, multi-destination travel patterns.

This combination of a low density city with a high degree of car dependency and the interplay of these factors with the changing demographic trends noted earlier are at the crux of planning for the future of Melbourne.

THE CHANGING POLICY CONTEXT

The late 1980s and 1990s marked a significant shift in the policy environment for Melbourne's urban development.

Some of the key changes were:

* Removal of the metropolitan planning function from the MMBW in 1985 and the subsequent devolution in 1989 of planning to the then 52 local councils which comprised the metropolitan area. There was no formal

metropolitan planning authority. The broad thrust of the MMBW planning strategy was maintained but little effort was put into implementation at the metropolitan level;

- The Planning and Environment Act 1987 which enshrined third party objection and appeal rights as part of the development approvals process;

- Legislation governing subdivision of land and buildings was reformed in 1988 providing, amongst other things, for the sale of property 'off the plan' with the added incentive of avoiding stamp duty. This allowed developers effectively to pre-sell a development and thus secure finance;

- The corporatisation or privatisation of major infrastructure utilities (water, sewerage, drainage, electricity and gas);

- The introduction of user pays and developer contributions schemes to assist in the funding of urban infrastructure and services;

- The amalgamation of local government authorities which, amongst other things, left local government as a relatively weak player in planning and development for several years as it coped with massive restructuring, compulsory competitive tendering of services and mandatory income reductions of 20% imposed by the State government;

- The simultaneous introduction by the State government of a new performance based system of planning control which sought to streamline and standardise the planning system; and

- The freeing up of regulatory restrictions on types of urban development, particularly medium density housing.

Taken together these public policy initiatives (with the possible exception of the objection and appeal rights of the Planning and Environment Act 1987) have 'freed up' the processes of urban development in Melbourne.

One of the immediate impacts has been a rapid escalation in medium density housing in all areas, but particularly the inner and middle suburbs. This has exposed a substantial pent up demand for a diversity of housing types and densities.

Deregulation, privatisation and a move towards user-pays regimes has meant that developers are less able to rely on government funded infrastructure being provided in fringe areas than in the past. Due to the reallocation of urban development costs in this way, large scale development on the urban fringe can now be more costly to initiate so only the larger developers are likely to be able to participate in this market.

There have also been changes in the nature of property investment over the last decade with increasing investment capital being made available through superannuation funds. Other factors affecting housing investment decisions which are peculiar to our jurisdiction include the tax benefits of negative gearing for those wishing to invest in rental property and the availability of retirement and redundancy lump sums which have provided capital for property investment. There has also been an increase in international capital and investor interest in property markets, aided by the lessening of restrictions on foreign investment.

PHYSICAL AND SOCIAL OUTCOMES

These changes in public policy and the broader economic environment have had a number of impacts on Melbourne's physical development and social patterns.

The inner city of Melbourne has benefited from the rapid growth in finance, property and business services, in line with international trends. The concentration of hotels, restaurants and entertainment facilities in inner urban areas has enhanced the role of the central city as a place for recreation and leisure consumption.

The inner areas of Melbourne have also become attractive residential locations with the result that inner city and inner suburban populations have increased for the first time in decades. For example, Port Phillip and the central area of Melbourne have each seen increases of about 4000 residents between 1996 and 2000. Another inner suburb, the city of Yarra, has seen an increase of 2500 in the same period. The contemporary image of the inner city has resulted in substantial redevelopment and increasing property prices. This location has proven particularly attractive to younger people due to the availability of rental accommodation, proximity to educational institutions, access to employment, particularly in the growth sectors of technology, information and hospitality, and 'lifestyle' attractions in the cultural and entertainment precincts.

This has changed the population profiles of many inner city areas, with a reduction in the number of families with children but a significant gain in non-family households. It has also led to a reduction in the number of cheaper accommodations such as boarding houses thus reducing housing choices for the disadvantaged and exacerbating the problem of homelessness.

As in other cities there has also been substantial redevelopment of former industrial and utilities sites. In this respect Melbourne has been particularly fortunate. It has avoided the inner urban blight experienced by many other 'developed' cities of the world as inner city industry relocated to greenfield sites leaving workers stranded and factories empty. This has been by accident rather than by design. Relocations and the departure from inner city sites fortuitously coincided with the beginning of the rediscovery of the inner city area and of its heritage values which are now much appreciated. This movement gathered momentum in the 70s and 80s giving these areas more widespread popularity.

With the explosion in the market for inner city living in the 90s many sites were prime for redevelopment.

Another feature of the housing market that has come from the public policy changes of the 1990s is the discovery of high rise living in apartment blocks. The ability to test the market through pre-selling has revealed a healthy demand particularly in and around the central city. Higher density living in apartments and townhouses has now become part of Melbourne's housing market in a way that was almost inconceivable as little as ten years ago. There is an increase in rental stock and a relatively new market niche being described as vertical retirement villages. These are essentially apartment blocks with 'live-in' management and health care but they seek locations near centres of activity. They cater for the post-55 age groups and emphasise security as well as availability of support services. They are a direct reflection of the aging of the population previously noted in this paper.

Parts of the middle and outer suburbs of Melbourne have also experienced a partial reinvention of their role. Some suburbs with large blocks and an aging housing stock are proving to be economically attractive for some medium density housing but this transition is at a much smaller scale in these suburbs than it is in the inner areas. For example, in the fastest growing fringe area of Casey only 5% of new housing starts are medium density. Compare this with a figure of 36% for the established middle ring suburb of Glen Eira and 96% for the city of Melbourne. These patterns suggest that current planning in Casey should provide a sufficiently robust structure to accommodate considerable 'retrofitting' of a more diverse housing typology over the next twenty years if it is to establish a sustainable community.

Another phenomenon which runs counter to tradition is the increased number of fringe estates which are making themselves attractive to the second and third homebuyers market. In the past these areas have always been seen as the natural location for young first homebuyers. It has been suggested that the increased development costs for infrastructure delivery in these areas has led some developers to build an increased margin into their cost recovery strategies by providing a more complete development package and marketing to the more lucrative second/third homebuyer.

GEOGRAPHICAL DIMENSIONS OF THE 1990s

There are two significant geographical factors that have emerged from these recent changes in housing development in Melbourne. The first is that the share of new housing built on the fringe has declined. In 1992 over 75% of new housing was built in new estates on the outer fringe of the metropolitan area and approximately 24% was built in established but not yet fully developed suburbs. (Depending upon definitions, some commentators put the fringe figure at 85%). For a brief period in the mid-1990s the proportion of new dwellings built in the inner and middle ring established suburbs reached 45% with the obvious relative decline in the outer areas. The situation appears to have stabilised a little and the proportion now anticipated for the next two decades is about 65% on the fringe and 35% in the established suburbs.

The second is that the distribution of more affordable housing also appears to be changing. It is no longer in the outer fringe estates or the 'run down' inner city but in a number of quite different locations which seem to have no distinctive characteristics in common other than that property prices have declined in real terms over the last 10 years. Whilst more research is required on this, it would appear that there is a new distribution of disadvantage overlaying the traditional Melbourne pattern of north-west versus south-east. Nevertheless this traditional pattern still dominates as exemplified in the distribution of employment. From 1981 to 1996 the share of metropolitan jobs located in the north-west fell from 49% to 37% while the share in the south-east increased from 49% to 61%. In contrast the distribution of the city's population between these two regions remained constant.

The geography of disadvantage is a very real issue for Melbourne. There appears to be a general understanding that the 'new economy' is resulting in a greater divide between rich and poor in all the developed economies of the world. Inevitably this is translating into spatially separated areas of affluence and areas of disadvantage. It is not a new phenomenon; it has been observed in cities for many decades. But a 1999 study showed that Melbourne had the greatest social disadvantage of any capital city in Australia and that a concentration of resources, knowledge, money and power marked clear geographical as well as the obvious social divisions, widening the gap between rich and poor.

This is clearly one of the major and most complex challenges facing planners in the 21st century. But it does not yet figure large as a force or social movement in the community. Indeed, one of the most disappointing if entirely predictable aspects of the last decade's changes in the development of Melbourne has been the emergence of a social movement the focus of which is almost entirely the **physical** form of new development and the protection of existing suburbs for existing residents. This is entirely understandable and probably inevitable in the face of the pace and amount of change that has recently been experienced in some of the established suburbs of Melbourne. It also reflects our cultural obsession with property ownership which naturally engenders spirited attacks on anything which may appear to have adverse impacts on residential amenity and property prices.

We have not yet seen the emergence in any coherent way of similar concerns among the community with the **social** impacts and implications of the changes underway in our urban environment.

It is also interesting to note that in the current debates about sustainable development and the triple bottom line of economic, social and environmental impact assessment, it is the social element that is least understood and attracts least interest. This may be the most fertile and ultimately most important area for consideration as we explore urban habitats for the 21st century.

CURRENT CONCERNS

Nevertheless, some evidence of community interest in such matters has recently been surfacing in an extensive round of community workshops undertaken as part of a process of developing a new metropolitan planning strategy for Melbourne. The evidence is largely anecdotal and of a general nature. However, what is remarkable is that similar issues have been identified spontaneously in workshops across the metropolitan area and in our regional towns and cities.

There is general acknowledgement that the gradual fragmentation of society from extended families to nuclear families to increasingly smaller household types is associated with a perceived loss of community. This finding has also been reported in a number of council surveys of their municipal populations.

Public safety is also a matter that has been regularly identified and given top priority.

The importance of public transport and the need for improved regularity and safety of services is also clearly high on the public agenda as is clean air and water and the maintenance of vegetation.

Interestingly these workshops have not given such a high priority, as a community concern, to medium density housing as the more activist group referred to previously.

What is most noticeable about these priorities (and it is by no means an exhaustive list) is that they focus on the public realm and support social interaction. As neatly summarised by one participant 'I simply want to be able to visit friends and go shopping. I need good public transport because I don't drive any more. I want to do it safely and I want to be able to breath clean air while I'm doing it.'

This surely is not too much to ask of an urban habitat in the 21st century?

LOOKING FORWARD

In attempting to distil the demographic trends, urban development outcomes and emerging concerns set out in this paper, there appear to be three major issues confronting the planners, designers and developers of urban habitat and two significant debates which our community will have to have and resolve.

The first issue is about aging in place and sustainable communities. One of the most widespread and powerful social stories in Melbourne today is about the aging relative who no longer wants to look after a large family house on a large block of land but who does not want to lose touch with their existing social networks and familiar neighbourhoods. They are struggling against the loss of community. But, they complain, there is no suitable housing in the area. With an aging population this is possibly the most potent argument in support of achieving greater diversity of housing types across all suburbs.

Furthermore, relatively few people entering the 60 plus age group in the next twenty years will have sufficient superannuation or pension funds to maintain existing lifestyles. Many will seek to capitalise on the family home and to purchase something cheaper so the difference can supplement their other sources of income. An incremental spread of housing redevelopment across all suburban areas will be needed to provide for these choices and to avoid market distortions. Unfortunately some distortions will be inevitable as those with high incomes chose the more attractive locations and in doing so push prices upwards. But more opportunities will be available for more people if we allow and encourage incremental redevelopment across the board.

Nor is this an argument against the more conventional strategy of increasing densities significantly around major public transport nodes. It is entirely complementary. We need both.

The second issue is about security and public safety. It has many facets. It arises from concerns with loss of family values, respect for the elderly and any sense of community behavioural norms; it appears to have been exacerbated with the advent of more widely spread drug addiction and associated personal assaults and theft of property; and it is a concern regularly expressed by older people and women when asked about their attitudes to the public realm.

Considerable work is being done across Victoria with police, councils and communities working on strategies to improve community safety. In physical terms it means that all our development planning, design and approval processes need to build in consideration and understanding of community safety principles. These will not be optional for the urban habitats of the future; they will be a mandatory starting point.

Part of the response to safety and security concerns will be resort to high rise apartment living with full time concierges and maximum security devices. Pressure will also build for more 'gated communities' which, in a sense, are just the horizontal version of the apartment block. The point of difference is that the apartment block still takes a relatively small ground level footprint. They can be planned and distributed in such a way as to maintain the pedestrian network and 'permeability' of the physical environment at ground level. Gated communities, by contrast, are extensive in ground level coverage and tend to distort accessibility networks. They are prone to walling off significant sized blocks and creating a no-go zone of substantial proportion which is not conducive to maintaining public safety in the vicinity. They should be strongly resisted.

The third issue is about public transport. Obviously volumes could be written about it. In the context of this paper, however, only two aspects require mention. The increasingly aging population will place greater importance on safe and secure public transport for accessibility. Whilst older people certainly **can** still drive, the question more are asking themselves (and others are asking on their behalf!) is whether they **should** still drive. Only the availability of regular and safe public transport services will encourage these age groups, and, indeed, all ages, to make better use of public transport.

Perhaps more importantly, public transport can be a social connection. Whilst the anglo-saxon tradition tends towards silent queues and silent trips it need not necessarily be so. For many from other cultural backgrounds and the more adventurous of anglo-saxon descent the trip itself provides valuable opportunities for social interaction. Public transport is not just about getting from A to B. It is a social experience in itself and it provides unimaginable quantities of material for subsequent conversations and social discourse.

This is not a trivial point. Social interaction and socialisation processes have long been the mortar of a cohesive society. The more our society fragments into smaller households coupled with our obsession for domestic privacy (what other city has so much opaque glass, sill heights of 1.6 m and flimsy screening devices to prevent 'overlooking'!), the more important it will be that we create opportunities for social interaction. This is not about compulsory conversation with the person next to you on the number 67 tram; it is about preserving and promoting safe and neutral territories in the public realm where social interaction and connectedness can occur. Such opportunities are vital and the public transport system can play an important, if previously unrecognised, role in providing such opportunities.

Finally, the two debates we have to have. The first is peculiarly Australian and concerns the future of immigration policy. As noted in this paper, with declining fertility rates, already below replacement levels, and with fewer people in the child bearing years the rate of population growth will decline in the next twenty years. This may be music to the ears of environmentalists who believe this country is already beyond its sustainable carrying capacity. But to others it heralds a declining economy and living standards. Strong arguments will be put for increasing immigrations levels to boost a flagging economy and increase demand for the housing and construction industry to name but one. And hence the need for debate and resolution. We regularly run up to this hurdle but usually within the confines of our adversarial political system and not on the basis of informed community conversations.

The second is one which many cities around the developed world are going to have in the not too distant future. It is about urban density and public transport. Despite some suggestions that heavily subsidised public transport can work in relatively low density cities, the bulk of evidence suggests that there is better prospect of sustaining good public transport with the sorts of potential patronage levels that can only be provided by densities of development with which we, in Melbourne, have traditionally been uncomfortable. Despite the fact that the clustering of higher density development around public transport nodes has been part of metropolitan strategy for the past twenty years, it has been honoured more in the breach than the observance. It is the classic urban argument where everyone agrees as long as the proposed higher density development does not affect them. But the related matter of traffic congestion must also be addressed. Despite the views of some residents, Melbourne is not a heavily congested city. Yet most of our transport planning and a disproportionately large part of the overall transport budget continues to be allocated to road transport at the expense of public transport. There may well be a case to be made for allowing greater congestion as part of a strategy of supporting public transport rather than constantly adopting the approach of trying to reduce congestion.

A difficulty we have had in addressing this issue is that debate tends to get polarised by ten second sound bites into public transport versus roads and private cars as though they were mutually exclusive options. They are not. And we need a more sophisticated public debate about the role of transport in our urban habitat and how we want to arrange it rather than having it arrange us.

CONCLUSION

In conclusion, urban habitats of the future, at least in the Australian context, will have to be responsive to changing demographic trends in terms of location and diversity of housing types in order to support socially sustainable communities. They will have to accommodate higher densities of physical development in order to sustain a safe and regular public transport service. And they will have to provide a safer public realm to maximise opportunities for social interaction as a basis for sustaining a cohesive society.

In Search for a Local Image in the Arab City: The Case of Kuwait City

Omar Khattab

In its earlier attempts to catch up with western modernity and civilisation, the Arab City has lost, whether intentionally or unintentionally, its local image. Since this image was greatly associated with backwardness and obsoleteness. And was regarded as unfit setting for modern activities and uses. At the same time western models conveying the desired *modern* image were ready for the pick. Hence architectural development mainly followed the prevailing international style with few exceptions, which addressed the history and tradition of the Arab City.

With the loss of local image in the modern Arab City there was a greater loss of identity. This has resulted in a separation between the Arab City's past and its present, as well as a greater concern for its future. This was apparent in the main cities of the Arabian Gulf states. Particularly in the case of Kuwait, reassert of local identity has become a matter of great importance especially after Iraq's claims in Kuwait and the second Gulf war. The search for a culturally and traditionally responsive architecture, which started during the late 1980s, has taken a boost by the political conditions.

This paper compares some examples of contemporary architecture in Kuwait to examine the effect of Western architecture on their design. At the same time it contrasts this with some other examples, which revive local design elements and traditions in contemporary practice in order to preserve continuity and reflect change. The aim is to document the effect of globalisation on the local architecture of Kuwait, and to investigate whether an attempt to portray a local indigenous image with some global ingredients can prove successful.

1.1 INTRODUCTION

The Arab City has experienced dramatic changes during the second half of the 20th century. Changes in the political, economic and cultural arenas especially in architecture. These changes are attributed to a number of factors such as political independence, rapid population increases and most important of all is the discovery of oil in the Arabian Gulf states. As a direct consequence the built environment of major cities in the Gulf States has gone under a rapid and gigantic transformation. In the newly thriving Gulf States the oil boom resulted in the flourish of the building industry. The enormous job opportunities in this field have attracted foreign experts and labour to the Gulf States to meet the increasing demand for urbanisation and modernisation.

Kuwait is one of the most important Gulf States. Kuwait City became the focus of international construction activities as many of the world famous architects competed for large-scale projects. In addition, the government turned to famous foreign architects for major building commissions, and so did most of the private sector too.

Although tradition should have been of great significance for contemporary architecture in Kuwait, it was not adequately manifested in the majority of buildings during the last few decades. This alienated the city's inhabitants and resulted in a loss of local identity. Which was anticipated by prominent planners and architects around 40 years ago (Kultermann, 1999).

1.1.1 Early Modern Architecture in Kuwait

The use of traditional local building materials and techniques, which were associated with backwardness and obsoleteness, was abandoned in favour of modern imported materials and techniques. Affluence due to the reinvestment of oil revenues in building construction has dramatically changed the methods of traditional design and construction. Especially that modern building technology and materials started to be imported from all over the world (Bosel, 1995).

Some buildings built between 1970s and 1980s in Kuwait were influenced by the early modern and corporate styles developed in the West from 1950s till 1970s. Early modern movement was characterised by simplicity in plan and form avoiding the sophistication of the neo-classical architecture at the beginning of the 20th century. These were mainly office complexes for both public and private influential corporations, which borrowed internationally recognised images of famous office buildings.

Corporate architecture aimed to convey a sense of permanence and authority and was characterised by the extensive use of glass and steel (Peel *et al*, 1992). In the case of Kuwait, fair-face concrete was extensively used in early corporate architecture as well as stone and marble cladding.

Examples on Early Modern architecture include the Kuwait Fund Headquarters, the High Courts complex, Behbahani, Al-Muthanna, Dasman and the Blue towers (Fig. 1.1). Design commissions of these buildings were awarded to internationally renowned firms such as the Architects Collaborative and Skidmore, Owings & Merrill of USA.

Kuwait Fund Headquarters

High Courts complex

Behbahani tower

Al-Muthanna complex

Dasman tower

The Blue tower

Al-Rashed & Al-Anjari tower

Office tower on Fahed Al-Salem street

Figure 1 Examples of Early Modern Architecture in Kuwait.

1.1.2   Neo-Islamic, Late and Post Modern Architecture

As an Arabic and Islamic country Kuwait was bound to experience some experimental work in Neo-Islamic architecture. This movement was characterised by the use of Islamic motifs and architectural elements in a different context, and came as a reaction to examples of Islamic architecture revivalism at the end of the 19th century and the first half of the 20th century. Like the use of the decorative element 'mukarnas' as a column capital in Al-Awadi towers (Fig. 1.2). In fact this phase was characterised by ambiguity between what is functional and what is decorative.

As Makiya (1990) explains that structure and non-structure were becoming ambiguous which is what they became historically in later Islamic architecture. "Is the 'mukarnas' a structural element or a decorative one? The inspiration behind it is constructional: the problem of making the transition from a rectilinear to spherical geometry, from a square plan to a dome" (Makiya, 1990, p. 79)

The use of Islamic geometric patterns on façades presented another way to achieve an image that can be associated with Islamic architecture. This is clear in the examples shown in Fig. 1.2.

Al-Awadi towers

Kuwait Airways Headquarters

Le Meridian hotel

Audit Bureau Headquarters

Figure 2   Examples of Neo-Islamic Architecture in Kuwait.

After the liberation of Kuwait in 1991 till now a number of large building commissions were completed in the efforts to rebuild the country after the Iraqi

Al-Ahli bank (left) and Kuwait Finance House

Hmoud tower

Al-Sahab tower

Al-Ahlia Insurance Company

Gulf Co-operation Council Fund

Al-Jahra Gate tower

United Real Estate tower

Kuwait Chamber of Commerce Headquarters

Figure 3   Examples of Late Modern Architecture in Kuwait.

invasion. Some of these large commissions were influenced by the growing trend in Western architecture towards Late Modern and Post Modern movements. The effect of international architecture became more apparent after Kuwait has opened even more to the West whose US-led troops have secured the withdrawal of Iraqi forces from Kuwait. The influx of foreign architects practising in Kuwait continued adding another dimension to the effect of global-isation on the Kuwaiti architecture. The use of high-tech materials and tech-niques gradually started to replace the extensive use of concrete and stone cladding.

At the same time some of these architects tried to reinvent traditional Gulf-Arab forms, introducing them into the contemporary development of Kuwait City. This is what consultants HLW International has tried in the Kuwait Chamber of Commerce Headquarters (Fig. 1.3). Where the idea of a traditional Caravanserai (traditionally is a resting place for caravans travelling across the desert) was translated into a modern idiom to create an urban landmark (AR, 1999). Some examples of Late modern architecture in Kuwait are presented in Fig. 1.3.

Post Modern movement in architecture became popular during 1970s and 1980s. The use of classical architectural elements such as Roman orders in an unusual setting, materials and colours characterised this movement. Although the columns and pediments of Post-Modernism went out of fashion in Europe after the 1980s to give way to new directions and concepts in architecture (Tietz, 1999), the spread of post-modernism can still be seen in buildings recently com-pleted in Kuwait. The well-known broken pediment of the AT&T headquarters, by Philip Johnson, which became the cliché of post-modernism (Peel *et al*, 1992) has appeared in Kuwait in residential, commercial and office buildings. Some examples of Post Modern buildings are shown in Fig. 1.4.

1.2   A NEW VISION

"A country is recognised by its architects and its history is built into it ... else the new would have nothing to do and nowhere to go" (Goodwin, 1997, p. 46). In addition to the previous examples influenced by the international modern image, there were a few attempts by some local and foreign architects to recog-nise and acknowledge the Kuwaiti architectural heritage. Among these is Saleh Al-Mutawa, a Kuwaiti architect whose dream was to revive the traditional archi-tecture of Kuwait (Al-Mutawa, 1994). He set to do just that in his practice using some local architectural vocabulary in a contemporary language of three-dimensional forms. "His architectural language makes an immediate impact on all who see it, but it is disconcerting to fellow architects. The proof of this is that trivial bits of pastiche have appeared on a few other works by architects who have in no way absorbed the ocean of experience, both intellectual and emo-tional, behind Al-Mutawa's creativity and merely borrow his ideas as if adding sugar to a glass of tea." (Goodwin, 1997, pp. 49–50).

Four apartment blocks using the pediment cliché of Post Modernism

Al-Mulla travel agents offices   Central Plaza shopping mall and offices

Office building in Sharq   Al-Usaimi office tower

Figure 4   Examples of Post Modern Architecture in Kuwait.

1.2.1   The Influence of Tradition

Al-Mutawa has identified a number of elements of old Kuwaiti architecture (Al-Mutawa, 1994). The most important of these are the courtyard, the *liwan* or passage surrounding the courtyard, the vestibule or bent house entrance, the *mastaba* or outdoor bench, and the high roof parapet. He also identified the *merzam* or roof gutter, the *jandal* or exposed wooden roof structure, the *badgeer* or wind tower, the *mindah* or supporting pillars around the courtyard as components of the Kuwaiti architectural heritage. The use of teakwood for parapets, doors and windows, and finally the specially decorated building corners were a common practice in old Kuwait (Al-Mutawa, 1994, pp. 23–26). Some of these elements are shown in Fig. 1.5.

In Al-Mutawa's architecture all of these traditional elements are employed in an active and somewhat contemporary fabric. According to Goodwin (1997) Al-Mutawa first felt and absorbed what is pertinent in the past through his childhood experience, then with an extraordinary empathy, he was able to transmute it into the living circumstances of today and the future.

1.2.2   Early Works

From the late 1980s till early 1990s Al-Mutawa has produced a large number of middle to large size apartment blocks around Kuwait. In addition to a high rise office building. The main characteristics that distinguished this phase are the extensive use of wooden elements, old and new. The punctuated fair-face concrete walls painted in white with square patterns. And the utilisation of some traditional Kuwaiti elements for the same function in the new buildings.

The use of historic concepts and sometimes building materials in terms of the present and without sentimentality, like the use of traditional drain spouts, also characterised his early works (Goodwin, 1997). The white colour used in all the buildings works as an effective background to dark brown or blue woodwork ornamentation and to the play of shade and shadows of recessed ventilation holes and various projections and cantilevers. This has rendered his buildings with a different image and gave them a special flavour than the rest of similar ones in Kuwait (Fig. 1.6).

1.2.3   Late Works

Since mid 1990s till now there was a continuation of Al-Mutawa's active contribution to the built environment in Kuwait by undertaking the design of public buildings. Among his latest projects is a restaurant/coffee shop and a commercial and hotel complex which is another tall element after his earlier office building. The latter seems to have been met with enough success to be duplicated in a next door site even before the inauguration of the first complex.

Court, liwan and mindah

Vestibule

Mastaba

High roof tops fences

Wooden windows, doors and parapets

Merzam

Jandal

Badgeer

Figure 5   Examples of traditional elements of architecture in Kuwait.

Salwa 1 apartment block

Salwa 3 apartment block

Salwa 5 apartment block

Salwa Complex apartment block

Al-Jabria apartment block

Al-Jabria apartment block

Al-Mubarakia shopping mall

Al-Mutawa office building

Figure 6   Examples of Al-Mutawa's early works.

The extensive woodwork in the commercial and hotel complex Salmiya Palace is interesting and variable in proportions which are generally based on traditional formats. The façades of Salmiya palace simulate decorated lacework especially at night when light comes through colourful glass panels (Fig. 1.7). The wide use of repetitive design elements, like in the façades of Salmiya palace, has characterised this late phase. The use of mahsrabiya or wooden lattice in Salmiya Palace does not quite recall a Kuwaiti tradition but rather an Arab one. Also the yellow paint of the fair-face concrete walls of the Bedaa restaurant/coffee shop presents a deviation from white colour used in almost all his buildings before.

Salmiya Palace hotel and shopping mall

Salmiya Palace complex duplicate

Salmiya Palace hotel entrance at night

Salmiya Palace mashrabia detail at night

Bedaa Restaurant and Coffee shop

Figure 7   Examples of Al-Mutawa's late works.

From a technical perspective, some studies have stated that Al-Mutawa's buildings have proven to be cheaper to cool or heat. "It is important to note that all
the elements together in a building by Al-Mutawa can reduce the usage of air
conditioning by as much as fifty per cent." (Goodwin, 1997, pp. 57–58)

1.3   DISCUSSION

Recognition and acknowledgement of tradition can take the form of interpreting
the essence, in meaning and function, of certain elements of architectural heritage and abstracting them in modern designs. It can also take the form of
reusing, or recreating, some of these architectural elements in a contemporary
design to convey a traditional image. Both approaches have been tried in current
architectural practice in Kuwait. In my opinion, Al-Mutawa's endeavours come
under the second approach. However the previously presented types of
contemporary architecture in Kuwait claim to be reflecting its rather brief traditional heritage in one way or another, they do not for sure reflect a local image.
    At the same time the various examples of Al-Mutawa's architecture
whether in residential, commercial or office development convey immediately
this lost local image. Not to say that this should be a prototype for the revival of
traditional Kuwaiti architecture. Or to justify revivalism in the nineteenth
century sense by emphatically copying buildings seen and gone (Goodwin,
1997). Also it is not out of nostalgia to a past that no man can recreate, or
attempt to create a museum city. On the contrary, it must not become a prototype, or rather a stereotype. But the essence and merits behind his work deserve
to be studied before one embarks on a design that needs to reflect a local, yet
contemporary, image of Kuwaiti architecture.
    After scanning some of the works of Al-Mutawa, one can define some
positive and negative aspects. Among the positive aspects is that his unique style
is the sole invention of his own form that mainly depended on his personal
development, cultural background and advances in modern engineering
(Goodwin, 1997). He has, also, shown how he can break the conformity to the
monotony inherent in most of the building sites in Kuwait through "the creation
of poetry of concealed disorder in his domestic buildings" (Goodwin, 1997,
p. 83).
    Yet there are also negative aspects in his work like the superficial use of
wooden beams under concrete roofs and balconies. Mixed style arches and
motifs, sometimes contradicting with each other, in the same building and same
façade. Nevertheless, a local image is reflected in Al-Mutawa's work. Whether
this image is a Kuwaiti image as he claims or not this is open to debate. Some
architectural critiques agree, while others criticise his work as pastiche and
stereotyped decoration of traditional forms (Goodwin, 1997).
    What is undeniable is that Al-Mutawa has developed a "style of practice"
of his own that is reflected in all his works since his early beginnings and right
down to his latest project. A style that is able to provoke architecture critique
whether praise or curse, approval or disapproval. But certainly his architecture

prompt the emotions of all viewers, in one way or another and cast a lasting impact on their minds.

While few foreign architects might have well understood the culture and heritage of Kuwait and produced traditionally responsive architecture, the majority did not. The latter group once even suggested that "tradition and innovation" are at odds and cannot be combined in architectural terms (Randall, 1985). Being a native of Kuwait is not the only advantage for Al-Mutawa. Like other native architects such as Hasan Fathy of Egypt and Charles Correa of India, who are both recipients of the UIA Gold Medal for their achievements, who did not only rely on their nationality but on deep understanding and inter-pretations of local tradition. Both of them used the previously mentioned approaches to tradition; i.e. the reinterpretation of inherited functions and mean-ings as well as the reuse of local architectural elements in contemporary fashion.

1.4 CONCLUSION

This is, by no means, a comprehensive view of contemporary architecture in Kuwait, since there are numerous other buildings and projects worthy of analy-sis. Also this paper did not aim to analyse the full architectural spectrum in the country, but rather to examine by example some of the current architectural trends. It aims to document an attempt to reflect a local architectural image in the design for contemporary architectural functions such as high-rise offices and hotels, which were considered unsuitable for this type of expression.

In the first part of this paper, the author looked at the attempts to mod-ernise Kuwait by adopting international global architectural images for the design of medium and large-scale projects. At the same time, the second part of the paper presented a small-scale attempt by a Kuwaiti native architect to reflect a local Kuwaiti image through his designs for modern contemporary functions.

We may agree, or disagree, that Al-Mutawa's architecture represent a revival of the traditional Kuwaiti architecture. We may also like it or not accord-ing to our personal architectural taste. In fact some Kuwaitis dislike his style while some foreigners admire it so much (Goodwin, 1997). But I believe that no one would disagree that his consistent and deliberate style is not influenced by the prevailing international style. It is more influenced by the neo-vernacular trend that prevailed during the second half of the 20th century. Al-Mutawa's architecture surely reflects a local image of some sort, if we don't take it as a Kuwaiti image. An image that certainly fits its local built environment context much better than other examples presented earlier. This image is undeniably local and in harmony with its physical, social and ecological settings.

1.5 REFERENCES

Al-Mutawa, S. A., 1994
    HISTORY OF ARCHITECTURE IN OLD KUWAIT CITY, (Kuwait:
    Al-Khat).

Bosel, S., 1995
    A PERSONAL VIEW OF ARCHITECTURE IN KUWAIT. In *Kuwait
    Arts and Architecture: A Collection of Essays*, edited by Fullerton,
    A. and Fehervari, G., (U.A.E.: Oriental Press), pp. 209–219.

Caravanserai Revisited, 1999
    In *The Architectural Review*, no. 3, pp. 20–27.

Goodwin, G., 1997
    SALEH ABDULGHANI AL-MUTAWA: NEW VISION IN KUWAIT,
    (London: Alrabea Publisher).

Peel, L., Powell, P. and Garrett, A. 1992
    AN INTRODUCTION TO 20TH CENTURY ARCHITECTURE,
    (Leicester: Magna Books).

Kultermann, U, 1999
    CONTEMPORARY ARCHITECTURE IN THE ARAB STATES:
    RENAISSANCE OF A REGION, (New York: McGraw-Hill).

Makiya, K., 1990
    POST ISLAMIC CLASSICISM: A VISUAL ESSAY ON THE
    ARCHITECTURE OF MOHAMED MAKIYA, (London: Saqi
    Books).

Randall, J., 1985
    SIEF PALACE AREA BUILDINGS, Kuwait. In *Mimar: Architecture in
    Development*, issue 16, pp. 28–35.

Tietz, J., 1999
    THE STORY OF ARCHITECTURE OF THE 20TH CENTURY,
    English-language edition by Youngman, P. (Cologne: Konemann).

PLANNING AND ARCHITECTURE

Tall Building Design Innovations in Australia

Brian Dean, Owen Martin, David Emery, and Peter Chancellor

1.0 ABSTRACT

This paper describes three examples of innovative tall building designs undertaken by the authors' firm in Australasia. These involve the design of building structures up to 88-storeys and as a multi-discipline group, the firm's innovations have embraced structural, M&E, façade and fire engineering disciplines.

The paper describes a number of innovative stability designs for tall buildings with minimum impact to the functional planning. The first case study involves a novel off-set outrigger stability concept, for the Aston Apartment building in Sydney. Further case studies are the 88-storey Eureka Place Apartment Tower currently being designed for Melbourne's Southbank and the 86-storey World Tower building currently under construction in Sydney.

2.0 CASE STUDY 1 – ASTON APARTMENTS, SYDNEY OFFSET OUTRIGGERS

Aston apartments is a slender 30-storey residential tower accommodating 145 serviced and owner occupied units. Its slender form and difficult positioning have required unique structural design solutions.

The reinforced concrete framed structure is 90 metres high and is only 13 metres wide in the north-south direction, resulting in a slender height-to-base ratio of 7:1.

An innovative wind resisting system using "offset outriggers" has been developed by Connell Wagner for this building, a first in Australasia. The "offset outriggers" shown in Figure 1 significantly reduce building deflections and core bending stresses. The offset system enables the outrigger arms to be placed across the full building width at locations away from the plane of the lift cores and mitigates some of the disadvantages of conventional outrigger systems, such as the outrigger arms obstructing occupiable and valuable floor space.

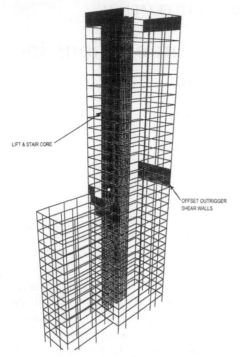

LIFT & STAIR CORE

OFFSET OUTRIGGER
SHEAR WALLS

Figure 1   Frame Perspective – Aston Apartments.

A conventional outrigger system requires the core and external columns to be directly coupled. Detailed structural analyses conducted have shown that for typical floor slab plan dimensions and thicknesses, there is little reduction in the effectiveness of the outrigger as it is offset further from the centre core.

The outriggers system for the Aston Apartments rely on the floor diaphragms to transfer shear forces to mobilise the perimeter columns of the building, as shown in Figure 1.

The offset outriggers couple the columns and core by development of opposing shear forces in the floor diaphragms at the top and bottom levels of the outriggers.

In the Aston Apartments the offset outriggers, consisting of two storey high shear walls 200mm thick, are located on the side elevations at mid-height and the top of the building as shown in Figure 1. One of the advantages of this innovative wind resistance system is that it has virtually no impact on the planning of the building and permits a very simple floor plate to be achieved. This has maximised the floor layout for the client and provided greater flexibility for the architect to meet functional and aesthetic design requirements.

The offset outriggers are very economical and efficient because the system utilises the axial strength of the perimeter columns to resist the load at maximum lever arm. The offset outriggers limit the drift of the building and reduce the bending actions in the core, minimising wall thicknesses and core reinforcement. Optimisation of outrigger locations on the Aston Apartments project enables core wall thicknesses to be limited to 200mm even in the lower levels of the building, resulting in a very simple and quick to construct core with no variation of thickness over the entire thirty storeys.

The outrigger walls effectively link perimeter columns of similar load and hence no problems are encountered with regards to differential shortening between core and columns which would normally otherwise be expected where the core and columns are rigidly connected by outriggers.

The effectiveness of the outriggers has seen the requirement for only a lightly reinforced core, which is required to resist only 20% of the total base bending moment.

3.0   CASE STUDY 2 – EUREKA PLACE TOWER, MELBOURNE

This project, under design at the time of writing this paper, involves an 88-storey apartment building to be located on the south bank of the Yarra River in Melbourne. With a height of over 300 metres, the building will become the tallest apartment building in Australia. Designed by award winning Architects Nation Fender Katsilidis, the building footprint comprises a central diamond section which extends over the full height of the tower. On each side of the diamond are offset rectangular tubes, which start at ground level but sequentially drop off at different heights up the tower, as shown in Figure 2. This architectural concept thereby takes maximum advantage of the stunning views over the city.

A major complexity for the project is the site geology which comprises 30m deep layers of silts and gravels with intermittent basalt flows, overlying siltstone bedrock.

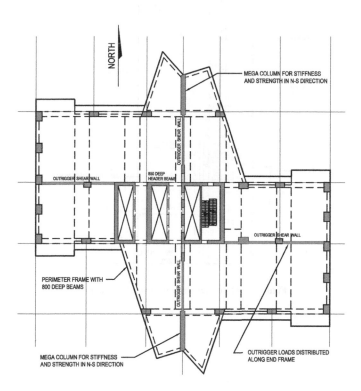

Figure 2   Typical apartment floor.

In consultation with the developer/builder, Grocon Constructions, a concrete structure was selected utilising high strength concrete up to 100MPa for the vertical elements to maximise the useable floor area. Post-tensioned floors are proposed for speed of construction and to minimise the structure self weight. The cladding comprises a curtain wall system with external balconies.

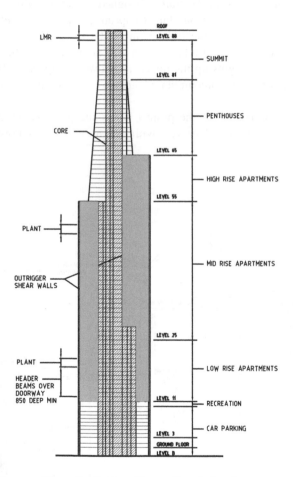

Figure 3   Outrigger/shear wall south elevation.

3.1.1 Stability

With a height to base width aspect ratio of 7:1, the building is relatively slender and a critical structural challenge is to resist the wind loading and control building accelerations in the most cost effective manner.

The structural stability is provided by a composite system comprising the following components:

- Coupled lift cores
- External tube beam column and frame incorporating two "mega columns".
- An east-west and north-south outrigger shear wall system linking the central core to the perimeter frame.

The extent of the shear wall outriggers are kept to a minimum and are positioned within the hotel/apartment wall layout to minimise loss of floor area. The outrigger locations have been optimised vertically up the building to maximise their impact.

A number of structural schemes were investigated for the tower using either composite concrete and steel construction or fully concrete construction.

For the floors, steel potentially offers the advantages of faster erection speed, however it is more crane dependent which can be a disadvantage in Melbourne, with significant downtime due to wind. Its other main advantage is reduced self-weight which is significant for such a tall tower reducing the column/wall sizes and the foundation costs.

Generally, in Australia, concrete is more cost effective, particularly for the vertical structure, columns and core. Concrete floor construction has been shown from previous projects to match the speed of steel in Melbourne. It is also more suited to apartment construction enabling lower floor to floor height, significantly reducing the lateral forces due to wind as well as reducing the cost of the curtain wall and vertical structure.

For Eureka, the floor spans are relatively long span for apartments (up to 11m) and these are best handled by post-tensioned concrete to provide a low deflection design.

With such a tall, slender building, a key structural challenge is to minimise the structural cost penalty associated with stability. Melbourne is a relatively low seismic activity area and consequently for tall buildings the wind forces govern the design. The highest wind forces are associated with thunderstorm activity, and generally emanate from the west.

The two main design criteria to be satisfied are, firstly the strength ultimate limit state and secondly the occupant comfort serviceability criteria, namely horizontal acceleration limits. The latter criteria is very important with a relatively tall slender tower having a minimum aspect ratio about 7 and a first mode natural frequency about 0.11Hz.

At the outset it was decided that we needed to mobilise the strength and stiffness of the outer tube as much as possible for structural efficiency, particularly as the central core structure is relatively small. However the architectural concept required that the building perimeter be as open as possible to take maximum advantage of the panoramic views. Consequently the outer tube by itself was relatively flexible.

We therefore examined several schemes with outrigger structures linking the central core with the outer tube. These included:

- Outrigger frames located at the plantroom levels and refuge level
- Outrigger shear walls located at discrete levels over the tower's height

For both schemes, multiple and single outriggers were investigated in each of the orthogonal directions.

A key feature of the design which emerged early in the process was the concept of a mega-column on the north and south corners sides of the tower diamond. The architect was keen to express this mega column as a bold element on the façade.

With such an unusual tower geometry, it was appreciated that a wind tunnel test would be required to verify the design forces and the horizontal accelerations at the top of the tower. Nevertheless, the preliminary design for market sales had to be undertaken prior to the wind tunnel test, and this was undertaken using the approximate dynamic analysis methods in AS1170 Standard.

The preliminary analyses using the ETABS computer programme indicated that discrete outriggers restricted to the plantroom levels did not provide adequate stiffness to the tower. To extend the outriggers into the residential levels required that they had to be located within the designated apartment walls and also required their width to be minimised to provide as much useable floor as possible. Furthermore, the analysis indicated the possible need for a roof top damper to increase the overall damping – see Section 3.1.2 below.

The stability concept therefore evolved to the scheme shown in Figure 3, comprising a central core linked to the outer tube and the two mega columns by way of two orthogonal shear wall/ outrigger systems extending over part of the tower height. Provision will be made in the shear walls for future openings enabling apartments to be linked in response to market demand as the sales progress.

3.1.2 Wind Tunnel Test

A 1/400 aerolastic scale model of the Eureka Tower was tested in the wind tunnel at Monash University in October 2000. The aim was to confirm the preliminary design wind forces and in particular the predicted accelerations at the top of the tower. The test used a linear mode aerolastic model pivoting about the basement level. Strain gauges directly measure the base moments and torsion of the building. The wind tunnel test results generally confirmed the overturning moments, except that cross wind response caused by westerly winds at about 250° was significantly higher than predicted by the approximate method in AS1170.2.

Regarding the predicted accelerations, the wind tunnel test indicated that these will be lower than the generally accepted limit, however there is a narrow band of wind direction for which the accelerations are considered marginal. It is therefore proposed that allowance will be made for a rooftop damper to provide additional damping.

A cost effective form of damper, presently proposed is to configure the rooftop water storage tank as a liquid damper.

CASE STUDY 3 – WORLD TOWER

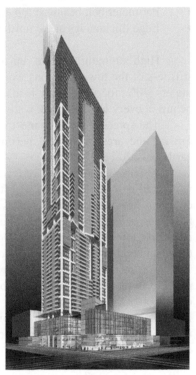

The World Tower has been designed by Connell Wagner, Sydney and is currently being constructed in Sydney's CBD at the corner of George Street and Liverpool Street. With a height of 260 metres and comprising 80 storeys, this will become Sydney's tallest building. The tower floor plan has an overall rectangular shape of dimension 55m × 29m with long span cantilevered wintergardens and balconies on many levels.

The development consists of a podium with 6 levels of basement carpark below 7 levels of retail/commercial, and low, medium and high rise residential levels comprising about 705 apartments. Intermediate 2 storey high plant and recreation/pool levels are located at quarter heights in the building at Levels 14, 37 and 60 with additional plant located within a 10m high beacon at the top of the building.

Designed by architects Nation Fender Katsalidis for developer Meriton Apartments, the tower is very slender with a height to base ratio of 9.5 and a core which is only 9 metres wide. A feature of the building is the expressed perimeter "superframe" of columns, and beams located on every third level.

The floors will be post-tensioned flat plates spanning 9m between the core and the perimeter columns.

4.1 Stability

The stability design comprises an innovative combination of structural systems which are outlined below.

The vertical structure of the tower comprises a central core of reinforced concrete shear wall elements and 20 perimeter columns. The central core is made up of 6 discrete core box elements which progressively terminate at low, mid, and high-rise levels.

The lateral load resisting system for the tower comprises:

- Coupled cores.
- 2 pairs of 8 storey high triangulated post-tensioned outriggers between core and perimeter columns centred about both mid-height plant levels.
- 2 storey 2 span spreader walls located at the ends of each outrigger.
- Outwardly spreading tower columns in podium levels.
- Inclined wind columns forming deep outrigger trusses in podium levels.
- Perimeter belt beams at top and bottom of mid-height plant levels.
- Edge thickenings at the north and south ends of the floors.

High strength concrete, up to 90MPa has been utilised to maximise the stiffness of the tower columns which when outrigged to the core resist approximately 70% for the lateral loads. The core elements are coupled by header beams at every level and link walls at discrete locations in plantroom levels.

The spreader walls connect each outrigger to 3 tower columns thus maximising the area of vertical perimeter structure mobilised to resist wind and earthquake forces.

A temporarily adjustable connection between outriggers and spreader walls will be provided to allow differential shortening between cores and tower columns, and then fixed after construction once most of the differential movement has occurred. The triangulated outrigger blades are located within apartments on inter-tenancy wall locations. Connection to spreader walls occurs in the plant levels where adjustment access is available.

The tower columns are outwardly inclined between Level 14 and 9 to provide an increased spread at the base of the tower. This had a threefold benefit of reducing the slenderness of the tower, enabling reuse of original tower pad footings built in 1990 for a previous commercial tower development, and also utilises the existing provisions for tower column penetrations in the as-built carpark slabs.

Increasing the overall base dimension of the tower by spreading the tower columns at the base reduces the tower slenderness and provides additional lever arm. This has a significant effect on reducing building sways and accelerations. Forces resulting from outwardly spreading tower columns between Levels 14 and 9 are restrained by post-tensioned tie and strutting beams built into the floors of Levels 9 and 14.

A wind tunnel test was undertaken at Monash University by MEL Consultants to enable more accurate assessment of wind forces and sway induced building acceleration. Lateral forces in the east-west direction, are governed by westerly winds on the larger building "sail" area whilst in the north-south direction earthquake forces are the governing design lateral load. The building's first mode natural frequency is 0.14Hz and the predicted ultimate limit state wind deflection at Level 80 due to a westerly wind is about 500mm which is height/520. The predicted 5-year standard deviation acceleration at RL=234 is less than 5millig and within acceptable limits.

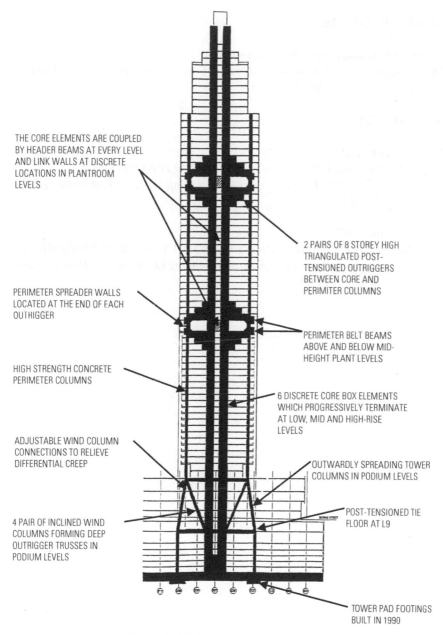

THE CORE ELEMENTS ARE COUPLED BY HEADER BEAMS AT EVERY LEVEL AND LINK WALLS AT DISCRETE LOCATIONS IN PLANTROOM LEVELS

2 PAIRS OF 8 STOREY HIGH TRIANGULATED POST-TENSIONED OUTRIGGERS BETWEEN CORE AND PERIMITER COLUMNS

PERIMETER SPREADER WALLS LOCATED AT THE END OF EACH OUTRIGGER

PERIMETER BELT BEAMS ABOVE AND BELOW MID-HEIGHT PLANT LEVELS

HIGH STRENGTH CONCRETE PERIMETER COLUMNS

6 DISCRETE CORE BOX ELEMENTS WHICH PROGRESSIVELY TERMINATE AT LOW, MID AND HIGH-RISE LEVELS

ADJUSTABLE WIND COLUMN CONNECTIONS TO RELIEVE DIFFERENTIAL CREEP

OUTWARDLY SPREADING TOWER COLUMNS IN PODIUM LEVELS

4 PAIR OF INCLINED WIND COLUMNS FORMING DEEP OUTRIGGER TRUSSES IN PODIUM LEVELS

POST-TENSIONED TIE FLOOR AT L9

TOWER PAD FOOTINGS BUILT IN 1990

Figure 4 World Tower – east/west section.

ACKNOWLEDGMENTS

Multiplex Constructions – Aston Apartments;
Michelmersh/Grocon Constructions/ Nation Fender Katsalidis – Eureka Tower;
Meriton Apartments – World Tower

REFERENCES

Dean, Brian K. and Webb, J. 1998
    RECENT TALL BUILDING DESIGN INNOVATIONS IN
        AUSTRALIA. *Fifth International Conference on Tall Buildings.*
        Hong Kong.

Owen, Martin. 1997
    EFFECTIVE USE OF CONCRETE IN MULTI-STOREY BUILDINGS,
        Concrete Institute of Australia Building Our Concrete Infrastructure
        Seminar, Perth.

The Emirates Towers

Hazel W. S. Wong

THE COMPETITION

Initiated in the mid 1990's as an invited international design competition by His Highness Sheikh Mohammed bin Rashid Al Maktoum, Crown Prince of Dubai and United Arab Emirates Defence Minister, the intention of the Emirates Towers project was to create a landmark development, comprising twin office towers to frame and to be at least twice the height of the 149-metre Dubai World Trade Centre, the latter being an existing focal point in the City built by his late father Sheikh Rashid bin Saeed Al Maktoum over twenty years ago. My proposed winning competition design (Fig. 1) realized this objective through the creation of twin slender towers, carefully located to pay homage to an existing landmark, and flanked at each tower base by low curvilinear buildings, in forms reminiscent of massive, shifting sand dunes, housing parking and service elements.

Despite subsequent market analysis that necessitated the conversion of the shorter of the two office towers into a hotel and the incorporation of a 2-storey retail component at the base, the overall design was sensitively and faithfully retained throughout the development.

Site Context

The site for the project, approximately 500 metres × 350 metres and 169,000 square metres in area is located along a major thoroughfare in the new bustling commercial and residential centre of the city. The design concentrates the buildings away from the highway in a central portion of the site. This way the project is distinctly set apart from the standard commercial development and the twin towers are positioned so as to create meaningful visual compositions from every viewpoint. The site has been organized around five primary structures – the office tower, the hotel tower, retail podium and associated parking buildings. A primary ring road conveniently accesses all components on the site. A series of terraced urban plazas are introduced at the tower bases with landscape elements including an 80 metre wide waterfall, lakes, fountains and causeway.

Beyond the immediate periphery of the building complex, the formal landscaped environment progresses into a natural park-like setting with gentle contours and lush vegetation buffering the roadways around the site perimeter completing the development which represents a major destination point and offers a sense of an oasis amidst an urban concrete environment.

Figure 1   Winning Competition Design.

The Concept

Both towers feature equilateral triangular cross sections evocative of the Islamic cultural vocabulary, representing the three heavenly bodies – earth, sun and moon. The circular drum at the base and the cylindrical feature at the top of each tower echo the concept of the circle as the 'timeless whole'. Conceived as pure sculptural forms, the buildings present a dynamic silhouette against the rapidly changing Dubai skyline. As in the poetic movements of a pas-de-deux, the slender triangular towers clad in aluminium panels with copper and silver reflective glass capture the changing light of the desert sun and show off their dramatic integrated illumination at nightfall (Fig. 2).

Figure 2   Pas-de-deux.*

Although the similarities of the two towers enhance their interrelationship and reinforce the expression of the tall building, subtle differences in the cladding and detail design highlight and reflect their differences in building type and programme.

At the base, intersecting planes of curvilinear and vertical elements frame grand stairs leading to the podium levels. The solidity of the Brazilian 'Kinawa' granite walls and the lightness of the stainless steel and glass entrances into the retail boulevard are juxtaposed to create a mediating scale between the towers and the streetscape (Fig. 3). Dune-shaped low-rise parking structures echo the desert landscape of the region.

Figure 3   Podium.*

The Office Tower

The Office Tower, rising above 350 metres (1148ft.), is ranked amongst the top ten tallest buildings in the world. The 52-storey tower has a total gross floor area of 68,500 square metres with a typical floor plate of 1334 square metres, a size, which reflects current market demands. An entrance ramp leads to a grand porte cochere and a voluminous sky-lit entrance lobby. The Entrance Lobby provides access to all office floors by means of 16 passenger elevators travelling up to 7 metres per second.

Interiors have been designed to provide thoroughly flexible and low maintenance office accommodation, which can adapt to changing occupant needs and to new business technology with ease. The floor-to-floor height of 4.5m allows the use of access flooring and a deep ceiling void to house state of the art I.T. equipment and building services. A highly efficient floor plate, devoid of any interior columns, provides close to 9 metres office depth at the narrowest points between core and building perimeter. Cherry wood wall panels, stone floors and decorative lighting in the typical elevator lobbies all contribute to a distinctive corporate office environment (Fig. 4).

The Hotel Tower

The 5-star, 400-room business Hotel Tower features a 31-storey glazed atrium overlooking the Arabian Gulf. The 52-storey structure, rising 305 metres, has a total gross floor area of 50,360 square metres. A total of 8 passenger elevators, including 4 panoramic, serve the 339 deluxe guest rooms and suites, 52 club executive rooms and 9 presidential suites. Guest rooms and suites line the tower sides facing away from the highway. The feature atrium with dramatic views of the Gulf through it's glass façade acts not only as a sound buffer, but provides visual transparency and becomes the hotel's window to the city (Fig. 5).

The lower levels are dedicated to Conference and meeting facilities, business centre and hotel executive offices. Distributed through the hotel are 10 food and beverage outlets catering to the needs of the guests including an 800-seat ballroom, restaurants serving various international cuisine and penthouse wine bar and fine dining.

Back-of-house and other support services are located in the podium levels immediately below the Hotel Tower.

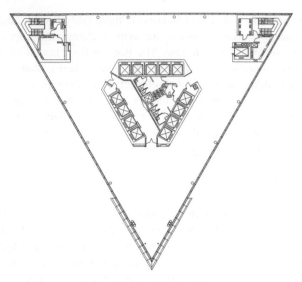

Figure 4   Typical Office Plan.

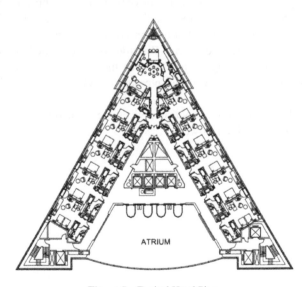

Figure 5   Typical Hotel Plan.

A 2-storey retail boulevard and a full level of parking connect the two towers at the base. Through the pyramidal and linear skylights, visitors are constantly aware of the tower structures above. Complimenting the corporate offices and elegant hotel, the boulevard takes on a lighter, fun-provocative and colourful palette of materials and finishes including ceramic tile flooring, copper wire mesh ceilings and timber wall panels (Fig. 6).

Figure 6   Retail Boulevard.

Parking Structures

Over 2000 car parking spaces and ancillary building services are provided in the two curvilinear buildings and in the lowest level of the podium. The stepped structures, reminiscent of massive shifting sand dunes, are clad in profiled pre-cast concrete panels and covered with greenery cascading off its perimeter planters (Fig. 7).

Figure 7   Parking Structure.

Structural Systems

Each tower is set on a 1.5 metre thick raft slab atop 100 friction piles of up to 47 metres deep and 1.5 metres in diameter. The structural system for both towers is only similar at the base and peak. In the Office Tower, steel transfers at level 9 distribute the loads from perimeter concrete filled steel tubular columns set 9 metres apart to three triangular legs. Three additional transfer floors and a tuned mass damper at the peak provide for stability under all load conditions. A steel and concrete hybrid solution achieves column-free office space and speed of erection. The Hotel Tower adopts an all-concrete solution for better acoustic performance and stiffness. Two visible prefabricated steel trusses support the 31-storey atrium glass wall and contribute to the structural integrity of the tower. Both tower peaks and the 44-metre tall spires containing three tuned mass dampers each are supported on a complex latticework of steel sections.

An additional 2000 piles of various diameters and depths support the concrete-framed multi-levelled podium and parking structures.

The Completion

Construction for the project was carried out from 1997 to 2000, and was delivered on March 2001 on time and on budget. Completed on the eve of the new Millennium, the Towers present a strong metaphor of sleek, modern technology reaching upwards and outwards into the future, yet firmly rooted in the cultural and environmental origins of it's past (Fig. 8). Regarded as a major architectural accomplishment in the region, the Emirates Towers clearly confirms Dubai as one of the leading urban centres of the world.

Figure 8   The Emirates Towers *

The Project Team

Client

His Highness General Sheikh Mohammed Bin Rashid Al Maktoum

Team

Lead Consultant	Hyder Consulting
Design Architect	Hazel W.S. Wong
Architect	NORR Group
Structural Engineers	Hyder Consulting
M & E Engineers	DSSR/TMP
Interior Design (Hotel)	Design Division
Project Managers	Turner International
Contractor (Office)	Nasa Multiplex
Contractor (Hotel)	BESIX – Ssang Yong

---

* Photographs courtesy of Turner Steiner International

PLANNING AND ARCHITECTURE

Two Towers

Phil Castillo

Two towers, Duetsche Post in Bonn and the MAX in Frankfurt, rethink the typology of the high-rise tower in relation to function, technology and user comfort. The primary material used in the expression of these towers is glass. Both buildings use glass with differing results. It is the one material that offers the opportunity for technological advancement, primarily in the development of the façade and the resultant effect on the energy systems and user comfort. Its qualities of transparency, opacity, reflection and refraction, allow for a varied architectural expression.

Duetsche Post is a 42 story, 162-meter tower sited as an extension of the Rheinauenpark forming an edge to the city. The plan is conceived as split, shifted oval with its primary orientation towards the Rhine. In addition to facilitating views toward the city, the aerodynamic shape minimizes negative wind effects

In plan, the split oval halves are separated by a 7.20-meter space. Connecting glass floors at 9 story intervals form skygardens that serve as communicating floors and elevator crossovers. The glass elevators of the low and high zones run in the center of the skygarden, providing views and orientation.

The typical floor plate of 1818 SM has a lease span of 6.80 m, allowing for the typical layout of cellular offices. Column spacing is nominally 5.80 m. Concrete cores provide lateral stiffness. The two halves are tied together with X-bracing at the skygarden levels so that the tower behaves as one structure.

The building has a twin shell façade. The outer shell is completely out of glass, enabling natural ventilation especially in the spring and fall. The outer shell protects from rain, wind and noise and allows for placement of the sun-shades. Glass from floor to ceiling optimizes daylight. The blinds in the interstitial space further protect the inner façade from direct solar gain. The inner shell is double-glazed with a low-e coating on the number 2 surface and has a series of operable windows allowing for natural ventilation of the offices. The result is the creation of two channels for air, the inner for ventilation and the outer to exhaust heat gain. Solar gain heats up the outer shell creating a convection current that draws the air up through the cavity

The concrete structure has an integral heating and cooling pipe system, taking advantage of the low energy characteristics of water and the thermal storage capacity of the concrete.

If comfortable temperatures cannot be achieved at the high and low exterior temperatures of summer and winter, an air displacement system along the façade mechanically assists in creating a comfortable environment.

Exhaust air from the offices is used to condition the skygardens with some mechanical assistance in special occupied zones. The east–west orientation of the skygarden allows for cross ventilation. A computerized building management system controls all of these components and selects the most effective operational mode, balancing the exterior and interior conditions. Cost comparisons show that the total cost of the climate systems and twin shell façade is comparable to a conventional system. Operating costs are reduced by 60%.

Lighting is an additional feature incorporated into the façade. A series of three cold cathode tubes, red green and blue, are synchronized to allow for a variety of color combinations.

The design for the MAX was selected in an invited competition. The shape, an incised ellipse, responds to its central and mid-block location in the high-rise bulk of Frankfurt next to the Commerzbank and new Rhein-Main Tower. This tower is 63 stories and 228 meters in height. Typical floor areas are approximately 8,493 SM.

Several strategies were used to make the building lighter, less material and more transparent towards the top. As the core areas diminish, the building sets back four times minimizing deep space. At the long axis, skygardens bring light into the interior plan areas allowing for the layout of cellular offices. Within the last segment of the tower they continue to setback behind the all glass enclosure of the skygardens thereby increasing the dematerialization. Within the top garden, light steel structures for elevators and stairs facilitate circulation between those special floors. The top floor is developed as a conference center with perimeter circulation reading as recognizable figure and skyline image. At night special lighting reinforces the towers figure by distinguishing between its solids and its voids.

The structure is reinforced concrete. Stiffness is provided in the core. Concrete columns, slabs and walls are intended to remain exposed.

The façade furthers the architectural intentions through reduction of the stainless steel spandrels until their elimination at the top. The façade also re-examines the nature of the high-rise enclosure, addressing shading, daylighting, natural ventilation, and increased vision while maintaining visual control of the tower façade. The typical façade module of 1.35 m consists of a narrow operable window behind vertical panels of perforated stainless steel and a fixed glass portion of triple glazed units with a low-e coating on the number 2 surface. Only a selective interior shade reduces vision and daylight.

The triple-glazed, heat-absorbing glass reduces solar transmission by 62%, yet allows 75% of the daylight to pass through. The interior shade reduces energy transmission by another 50%. This results in the room comfort being primarily affected by interior loads. The perforated stainless steel panels in front of the operable windows also serve as sun and weather protection. From the inside they give an interesting modulation to the views, form the outside they reinforce the verticality.

Again the offices are naturally ventilated. During exterior temperature extremes, fan coil units in the floor, which distributes the air as a displacement system, provide mechanical assistance. Low air changes of 1.5/hour are achieved through radiant cooled or heated integral piping in the concrete structure again using the efficiency of water as energy carrier and concrete for thermal mass.

These buildings represent the only way architecture can be new and responsible, not only relying on form and aesthetic. Responsible architecture can control its environment through design, not solely through added technical and mechanical systems.

The goal here is to include the idea that the skin of the building can modulate its own climate through daylight, natural ventilation and solar energy as essential components in commercial design. The result is a building with high technology and low energy.

Form, space, function, materials, construction and technology all enforce and support each other in a totally integrated design.

PLANNING AND ARCHITECTURE

The Citigroup Centre At No. 2 Park Street, Sydney

Mike Haysler and Robert Facioni

INTRODUCTION

The Citigroup Centre rises 250m above street level and is one of the tallest buildings in Australia. It consists of 4 levels of underground parking, and a 6-level retail podium with a 45-level office tower over.

Hyder worked with architect Crone & Associates and contractor Multiplex Constructions from concept design through to completion ensuring an optimum structure that could be built to suit a tight construction program and meet the target budget.

This paper covers the history of this site, a notorious black hole for many years, features achieved working on a design and construct basis and some of the technical challenges facing the structural engineer.

The successful completion of this prestigious project prior to the arrival of international visitors to the Sydney Olympics is a testament to the success of design and construct delivery method.

HISTORY OF THE SITE

The Citibank Centre is located where the Waltons Department Store once stood. The site was notorious as Sydney's biggest remaining hole in the ground at the centre of the city opposite the restored Queen Victoria Building and Town Hall.

Hyder under its former name of Wargon Chapman Partners had a long association with the site working on various schemes with architects Crone and Associates.

The original development proposal for the site prepared in 1982 consisted of a 33-storey building constructed over a 6-storey retail centre. In 1984 with further consolidation of the site a design for a 45-storey office tower was developed.

In 1987 following the purchase of the adjacent Hilton Hotel and consolidation of all the properties between the hotel and Park Street, the Bond Corporation proposed the landmark 102 level 'Skytower' development. The wind resisting structure for this scheme comprised an externally braced steel frame with the structural system highlighted in the curtain wall façade. The structural steel core played no part in the wind-resisting frame.

There were some perceived urban planning issues with Skytower and in 1988 the design was changed to an 85-storey tower with podium level retail. The structural system for this scheme was similar to Skytower but the external steel bracing frame was no longer expressed in the façade. This scheme did not progress due to the decline of the Bond Corporation and in 1989 Kumagi Gumi purchased the site. Japanese architect Kisho Kurokawa worked with the Crone team to develop the Park Plaza scheme with a 210 metre high 48-storey tower over 9-level retail centre for international retailer Sogo. This scheme included a concrete core and composite steel frame

1984

1987

1989

Contractor Multiplex started on this scheme with excavation and shoring substantially completed and foundations started. The School of Arts was underpinned and a tunnel constructed underneath which was to be a link to the proposed Park Street Tunnel. Work also included the temporary support of the Park Plaza monorail station. This scheme again stalled with the economic downturn.

Multiplex continued their interest in the site and looked to progress the development. The current scheme was in its infancy in 1992. Hyder worked closely with Multiplex and the Architects Crone & Associates in fits and starts until the scheme finally got the green light in 1998.

DESIGN AND CONSTRUCT FEATURES

Working from early concept with the Developer and Contractor, Multiplex Constructions, Architect and Services Consultants led to the incorporation of features to improve buildability and speed of construction, to minimise the intrusion of structure and maximise lettable area. The structure was also developed to integrate with the building services requirements.

The primary structural feature to develop from co-operation between the design team was the minimisation of column transfers. This obviously simplified the structure and made the construction quicker and easier. The only significant column transfers are at the corners of the tower.

As there was a high level of repetition a lot of time was spent optimising the typical tower floor plate. To tie in with the façade and achieve the maximum lettable area the column sizes were kept to a minimum. This was achieved by using 90 Mpa concrete and 8% reinforcement at the lower tower levels immediately above the podium. As the columns were relatively large through the podium (1200mm square) prototypes were made and temperature differentials checked to ensure there would be no problems associated with curing. Various curing methods were tested on the prototypes with the most effective, leaving the formwork in place for a minimum of 7 days, being adopted. Cost studies showed that the high strength concrete was cheaper than using high reinforcement ratios so the 90 Mpa concrete was used up to the 34th level with the reinforcement ratio dropping to 1%.

A column free space was considered essential for a premium grade office in the centre of Sydney. The use of post-tensioned band beams spanning up to 14.5 metres was found to be the optimum. Band beams at relatively close centres ranging from 4.0 to 5.6 metres were adopted to minimise the slab weight with typical slabs being only 120mm thick. This is the minimum practical depth to meet fire resistance and acoustic criteria.

Changes to the requirements for formwork caused an increase in the cost of conventional systems such as table forms that would have normally been adopted for this type of floor. Various alternatives were considered before adopting a metal decking system as permanent formwork for the slabs. This reduction in conventional formwork meant that a 4-day floor cycle could be achieved with resulting benefits in construction time.

There were significant benefits to be made by minimising the floor to floor height. This would allow the maximum number of floors within the approved building height, whilst keeping FSR in line with Council approval. It would also reduce the height of curtain wall façade per floor with significant cost saving. The most significant reduction was achieved by notching the beams at the perimeter and at the connection to the core to allow air conditioning ductwork to pass under.

For commercial reasons and to meet City Councils requirements for substantial completion prior to the Olympics fast tracking was essential. The onus was on the design team to produce construction documents early as excavation for basement carparking had already been completed prior to the main project starting. The design progressed in stages sufficiently in advance of the construction to allow co-ordination between disciplines, optimisation of the structure and resolution of buildability issues.

STRUCTURAL FEATURES

The concept design included steel truss outriggers at the mid-level plantroom connecting the core to the perimeter columns as part of wind resisting frame.

The wind frame was optimised by utilising wind tunnel testing of an aerodynamic model in which the wind induced loads were measured using a very sensitive, high frequency Base Balance. Initial sizing of structure, core wall thickness and steel member sizes was done manually. A finite element computer model was then set up to confirm the fundamental natural frequencies at 0.152, 0.218 and 0.393 Hz.

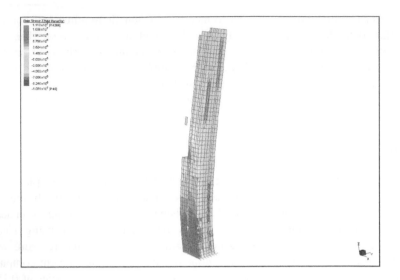

The wind tunnel testing was carried out using these parameters and also with higher and lower natural frequencies to determine the effect of varying the stiffness and allowing optimisation of the structure. This showed that the outrigger trusses could be omitted with the consequent benefit of time and cost saving and uninterrupted space in the mid-level plantroom. The wind tunnel testing gave peak deflection of 295mm and peak acceleration of 23mg at the highest occupied floor. This was within the acceptance criteria of the International Standard ISO 6897 for the response of occupants to low frequency motion.

Three railway tunnels run through the site. The podium area over these tunnels was separated from the rest of the structure. Elastomeric bearings were used to reduce the transmission of noise and vibration from the trains. The podium columns over the tunnels were on large spread footings so as not to overload the tunnels. Special care was taken with excavation and construction in the vicinity of the tunnels.

The site included the heritage-listed School of Arts. This was restored to its original splendour working in conjunction with specialist heritage architect Howard Tanner and Associates.

The existing Monorail station was incorporated into the new structure. It was necessary to move one of the monorail columns without disruption to the running of the system.

Computer modelling of the 37-metre high structural steel spire at the top of the building showed it to be potentially responsive to wind excitation,

with resultant fatigue problems. Specialist advice was sought from Professor Bill Melbourne. Chain dampers were installed in the spire. These dampers have been tested after construction and shown to be effective.

CONCLUSION

The involvement of the Contractor from the early stages of design is an essential part of a successful fast track project. A condition of the development approval from the City of Sydney required that the building be substantially completed prior to the Olympics. The Citigroup Centre, one of the tallest office buildings in Sydney, was completed 2 months ahead of the planned 26-month construction program. This is a testimony to successful co-operation between the designers and builders.

Liberating Urban Architecture (The Merging of the Virtual and the Real)

Peter Pran

Visionary creativity and innovative thinking in architecture and engineering are liberating us and allowing us to develop outstanding new solutions for buildings and urban design. Our close architect-engineer-developer collaboration and trust makes everything possible.

Our work and design visions are on the edge, celebrating complexity, layered meanings and instability ushered in by the 21st century. Today, the real and virtual worlds are rapidly crossing paths, exploding our perceptions of where and how we live and work. Boundaries of all kinds are breaking down, providing opportunities for change in a tectonic realization of existentialist architecture. The job of predicting and charting a course for the future while striving to re-invent reality is destined to become a global struggle. Our goal should be to move aside whatever stands in the way of innovative thought, to articulate liberated buildings and spaces that make lives richer and more meaningful, and to define a vision for everyone as individuals. Based on progressive social, cultural and political ideals, we are in a position to take advantage of interaction across all media, and to express the full complexity and equality of all people. This kind of creative collaboration, viewed on a universal scale, reflects the marvelous potential we have available to us through the integration of our professions.

As the mind is liberated by the computer's borderless realm of possibilities, all things tangible will also begin to reflect an increasingly open world of opportunity. Architecture will soon outgrow its dependency on the rules and regulations that historically have driven the creation of form and space, and will be able to address a demand for more sensual and multi-faceted environments.

During the 21st century, the greatest freedom and the greatest dilemma of human existence are likely to unfold. How will society evolve to accommodate global electronic connectivity while maintaining a physical sense of place and belonging? What kind of environment will incorporate transient communication nodes, marking only moments in space and time, into settings of geological and cultural permanence? Architecture will begin to reflect the influence of these forces, along with digital technology, to introduce a new spectrum of places for living and working beyond the current boundaries of design.

LIBERATED DESIGNS REINVENT CITIES

The last two decades have benefited from progressive urban developments initiated by political leaders in notable cities. Specifically, such as in the work of
Mayor Pasqual Maragall in Barcelona, and President Francois Mitterand in
Paris. At the same time, the support of many private institutions and clients has
helped to sanction and spread modern architecture – establishing progressive
architectural and democratic socio-political thought as integral parts of the same
trend into the future. The effect has been to start to awaken people to a new perspective, providing opportunities to live better lives through innovative and
compassionate design. A society's worth must be measured by how it treats
those that are worse off. Our society needs more such public figures and leaders
in private business that stand up for architecture and advanced urban living,
recognizing and promoting the potential of architecture to revitalize our cities.

Contemporary modern designers have decisively won the 20-year battle of
modern architecture versus repeated traditionalism. This is important to our
society, as we continue to reinvent ourselves and provide an authentic architecture for now and for the future.

One exploration of what a specific new urban habitat can be and what life
it can achieve is the Seoul Dome in Seoul, Korea by NBBJ and SWMB (Fig. 1).
This project shows how a mixed-use sport/entertainment/cultural program transformed a collection of separate functions into a coherent interlocking, pulsating
whole. A new entity was created that allows for dialogue between its uses,
emerging as a profound new urban statement that reaches out and proclaims a
change to the city forever. In this way, the Seoul Dome provides a focal point
for people from all walks of the city to come together and infuse their community with new energy and purpose.

It is intriguing to compare the spirit of this building with one from three
decades ago – the National Gallery in Berlin, Germany, (Fig. 2) – which this
author worked on as the project designer for and with Mies van der Rohe. As if
looking to the future, or perhaps recognizing a constant universal theme, the
floating, minimalist, endless planes of the roof and the podium extend out and
give an architectural and cultural coherency to an otherwise idiosyncratic, partly
broken and traditional city plan. This project and the Seoul Dome explore movement and stillness; equally needed human/urban values that rarely have the
chance to co-exist. But architecture makes it possible.

The new center and expansion for the City of Lille, France (Fig. 3) in its
thoroughly modern concept, is another successful major urban design. This
project recalls the modern authenticity of the Weissenhofsiedlung Building
Exhibit that was held seventy years ago. Moving away from stagnant, rule-rigid
urban development, it acknowledges the many overlapping aspects of human
existence and demonstrates the power of 100 percent modern design to endure
as successful architecture.

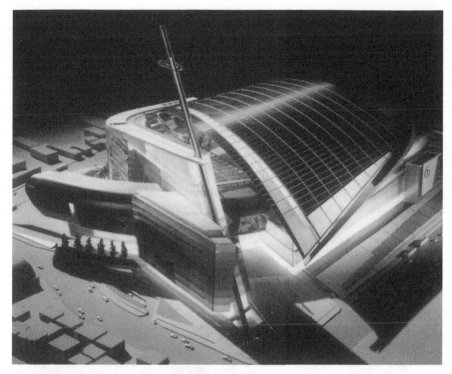

Figure 1 Seoul Dome, Seoul, Korea.

Figure 2 National Gallery, Berlin, Germany by Mies van der Rohe.

COURAGEOUS BUILDING STATEMENTS PRESERVE A PIONEERING SPIRIT IN CITIES

In New York City over 40 years ago, three authentic modern architectural masterworks were created: the UN Headquarters along the East River by Le Corbusier and Oscar Niemeyer; the Guggenheim Museum by Frank Lloyd Wright; and the Seagram Building by Mies (Fig. 4). Ironically, New York City's two largest recent urban developments, the Battery Park Residential Area, and the large Westside Development (Fig. 5), show a regression back to a traditional design formula from the 1930s. Obviously the city and we cannot rest on the laurels of past great achievements. Always we must renew our thinking and strive to move forward, finding ways to address the restrictions that recent developments have had to confront, such as strict zoning laws, input from conservative community groups, and the city's prohibitive involvement – while keeping a commitment to a renewed modern architecture. Even before design had begun on the Westside Development, every shaped massing step had been pre-determined in a traditional style. At this rate, the only freedom left to the architects was to select the color of the glass and the stone for the façades.

Figure 3   Lille, France.

This year, Ada Louise Huxtable, brilliant long time architectural critic for the *New York Times*, pointed out that restrictive new zoning laws proposed this year 2000 present a clear further danger to the creation of spirited high-rise additions in New York. This is very much in contrast to the earlier and middle parts of the last century when New York was recognized as a birthplace for visionary high-rise design concepts. Similar restrictions can be seen in new, higher buildings in downtown San Francisco and other major cities where zoning laws and regulations dictate the overall design. Contextual often becomes a catch phrase for copying the buildings next door, leading to a serious misconception about the value of new contemporary design for enhancing and enlivening the environment.

Figure 4   Guggenheim Museum by F. L. Wright; UN Headquarters by Le Corbusier and Niemeyer; Seagram Building by Mies van der Rohe.

With a longer perspective, Berlin serves as an example of what could happen in New York. Ten years ago, the building director and the building committees for a new city plan laid down rules to make new construction look 'like the old Berlin,' and set the city back substantially in its overall urban approach. Despite this severe hurdle, though, a few leading architects have succeeded in breaking away and have introduced important architecture; the work of Jean Nouvel, Daniel Libeskind and Frank Gehry, along with certain architectural work of the new Potzdammer Platz, have provided superb buildings that help to lead the way to new concepts. They are a reminder to us of the need to promote visionary thinking.

SOARING TOWERS ARE THE MODERN CITIES' WAY OF PROCLAIMING HOPE FOR THE FUTURE

Towers have the distinction of allowing people to experience cities from above. In a majority of the largest European cities, however, this is not an option because high-rises are unfortunately considered inappropriate in most areas. Similarly, many major cities in the US have in recent times put into law certain height limitations of 30, 40, or 50-stories. These regulations are not reasonable for the most part and need to be changed or partly eliminated. But it is possible to look to Asia for inspiration where one notices relaxed rules with new buildings. In Hong Kong and Singapore (Fig. 6), for example, the cities and the clients are very open to new innovative architecture in high-rises, with very few height restrictions. The constantly emerging skylines that can be seen are vigorous and fresh.

Figure 5   West Side development, New York.

Figure 6 Hong Kong and Singapore.

Of course there are many places and areas where high-rises are not meaningful to construct but, in the larger cities, they make sense by achieving a denser, more energetic city core with shorter travelling distances, more efficient energy use, and more dramatic urban settings that attract social and cultural centers.

The resistance that Norman Foster's 80-story Millennium Tower in London (Fig. 7) faced is another instance of a lost opportunity to strengthen and enrich the identity of a city. The explanation supplied by city bureaucrats and a large percentage of the city's population was that the tower would overpower the neighboring cathedrals and other large public buildings. Of course everybody wants to protect our heritage of outstanding historic buildings, and to respect their relationship to planned new buildings. But, in general, it is not right for one or two hundred year old buildings of limited height to continue to dominate a modern city's skyline for the foreseeable future.

Mies brought compassion and dedication to the development of the most innovative high-rise concepts, including their poetic urban articulation, as seen in his Chicago Federal Center (Fig. 8) and his Toronto Dominion Centre (Fig. 9). Both were at the vanguard of design with powerful and minimalist pure, fluid urban statements. This author had the great fortune to work on both of these masterful buildings, as well as on SOM's Sears Tower (Fig. 10). The Sears Tower involved the work of the renowned structural engineer Fazlur Khan, Structural Partner, and Bruce Graham, Design Partner, in the schematic phase.

Figure 7   Millennium Tower, London, England by Norman Foster.

Two Union in Seattle (Fig. 11), a 56-story office building by NBBJ, represents one of the most outstanding high-rises internationally in terms of architectural achievements. The advanced structural engineering that SWMB provided brought substantial savings of US$8 million to the construction cost, and showed that well-designed and constructed high-rises are a viable development option.

Figure 8   Chicago Federal Center.

Figure 9   Toronto Dominion Centre.

Figure 10   Sears Tower.

Figure 11   Two Union Square.

In our many exploratory, creative, innovative high-rise urban developments, our designs grow from site-specific information that gives a vision of a new architecture and a new city. A number of Ellerbe Becket projects that our design team created have this effect, as seen with the Samarec Headquarters in Jeddah, Saudi Arabia (Fig. 12), and the Portofino Apartment Tower in Miami, Florida (Fig. 13), both which have a daring sail-shape that becomes the entire building. In the Jeddah project, there are also tie-backs through the tall atrium to an adjoining rectangular building wing, which makes the structural system viable. The Karet Office Tower in Jakarta, Indonesia (Fig. 14), with its tight budget, achieves one move with an exuberant curved shape at the top that ingeniously includes the mechanical systems, as well as a floating restaurant.

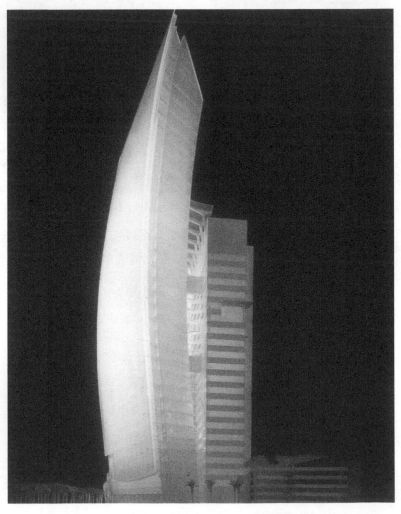

Figure 12   Samarec Headquarters, Jeddah.

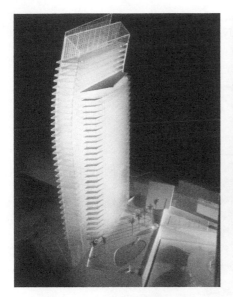

Figure 13 Portofino Tower, Miami.   **Figure 14** Karet Office Tower, Jakarta.

The 50-story Graha Kuningan in Jakarta, Indonesia (Fig. 15) is further an important case of architectural and structural solutions being in full harmony. Still under construction, it is the work of our design team in Ellerbe Becket in SC and DD, and of the leading designers continuing in NBBJ in CD. With a double-decker elevator system and a sky-lobby on the 34th and 35th floors, this elegant, soaring, slim tower allows for a small core and over 85% floor efficiency. The low-rise building wings connect with the surrounding neighborhoods and the fabric of the city. LERA developed the most advanced structural system for this building, and saved the client US$6 million. The structure uses mixed composite steel and concrete, with composite columns. On the 34th and 35th floors, it has steel outrigger trusses from the core, and belt trusses around the building. Along the perimeter of the building floors, widely spaced thin columns allow for maximum panoramic views and emphasize a clear advantage of the high-rise perspective.

Here are further examples of exuberant design schemes by our design teams in Asia, showing the possibilities of a soaring curved edge, and the ability to be completed within a tight budget. This is the 80-story Mixed-use Tower in Jakarta, Indonesia (Fig. 16) by NBBJ; the TNB Headquarters in Kuala Lumur, Malaysia (Fig. 17) by Ellerbe Becket and Focus Architects; the El Presidente in Manila, Philippines (Fig. 18) by NBBJ; and the 50-story Nihon TV Headquarters in Tokyo, Japan (Fig. 19) by NBBJ and Ishimoto. The TNB has varied atriums "walking up" the entire height of the building along the curved edge, while the tight site for the El Presidente calls for one metal/concrete façade and another consisting of an open, twisting all-glass curtain wall. In other words, these projects demonstrate highly creative means for defining and contributing to their demanding contexts.

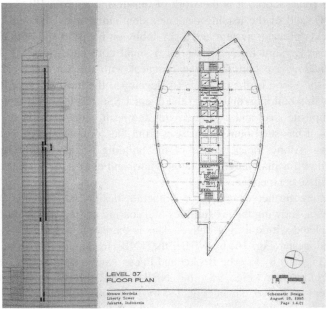

Figure 15   Graha Kuningan, Jakarta.

Figure 16 80-story Jakarta. Mixed-Use, Jakarta

Figure 17 TNB Headquarters, Kuala Lumpur.

Figure 18 El Presidente, Manila.

Figure 19 Nihon TV Headquarters, Tokyo.

Continuing in the same vein, the Stulang Development (Fig. 20) in Johor, Malaysia by NBBJ and Focus Architects, shows an evocative stepping, moving urban complex with apartment towers, performing arts center, and a hotel whose waterfront orientation faces Singapore. In this case, a single but complex architectural expression combines a variety of distinct functions and brings them to life as a unique new urban entity.

Reaching into the 21st century, the Kwun Tong Town Centre in Hong Kong (Fig. 21) by NBBJ shows a dynamic and poetic urban complex. It consists admirably of an 88-story office tower (with a homage to Brancusi), five 50-story apartment towers, a horizontal, curved hotel, a retail center, an arrival hall from the adjacent metro station, parking and the bus terminal below. The lower floors connect with the existing urban setting.

In summary, what these projects illustrate are the many successful urban high-rise projects that we have developed over the last ten years. They are recognized as standing in the forefront of the most innovative and creative towers internationally; and are distinguished for having a poetic quality that makes them memorable anywhere.

Towers essentially are vertical cities. Not only does innovative engineering make them possible, but also gives us the chance to explore and expand this concept, and to devise new never-thought-of resolutions. Downtown residential high-rises introduce activities around the clock, and make many previously empty downtown urban areas come alive again. We all recognize the incredible energy that a dense vertical city like New York has – energy with the power to attract the necessary concentration of public institutions that allow a world cultural center to blossom.

Parallel with the advancement of modern design, it is important to follow the development of ideas in structural engineering. The attached graph by Skilling Ward Magnusson Barkshire, showing Structural Steel Weight versus Number of Stories in High-rises (Fig. 22), illustrates the most successful high-rises in terms of structure in recent years. It is interesting to note that what we considered cutting-edge in 1970 is now less in the forefront, and can be embellished in many ways. In architect-engineering collaborations that encourage more daring accomplishments, structural engineers' constant search for improvements lead to better results in design. Parallel to this study and information, however, we also need to review the varied construction costs in different countries.

VERY RESTRICTIVE ZONING AND RULES OFTEN UNDERMINE INNOVATIVE ARCHITECTURE

Returning to the subject of restrictive urban zoning, it is clear that the Guggenheim Museum in New York could only have been built by breaking all the existing rules in New York City. So thanks to Mr. Moses' insistence on breaking the zoning rules, it got built.

It is meaningful to have some zoning laws ensuring that sun and daylight reach the streets, but the resulting FAR rules and "Sky Exposure Plane" in New York City (Fig. 23), as well as a number of other cities, negatively lead to mostly undesirable "wedding cake" building masses that often undermine many great architectural visions and concepts. Buildings that go vertically straight up from the street – and are likely the best architectural solutions – are not allowed or accepted based on unreasonably narrow criteria.

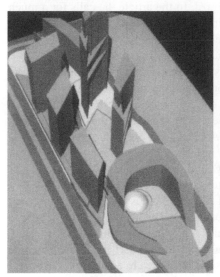

Figure 20   Stulang Development, Johor.

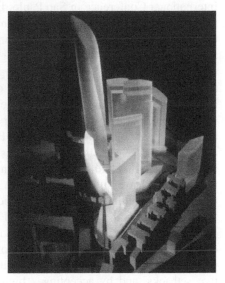

Figure 21   Kwun Tong Town Centre, Hong Kong.

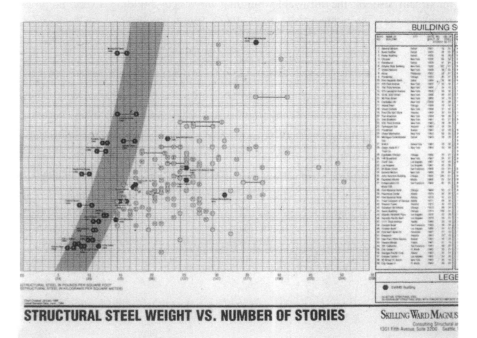

Figure 22   Chart: Structural steel weight vs. number of stories.

Unfortunately, applying for zoning change can take years which leads most developers to avoid the hassle. Valer Mocak, Ph.D. Architecture, San Francisco, wrote an excellent article titled "*A Quest for Sanity in Skyscraper Design*" for the 1997 Council on Tall Buildings and Urban Habitat's International Conference in Sao Paulo, Brazil. In the article, he calls for a much more free and open interpretation of zoning laws in order to allow for a substantially more creative and bold architecture. Very pointedly he asks, "Why are so many pre-zoning cities like Venice, Prague, Vienna, Budapest so beautiful?"

To consider the role of tall buildings realistically, and to reconsider revising the interpretation of the zoning laws, one can as examples suggest a greater length of waterfront as an alternative to one or two smaller sites on the waterfront, or consider how a tower might serve a large segment of the city rather than a specific small site. If you follow the FAR and "Sky Exposure Plane," you do not automatically get attractive buildings. Instead, it is often quite the opposite. It becomes clear that with very strict zoning rules, building shapes are often predetermined and leave little artistic freedom for the architects and engineers.

A few buildings, such as Raymond Abraham's Austrian Culture House on 52nd Street (Fig. 24) and Christian de Portzamparc's LVMH Building (Fig. 25) on 57th Street, are exceptions to the rule, achieving high quality architecture despite the zoning laws. The wealthy owner of the LVMH made this freedom possible by allowing for a reduction in usable floor area to achieve all the attractive setbacks, and by accepting a higher cost. Unfortunately for the value of design, the majority of developers would turn the other way and require the maximum use of the allowable building envelope, building every possible square foot and reducing artistic freedom to a minimum.

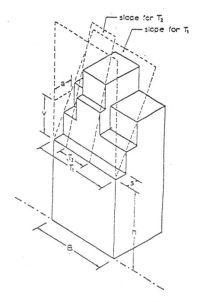

Figure 23   "Sky Exposure Plane" zoning for New York City.

It is the hope that the few new innovative, successful buildings in New York will ultimately set a standard that other developers cannot ignore, and will make them feel an obligation to rise to the occasion. It is essential for architects to team up with progressive clients that are open to and look for innovative design. And collaborations with great clients are real highlights in our professional careers.

As another example of what can be achieved against the odds, the New York Psychiatric Institute (Fig. 26), by our design team at Ellerbe Becket, transformed a major Manhattan site into a liberated and fluid building that presents different faces to meet the specificity of different site directions. Its movement expresses a metaphor for the movement of Hudson River and the cars on the Westside Highway, while also giving the entry along Riverside Drive a welcoming gesture.

Figure 24   Austrian Culture House, New York City.

Figure 25   LVMH Building, New York.

Figure 26   New York State Psychiatric Institute, Manhattan.

ENERGY CODES CAN WORK AGAINST ARCHITECTURE

Everybody agrees in general with the importance of saving energy, but energy codes generally kill floor-to-ceiling glass exterior walls in new construction cases. Mies' floor-to-ceiling glass apartment and office buildings that provide magnificent panoramic views and a feeling of suspension can rarely be built any longer. This is a very big loss. The 860–880 Lake Shore Drive Apartments in Chicago provide an important reference. Clearly their message is that saving some energy is always more important than creating great architectural spaces with floor-to-ceiling glass and wonderful panoramic views. It is as if the wonderful transparency of glass is sacrificed at the altar of energy saving (tinted, energy saving glass often looses it transparency). What we really need, however, is a balance of all these concerns. The final architectural result should always be given the highest priority.

CHANGING CITIES THROUGH CULTURE

With one building in Bilbao, the Guggenheim Museum by Frank Gehry (Fig. 27), the entire city has been reinvented and given international prominence. The architectural vision attracts large numbers of visitors and has made the city renowned everywhere. It shows that there is no excuse for any city to be defined by mediocre buildings, and be deprived of or deny the vision, collaborative strength and courage of outstanding architects and architecture. It also totally liberates architecture, showing the potential to achieve never before seen or even imagined designs.

Figure 27 Guggenheim Museum, Bilbao, Spain.

Daniel Libeskind's Museum in Berlin achieved an unforeseen penetrating thought process and discourse. Reflecting the force field of the city with its layered meanings, such as its personal history of the Jewish people, and the void spaces that symbolize the Holocaust victims, we are not allowed to escape awareness of the most excruciating evil while at the same time the building miraculously gives us a constructive dialogue and great hope for the future.

Creations like Wolf Prix' excellent Cinema complex and Itsuko Hasegawa's almost ephemeral but powerful lyrical statements in her domes, further transform our cities to new levels of cultural integrity, social awareness and innovative architectural experiences. Ultimately, innovative and profound buildings such as these change our entire world and consciousness, and prevent us from ever returning to the "before time."

Looking to the future, it is likewise important for architects and artists to interact, provoke and merge ideas with each other, as with Jenny Holzer's work and Richard Serra's work (Fig. 28) in Gehry's Bilbao Museum. Simultaneously, the virtual museum is already here and ready to add its influence; as we may now enter museums all over the world through cyberspace. Here is a range of our decisive responses to varied programs and varied urban settings toward a new urban habitat.

As a new aspect of urban expressions, is the notion of the endless building; the "endless horizontal building" seen in our AA/NW Terminal at JFK International Airport, by our design team at Ellerbe Becket, and the "endless vertical building" in Jean Nouvel's round Tower at LaDefense outside Paris. Both buildings have no beginning and no end.

Figure 28   Richard Serra artwork in Bilbao Museum.

The Vulcan Building in Seattle (Fig. 29) by NBBJ Design interacts with its historic neighborhood as well as with the urban forces of the city; it shows architecture as an interactive art, that participate and transforms scale, masses, economy, land lots and cultural content. The Manggarai Train Station/ Transportation Centre, Jakarta, Indonesia (Fig. 30) reaches into the surrounding communities, while giving these a new invigorating identity and character. The Gateway+Maintenance Building at Dallas Fort Worth International Airport, Dallas, Texas (Fig. 31) expresses the anticipation of flight before one flies, and define and celebrate the arrival and departure to/from the airport. The Staples Center arena (sports/conference, entertainment center) in Los Angeles, California transforms an important urban district downtown, and brings in life and excitement, and thereby changes the city itself (similarly to the concept of the Seoul Dome, described earlier).

Figure 29 Vulcan Headquarters, Seattle.

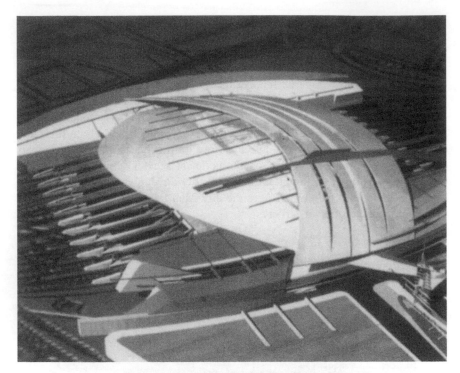

Figure 30   Manggari Trainstation, Jakarta.

Figure 31   Gateway-Maintenance Building, Dallas Fort Worth International Airport, Dallas.

There is a move toward global architecture, that is simultaneously site and place specific, such as in the Reebok Headquarters near Boston in Massachusetts (Fig. 32), and the Telenor Headquarters, Oslo, Norway, both by NBBJ, where the work place is transformed and the progressive companies are given a new, never-seen-before identity.

APPROACH – PERCEPTIONS/INTERSECTIONS

The first step is our approach: How can we change the program? The second move: How can the different program elements inform each other and create conditions and solutions possibly not thought about before, organized anew? From an inside thought process, it becomes a new architecture, complex and diverse. We often see a reduced vision and a general lack of will to confront urban design problems unfortunately; the pressure is often on to demand normalcy.

We embrace and welcome experimentation, diversity and complexity, which gives depth and insight; an architecture that is indeterminate – and has both movement and stillness, multi-layered; an architecture that is democratic and has an open mind.

Figure 32   Reebok Headquarters, near Boston.

PLANNING AND ARCHITECTURE

Aurora Place Commercial Office Tower – 88 Phillip Street, Sydney

Rocco Bressi

1.1 INTRODUCTION

Aurora Place is a landmark mixed use development situated in the core of Sydney's Central Business District on the site of the former State Office Block.

The project, completed in December 2000, comprises a 44 level office tower, 18 level residential building and supporting retail facilities. The 4,262 square metre site is bounded by Phillip Street, Bent Street and Macquarie Street.

Aurora Place has been developed by Lend Lease and East Asia Property Group, the extraordinary design is the work of world renowned architect Renzo Piano. Bovis Lend Lease is responsible for the project management of the design and construction.

The office tower at Aurora Place has a net lettable area of 49,500 m^2 and has attracted several pre-eminent companies as tenants such as ABN-Amro, Minter Ellison, Challenger International and the Executive Centre. The office tower not only boasts a prime location, landmark design and quality, it also rates as one of Sydney's most efficient and effective office towers.

Figure 1   East view of Aurora Place Development.

The residential building at Aurora Place is known as Macquarie Apartments and comprises 62 luxury residences, all of which enjoy unobstructed views across the Royal Botanic Gardens to Sydney Harbour and the Opera House.

This paper deals primarily with the structural design aspects of the commercial office tower.

1.2   BASIC BUILDING STATISTICS

The following statistics summarise some of the key features of the commercial tower:

- Number of commercial levels – 38
- Dedicated plant levels – Levels 3, 21, 42 and 44
- Number of basement levels – 4
- Floor to floor height of commercial floors – 3,720 mm
- Ceiling height to office levels – 2,700 mm
- Approximate NLA commercial floor areas including wintergardens (square metres) – 1,285 LR, 1,358 MR, 1,435 HR
- Core areas (square metres) – 354 LR, 324 MR, 260 HR
- Overall building height (to top of sail) – 200 metres
- Core slenderness ratio (height/depth) – 21
- Maximum lateral east-west sway movements – 300 mm
- Tower building frame lowest natural frequency – 0.18 hertz
- Stability frame – No outriggers used, building stability relies on combined frame action on core, floors and columns
- Average overall axial shortening of tower frame due to creep and shrinkage – 75 mm
- Volume of concrete used – 35,000 cubic metres
- Tonnes of reinforcement used – 6,200 tonnes
- Tonnes of post-tensioning used – 350 tonnes
- Tonnes of structural steel used – 650 tonnes
- Overall construction time including site demolition – 4 years

1.3   DEMOLITION OF THE STATE OFFICE BLOCK

As sites become scarce for tall buildings, these types of projects usually involve consolidation of land which often require demolition work.

This preliminary work is almost synonymous with tall building projects and are often complex.

The Aurora Place project was no exception. The site to be cleared involved the "de-building" of a 31 storey building, a 12 storey building and a 9 storey building. Known as the NSW State Office Block, the buildings were completed in the mid 1960's.

Obtaining information on these buildings is often difficult and often requires a significant level of exploratory work to analyse and understand the structures to be demolished.

Preparation and detailed planning with an integrated team of structural engineers and the builder was the key to the success of this "de-building" project.

Figure 2   State Office Block tower demolition.

The main building included a large central core of 6 lift banks, including in total 18 lift shafts.

The floor plates included steel beams and composite columns with a slab on ribbed sheet metal which had no shear studs attachment to the beams.

The integrated team evolved a method of "de building" that produced a 2.5 day cycle per floor.

This was an outstanding achievement and saved 4 months on the overall project program.

In the process, 98% of the base building material was recycled and the de-building work was completed with an outstanding safety record.

1.4   SITE GEOLOGY AND FOUNDATIONS

As the site was formerly occupied by the State Office Block between two and three levels of basements had been previously constructed over the entire site. The proposed development required that the excavation be extended into the Hawkesbury Sandstone by a further 10 metres to allow for the construction of two additional basement levels.

Maximum tower column working loads are in the order of 40,000 kN and are supported on reinforced pad footings. The central core having a total working load in the order of 730,000 kN is supported on a 1.5 metre thick continuous core raft projecting 1.5 metres beyond the external perimeter wall lines.

Founded onto Class II and III sandstone, the design bearing pressures vary between 3.0 MPa to 6.0 MPa as recommended by the geotechnical investigation work carried out by Coffey Partners International Pty Ltd.

The existing reinforced concrete basement walls of the State Office Block building were retained and underpinned along Bent Street to maintain support for the roadway and high voltage electrical cables. Temporary support of the other boundaries was achieved using conventional methods such as anchored soldier piles and shotcrete walls.

Figure 3   Core raft construction.

Figure 4   Axial core stresses at raft interface.

1.5   BUILDING CODES AND REGULATIONS

The structural design of the building has been carried out in accordance with the relevant SAA Codes and the Building Code of Australia. Fire engineering principles were adopted where appropriate dispensations could be sought by Sydney City Council and other relevant authorities

From a structural engineering viewpoint, fire engineering design principles were applied to the composite structural steel floors, roof and sail elements above Level 41 to assist with the deletion of traditional fire rated steel construction. Active sprinkler systems were used throughout the building.

Figure 5   Level 42 plantroom needle columns and steel framing.

1.6   STRUCTURAL FRAMING DETAILS

The structural elements of the commercial office tower can be considered as two basic components, the primary building frame and the secondary structural façade support elements attached to the building frame.

The primary building frame is mostly constructed from a combination of reinforced and post-tensioned concrete. Composite structural steel has also been used for the accelerated building frame structures erected to Level 3 and the plantroom roof elements.

The secondary structural elements are constructed from structural steel and interact directly with the projecting external façade components. These elements are termed fins, tusks, sails and mast. Dog-bone mullions manufactured from aluminium are used and span between the cantilevered secondary steel support members. The externalised glass panels are supported directly from the dog-bones with silicon only.

1.7   BUILDING FRAME AND LATERAL STABILITY

The reinforced concrete core works integrally with the floor plates and columns to form a combined moment resisting frame. The distribution of loads to the core, floors and columns has been determined by finite element analyses. The wind loads applied to the building have been verified by aeroelastic wind tunnel model tests carried out by MEL Consultants at the Department of Mechanical Engineering, Monash University.

The frame action of the core, slabs and tower columns contribute to the lateral stability of the building. This requires that the floor-to-core connection be designed to resist the applied frame moments. Each floor acts as a mini-outrigger eliminating the need to adopt a centralised floor to floor outrigger system that would normally occupy valuable space within the building.

Figure 6   Core stresses.

A 320 mm thickened slab around the core enhances the floor to core moment connection. Two layers of 20 mm diameter screwed couplers are placed continuously at 150 mm centres and splicing with top and bottom reinforcement to provide a continuous connection for the thickened slab to the core.

Figure 7   Core jumpform construction.

The core provides 70% of the lateral load stiffness and the remaining 30% is taken by the combined frame action of the slabs and tower columns.

The lozenge shaped reinforced concrete core has a maximum width of 9.5 m. Permanently anchoring the 1.5 metre thick core raft into the bedrock provides base fixity of the core structure. Along the perimeter of the outer core walls, ten permanent rock anchors each of 8000 kN working load capacity were drilled through the core raft and anchored approximately 16 metres into the sandstone bedrock. The ground anchors ensure that no net tension results under the raft when the most adverse lateral and eccentric load conditions are applied to the building frame.

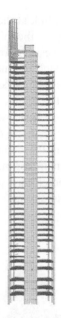

Figure 8   Finite element model cross section.

A view of the broad elevation of the core shows how corbelling out of the main walls above the lobby level and carpark entry transfers the northern and southern ends of the building. The corbelling extends the core by 10 metres approximately in the north and south directions along its longitudinal axis.

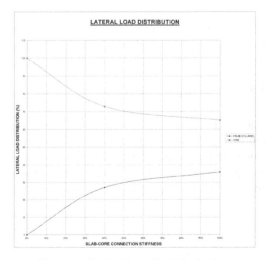

Figure 9   Graph of lateral load distribution.

To form a concierge through link at the lobby level an opening was made through the main walls adjacent to the high rise lift lobby. The structural opening through the eastern and western core walls is 5.0 metres high by 5.9 metres long. To transfer the truncated perimeter wall loads 3.0 metres deep by 1.0 metres wide reinforced concrete lintel beams are flinched to the core walls above the openings. A 1,200 mm thick reinforced concrete plate spans between the flinched beams to support the secondary walls located above the concierge.

Verticality of the core under eccentric dead and live loads was maintained by ensuring that the stress levels in the outer walls were similar on both sides. This was achieved by tuning the main perimeter wall thicknesses, resulting in a 500 mm base wall thickness on the eastern side and a 400 mm base wall thickness on the western side. Concrete strength for the core walls varied between 60 MPa and 32 MPa.

Figure 10   Aurora tower construction.

Splitting the jump form system into two segments enabled construction of the core to be carried out more efficiently. Screwed coupler bars are used to join the construction joint match lines of the core segments.

1.8   BASEMENT AND LOBBY FLOORS

For the commercial tower up to four basement levels extend out to the perimeter of the site. A combination of reinforced and post-tensioned band beams supporting reinforced slabs have been used to frame out the basement carparks, ramps, loading docks and plant areas. Earth pressures below the ground level are resisted by concrete retaining walls braced by the diaphragm action of the basement floors.

1.9   ACCELERATED TOWER CONSTRUCTION

Utilising composite structural steel framing, Level 3 floor plate of the tower and the twelve perimeter columns were constructed under accelerated conditions to facilitate a work front above and ahead of the final site excavation and prior to the construction of the basement and lobby floors. Prefabricated reinforced cages were fixed inside 12 mm thick tubular steel column form liners measuring 1,250 mm in diameter. The tubular steel forms were then filled with high strength concrete pumped from their base up to a height of 25 metres.

Basement slabs were connected to the tubular steel liners with internal and external stud attachments and lintel flange plates coinciding with the relevant floor levels.

Figure 11   Corbelled south core above lobby level.

The typical floor table formwork system was then introduced. The enhanced structural capacity provided by the composite beam action alleviated the need for back propping of the formwork system below Level 3.

Figure 12   Accelerated Level 3 structural steel framing.

1.10   TYPICAL OFFICE FLOORS

The typical office floor slabs span from the core a distance of 10.8 metres to the west and 12.0 metres to the east to the perimeter beams. Six tower columns, spaced at 10.8 metre centres, support the floors each side of the core and are located along the curved east and west building perimeter. The floor plates cantilever at the four corners of the structure. The south-east corner of the tower structure gradually extends and leans eastward at every level. Summation of this floor to floor incremental offsets results in a 5.0 metre change in span of the edge beam between Level 3 and Level 41.

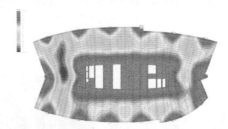

Figure 13   High rise floor deflections contours.

Notched post-tensioned band beams spanning between the core and the perimeter beams have been designed to a minimum depth to optimise services clearances and are typically 470 mm deep and 600 mm wide. The beam-ends are notched 150 mm for a length of 1.3 metres. This allows the underside attachment of the live end anchorages and stressing of the post-tensioning cables. The floor beams are spaced radially at 2.7 m centres and support a 120 mm thick mesh reinforced concrete floor plate. The floor to floor height is typically 3.72 metres. Perimeter edge beams are 880 mm deep × 400 mm wide and cantilever northwards and southwards to support the wintergardens and projecting façade elements. The edge beams and attached facade elements (fins) cantilever in some instances in excess of 10 metres beyond the first internal column.

Figure 14   Finite element model of typical floor soffit framing.

Located continuously for a 2.0 metre width around the perimeter of the core, a 320 mm thickened slab section provides connection of the band beams to the core. The thickened slab section transfers the plate bending moments and in-plane diaphragm actions generated in the floors to the core walls. Circumferential post-tensioned cables are located within the thickened slab adjacent to the core. The curved plan profile of these cables assists in applying extraneous in-plane forces from the floors to the core similar to the concept of placing an elastic rubber band around the perimeter of the core. These cables also restrain the forces generated by the cantilevered and corbelled core walls located on the northern and southern ends of the building.

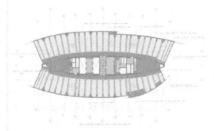

Figure 15   High rise floor framing plan.

Figure 16   Aerial view of the core and floor construction.

The typical floors are generally designed for the following loads:

Live loads	3.0 kPa
Partitions	1.0 kPa
Services and ceilings	0.5 kPa
Raised computer floors	0.6 kPa

Figure 17   Typical floor band connection to the outer core wall.

Structural provision at four locations between slab bands has been made for additional penetrations through the floors to take the loads imposed by tenant specified interconnecting stairs.

1.11   PLANTROOM FLOORS AND ROOF

Double height plantrooms are located at Level 21 and Level 42 and the floors have a similar structural configuration as a typical floor. The required enhanced live load capacity of 7.5 kPa for equipment is obtained by increasing the overall structural depth to 500 mm and providing additional post-tensioning cables and reinforcement in the bands and slabs. Level 3 and the roof slab at Level 44 are supported from composite steel beams and profiled steel sheeting. The surfaces of external areas and wet areas are cast integrally with falls to assist with drainage and waterproofing.

1.12   STRUCTURAL ANALYSIS

Bovis Lend Lease structural engineers used Strand7 finite element program to perform numerous structural analyses on the building frame of the commercial tower.

A rigorous structural analysis of the tower building frame was carried out to comply with Clause 7.8 of AS3600, Concrete Structures Code. The analysis of the building frame takes into account the relevant material properties, geometric effects, three-dimensional effects and interaction with the foundations.

The principal aim of the analysis was to effectively predict the structural behaviour of the unique shape of the tower frame subjected to various static and dynamic loading conditions. These loading conditions were generated from combinations of superimposed dead loads, live loads, wind loads and seismic loads.

Static load combinations complying with the relevant Australian Standards provided realistic predictions of the actions of the building frame, particularly due to applied lateral loads and eccentric gravity loads.

The dynamic response of the structure also needed to be assessed to evaluate the natural frequencies of the tower frame and consequently the response of the structure in terms of perceptible building accelerations.

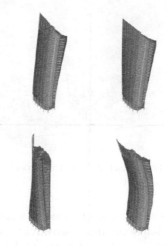

Figure 18   Natural frequency modal shapes.

A finite element linear buckling analysis enabled the determination of the tower column effective lengths, which then assisted with their final detailed design.

The columns measure 1,250 mm diameter up to Level 3, then change to a reducing rectangular shape through the tower. The low-rise column measurements are 1,200 mm × 800 mm, the mid-rise columns are 1,200 mm × 650 mm and the high-rise columns are 1,200 mm × 500 mm.

A suitable mesh grading to model the floor in plate elements was developed. Five plate element properties were used per level to account for the end notching of the beams and slab thickening around the core. Strand7 allowed the use of different membrane and bending thicknesses for each plate property. Beam elements were used to model the perimeter beams. The self-weight of the structure could be determined by assigning plate densities. Superimposed floor loads and live loads were applied as face pressures to plate elements.

Plate elements were used to simulate the façade of the building. The façade plate elements were a Quad4 Plate/Shell type, each node connected to the structural beam elements. The lateral wind pressures were then applied to the face of these plate elements.

The linear static, non-linear static, natural frequency and linear buckling solvers were used to evaluate the structural behaviour of the tower frame. Linear static analyses were used to predict the structural behaviour of the structural elements for a combination of lateral loads and eccentric vertical loads.

Figure 19   Aero-elastic model of Aurora tower.

The moment distribution of the lateral load between the core and the frame was also evaluated. It was important that a sensitivity analysis be carried out to account for the varying stiffness of the slab-core and slab-column connection.

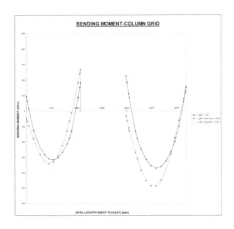

Figure 20   Typical floor slab bending moments.

The sub-modelling feature in Strand7 was used to study the detailed actions of the structural components in critical areas. This was achieved by using a much finer mesh in these areas compared to the coarser mesh of the global model. The sub-modelling feature allowed the bending moments, shear forces and axial forces to be found for the subsequent design of each structural component using the results of the global model.

A sensitivity analysis was also performed on the model by varying the dynamic and static moduli of the structural sections. This enabled the range of possible natural frequencies for the tower to be determined.

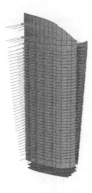

Figure 21   Westerly wind pressures apply to global FEM model.

The occupancy comfort was gauged by a study on the accelerations of the building under dynamic loading conditions. For the lowest natural frequency, the calculated building frame accelerations were compared to maximum peak recommended values for a mean wind return period of 0.5, 1, 5 and 10 years as set out by AS1170 Part 2, Wind Loading code.

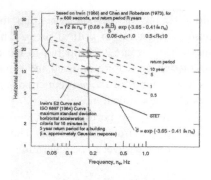

Horizontal acceleration criterion for occupancy
comfort in buildings.

Figure 22   Building accelerations for occupancy comfort.

The base finite element model has the following statistics:

Number of nodes	36,097
Number of beam elements	4,840
Number of plate elements	38,566
Number of equations	215,856
Run time	7 hours approximately

The model was run in excess of 160 times to evaluate the sensitivity and effect of different structural parameters relating to the building frame.

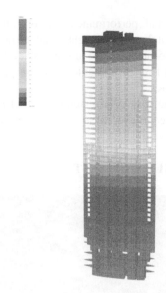

Figure 23   Lateral deflection of core elements.

1.13   CONCRETE QUALITY

For the project, the specified concrete 28-day strengths varied between 32 MPa and 80 MPa. High performance concrete was used to control long-term differential elastic axial shortening, shrinkage and creep between the columns and core walls. All concrete mixes were super-plasticised and the maximum 56-day shrinkage was limited to 600 micro-strain.

By way of trying to equate the overall axial elastic shortening, shrinkage and creep between the core and tower columns, the concrete shrinkage specified for the concrete used in the columns was limited to 450 micro-strain compared to 600 micro-strain used for the core.

1.14   FAÇADE SYSTEM

The aluminium and glass curtain wall system spans from floor to floor and is supported by the edge beams with cast-in anchorage brackets. The double glazed façade system is designed to cope with the anticipated building movements determined by structural modelling.

The external layer of glass is chemically treated with a ceramic silk screen frit that helps to reduce the effects of direct solar heat gains and re-radiated heat gains. The ghost-like appearance is created by grading the frit around the perimeter of the vision panels from 80% at the body of the building to 40% at the edges of the fins and sail. The frit also modulates the transparency of the vision panels and hides the column and spandrel structures. High-energy efficiency, maximum thermal performance and optimum visual comfort are achieved by using a moderately reflective glass and a low-E coating.

The wintergardens located on the northwest and southeast ends of the tower have clear glass and framed operable windows protected by cantilevered aluminium louvred sunshades.

The projecting façade glass that extends beyond and above the building footprint is supported by cantilevered structural steel framing attached to the edge beams.

1.14 FINS, TUSKS, SAILS AND MAST

Figure 24   Mast and west sail.

These elements are fabricated from structural steel and attached to the concrete building frame using cast-in embedments and bolts. The wind forces to be carried by these projecting elements have been determined by wind model testing as part of the aeroelastic studies carried out by MEL Consultants.

Figure 25   Façade fin cantilevered steel framing.

The cantilevered sail needles that project above Level 44 are laterally restrained by composite structural steel floor systems incorporated in the two top levels of the tower. The cantilevered needles back span through the high rise plantroom between Levels 42 and 44. Roll formed and flat steel plate sections are profiled and continuously butt welded to form the tubular and tapering needle sections. Being of grade BHP-300PLUS, the plate thicknesses used varied between 12 mm to 40 mm. Cantilevering in excess of 30 metres, the tallest needles located on the northwestern end of the sail weigh approximately 28 tonnes. The needles reduce in height and plan area as the sail extends southward.

Figure 26   West sail finite element framing model.

About their major axis the needles are required to resist ultimate wind pressures up to 3.6 kPa normal to the sail. Acting simultaneously, significant lateral load on the needles will develop produced by wind shear flow on the sail plus wind drag on their projected faces thereby causing potential in plane sway movements of the sail. These movements are significant and will cause racking of the glass façade panels if not controlled either by increasing the bending stiffness of the needles or by providing in plane bracing of the sail (or both). To minimise the width of the needles, three sets of post-tensioned stainless steel wire bracing was installed. Offset and curved 168 mm diameter pipe transoms span between the needles and are used to fix welded outriggers that support the 1,350 mm wide by 3,700 mm high facade panels.

To eliminate in plane shear forces developing in the glass façade panels due to the potential racking of the sail, the connections of the transom outriggers to the dog-bone mullions were designed so no restraint against in-plane movement would develop.

Figure 27   Aerial view of south fin projections.

A continuous 323 mm-diameter pipe that also supports a continuous access way for maintenance and cleaning of the west sail links the tops of the needles.

Figure 28   Needle fabrication.

The tapered tubular steel sections used to fabricate the mast have a wall thickness of 12 mm. The mast diameter varies from 1,450 mm to 150 mm and the mast extends 99 metres above the crow's nest located at Level 34. Outriggers connected to cast in plates set flush in the core provide support for the mast.

Figure 29   West sail needle erection.

1.15   PIAZZA CANOPY

Ove Arup and Partners were appointed as specialist-engineering consultants to develop the concept for the canopy as a thin layer of glass with no secondary structural members supported by a "spiders web" cable net, slung between the two buildings.

Figure 30   Model of Piazza and canopy.

Having a maximum span of approximately 30 metres, the glass canopy has a plan area of about 650 square metres and is slung between the residential tower and office tower directly above the linking piazza. The cable net is formed into an anticlastic surface to ensure structural resistance to both downward and upward loadings and provides support for the frameless suspended glass.

The cable net is typically formed of 18 mm diameter, high tensile, and stainless steel rods connected at each intersection via stainless steel cast nodes. The bars specified were cold worked grade 316 stainless steel with a yield at 0.2% strain of 530 MPa. The glass is mostly 16 mm thick laminated, toughened and patch supported at each corner with stainless steel cast spiders. Varying diameter and length, stainless steel hangers, makes the link between the glass plane and the cable net.

To determine more realistic wind forces on the canopy surfaces, a 1 to 400 scale model of the canopy was pressure tapped and tested in the Monash University boundary layer wind tunnel. Surrounding buildings were modelled to at least 500 metres from the canopy.

The cable net is a structure that derives its strength and stiffness purely from its form; therefore, considerable design effort was spent to ensure the anticlastic shape was stiff for both upward and downward loads.

Figure 31   View of canopy and lobby entrance.

To determine the form and shape of the canopy the following loads were considered:

- Self weight of the rods, hangers, castings and glass
- Live loads though not critical for cable net design
- Wind forces as derived from the wind tunnel testing
- Prestress applied to the rods forming the supporting net
- Temperature differentials for a range of plus or minus 25 degrees
- Seismic forces though found to be negligible for such a light structure

For the main backstay rods between the "puntone" or bowsprit struts and the building anchorages, a higher grade of stainless steel bar was required to resist the large forces. In this instance the bar grade has an UTS of 1000 MPa.

All connections between rods within the net and at its boundaries were made using stainless steel investment castings. The casting material is grade SAF2205 that yields at 0.2% strain of between 450 MPa and 550 MPa.

Spiders with one, two, three and four legs were developed to connect the glass to the slender circular pipe hangers. Architecturally shaped, the spiders had a requirement to rotate at the neck to accommodate the varying angle between the glass plane and the vertical hangers. Once in place the neck is locked to resist bending moments from out-of-balance loads on the glass panels.

Figure 32   Piazza canopy spider and hanger connections.

The primary design action for the hangers was mainly for the downward tension loads developed by the dead, live and wind loads. However, upward wind loads are sufficient to induce compression in the hangers.

In order to verify design and as a check on the manufacturing process, three of each type of castings was load tested. The load testing included proof load testing to 1.25 times the calculated working load and then testing to destruction.

Due to the irregular shape of the glass and the point supports, non-linear finite element analyses were used to size the glass. Typically the glass comprises of one sheet of 8 mm thick toughened glass and one sheet of heat strengthened glass that is laminated with a pvb inter-layer. Glass panels with dimensions greater than about 2,000 mm had the thickness of each layer increased to 10 mm thick.

Erection of the cable net and the irregular and warping glass plane in the correct shape was essential to the structural performance of the canopy.

To erect the canopy, six stages were required as follows:

- Erection of the cast-in plates on the commercial and residential towers
- Erection of the cable net to a snug tight condition
- Stressing the net
- Erection of the droppers
- Erection of the glass panels
- Sealing and setting the joints between the glass panels

1.16   BUILDING MAINTENANCE UNIT

The Building Maintenance Unit (BMU) located on top of the building is one of the largest in the world with a total weight of one hundred tonnes. The BMU services almost all of the façade area of the building. A gantry cradle and a davit cradle system service the areas inaccessible to the BMU.

When the BMU is parked it is hidden within the building structure, and cannot be seen from ground level. When operating, the entire BMU rises five metres from its parked position and then the five stage telescopic jib can extend out forty-seven metres to reach all around the perimeter of the building. The jib also luffs up forty-five degrees from the horizontal and luffs down twenty-six degrees from the horizontal to bring the cradle as near as possible to the building façade.

Figure 33   Aerial view of BMU and sloping roof.

Operation of the BMU is simplified by the installation of a programmable logic controller (PLC). The PLC, through various counters, limit switches and sensors, determines the speed and movement of the BMU functions to give the operators the safest and most efficient ride. The acceleration of the BMU functions is also controlled to prevent any "whipping" action in the jib and thus provide the operators with a much more comfortable ride.

Other functions are also included on the BMU such as; a touch screen monitor to provide detailed information about the BMU condition, glass handling facilities on the cradle, and telephone communication between the operators in the cradle and the Building Superintendent.

Figure 34   BMU in operation.

All static and dynamic loads generated by the BMU are transferred to the tower columns and core walls by one storey deep steel trusses located within the roof plantroom space. The same trusses also assist with the strutting of the western façade cantilevered sail needles.

Figure 35   Completed Aurora Place project – December 2000

1.17   PROJECT CONSULTANT TEAM

1.17.1   Joint Venture Partners

Lend Lease Developments
East Asia Property Group
Mirvac

1.17.2   Project Management and Construction

Bovis Lend Lease

1.17.3 Architecture

Renzo Piano Building Workshop – Overall concept and façade documentation
Bovis Lend Lease – Building co-ordination
Gazzard Sheldon – Detailed documentation

1.17.4 Structural Engineering

Bovis Lend Lease – Overall structural concept and detailed design of the commercial tower
Taylor Thomson Whitting – Detailed design of the apartment building
Ove Arup and Partners – Secondary attached structures and glass canopy

1.17.5 Façade Engineering

Arup Façades
Permasteelisa

1.17.6 Wind Engineering

MEL Consultants – Professor Bill Melbourne

1.17.7 Geotechnical Engineering

Coffey Geosciences

1.17.8 Mechanical Engineering

Ove Arup and Partners
Environ

1.17.9 Other Building Services

Bovis Lend Lease

Integrated Engineering

Duncan Michael

I believe in Integrated Engineering. I want to develop it and promote it. My talk will be as rational and balanced as I am able but you should be aware that I am more than a commentator.

In my youth, systems looked to me like the best way forward and I discounted the personal differences, especially the interpersonal effects. Now I observe people as hugely and compulsively social in just about everything they do, integrating themselves into their zone of society. Being stubborn I have not so much rejected systems, as attached them to all of life. Then integrated engineering is for me, a good, and a part of the expression of human society.

The Urban Habitat and one particular species in that habitat, the Tall Building, are a massive case where integrated engineering and society are well and truly locked into each other, for better for worse.

Let us look at the engineering that is invoked for a Tall Building, in its design, its construction and its operation. There is even at a glance a vast amount of engineering and within it a huge diversity of functions. Just about everything that goes towards a Tall Building is engineered whilst the technologies that are used to affirm the answers are very specialised and getting more multiple all the time.

At this point one has a strategic choice; does one set out to integrate, disintegrate or simply ignore the issue and let events produce their own results. There is a case to be made for all three strategies. The integration route, the law and order approach, is my instinctive style and so it is difficult to articulate since it is so self-evident, to me. Clearly the integration route has to be in the design from the start if it is to deliver the benefits of its potential. These benefits include balance between the parts, prevention of imaginable problems and simplicity for operation. The benefits are often described emotionally as integrity, harmony and intellectual rigour.

Many people, indeed many in the audience, will argue strongly for a more relaxed approach. They will point to cases where the integration has taken over and become the aim making the engineering, indeed the project, badly compromised because of the diverted focus. They will say that today's specialists are so expert that it is counterproductive to make presumptions to impose on the experts other than to say what you want functionally. They are the free market believers. They do say that integration blunts brilliance, that the creative energies used up on integration would be better fed into the itemised designs themselves and that integration plays to the administrators and shackles the stars. They sometimes say that integration slows down progress, celebrates the ordinary and is socialism in disguise. Laissez faire is their literal motto.

In a third school the explicit case for disintegration can get made, with some odd outcomes. The layered approach for zoning the engineering gets promoted as part of integration, but that is really disintegration. I will live in my territory, you will live in yours and so there can be no problems; that is disintegrative. It is apartheid. Then you also get the deconstructivists, still having fun. The paradox here is that their apparent disintegrations are feasible only via the most inventive and integrated engineering available. You may note that I sound not very enthusiastic about this third school. It is their rhetoric that I get stuck over, the mismatch between the words and the action. I concede that they have been exploring new territory, though I see it being as fruitful as climbing high mountains. I can greatly enjoy fun construction, construction as a sport, without the smoke of disintegrated language. The trouble is, clients need a better story than mine, if they are to be convinced to join in.

These different approaches are often masked. Engineers do not usually march into a room and say to hell with all this tedious integration or let's disintegrate this one or I am going to integrate everyone's work and I can block until you all agree. The approaches often arise from attitude and get proffered as policy. If you are by nature a divide and control person then integration will continually make you uncomfortable. If seamless operation of the building is what you see as ultimate perfection, if team work appeals, then integration will be for ever beguiling you. If you take a long view and are fearful about the problems of maintenance and the inevitable alterations then too much optimisation from initial integration will not attract you. A general rough robustness will be what you seek, rather than an over-sensitive, highly interdependent, fully integrated composition.

My topic so far has been about engineering integrating with itself or not, self integration, an important topic to engineers but scarcely the whole story. If the case for self integrity is worthwhile then so much more is the case for integration of engineering with everything else.

In building design the great adjacency to the engineer is the architect. The relentless expansion of technology has moved the balance of the inputs of the architect and engineer, so that the traditional and organisationally recognised relationship is no longer the best method of working. Now and even more so in the future every item and issue will be sustained by an engineering input and by an architectural input, so that debate about which bits are architecture and which ones engineering is losing its meaning, and has become pointless, even if it ever was worthwhile. A better description for the work of today would be as a matrix, with the items across the bottom and the expertises as the up axis. Most of the intersections would have a Yes on them and you could score them on a number scale.

I lack the words here to sketch out the extra-engineering relationships for the unintegrated and the disintegrated approaches. The architect/engineer integration clearly is dominant in the building design area. It also extends into the money and time zones and touches the construction zone.

Along with construction method, the money and the time become dominant issues for the builder and the construction. In most countries the integration of the builder's work and the designer's work is at best partial. Certainly the potential of that interface is in no way fully realised. The exploration of that integration of design and construction could be the big theme for the progress of building work through the next decade. The linkage of design and construction has been particularly unproductive in the UK. Through the last 7 years the UK Governments have actively sought to crack the problem. [Symptoms: expensive work, often late, often exceeded budgets, bad relationships, lots of expensive litigation, little progress by industry. Not a technology issue, not a trades union issue.] Government is now trying to impose a revolution of single point procurement for all public works and is seriously coaching the players in the industry to reform into new teams complete with new formations, new tactics, new clothes and new language. It is studying how early in the process the single point, the prime contractor can be given his head. It is accompanied by lots of razzmatazz like conferences, Treasury papers, demonstration projects, ministerial appearances and explicit threats to the doubters and the disagreers. It will change things, some for the better. There is a parallel thrust in the commercial sectors driven by serial customers. All in all it is a once-in-a-generation initiative and it could just be a success.

For the engineering to be fully integrated, the engineers must in addition to integrating with the architects and the builders and amongst the engineers themselves, integrate also with the owners, the users and indeed with society since the building affects the indirect parties, the public, like neighbours or like tourists if you are lucky, like protesters if you are not. The energy needed to deliver thoroughly integrated engineering may seem excessive, given the gentle subtlety of the result, given the indirect reward and given the limited punishment if the integration is only partial. I would resist this low-added-value assessment. Dedicated practitioners become amazingly adept at the communications and relationships which go with integrated engineering. And in any case the value-cost issue of the engineering is not a precondition of working. It is a very limited issue and one that can be taken or left, depending on the objectives for the project.

It is desirable and often necessary to consider the green issues in our environment. The current version is "sustainable", that is, how much of today's consumption can we retain, can we get away with, can we sustain, over a very long period. Since integration offers processes that can be defined and measured and allows demonstrable optimisation to be exercised, sustainability goes very happily with integration.

Engineering and sustainability are not much linked in the public mind, nor indeed in the thoughts of the decision-makers. This is bad, for engineers obviously, but also for everyone since it is the engineers who will deliver so much of the means to sustainability. The word engineering is slipping out of fashion, which in our new common, global, instant information style is quite fast. The brave word is technology, with its presumption of new, whilst engineering is attached to old. It is a cultural problem and is very pervasive. The degree of the problem varies between countries but it is there, to a serious degree, in most if not all places. The term engineer can be (quite properly to a lexicographer) used to denote the inventor of fibre optics or the maintenance man for domestic equipment. Engineering as a concept and as an image gets crowded out by its immediate neighbours like science, architecture and styling. This optical indistinction depresses many of the norms of acceptance as varied as nationally recognisable charismatic personalities, recruitment of the brightest youngsters and typecasting in TV serial dramas. To change this misfit between perception and actuality would need a big effort, including some new language. Whether the arriving e-world will aggravate or redress, I am not clear yet. In the meantime the reality of engineering gets greater and greater day by day. You could say that engineering has become so all pervasive, so big, that it is hard to stand far enough back in order to see it. Maybe we should be glad to be so broad, to be so inclusive, the inn where diversity is welcome.

It is interesting to look at architecture and reflect on it in terms of integration. Buildings can be designed from the inside out, function being the banner. Integration will inevitably be sought. And buildings can be designed from the outside in or indeed from the outside and not many millimetres further. Architecture remains largely personality driven despite the fact that any one of our great buildings may have called up the inputs of 1000 talented and different designers and one of our lesser buildings used 100 such people. Nonetheless one can instantly spot the work of Architect X, like you can call out the names of the makers of paintings, poems, music or sculpture on first sight. The imposition of the personality on the sensation of their works happens much less amongst engineers and you can debate whether that is good or bad. It is great fun to assess name architects on an integration index.

You may be saying, if you can be so clear and dogmatic, Duncan, then why are you not applying it all in Arup? Then I will end with a lovingly critical look at Arup as integrators of engineering both intra Arup and extra Arup. The philosophical basis of Arup is expressed in the Aims, as we call them: excellence in everything we do; fairness and honour in all people issues; and sufficient prosperity. And that is all, no jargon, no manual on how to do it and no engineering prescription, just three attitudinal Aims. The social set-up in Arup has developed to be conducive to team work with soft organisation, vague titles, self managed time and long term employment, to the extent that we scarcely notice or celebrate our own culture. Of course the commissars keep arriving and with powerful logic want to de-disorganise us, shaking their heads in disbelief. Some of their efforts stick, which is good, since we do want to survive. We imported a fashionable computer bug, which ran rapidly through our world intranet. It took other organisations a couple of days to clean their bug out. It took us more than a week because of our accumulated system. We are now imposing more rules, commonalities, gates and controls. Will we be better for that? It is an open question. In the perpetual struggle between the creators and the managers, each with their camp and followers, our three little Aims have been a sufficient common ground to share and to sustain our own way. It shows in the many lovely and exciting bridges that we have designed, always with an architect to share the thinking and a builder also.

The urge to do better was very strong in the early Arup days. Ideology defined the world at the time whether as UNO or as capitalism versus communism or as sociology in education and health. It was a naive period, the best of times. So Arup set up Arup Associates (hiding it in the name The Building Group, not to alienate our many architect collaborators), to be a stable team of all the professionals for the design of buildings working together and abandoning all other working liaisons. The theory was that a team's outputs would get better and better if the team was stable and had as its diet a variety of work. It has been a hugely successful experiment, as measured by its building designs, showing what cooperation, articulation and shared attitude can do. Later the subset concept of all the engineering being offered by a stable and exclusive Arup team, to independent architects was developed. We call that Building Engineering and it is our No.1 product and preferred service. Round these methods we have gathered a range of best specialists to share throughout Arup, their service giving us the confidence to raise our ambitions for the excellence of the work and moving us forward in competence.

Now all at Arup is not perfect, by no means. These methods have limitations which we recognise and try to remember and compensate for. One obvious issue is stable teams; they will grow old and they may get stale too soon. Another one is highly differentiated pay in a team culture; does the cost of the resentment in others exceed the value of the star's input, judged against our Aims? However that is all detail. The key question is what next for Arup. You cannot dine out for ever on yesterday's new idea. Today is not the time to launch on you our next serious step change; since we cannot know its success until we do it we may not launch at all, but simply work our way into a better world. What we do know is that there are more ideas and unfinished sequences available than we can see how to harness. The potential is not limited by the supply of new ideas.

As a symbol of the Arup commitment to integrated engineering, we have put a few million sterling into a charitable Foundation whose purpose is to promote integrated education for the built environment. It has focused so far on inventing and funding new Masters degree courses, one at Cambridge for architects and engineers called Interdisciplinary Design for the Built Environment and one at London School of Economics for sociologists, economists, engineers and architects called City Design and Social Science. In both courses the gathering of different professions to work together and demolishing the boundaries were built into the teaching and the methods as a condition of Arup support. Both courses are now past their early years and operating successfully, attracting superb young people from around the world. The Foundation is now preparing to invent and fund its next serious initiative somewhere else in the world.

If I appear impassioned, well, I am. Arup is in many ways a dream come real every day. I hope that I have made a case for admitting Integrated Engineering. I thank you for this opportunity to be with you and I thank you for receiving me so attentively.

BUILDING SYSTEMS AND CONCEPTS

Structural Innovation

William F. Baker

INTRODUCTION

The very existence of skyscrapers is a testament to innovation.

Mankind had always lived close to the ground. During the thousands of years that preceded the birth of the skyscraper, life was limited by a reality of stairs and ladders. It was only after Elisha Otis invented a safer elevator that architects and engineers started a race to the sky with the 12 story Home Insurance Building – taking office buildings from the historical range of 5 or 6 stories to the 102-story Empire State Building in merely 40 years.

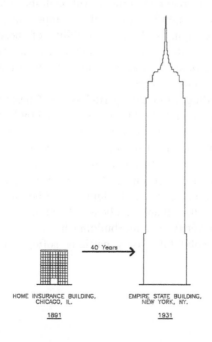

Figure 1   Skyscraper Development.

The invention of the skyscraper came shortly after major breakthroughs in railroads, bridge building, shipbuilding, and manufacturing. While these nascent technologies created the economic foundations for modern cities, the skyscraper created an entirely new kind of city. The skylines of Chicago and Manhattan were unlike anything the world had ever seen.

Technical innovations in systems and materials for these first towers are now subjects for historians. Although innovations that have followed the first skyscrapers have not been as sociologically profound, they represent levels of creativity and inventiveness that equal and, perhaps, exceed the original breakthroughs.

REBIRTH

The rebirth of skyscrapers in the 1960s, after a hiatus of over thirty years, came with structural innovations that transformed the industry. Gone were the interiors filled with columns and frames resisting the wind (with substantial help from the "non-structural" cladding and interior partitions). Instead there was a whole new structural vocabulary of framed tubes, braced tubes, tube-in-tube, bundled tubes, frame-shearwall interaction, outrigger systems, etc.

This cloudburst of new systems was made possible by the advent of computers and engineering pioneers such as Fazlur Khan, Hal Iyengar, William LeMessurier, Leslie Robertson and others. Although the conceptual foundations of these systems were straightforward, earlier computational methods were not adequate for use in design. At last, the viability of these systems could be demonstrated by utilizing the mainframe computer. In addition to validating the systems, important parametric studies were done to establish the applicable height range for the various systems.

A seminal building from this period is the Chestnut-Dewitt apartment building (arguably the first tubular building) designed by Fazlur Khan and architect Myron Goldsmith. This simple, elegant tower – now dwarfed by its neighbors – was a major development in modern architecture. The integration of the tubular structural system and Miesian architecture was complete and seamless. Architecture and structure were one and the same, inseparable. The computer was able to verify that this concrete building could be viewed as one complete continuous structure rather than merely a collection of columns and beams. Designers could now verify that the building behaved as a three-dimensional system much like a solid tube, only partially softened by the openings for windows.

Figure 2   Chestnut-Dewitt Apartment Building, Chicago IL.

If concrete was the material of choice for moderate height residential buildings, steel dominated the construction of super tall buildings – so much so that if construction of skyscrapers had stopped in 1974, the tallest buildings would have all been made of steel. The Sears Tower was a steel bundled tube; the World Trade Center and the Standard Oil towers were steel framed tubes, the much earlier Empire State Building was a steel frame and the John Hancock Center was a braced steel tube. All of the newer towers had long, column free, lease spans and, interestingly, were of similar slenderness ranging from a height to width ratio of 6.0 to 1 for the Standard Oil Tower to 6.6 to 1 for the John Hancock Center.

The first practical use of boundary layer wind tunnels for buildings occurred during this time. These tests often utilized expensive aereoelastic testing to measure building motions. Criteria were needed for these motions, so moving room tests were done for the John Hancock Center and the World Trade Center projects.

Speed of construction is critical in the economics of a skyscraper. Therefore, innovations in construction technologies were essential. Structural steel could be fabricated off site in ways that minimized field connections and thus speed the construction. Welding and high-strength bolts replaced the rivets of an earlier age. Although none of the newer buildings could match the incredible 18-month construction schedule of the Empire State Building (built during the Great Depression), these new buildings were efficiently assembled like great machines.

Profiled metal deck was a major addition to the industry. It increased the speed of construction by providing economical permanent formwork for the floors that could be used as a working surface for the ironworkers and could later function as reinforcement for the concrete slab. The flutes of the metal deck could also be configured to accommodate electrical and communication wiring for the tenants' workstations.

As essential as speed was to skyscraper construction, equally paramount was cost. The generally accepted measure of a structure's cost was the total steel tonnage divided by the total framed area. Expressed in pounds per square foot of floor area (psf), the economy of these buildings was made evident by the substainal reduction in the steel used. While the Empire State Building weighed in at 42.2 psf the substantially taller Sears Tower was only 33.0 psf.

RECENT TIMES

Unlike the generation-long interruption of projects between the time of the Empire State Building and the John Hancock Center, the construction of very tall buildings has not stopped for the last forty years (although there have been periods of slowing down).

In the era from the mid-1980s until today, there have been a series of innovations that have once again changed the vocabulary of tall building design.

One such innovation was a major shift from all steel construction to systems that utilize the benefits of both steel and concrete in the primary structural systems. The industry has weighed the strengths and weaknesses of these two primary structural materials and now uses each to its advantage.

Steel is excellent for framing long span office floors; it is lightweight, can be easily modified by tenants and can be quickly erected. Concrete, on the other hand, is very cost effective in carrying the weight of the tower, and the mass is beneficial in reducing building motions. New formwork systems and the ability to pump concrete to great heights have also greatly increased the speed of concrete construction. In addition, the inherent damping of a concrete lateral system is generally higher than a steel system.

The resulting merger of steel and concrete into composite buildings has resulted in seemingly endless combinations of the two systems. Perhaps the most common composite systems use reinforced concrete cores and steel floor framing. The perimeter framing varies from reinforced concrete frames or tubes to reinforced concrete mega-columns or even to perimeter steel frames. This results in an extremely large variation in the manner that the cores can be arranged or how the building might be expressed architecturally. The Jin Mao Tower in Shanghai is an example of a highly integrated super-core.

Figure 3   Jin Mao Tower, Shanghai, China.

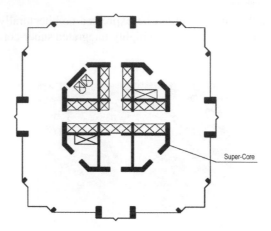

Figure 4   Super-Core, Jin Mao Tower, Shanghai, China.

Recent years have seen a major increase in computer power, with mixed results. The benefits of this enhanced computational power include the ability to perform more extensive behavioral studies to better understand the nuances of a given structural system. It is also much easier to accurately compare alternative structural solutions for a particular building.

Recent advancement in the application of optimization techniques to tall buildings has also resulted in more efficient buildings that use less of our limited resources. It is now possible to use optimization techniques to "tune" a building to achieve a better behavior under wind loading.

The "downside" of such advanced computational horsepower, however, is that it allows the engineer to make a building "stand up" without having a rational structural system. Although super-tall buildings still tend to have rational systems, many of the mid-rise towers often employ a complicated set of ad-hoc structural systems. It could be argued that architecture and the efficiency of many buildings, particularly in the late 1980s and early 1990s suffered from the undirected freedom permitted by the computer. In an earlier time, the computational limitations on the structural engineer necessitated straightforward structural systems that could be calculated with the available tools. The resulting architecture often benefited from this discipline.

In addition, optimized structural systems favor more efficient load paths and tend to eliminate less efficient ones. While this can greatly reduce the quantities of steel and concrete, it often results in a building with less redundancy. It is very important to study the robustness of an optimized structure and give special consideration to fracture critical members and similar issues.

Major advancements have occurred in wind engineering and motion control. Tall buildings are generally not controlled by the strength of the lateral load resisting system but by limiting the motion to within acceptable limits. The wind related dynamics of a tall building have three components: background static, background dynamic and resonant dynamic. The background static and background dynamic wind forces are independent of the building behavior and are a function of the surrounding environment and wind climate. Although these components are important, it is the resonant dynamic behavior that is critical to the structural design of a tall tower. The resonant dynamic forces and motions, particularly in the dominant across-wind direction, are related to vortex shedding and the accompanying pulsating forces.

The design team has several approaches available for controlling the resonant dynamic behavior. The natural dynamic behavior of the tower can be modified to optimize the response. This refers to modifying any or all of the basic structural properties of the building: mass, period, mode-shape and inherent damping. The design team may choose to add artificial damping to the tower through special devices and machines. Most importantly, the design team can actually decrease the dynamic forces imposed on the building by judiciously shaping the building in plan and profile.

Examples

The successful design of very tall buildings requires extremely close cooperation between the architect, structural engineer, building services engineer, specialty consultants and contractors. In order to demonstrate how this cooperation and interplay can be manifested in a skyscraper, two recent projects will be briefly reviewed.

The Samsung Togok project is a very tall residential tower in Seoul, Korea. Because it is a residential building, the motion criteria are the most restrictive. The span from the core to the perimeter is very short because of the natural lighting requirements of a residential building. It will be the tallest all-residential building in the world when it is completed. The integrated architectural and structural solution is a floor plate of three leaves that brace each other against the lateral forces of nature. The shape of the building greatly reduces the resonant dynamic forces on the building to such an extent that the motion criteria are easily met using a conventional structural system.

Figure 5   Samsung Togok Project, Seoul, Korea.

The important components of the building shape are the heights of each of the wings and the shape of the floor plans. The different height wings create different floor plans that have different vortex shedding behavior. This prevents the wind forces from becoming organized in a manner that can lead to large dynamic building responses. The elongated shape of the floor plate at the top of the structure also prevents well-defined vortices from forming, while the sawtoothed edge avoids creating an airfoil that would have created "lift" forces for certain wind directions.

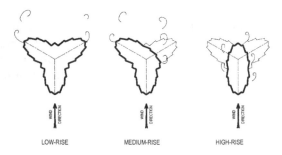

Figure 6   Samsung Togok Project: Vortex Shedding Behavior.

Another proposed building that incorporates many of the latest innovations in structural systems and wind engineering is 7 South Dearborn in Chicago. The design uses the latest innovations in materials, systems and wind engineering to create a tall, slender and economical tower. The proposed tower is very slender, with a height to width ratio of 8.5 to 1. It avoids the overly deep lease spans of earlier towers to create conventional office floor plates and uses a relatively small site of less than one acre for a building that would be the world's tallest.

Figure 7   7 South Dearborn, Chicago, IL.

The lateral load resisting system is termed a "stayed mast" that uses an outrigger system at the mid-height to brace the high strength concrete core that extends from the foundation to 1550 feet (472 meters) above grade. A "virtual" outrigger on the parking stories above the ground floor lobby augments this straightforward system. This system is the latest development in super-core systems.

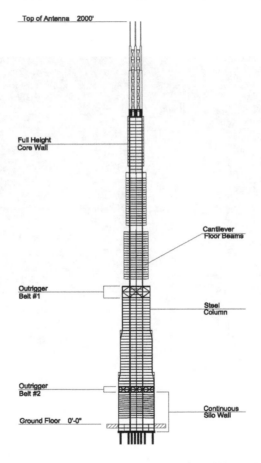

Figure 8   Structural System for 7 South Dearborn, Chicago, IL.

Structural steel is utilized for the floor framing and perimeter columns in the office floors due to advantages in tenant use and speed of construction.

Cantilever floor construction is used in the residential floors for reasons of construction, core strength, and improved serviceability. This cantilever construction takes full advantage of the speed that concrete core wall construction permits by allowing for multiple floors to be constructed simultaneously without needing to proceed in the one-floor-at-a-time method normally required. It also maximizes the gravity loads on the core, thus increasing the strength of the concrete core. In addition, this type of construction also eliminates problems of differential creep, shrinkage and elastic shortening movements between the core and the perimeter for these upper floors. Finally, this system eliminates the racking that often contributes to motion perception in tall buildings. It also creates a column free perimeter and spectacular views.

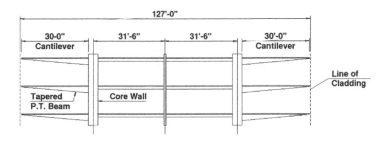

Figure 9   7 South Dearborn Cantilever Floor System.

Wind engineering and motion control have been addressed by the proper shaping of the building and by controlling the period, mass and mode shape of the tower. The tower was tested in the wind tunnel several times during the conceptual design phase in order to assess the wind behavior and to evaluate refinements in the building shape. The "wind signature" of the building was studied by analyzing the autospectra curves. This permitted judicious modifications of the building's dynamic properties. It also gave insight into the effects of various shapes. One series of tests was done to evaluate the effect of the distinctive notches on the building behavior. The effect was dramatic: the notches reduced the wind forces in critical wind directions by up to 20%.

The resulting building is new, exciting and unique. One of the wonderful results of new technology is the opportunity for new architectural interpretations of the skyscraper.

Damping devices

Although the 7 South Dearborn design met the strict residential motion criteria with only inherent structural damping, studies were done on the use of supplemental damping systems. Several innovative systems for motion control currently exist. These include mechanical or liquid tuned mass dampers, piston dampers, and visco-elastic dampers. An interesting system that was reviewed for the 7 South Dearborn project was a tuned liquid column damper, a type of tuned mass damper that uses the water in a tank as the mass, a water column at each end as the spring and an orifice to dissipate energy. This type of damper will undoubtedly be used more on buildings with less optimal architectural massing and structural systems.

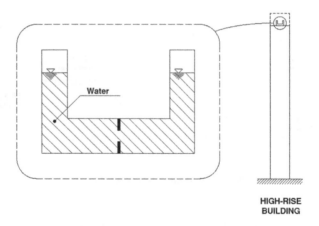

Figure 10   Tuned Liquid Column Damper.

Artificial damping devices should be used with caution. The reliance on these systems for issues of strength should be carefully considered and generally avoided. The actual track record for these systems is very limited, and long-term maintenance is a serious issue. It is not unusual for even famous buildings to undergo periods of neglect, as they become older.

There are times where the use of damping systems for improved serviceability would be appropriate. Even then, it is important to evaluate the total cost of these items, including the cost of occupying space that is very expensive to build and the cost and practical concerns of maintenance.

CONCLUSION

Structural engineering innovations and their influences on skyscrapers continue to be significant in changing the nature of tall buildings. It is important to note, however, that the most profound structural advancements occur through the collaboration of architects, structural engineers, building services engineers, specialty consultants and contractors. The time and money that it takes to construct a tall building require an efficiency and optimization that exceeds any other type of building.

The current level of innovation is quite high and is focused on composite systems, materials, wind engineering/motion control and optimization approaches. The future of tall buildings looks bright, as each of these fields is just beginning to be mined for their full potential. As in the past, skyscrapers of the future will change in response to these influences: contractors will find new ways to construct, engineers will continue to invent new systems, and architects will continue to create new buildings that embrace and benefit from these technical influences.

Progress in Construction Robotics in the U.S.A.

Miroslaw J. Skibniewski

ABSTRACT

This review discusses examples of the latest developments in robotics technologies in the United States that may be of interest to construction robot systems developers. Both theoretical, far-reaching research such as collaborative robotics as well as hardware systems oriented R&D is included in the paper. Examples of the latest industry initiatives and consortia activities are also described.

INTRODUCTION

The population of robots nearly doubled over the last decade in North America alone, and they are becoming increasingly important in applications ranging from quality control to space exploration, surgery to the service industries. The industry recently has seen the emergence of new types of devices, including tiny micro- and nano-robots and robots with multiple arms or legs.

Popular attitudes toward robots have changed over the years. The fear that robots would replace workers has disappeared. Instead of displacing large numbers of employees in manufacturing, robots have brought about a more highly-trained workforce better capable of running robots and computers. Research conducted by the author on behalf of CSIRO-Australia as early as 1992 indicated positive attitudes among Australian construction workers toward automation and robotics. There are many more trained people in robotics now, but some new challenges for robotics researchers are better human-robot collaboration interfaces, robot mobility and navigation in unknown surroundings, and better robot intelligence for service industries such as construction and for public transportation.

The number of robots per 10,000 manufacturing employees skyrocketed from 1980 to 1996. For example, it increased from 8.3 to 265 in Japan, 2 to 79 in Germany, 3 to 38 in the United States and zero to 98 in Singapore. In roughly the same time, the world robot population surged, going from about 35,000 in 1982 to 677,000 in 1996 and an estimated 950,000 this year. In the five years from 1992 to 1997, the robot population in North America increased 78 percent, from 46,000 to 82,000.

Robotics applications in the construction industry have been researched, explored and prototyped for the last 20 years. Major industry effort in Japan resulted in a number of robotic technologies for building construction, including automated building construction systems designed and produced by companies such as Obayashi, Shimizu, Taisei, Takenaka, and others. As a result, automation and telecommunication technologies of today are far different from those of the last two decades.

Innovative technologies emerging over the first decade of the 3rd millennium promise to affect virtually all aspects of everyday society, including health care, communication, manufacturing, transportation and, of course, construction. Robotics in the construction industry are in the large part based on advances in robotics developed first for other industries, and this trend is expected to continue.

EMERGING CORE TECHNOLOGIES

Engineers and scientists have already envisioned superbly efficient cars, "smart" offices equipped with wireless sensors, nanotechnology and a new class of miniature devices that dramatically speed up medical and biological testing, and electronic gear that runs for days on a single charge.

Nanotechnology will lead to ultra-compact video and audio equipment with far more storage capacity. More compact and versatile robots will enter the workforce in a number of industries. While it is impossible to predict exactly what will happen, some future trends appear to be obvious. Fuel-efficient cars will likely travel about 80 miles on one gallon of gasoline and use a "hybrid" drive system: an electric motor powered by batteries or fuel cells, supplemented with a traditional fuel engine only when needed for acceleration and high performance. However, the success of hybrids will depend largely on the perfection of a control system that switches automatically from one drive to the other. With two propulsive sources, the use of both will need to be optimized automatically to make it all happen smoothly. Hybrid fuel technology may have a profound impact on the future of mobile robotics much needed in automated construction of roads and in transportation of building materials.

The automotive leap will also depend on the refinement of fuel cells, which generate electricity through electrochemical reactions between hydrogen and oxygen. In the coming decades, new types of fuel cells will likely evolve. Some visionaries predict that, before the middle of the next century, automobile fuel cells will generate enough electricity to supply an average-sized building.

Another near-term innovation will be the proliferation of very small, wireless electronic sensors the size of computer chips that will improve safety and make life easier in the workplace and at home. In general, most objects that surround us will be considered 'smart.' Some sensors will monitor the air for carbon monoxide and smoke. Others, linked via radio signals instead of wires, will automatically adjust the lighting and temperature. Employees may wear electronic cards or badges that specify their whereabouts. The badges might be programmed with a wearer's atmospheric and other personal preferences, much like a system designed for Microsoft's Chairman Bill Gates' home in the state of Washington, USA. Such innovations will become commonplace in the 21st century. There will be thousands of these embedded chips, in the doorknobs, light bulbs, tables, chairs, all networked together. Similar sensors and electronic chips will be imbedded in structures such as bridges, road surfaces, foundations, and building walls to continuously monitor and report their condition and performance.

Advances in electronics and computers will have a direct impact on business travel, as improvements to videoconferencing make the long-distance medium much more realistic, perhaps even three-dimensional. Wireless Internet connections also will be refined on a commercial basis, and new computer software and hardware will make cellular phone communications more seamless. There will be no need to plug a computer into a telephone jack anymore. It will be possible to drive under an overpass, or inside a tunnel, and not have one's cell phone fade out. And there will be no need to worry about airplane flutter every time an airplane flies overhead.

Meanwhile, innovative electronics and software will increase the operating efficiencies of equipment such as laptop computers and cellular phones so that they run longer on a single charge. These devices will use knowledge of what is running on the system and then intelligently shut down the parts that are not needed. In doing so, power will be conserved and batteries will last longer.

The higher efficiency will make wireless communications more practical as well. Global Positioning Systems already are available that use satellites to pinpoint an exact location of physical objects such as people and equipment. New software advances are making satellite-imaging systems very user-friendly that they might soon be accessible to everyone from farmers to real estate developers and to construction contractors as well. The software will handle the most mathematically rigorous work so that people with minimal technical skills can operate these systems. At the same time, the satellites themselves are improving and will soon be providing more data that can be used in numerous applications, including land development and transportation planning.

In other research, scientists expect to be able to merge biotechnology with computer science, creating a new class of miniature sensors that will be imbedded in a human body to deliver a variety of information useful in determining biological performance of human organs during the conduct of work. This experimental concept is called MIBBS, for 'micro-scale integrated bio-separation and bio-sensing systems,' and is currently at a development stage at Purdue University.

These and other developments are likely to lead to better and smaller robots for a nearly unlimited range of applications, from medicine to household maintenance, and to construction related uses.

PROGRESS IN RESEARCH LABORATORIES

A number of R&D projects are currently underway throughout the United States that are of interest to the construction industry community. Several example projects are described below.

Progress in Hardware Research

Scientists and engineers have raised the possibility that machines constructed at the molecular level (nanomachines) may be used to cure the human body of its various ills. Nanorobots are nanodevices that will be used for the purpose of maintaining and protecting the human body against pathogens. They will have a diameter of about 0.5 to 3 microns and will be constructed out of parts with dimensions in the range of 1 to 100 nanometers. Although it is now difficult to imagine any potential applications of nanorobots in the construction industry, microrobots have already been prototyped and tested in a variety of tasks in the maintenance of built infrastructure. For example, trenchless technology is now becoming a mature application of this concept in the industry, including a larger scale of semi-automated devices. New pipelines may be installed or pipelines may be rehabilitated without the need for cut-and-fill excavation. Trenchless technology minimizes the cost of pipeline construction by avoiding excavation, minimize pedestrian and vehicular disruptions, retain landscaping, and reduce contractor's liability insurance premiums.

Legged Locomotion

Purdue University Robot Vision Laboratory has developed a vision-based quadruped walking robot system. The goal of this project was to design a legged robotic system that can sense, control, and plan its own locomotion in response to the conditions in the environment. Over the years it has become increasingly evident that all of these capabilities must work together for a legged robot to accomplish empirically interesting tasks.

The progress made so far in the design of legged robots has been mostly in the areas of leg coordination, gait control, stability, incorporation of various types of sensors, etc. This progress has resulted in the demonstration of rudimentary robotic walking capabilities in various labs around the world. The more stable of these robots have multiple legs, four or more, and some can even climb stairs. But what is missing in most of these robots is some perception-based high-level control that would permit a robot to operate with a measure of autonomy. Equipping a robot with perception-based control is not merely a matter of adding to the robot yet another module; the high-level control must be tightly integrated with the low-level control needed for locomotion and stability.

Mobile Robot Navigation

In the past decade, the Purdue University Robot Vision Laboratory has developed two unique reasoning and control architectures for vision-guided navigation by indoor mobile robots. Both of these systems use model-based vision to help a robot determine its location with respect to the environment. Both of these systems are fully operational and demonstrations of these are given routinely in the Laboratory.

The first of these architectures, called FINALE, allows a robot to navigate at approximately 8 meters per minute in the presence of stationary and moving obstacles. FINALE requires simple geometrical models of interior space. These models can be created simply by knowing the height of the ceiling and the locations of the principal corners in a hallway. The system can function even when some of these features are missing from the model map.

The second architecture, called FUZZY-NAV, uses neural-networks to keep a robot centered in the middle of a hallway and to generate steering commands when the robot needs to turn for either course correction or for avoiding obstacles. FUZZY-NAV needs only topological models of interior space. While the models are easy to create for FINALE, they are even more easily created for FUZZY-NAV. FUZZY-NAV is an extension using fuzzy logic for supervisory control of an earlier system, called NEURO-NAV, that also came out of this Purdue University laboratory.

Caisson Construction 3-D Modeling System

One of the critical links to the success of robotics in construction applications is the human-machine interface. This project is conducted at the National Robotics Engineering Consortium at Carnegie-Mellon under the sponsorship of Kajima Corporation. The researchers are developing a 3-D sensor system and graphical display to assist caisson construction equipment operators in digging a 42-meters deep caisson. The results of the terrain mapping are to be displayed in graphical form to human operators. These human operators are responsible for tele-operating excavating machines that are inside of the caisson. The human operators will be able to use the terrain-mapped display of the caisson to locate potential problem areas within the caisson structure and to determine what areas are stopping the caisson from sinking into the earth.

Automated Underground Mining

Another project underway at the National Robotics Engineering Consortium involves a system prototype involving a variety of excavating and transport machines configured to perform an underground mining operation.

Coal is mined using an ensemble of various machines. A continuous mining machine uses a spinning cutter head with "teeth" to break coal away from the face. The coal is deposited in shuttle cars for haulage to a permanent conveyor system and out of the mine. A roof-bolting machine drills holes in the roof and inserts bolts to stabilize the roof and minimize the risk of collapse. Sometimes the miner and the bolter are combined into a single machine. This project concentrates initially on automating the functions of a continuous mining machine and roof bolting units.

New Tactile Sensors

Scientists and engineers at Harvard University are currently involved in a Tactile Sensing Project involving development of new sensors, and integration of sensed information with real-time control. One new sensing technology already developed can localize high-frequency vibrations on the fingertip. They are working to use this sensor to measure friction at a single touch. They have also developed a fabrication technique for tactile pressure array sensors that results in a flexible, easily mounted sensor with very high sensitivity. These sensors are used with the two-fingered direct drive hand in the laboratory to study the role of tactile information in dexterous manipulation. The second sensing technology developed is based on deformable tactile sensors. Using this technology, one can reconstruct the deformed shape of the sensor on-line. This sensor is currently being used in manipulation tasks and in the development of a tool for minimally invasive surgery. It is expected that similar sensors may be utilized in a variety of non-destructive structural testing in the future.

Wireless Communication with Heavy Mobile Equipment

A vision of a job site where bulldozers "see" by sharing laser vision and terrain maps in real time has been created. In such a vision, a technician several miles away from the site can diagnose an earth-moving machine's malfunction. This communication all takes place without wires, antennae or complicated computer hardware.

Communication protocols differ between countries. There is also a significant problem of communication packets locating highly mobile machines that may move from site to site. There are also unique propagation problems at sites, not just the uneven terrain but the special problems of terrain changing shape significantly in real time. Add to this the complexities of multi-path interference. There are also a number of miscellaneous issues to be considered, for example:

- How do you integrate the pieces into a single system as you do with your wired computers?
- How do you deal with robustness?
- How do you deal with security?

Finally, although the construction industry is one of the largest in most developed or developing economies, it is not large enough for a solution to its unique requirements to emerge spontaneously.

Caterpillar Corporation of Peoria, Illinois and the Information Networking Institute at Carnegie-Mellon University are working together under the university's Wireless Research Program to develop a mobile wireless communication system for construction equipment. Such a system assumes no availability of public networks and it is to benefit the construction and mining industries by supporting their field communications. This type of a system must provide a high-speed communication between earth-moving machines and from machines to the outside world. It needs to support field service by providing remote access to machines and giving field technicians direct access to up-to-date service manuals and data. Additional requirements include allowing dealers to do remote diagnosis of a machine's condition and enabling Caterpillar to field new, high-speed applications.

Many corporate and government owners of construction equipment fleets have unique needs. The owners often require a single site system that is easy to install, operate and maintain. The dealers want a simple technician interface – a rugged laptop as easy to use as a cellular phone – one that only needs turning on. This demanding set of requirements raises several issues. First, the system has to handle a wide variety of communications from data and file transfers to real-time data, voice telephony, and, someday, video, radar and laser images. In addition, the system needs to handle emergency stops and deal with hostile terrain. The system also has to work with wireless technology anywhere in the world.

The researchers have begun to meet these challenges. Using an IP-based architecture – basically a sophisticated "Internet in the sky" – they dealt with the issue of moving vehicles. They developed new processes for simple but high-speed networks in open-pit mines and quarries. They created a patented concept of intelligent protocols that use knowledge of the terrain and vehicle position to enhance reliability and performance.

After demonstrating their first year's work in Pittsburgh and at Caterpillar's Proving Grounds in Peoria, the researchers continued their efforts. They analyzed computer simulations of Caterpillar machines working in mines and quarries, and developed new protocols.

In a subsequent demonstration in Pittsburgh, the Carnegie-Mellon campus became the location of a supposed dealership and the neighboring Oakland area was used as the dealer's territory. Technology Park, two miles from main campus, was used to simulate a remote major construction site. Since remote mines and major construction sites are the kinds of locations where the largest Caterpillar machines demand the most sophisticated applications, it was an ideal way to test wireless data services.

The demonstration showed successful ad hoc networking, significant protocol improvement using terrain knowledge and transparent use of multiple networks – all while the networks were stressed with heavy communications traffic.

When Caterpillar implements this system into a fleet of "communications ready" vehicles, the mining and construction industries will be thrust into a world of high-speed, mobile communication technology. In addition, the products are being explored for NASA to apply to multiple spacecraft, space stations and clusters of surface robots as well as for the U.S. Army to use in mobile battlefield equipment. Other companies and agencies involved in this project include Bell Atlantic Mobile, Beckwith Machinery Company, Cisco, DARPA, EFData Corporation, Intel, Lucent Technologies,

NASA, Glenn Research Center, Trimble Electronics, and the National Science Foundation.

Autonomous Rover Technologies

The primary goal of this project being conducted at the Carnegie-Mellon Robotics Institute is to research and develop the enabling technologies for autonomous planetary robot perception, position estimation, navigation, and integrated exploratory science from a robot, and validate such technologies through aggressive and rigorous field experimentation.

The recently executed research objectives include:

- Navigation and science from panoramic imagery: Prior research in wide field imaging developed teleoperated remote viewing and demonstrated its merits for robots, but fell short of the scope and benefits possible for automation with wide imagery. The attractive opportunity generated by capturing lateral and longitudinal views from a rover simultaneously has not been exploited. The team researches techniques for autonomous visual deduced reckoning, landmark based navigation, and scientific characterizations using panoramic imagery.
- Advanced radar perception and safeguarding: Sonar, stereo, and laser have dominated robot perception research, but each has liabilities and downfalls for application in space. Radar holds the prospect for modeling, safeguarding, and navigation from a space robot with advantages of operating in and through dust, in vacuum and atmosphere. The team investigates the merits of ultra high-frequency radar to detect objects, map terrain features, and even profile shallow subsurface geology in substantial dust accumulation during long traverses.
- Science data classification from multiple sensors: No "perfect" sensor or classification methodology exists for robustly distinguishing interesting science observations, like evidence for life, geologic anomalies, fossils, and meteorites among other rocks. The team has been developing a principled framework within which output from a variety of sensors and multiple classification algorithms are used to confirm or deny the detection of a scientific object of interest.
- Advanced rover autonomy: Extensive research has gone into obstacle detection and avoidance methods for autonomous robots. However, these methods largely rely on knowledge of robot characteristics (such as sensor coverage and mobility). Providing a robot with health monitoring and error recovery capabilities will allow the robot to notice that its turning radius has

increased and incorporate this into its planning allowing a mission to continue even though a malfunction has occurred.

The researchers have been developing a general health monitoring capability capable of detecting failures in the drive, steering, and sensor components of the vehicle. An error recovery capability is also under investigation, which will use the error diagnosis to modify obstacle detection and avoidance behavior.

Cognitive Robot Colonies

The primary objective of this work conducted at Carnegie-Mellon's Field Robotics Center and co-sponsored by Defense Advanced Research Project Agency is to uncover the basic principles that will best govern a group of robots trying to do useful work in difficult and hazardous environments, including construction sites. The foundation of the research begins with the idea that robot existence must be modeled probabilistically. Robots, like humans, are subject to physical laws and can be damaged or destroyed by both random and intentional events. In the extreme environments posed by space exploration, military operations, fire-fighting, nuclear cleanup or dangerous construction, the likelihood that robots will be injured is amplified. In many situations, the danger posed is so great that a single robot expected to perform adequately in these scenarios must be designed to mitigate every conceivable circumstance. Clearly, this task is either very difficult or impossible for most operations.

A promising approach is to use tightly coordinated groups, or colonies, of smaller, simpler robots to perform tasks in these dangerous locations. The fundamental advantage of this approach is redundancy of systems and mechanisms. If managed properly, the loss of a robot, although painful, will not be catastrophic and task execution capabilities will degrade gracefully across multiple robot failures. While very promising, the implementation of these ideas into a working system is fraught with the difficulties inherent in any highly redundant system. Specifically, the work at Carnegie-Mellon is focusing on the following major problem areas:

- Behavioral Strategies for Colonization: What is the "critical mass" of agents necessary to form an effective colony? Once formed, what kind of internal, environmental, and task-level cues can be monitored to maintain the optimal colony through pruning and growth behaviors? Are there times when it is more efficient to disband the colony and work at the level of the individual robot?

- Group Learning Algorithms: Can the sharing of experiences amplify the effectiveness and minimize the convergence time of the learning processes within a robot colony? What is the proper mix of individual and group learning strategies? Can effective learned behaviors be distributed as a commodity within the colony?

- Protected and Secure Resources: What is the best way to distribute data and processes within the colony to insure that loss of robots will minimize loss of learned knowledge and data? Once losses are realized, what is the best way to redistribute existing information to make it more secure? What are the most efficient and secure methods of routing critical information and data within the colony and to the outside world?

- Task-Level Reprogramming: Can we obtain an efficient system for doing real work in hazardous locations using behavioral seeds from individual robots to grow a task-specific organization? If so, what is the optimal matrix of basic behaviors and the learning algorithm required to foster this system?

- Scalability: Robot colonies that rise as a dynamic response to general task-level requests will need to have great variability in their computational, power consumption, communications, and physical capabilities. Thus, robot system architecture must scale to problems at both ends of the spectrum. Can we make a colonization architecture that supports ten robots in one task and ten thousand for another?

- Although the focus of this work is fundamental, the ultimate measure of success of any robotic system should be evaluated in terms of performing useful work in a practical job scenario. For this reason, one may test the validity of the proposed concepts with respect to the task of Distributed Mapping of Urban Environments.

The unique feature of the proposed distributed mapping system, and the eventual metric of its success, will be its ability to pursue this task when faced with multiple robot failures. The initial demonstration, tentatively scheduled for the fall of 2001, will be to deploy ten small robots into a "mock-up" of an urban facility. These robots will form a colony whose sole purpose is the generation of a map of this area. After an initial period during which basic distributed mapping operation is demonstrated, project sponsors will be asked to "disable" robots of their choice and observe the reaction of the colony to this loss. This process will continue until critical mass is lost and the colony is unable to function in terms of its primary mission. Thus, observers will be given an "on-line" demonstration of how our system adapts to multiple and catastrophic failures.

The list of major research tasks includes the following:

1. Formalize the core components of free market architecture
    *   Implement architecture and evaluate performance in simulation.
    *   Extend architectural capability to accommodate exploration of partially known worlds.
    *   Extend architectural capability to accommodate robot death.
    *   Evaluate architectural performance on real test bed with four robots.
    *   Extend architectural capability to accommodate multiple roles.

2. Map Reconstruction
    *   Robust matching of features between images taken at arbitrary positions by different robots, and recovery of robot poses and scene geometry.
    *   Recovery of dense depth maps from images from robots at arbitrary positions.
    *   Generation of texture-mapped realistic environment model from the reconstruction for presentation to users.
    *   Integration of multiple maps into a combined representation of the environment.
    *   Use of the geometric representation for reasoning about occlusions and view planning.

3. Communal Learning
    *   Develop communal learning techniques in which the colony shares experiences (data gathering) and the computational burden of learning.
    *   Prove that concepts can be learned by the colony that are beyond the capabilities of the individual.
    *   Show the benefit of having learned information preserved across the colony.
    *   Also show that the colony can reuse and build upon previously learned skills and behaviors. In situations, where new generations of robots replenish those consumed, this will lead to a progressive society where new generations can advance beyond the previous generation.
    *   Develop a "buddy" system for insuring fatal state information survives. Communal learning can then learn fatal situations and actions. In many situations, robot sacrifice will be required. Techniques will be developed that will ensure that the most information is gained for the colony from any sacrifice.

- Fit communal learning into the free market architecture. By making experiences, computation, and new or refined behaviors/information a commodity, communal learning will be part of distributed market architecture.
- Show that communal learning can lead to more accurate and meaningful negotiations. This will be done by learning transaction confidence and expected costs.
- Apply communal learning to: mobility models, behavior parameters, transaction models, maps and enemy tactic models for individuals and teams.

4. Robot Test Bed
    - 30 cm/s reliable obstacle navigation.
    - Integration of Firewire hardware on each platform.
    - Integration of audio hardware on each platform.
    - Data Abstraction Module for image acquisition (both streaming and static forms) and audio clips.
    - Multi-link routing module that enables point-to-point communication through multiple intermediaries to compensate for limited range in our wireless hardware.
    - Augment sonar-based obstacle avoidance with image-based approach.
    - Characterize odometric data (linear and angular drive vs time).
    - Develop network resource manager that empowers robots to obtain information and physical services from others.
    - Design and integrate mounting brackets for audio and imaging hardware.

Related future work may also include determination of the collective intelligence of a colony of robots in the performance of given work tasks, based on theoretical foundations presented in the forthcoming book by Szuba (see References).

Development of Standards for Automated Construction

Researchers at the National Institute for Standards and Technology are currently involved with three projects related to construction site automation:

- Non-Intrusive Scanning Technologies for Construction Status Assessment. This project is intended to initially benefit large earthmoving projects and aims at enabling the use of scanning technology by the industry for near real-time, inexpensive assessment and tracking of construction status. The challenges include a statistically based calibration of laser device and traceability to NIST measurement standards, development and evaluation of advanced computational models/ algorithms for generating 3-D surfaces through mathematical optimization, automatic spatial registration of multiple point clouds, development of a mobile platform and spatial registration of platform, and development of automated object classification and recognition methods.

- Real-time Construction Components Tracking. The objective is to provide infrastructure necessary for the industry to achieve real-time identification and position tracking of manufactured components on a construction site. The challenges include the development and integration of advanced measurement systems for high accuracy tracking of position and orientation for manufactured objects on a construction site, defining interoperability needs for field sensors and wireless communication systems for part tracking, and developing methodology for defining fiducial points for unique object position and orientation determination.

- Site Measurement System Interoperability and Communications Standards. The objective is to enable the use and integration of multi-functional, heterogeneous sensor arrays by the industry to achieve cycle time reductions in construction. The challenges include the development of data representation/interoperability standards and communication protocols that industry will accept and implement, the design of standards/protocols that work well over wireless networks, facilitation of sensors/user mobility about sites, interoperability assessment and field testing of standards, and determination of accuracy and reliability of field sensors.

STATE OF PRACTICE ON ROBOTICS FOR CONSTRUCTION-RELATED APPLICATIONS

As determined by researchers over the past two decades, practical use of robotics in challenging environments such as a construction site depends on the ability of robots to move around, sense the environment, process the data and information received and reason based on available information – all with sufficient levels of redundancy to ensure reliable field performance. A number of current and recent projects conducted in American laboratories address various aspects of these prerequisites.

The All-Terrain Robot Line from I-Robots, Inc. features four-wheel drive, differential steering, pneumatic knobby tires, long running time and a wide range of optional equipment and accessories. The ATRV family of robots are capable of performing work tasks in heavy terrain. These robots are built around RWI's FARnet™/rFLEX™ Robot Control Architecture, and the versatile MOBILITY™ motion control package. The system includes a proprietary Robot Software Development Environment. Applications are developed once and may be run on any robot in the ATRV family without re-coding or porting. In addition, I-Robots, Inc. supplies wireless communications based on RS-232 standard, on-board computers, scanners, vision systems, joystick controls and navigation sensors for heavy terrain locomotion such as accelerometers and compass-tilt sensors, for all ATRV outdoor robots.

The NASA Space Telerobotics Program is an element of NASA's ongoing research program, under the responsibility of the Office of Space Science. The program is designed to develop telerobotic capabilities for remote mobility and manipulation, by merging robotics and teleoperations and creating new telerobotics technologies. Space robotics technology requirements can be characterized by the need for manual and automated control, non-repetitive tasks, time delay between operator and manipulator, flexible manipulators with complex dynamics, novel locomotion, operations in the space environment, and the ability to recover from unplanned events. To meet these needs, the program is focused on the following goal: To develop, integrate and demonstrate the science and technology of remote manipulation such that by the year 2004, 50% of the EVA-required operations on orbit and on planetary surfaces may be conducted telerobotically. The Space Telerobotics Program consists of a wide range of tasks from basic scientific research to applications developed to solve specific operations problems. The program is focused on three specific mission or application areas: on-orbit assembly and servicing, science payload tending, and planetary surface robotics. Within each of these areas, the program supports the development of robotic component technologies, development of complete robots, and implementation of complete robotic systems focused on the specific mission needs. These three segments align with the application of space telerobotics to the class of missions identified by the potential space robotics user community. The objective of each of these program areas, from base technology development through systems applications, is to provide the technology for space telerobotics construction and related applications with sufficient validation that the designers of future spacecraft can apply the technology with confidence.

Turning to terrestrial disaster recovery sites, the Chernobyl nuclear power plant unit-4 in the Ukraine experienced a powerful explosion on April 26, 1986. Under extreme conditions, the Chernobyl Sarcophagus was constructed within six months of the explosion. Last year, RedZone's Pioneer system entered the damaged Chernobyl Nuclear Power Plant Unit-4 reactor to sample and characterize environmental and structural conditions of the Unit-4 shelter, which was built six months after the April 26, 1986 explosion. Pioneer's goal is to provide data to aid in stabilization and remediation planning. Pioneer conducts structural assessments including:

- Visually surveying structural elements, walls, floors, and ceilings for damage, cracking, or other deterioration.
- Taking samples of concrete to determine the physical properties or degradation due to prolonged exposure to high radiation fields and other hostile environmental conditions.

Environmental assessment includes tasks such as:

- Measuring the gamma field intensity and neutron flux, and
- Locating and estimating the volumes of fuel containing material (FCM).

Pioneer includes mobility platform and sensor/tooling packages, such as a coreborer, 3D mapper, manipulators, and environmental sensors that measure temperature, humidity, and radiation levels. Pioneer will be equipped with shovel and arm. Two of Pioneer's sensor/tooling packages include a plow bucket for pushing and lifting gravel and other debris, as well as a fully electric manipulator that can pick-up and move a variety of objects.

Houdini produced by RedZone of Pittsburgh is a tethered, hydraulically-powered, track-driven work platform with a folding frame chassis that allows it to fit through confined access ways as small as 22.5 inches (0.57 meters) in diameter. Modeled after bulldozers and backhoes from the construction industry, and hardened to work in nuclear and other hazards, Houdini features a rugged, intuitive work platform, rugged design and sturdy construction that make it well-suited for the rigors of waste mobilization and other heavy work tasks. The robot features low-voltage servo-valving, spark-proof hydraulic operation, environmentally safe hydraulic fluid, and self-collapsing capability (under gravity) for fail-safe removal. Houdini can travel on, over, and through materials ranging in consistency from concrete to sludge to liquid. The robot system is fully submersible, allowing material handling to be performed on tank contents. Houdini also provides simple material handling similar to conventional earth-moving equipment, such as bulldozers and backhoes, in conjunction with a variety of dislodging, retrieval, and conveyance technologies. Originally designed to support the retrieval of nuclear waste from the U.S. Department of Energy storage tanks, Houdini has already been successfully used for that purpose in numerous tanks at the Oak Ridge National Laboratories. Houdini was developed by RedZone in cooperation with Federal Energy Technology Center and Fernald Laboratory.

Another product of RedZone Robotics, Inc., Rosie, is a heavy-payload, long-reach remote work vehicle, used in decontamination and decommissioning applications. A rugged mobile platform with advanced tooling and controls, Rosie combines the ability to dismantle an entire structure with the dexterity to decontaminate selected building components. Rosie deploys a wide range of tools and sensors from her heavy manipulator or locomotor deck. Variable-speed motion control allows an operator to position tools quickly to perform desired work tasks. In addition, Rosie's work envelope enables floor-to-ceiling reach. A single operator stationed at a control console controls the robot. Primary system functions, such as boom, locomotor, system power, tether, and cameras, are operated using switches and joysticks on the desktop. Less frequently used functions and status information are accessed through a touch screen. Three video monitors with quad-splitting capability provide views of all on-board cameras. To maximize productivity, the operator can adjust the display location of any camera view. Additional video monitors can also be provided. Modular design of the robot allows expedient maintenance and deployment of alternate tools, such as a mechanical scabbler, sealant spray, reciprocating saw, plasma torch, pulverizer, wet/dry vacuum, and dexterous manipulator(s). At Argonne National Laboratory, Rosie works to dismantle the CP-5 experimental reactor. Though radiation levels at the work site have been higher than anticipated, this robot continues to be a reliable and flexible tool; maintaining a heavy and consistent workload, while preventing radiation exposure to human workers.

ACADEMIC, GOVERNMENTAL AND INDUSTRIAL INITIATIVES

The construction industry faces special challenges in reaping the full benefits of the automation and information technology revolution that has brought and continues to bring rich rewards to many industries. These include low R&D investment, fragmentation, and its unique project-oriented character. Purdue University and the Construction Industry Institute developed a web site featuring the latest developments in advanced construction technologies <http://www.new-technologies.org/ECT/>. The site serves as a vehicle for information transfer from research laboratories and technology developers in industry into the hands of actual and potential end users. A number of other web sites exist which provide relevant links to information sources related to professional and trade organizations in the construction industry, and to web-based project management services and other information technology resources.

The American Society of Civil Engineers established the Civil Engineering Research Foundation, which along with its affiliated "Innovation Centers" sponsors and organizes annual conferences and conducts an annual award scheme aimed at promoting the most promising innovations in design and construction. Additionally, ASCE operates a Field Sensing and Robotics Committee which organizes biennial conferences devoted to "Robotics for Challenging Environments." An organization based at the University of

Michigan sponsors an annual construction innovation NOVA award based on practical impact of the innovation on construction industry practice.

The FIATECH (Fully Integrated and Automated Project Processes) Consortium has been established as a Construction Industry Institute affiliated, collaborative, not-for-profit research consortium conducting leveraged research and development of Fully Integrated and Automated Project Processes. Achieving this objective requires a breakthrough change in technology intensive processes. The first task involves a seamless integration and management of project information within the context of an entire facility life cycle and enterprise-wide resource planning system. The second tasks will bring about wireless data from the construction site into the project management information loop. FIATECH will conduct leveraged R&D in partnership with suppliers, with firms in the software/information technology industries, and with the public sector. Likely public sector partners include the National Institute of Standards and Technology, The U.S. Department of Transportation, The Army Corps of Engineers, and the General Services Administration.

CONCLUSION

The above overview of automation related activities in the United States indicates high volume of robotics research and development on behalf of industries outside of the construction sector, but with similar work characteristics as those typical on construction sites. Still, as opposed to other countries such as Japan, little robotics R&D activity takes place aiming directly at on-site applications. Historically, most robotics R&D are aimed at tasks related to disaster response, maintenance and repair of built facilities.

However, there is certainly an information technology revolution taking place in industry, particularly in the area of sensor-based integrated data acquisition methods, metrology and Internet based multi-media communications, with direct impact on construction procurement, project planning, design and construction efforts (see References). It is in this aspect of construction site and office automation that the United States is leading the way and it is likely to retain its leadership in the foreseeable future.

REFERENCES

Abduh, M., 2000
UTILITY ASSESSMENT OF ELECTRONIC NETWORKING
TECHNOLOGIES FOR DESIGN-BUILD PROJECTS, unpublished
Ph.D. dissertation, School of Civil Engineering, Purdue University,
West Lafayette, Indiana, July.

Cheok, G., Lipman, R., Witzgall, Ch., Bernal, J., Stone, W., 2000
NIST CONSTRUCTION AUTOMATION PROGRAM REPORT
NO. 4: NON-INTRUSIVE SCANNING TECHNOLOGY FOR
CONSTRUCTION STATUS DETERMINATION, NISTIR
6457, Building and Fire Research Laboratory, National
Institute of Standards and Technology, Gaithersburg,
Maryland, January.

Ciesielski, C., ed., 2000
PROCEEDINGS, INTERNATIONAL CONFERENCE ON
AUTOMATIC DATA COLLECTION TECHNOLOGIES IN
CONSTRUCTION, Las Vegas, Nevada, March 28–30.

Field Robotics Center, 2000
CARNEGIE-MELLON UNIVERSITY, COGNITIVE COLONIES
PROJECT, http://www.frc.ri.cmu.edu/~sthayer/sdr/ (July 30).

Johnson, D. and Bennington, B., 2000
MOBILE INTERNET PROTOCOL PERFORMANCE AND
ENHANCEMENTS OVER ACTS, Proceedings, Advanced
Communication Technology Satellite (ACTS) Conference 2000, Sixth
Ka-Band Utilization Conference, NASA Glenn Research Center,
Cleveland, Ohio, May 31.

Leger, P., 1999
AUTOMATED SYNTHESIS AND OPTIMIZATION OF ROBOT
CONFIGURATIONS: AN EVOLUTIONARY APPROACH, doctoral
dissertation, Robotics Institute, Carnegie-Mellon University,
December.

Neil, C., Salomonsson, G., Skibniewski, M., 1993
ROBOT IMPLEMENTATION DECISIONS IN THE AUSTRALIAN
CONSTRUCTION INDUSTRY, CSIRO Research Report DBCE DOC
93/37 (M), Highett, Vic., March.

Nof, S., ed., 1999
    HANDBOOK OF INDUSTRIAL ROBOTICS, 2nd Edition, John Wiley &
    Sons, New York.

Szuba, T., unpublished
    COMPUTATIONAL COLLECTIVE INTELLIGENCE, Wiley Series on
    Parallel and Distributed Computing (forthcoming).

Wetzel, J., Bernold, L., Stone, W., eds., 2000
    PROCEEDINGS, ROBOTICS 2000 – The 4th International Conference,
    Exposition, and Demonstration on Robotics for Challenging Situations
    and Environments, Albuquerque, New Mexico, American Society of
    Civil Engineers, February 28 – March 2, 2000.

Kidney Stones and The Structural Design of Tall Buildings

Gary C. Hart

1.1 INTRODUCTION

The author went to his doctor to report his latest Kidney Stone attack. Prior to the visit he had to take a blood test, have an X-ray session and then the latest Cat Scan imaging session. The doctor showed him the images from the sessions to impress upon him how "high tech" he had become. Then he discussed the options available to the author to stop kidney stones.

This visit to a modern "high tech" doctor and the doctor's process to meet the author's needs was enlightening. It was clear to the author that the doctor took three basic steps: First, he explained the results of his analysis in terms that the author understood, even though chemistry was the author's worst subject in college. Second, the doctor presented the options for stopping future Kidney Stone attacks. Third, the doctor, when pressed by the author, who knows probability and reliability theory, stated that there was NO guarantee that if the author followed all of the doctor's orders that those dreaded Kidney Stones would not come back again, i.e. they would stop! These three basic steps illustrated the fundaments of communication that must take place in this scientific and high tech age. What the doctor really did was to provide the bridge between the science of medicine and non-doctor descriptions of the options, and then very importantly quantified, based on his experience and the analysis of the author's case facts, the probability of success for different options that could be selected by the author. That is, science the author could not understand, options the author could understand, and then quantification of success for each option based on scientific analysis. Finally, the author had to make the decision on which option to take – not the doctor!

This is a Tall Buildings conference paper, so what does it have to do with Kidney Stones? Wait a minute and the answer will come.

Structural Engineers are professionals in this "high tech" age just like doctors. They work for clients that really have no technical training or experience to evaluate their work except when the dreaded Kidney Stone, or the Design Basis Earthquake or Wind occurs. In the structural engineers natural hazard world, it hopefully will never occur. (Some would argue that construction cost per square foot is a good measure, but then they really do not know if the great low cost building was the result of bad or good structural engineering and they really do not want to find out!). Fortunately for structural engineers, the theory of structural reliability was introduced in the 1960's and this introduction brought order to structural engineering. It enabled the structural engineer to communicate with his or her client in the same way that the doctor of the author communicated with him.

To understand the importance of the structural reliability theory and why it must form the foundation for a structural design criteria consider the following. The structural engineer that is about to design a tall building is in a unique and an especially difficult position relative to other building design professionals. The fundamental difficulty is the immense challenge faced in representing nature and its force on buildings, and also the prediction of the performance of the building due to these forces of nature. This is followed by the checking for human errors when structural engineers, usually under severe time pressure, make many complex calculations and decisions. The representation of nature and its impact and response of buildings demands the development of mathematical models that are ever increasing in both their accuracy and their complexity. For example, the author just finished a book for the publisher John Wiley and Sons entitled *Structural Dynamics for Structural Engineers*. The book started out to be about 200 pages in length and ended up being over 400 pages long. This result, even with the author's desire, was to minimize the mathematics of structural dynamics and to emphasize the physical feeling for structural dynamics. Therefore, one can imagine the shock to the author when it was discovered that the book contains over 4,000 equations! So unfortunately the world of equations is essential and because of the fantastic creativity of the architect each building has a unique optimal design.

What is the reluctance to equations and the sophisticated structural engineering theories that are available today? To answer this question consider the following. The author has had the privilege of knowing rather well many great structural engineers from around the world that are now still living and are near or past 90 years of age. Without exception they trace the modern development of structural modeling of structures from the late 1930's and 1940's. In the United States some would state that it started with the teachings of Timoshenko in the United States. Therefore, modeling the forces of nature and building response to these forces is a rather young and fast changing science. Therefore, from a quality control and business profit perspective it is a challenge.

As noted above, structural reliability theory, which must form the basis of modern structural design criteria, demands and uses these complex theories and corresponding equations. However, the reward is that it is the communication bridge that is used by doctors. On one end of the bridge structural reliability theory demands that the performance of the building be described to the lay person – in non-structural engineering words. For example, the behavior of the building will be such that the computers will not fall off the table due to an earthquake motion in the next 10 years. Or for example, the motion of a building's 35th floor in the wind will be such that the occupants not feel any building motion in the next 5 years. Or, for another example, the steel bar in that beam on the 30th floor near the nursing room will stretch in earthquakes that will occur over the next 30 years, but after the earthquakes are over the length of the steel bar will be the same as it was before the earthquake.

The structural engineer like the doctor can explain the possible responses that the building can experience when subjected to, for example, nature's wind and earthquake loads. Then the structural engineer can identify the structural engineering options that are available for the design of the building that can eliminate the undesirable building responses. Finally, the structural engineer can quantify using what is often called a risk analysis the real probability or chance that these undesirable building responses will occur.

1.2 THE LIMIT STATES

Modern structural engineering for tall buildings demands that the structural engineer and the building owner discuss at least three basic options or as in the vocabulary of Structural Reliability limit states.

The first limit state is the classic limit state that is considered in all building codes. It is life safety protection. This limit state must be quantified in terms of element performance and usually in today's modern structural engineering the strain in the steel or the concrete of the building. The goal here is to insure that no persons will die in the building as a result of the collapse of one or more members in the building for the largest earthquake or wind that will occur during the life of the building. We want a building design that provides a life safe behavior. This limit state must be discussed with the building official and the members of the community because the people are in the building and are being exposed to a risk often not of their own choosing – they may have to work or bank in the building. Because the structural engineer does not know with certainty what the largest earthquake will be in the building during the life of the building the discussion must be carried out in probabilistic terms. This is not unlike the terms that each person over 50 must address when asked how long he or she will live. The age of death and the associated estate planning must be carried out in probabilistic terms.

The second limit state is a damage control limit state using the relative displacement between the floors of the building. The goal here is to provide a cost-effective design when the costs associated with the consequences of different between floor displacement magnitudes are balanced with the cost to reduce the response to different magnitudes. This is a business decision and not a life safety decision. It does not require public input and is no person's business but the building owners. Unfortunately, all too often the structural engineer can not perform such a cost benefit analysis and therefore makes the decision without the owner's knowledge.

The third limit state is another damage control limit state that this time uses the acceleration of the floors in the building. Simply speaking the acceleration on the floor multiplied by the weight of an object or content on the floor must be small enough not to cause the object to tip over and break. The acceleration is also critical in the occupant's perception of the motion of the floor and the negative cost impact of feeling this motion. These acceleration limit states are not life safety decision limit states. The cost to reduce the response must be balanced against the damage for each response level. Again this is a cost-benefit analysis that is called a risk analysis.

1.3 THE LIFE SAFETY LIMIT STATE

Because most structural engineers prefer to perform a three-dimensional elastic analysis of the building, the life safety approach involves two levels of earthquakes and winds.

The life of the tall building, and therefore the time it is exposed to nature's forces, is assumed to be 100 years.

The elastic analysis design earthquake or wind is called Level I and it is used as input to an elastic structural analysis computer program.

LEVEL-I:

An earthquake ground motion represented by an acceleration response spectrum or a set of wind forces on the building that have a 50-percent probability of being exceeded within a 100-year period.

The non-linear analysis design earthquake or wind is called Level II and it is used as input to a non-linear structural analysis computer program.

LEVEL-II:

An earthquake ground motion represented by an acceleration response spectrum or a set of wind forces on the building that have a 2-percent probability of being exceeded within a 100-year period.

All buildings shall have a site-specific earthquake ground motion study. The study shall account for the regional seismicity and geology; the expected recurrence rates and maximum magnitudes of events on known faults and source zones; the location of the site with respect to these; near source effects if any; and the characteristics of subsurface site conditions. A review of the site-specific earthquake ground motion study shall be performed by an independent State of California Licensed Professional Geotechnical Engineer experienced in methods used to perform a site-specific ground motion study.

All buildings shall have a site-specific wind study as defined in the 2000 IBC. A review of the site-specific wind study shall be performed by an independent Licensed Engineer experienced in methods used to perform a site-specific wind study.

BUILDING SYSTEMS AND CONCEPTS

Structural Standards Globalization – An Asia Pacific Perspective

L. K. Stevens

1 INTRODUCTION

1.1 General

A Standard may have quite different functions depending on its objectives and regulatory standing. It may primarily be an advisory source of information on currently accepted practice for use by designers or it may serve as a mandatory whole or part of an enforceable National Standard, Building Code or Building Standard Law of a building regulatory system.

There are several hundred bodies concerned with the development of Standards across the world. Some of these are national government controlled bodies while others are associations of engineering or technical institutions with varying levels of accreditation and authority.

No Standard can be solely a technical product but will reflect in some way the national concerns and aspirations relating to economic, cultural, geographical, climatic and political factors as well as current and previous historical alliances and technological development.

As a result, there has often been a concentration on local or regional issues in the development of national Standards particularly in long established industries such as building and engineering. This has led to substantial differences both in broad philosophical content as well as in detail, for reasons which may now be irrelevant. For example, slightly different rail gauges may have seemed logical for unconnected communities but are now a serious and expensive problem. Similarly, different electricity supply voltages and the multiplicity of power point sockets now appear as quite indefensible anachronisms.

The recent growth of international interaction in communications and particularly in information technology has made it imperative to move rapidly towards much greater levels of standardization with establishment of strict technical protocols. In the building and engineering industries, where Standards have been evolving gradually for the past 100 years, progress has been much slower. The conservative nature of these industry structures and the very real differences in local or regional conditions have inhibited any rapid implementation towards the adoption of identical standards or progress towards alignment. In particular, until comparatively recently, there has not been general acceptance of any general strategic guidelines for achieving such adoption.

1.2   international Organization for Standardization

The formation of the International Organization for Standardization (ISO) some 50 years ago has provided the opportunity for development of internationally acceptable bases for national or regional Standards.

In the building and construction industry ISO has concentrated on preparing broadly based Standards as guidelines. Few of these ISO Standards are in a form which can be adopted as operational Standards or used as regulatory documents. They are however intended to provide a consistent framework for development of national or regional Standards. Such Standards can therefore reflect not only the relevant technical requirements but also the economic conditions of a particular country, as well as the nature, type and conditions of use of a structure, together with the properties of the construction materials, during its design working life. A considerable level of success has been achieved in progress towards development and application of these guidelines particularly in Europe.

1.3   European Committee for Standardization

Another significant association, but with a more restricted regional scope than ISO, is the European Committee for Standardization (CEN) which was established by the European Commission. CEN is currently producing a suite of 10 Structural Eurocodes with the objectives of establishing "a set of common technical rules for the design of buildings and engineering works which will ultimately replace the existing differing rules in the various Member states".

These Eurocodes have drawn heavily on relevant ISO Standards and are consistent with the broad principles adopted by ISO for structural Standards. However, Eurocodes are clearly not suitable for general world wide adoption because of their focus on the specific regional requirements of the European Union.

In particular, the Eurocode Standards are not suitable for adoption for the Asia Pacific region because of the widely different technical and other conditions which exist between its economies.

2   ASIA PACIFIC STRUCTURAL STANDARDS

2.1   General

The Asia Pacific region has many major different characteristics from other parts of the world. Its population is still rapidly growing and it has substantially different standards of living and requirements in the housing, building and engineering industries. Technological and educational development varies widely and types and quality of resources are distributed unevenly.

Possibly because of these significantly differing characteristics, there has not been until very recently any organization concerned with the development of Structural Standards specifically for the Asia Pacific region which are consistent with international practice.

2.2   Asia Pacific Economic Cooperation (APEC) Forum

The Asia Pacific Economic Cooperation (APEC) forum in considering Technical Barriers to Trade, has recognized the importance of the adoption of International Standards and conformity assessment systems for improving efficiency of production and facilitating the conduct of international trade. As a result, a decision was reached by APEC in 1996 that "where technical regulations are required and relevant international standards exist or their completion is imminent, Members shall use them as a basis for their technical regulations except where such international standards or relevant parts should be ineffective or inappropriate means for fulfilment of legitimate objectives pursued, for instance because of fundamental climatic or geographical factors or fundamental technological problems" (Agreement on Technical Barriers to Trade).

2.3   Technical Group TG1

In accordance with the above agreement, APEC has set up three Technical Groups in the building industry field, including housing and timber products.

Technical Group 1 has been made responsible for Structural Standards with the objectives: "To provide a basis for the creation of aligned Standards by facilitating the timely production of International (ISO) Standards for loading and material specific design Standards that fully reflect the technical requirement of APEC Member economies and enable the removal of Technical Barriers to world wide trade".

The term "alignment" used in the above may be considered to be effectively synonymous with the ISO term "harmonization". Harmonized Standards are defined by ISO as being "Standards on the same subject approved by different standardizing bodies that establish interchangeability of products, processes and services or mutual understanding of test results or information provided according to those Standards". This provides for each Member economy to develop its own Standards provided they satisfy the conditions of alignment.

Benefits in addition to the removal of barriers to trade, include the following:

- facilitation of design operations between different countries;
- sharing of well established and tested knowledge;
- reduction of opportunity for mistakes in application;
- reduction of cost and problems in evaluation and certification of products;
- increase in confidence in quality assurance of products;
- facilitation of realistic and consistent assessments of comparative risks implicit in Standards;
- facilitation of education in common engineering principles thereby increasing mobility and opportunities for graduates;
- introduction of performance based Standards facilitating innovation in design and product development.

2.4 Informal Network

As an essential adjunct to TG1, an extensive informal network of technical experts from the Asia Pacific region has been operating under the direction of Doctor Lam Pham of the Commonwealth Scientific Industrial Research Organization, Australia. The membership is drawn from all APEC economies and currently includes members from Australia, Canada, Indonesia, Japan, Korea, Malaysia, New Zealand, Papua New Guinea, Peoples Republic of China, Philippines, Singapore, Thailand, United States of America and Vietnam.

The objectives of the Informal Network are to facilitate the development of aligned Standards in building and construction and to provide advice to Technical Group 1 for consideration as inputs to ISO through the member economies of APEC.

The initial focus of the Informal Network has been directed at the development of General Design Requirements and Loadings including particularly Wind Actions, Seismic Actions and Load Combinations. The emphasis has been on the development of performanced based specifications to enable greater freedom for designers and product manufacturers to achieve desired outcomes as an alternative to prescriptive regulations.

3 ACHIEVEMENTS OF APEC TECHNICAL GROUP 1 AND THE INFORMAL NETWORK

3.1 General Design Requirements

ISO 2394, "General Principles or Reliability of Structures" has been adopted as the basis for the verification of the reliability of structures. A major significant consequence of this decision is the acceptance of the Limit State Design philosophy with the adoption of a performance based approach..

As as an extension of ISO 2394, a proposal has been made to ISO to establish a New Work Item for the development of an ISO Standard for the General Framework for Structural Design. This is intended to be a guideline for the future development of all international Structural Standards but will be particularly useful for the member economies of APEC in developing their own Standards.

This proposal was accepted by ISO Technical Committee TC98 in December 1999 and Standards Australia International has now formally submitted the proposal to ISO.

Included in the General Requirements scope of TG1 has been the decision to adopt ISO recommendations on Notation, Terminology and Units which are also areas of concern for the production of Standards by APEC economies.

3.2 Wind Actions

ISO 4354, Wind Actions on Structures, is in need of revision and a proposal to ISO/TC98 has been accepted for a New Work Item to carry out this revision. This proposal has also been formally submitted to ISO by Standards Australia International.

Such a revision is most opportune for APEC Member economies since many current wind standards are in need of updating or are in process of revision.

3.3 Seismic Action

It has been agreed that the highly seismic regions of the Asia Pacific region require specialized knowledge from experts intimately associated with the region and that Standards prepared for other regions may not be applicable or appropriate for adoption without due consideration of local conditions and requirements. It has therefore been agreed to participate actively with ISO in the development of the next generation ISO Standard on Seismic Action.

4 CONCLUSION

Considerable progress has been made towards development of proposals for the alignment of Structural Standards in the Asia Pacific region in agreement with the principles established by the International Organization for Standardization. But even more importantly, the Asia Pacific region, through the Asia Pacific Economic Cooperation forum and the Informal Network has become a significant contributor to that body and to the development of international Standards which truly reflect the requirements of a very large proportion of the world's population in a major growth area.

This contribution has been recognized by the ISO/TC 98 meeting in Oslo in December 1999 where the following motion was recorded: "ISO/TC98 recognizes and appreciates the initiative taken by the APEC Informal Network to increase the use of Standards prepared by ISO/TC 98. This initiative should be used as an example to encourage other members of ISO currently not participating" (Resolution 139).

5 ACKNOWLEDGEMENTS

The following organizations and people have contributed significantly to the developments described above and their part in this exercise is gratefully acknowledged.

- Department of Science, Industry and Resources. Commonwealth Government of Australia.
- Doctor Lam Pham, CSIRO, Australia.
- Professor D.S. Mansell, CSIRO, Australia.
- International members of TG1 and the Informal Network.

Greening Office Towers

Christoph Ingenhoven

1.1 THE OFFICE INGENHOVEN OVERDIEK UND PARTNER

The Ingenhoven Overdiek und Partner Office in Düsseldorf has been under the leadership of Christoph Ingenhoven and Jürgen Overdiek since it was founded in 1992 as a shared architectural practice.

Christoph Ingenhoven, born 1960. Studied architecture at the RWTH in Aachen and at the Academy of Arts in Düsseldorf. Jürgen Overdiek, born 1954. Studied architecture at the RWTH Aachen and Villa Massimo Stipendium Rome. Today, there are 150 staff working in the office (architects, interior architects, designers and model makers).

The projects are mainly offices and administration buildings, commercial and insurance company headquarters, high rise buildings in Germany and abroad, department stores and infrastructure projects (airports, railway stations), town planning and landscape architecture works.

The aim in all the projects is to work closely with engineers and other specialists throughout the whole of the design process. Internal questionnaires are used to measure the efficiency, the ecological consciousness, the economy of resource usage and the buildability of every project. This creates an architecture that is characterised by technological innovation and produces buildings appropriate to people's working and leisure needs. It does not spring from artistic requirements alone but is rather a reflection of the architect's responsibility for the environment.

2.1 HEADQUARTERS FOR RWE AG ESSEN, 1991–1997

The construction of the 163 metre high cylindrical high-rise tower which houses the RWE AG represents the completion of the first stage of the competition decided upon in 1991 for the headquarters of the RWE AG/RAG. The ecologically oriented building will be completed at the end of 1996.

The RWE tower is free-standing, built behind the curved slab block. Areas of higher density keep the interior of the site free for a generously proportioned park. Access to the high-rise is through a public plaza extending outwards. The newly created streetscape is interrupted at this point and bridged over by a roof-level loggia with photo-voltaic lamellae. The 30 floors of the building, including a roof garden, are accessible from the lobby through lifts located outside the tower itself.

Ingenhoven Overdiek und Partner, Düsseldorf

The RWE AG company headquarters can be described as an ecological high rise development. The building incorporates innovative systems for optimisation of the building services.

The office zones are naturally ventilated by a full storey-high double skinned glass façade and by openable windows. Ventilation is optimised by the incorporation of a newly developed integrated "fish's mouth" element, with sun shading and anti-glare blinds to make the most of daylight and natural ventilation. Control panels ensure that the building environment can be adjusted to suit individual demands and requirements. Roof elements contain the necessary building services for air and space conditioning such as 2 component lighting, cooling and acoustic absorption panels, in addition to smoke alarms and sprinklers.

The element is freely suspended in order to be able to use the energy storage qualities of the concrete floor.

Even the interior architecture and furnishing follow the integrated architectonic philosophy and forms, and are found recurrent. applications in the various usage zones.

Figure 1   RWE AG Highrise, Essen – photo: H.G. Esch, Köln.

2.2 WAN XIANG INTERNATIONAL PLAZA, SHANGHAI, 1995–2003

The high-rise project for the Wan Xiang International Plaza Company in Shanghai was awarded 1st prize in an international competition in 1995.

The site is in the centre of Shanghai on Nanjing Road in the immediate vicinity of the People's Gardens. Approx. 1.5 million passers-by flow daily through the largest commercial street in China. A spacious public plaza is planned on Nanjing Road. From the plaza, you can reach the adjacent department store and the offices in the high-rise over a transparent multi-level access area that runs on the diagonal.

The 53-storey and 288 m high building, based on a triangular plan, was developed for hybrid use incorporating shopping areas, flexible office space, restaurants and rooftop gardens. Its architectonic image is created by the structure of diagonal struts parsing up the front of the façade.

Figure 2   Wan Xiang International Plaza Shanghai – Illustration: Peter Wels, Hamburg.

2.3   HQ3A/4B, CANARY WHARF, LONDON, PROJECT 2000

The two high rise buildings on Heron Quays are part of the world financial centre at Canary Wharf. The plan shape creates highly efficient office and administrative zones and provides extensive space for financial trading offices.

The multi-skinned façade design allows natural ventilation in the office areas. This results in a reduction of the required storey height in comparison to conventional designs.

Figure 2   HQ3A/4B, Canary Wharf, London – Illustration: Peter Wels, Hamburg.

The Urban Habitat After Globalization: Cities in the Age of Climate Change and Fossil Fuel Depletion

P. Droege

1 INTRODUCTION

Two daunting challenges confront the world's cities and city regions well within this coming generation, affecting the global urban system and human civilization as a whole: fossil fuel depletion and man-made catastrophic climate change. If these are not swiftly and effectively met their impacts will deeply affect all industrial, world and mega-city systems – and hit hard the fast-growing, major urban agglomerations of the developing world, along with their economies. Tall buildings carry a special significance in this context.

Since the 1970s and 1980s the prospects of fuel depletion have only slowly begun to enter general urban planning and development frameworks, largely as energy efficiency and conservation issues. In terms of climate change communities have only during the 1990s begun to recognize that all greenhouse gas (GHG) emissions are directly or indirectly generated locally, through acts of agency, production or consumption. This has provided a boost to the role of local places in the debate since GHGs can be allocated and made understood locally and hence form the basis for specific policies, programs, plans and projects.

A number of organizational and cultural barriers mitigate against swifter, wider change. Among these are the subsidiary regard in which cities are held in the traditional hierarchical frames of international arrangements dealing with globally encountered challenges. Another is posed by the short planning horizons and political uncertainties that prevail on the local level. In terms of policy development, measurement techniques and planning reality an extraordinary, even paradoxical gulf exists between the global nature of greenhouse gas impacts and fuel depletion prospects, and the local reality, that represents both final impact and original source of globally experienced changes.

Despite the significant hurdles, energy issues have begun to take center stage in the reality of an increasing number of cities and towns around the world. The leaders of these pioneering communities realize that because of the central significance of cities in national economies, and their utter dependence on relatively short-term fossil fuel supplies – and the devastating effect of their burning on human health and global climate – the speed and magnitude in which renewable energy strategies are being introduced will be of crucial importance to the future of global civilization and local cultural settings alike. As socially, politically, economically and culturally significant settings cities face increasingly intense local action, in their communities' search for improvement of the local environment, and in a rising movement to combat global warming well before that time. Business, industry, science, technology and governments are being challenged to respond and deliver solutions. It is here where a growing number of new urban action and development initiatives are being readied to link local agendas and national frameworks to international challenges and resources.

The challenges confronting a rapidly urbanizing human civilization are unprecedented. As long as the current path is pursued there is the dilemma of a massive investment in obsolete infrastructures, since any urban area or long-term urban systems element that is intrinsically dependent on fossil fuel will be rendered dysfunctional within only a few decades. All basic urban communication infrastructures, both traditional – such as roads, rail, air and sea ports – and advanced have been nurtured in a world of total fossil fuel dependency. Indeed, the Internet, an increasingly vital global and largely urban network of networks, relies largely on fossil-fuel operated hardware, conduit-based webs, and wireless communication carrying media. If global communications are to be sustained beyond the middle of this century, they must be powered by distributed, ubiquitous and redundant renewable power supply systems.

2   MAJOR URBAN ENERGY ISSUES

2.1   Fossil Fuel Depletion and Cities

The world depends on fossil fuel. Fossil fuel use increased five-fold over this past half century, from 1.7 billion tons of oil equivalent in 1950 to 8 billion tons in 1999, providing 85% of the world's commercial energy. The majority of this energy is used either within cities, or for transport from and to cities.

Most contemporary fuel sources are due to expire well within this century as widely available and relatively peacefully contested sources, and much of this reality will become globally pervasive within the next thirty to fifty years (see

Figure 1). Even conservative industry, national and international governmental sources estimate that oil will expire by 2050 – the depletion of the more easily accessible supply sources is likely to take place already in the 2020-2030 time frame. Given rising use rate scenarios natural gas is likely to evaporate by 2040. The logistically and environmentally most problematic source of global fuel supply, coal, is expected to expire well before 2100 – provided no large scale efforts of energy fuel substitution is pursued to stretch its deposits. Uranium is expected to be depleted by the mid-2030s, even if the daunting cost, security, safety, disposal and public acceptance dilemmas were not to weigh so heavily against its application. Significant growth rates in uranium use are not anticipated, given the risks involved and a waning willingness among many governments to pursue nuclear fission further. Nuclear fusion, more than half a century after its arrival as a great Modern dream, still struggles with enormous conceptual and technical flaws, and the problems associated with huge financial, environmental and political costs. These include the specter of wide-spread public resistance to a new generation of nuclear super-reactors.

Reserves and future availability of oil and natural gas.

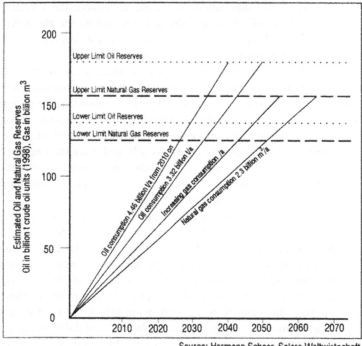

Source: Hermann Scheer, Solare Weltwirtschaft

Figure 1 The prospects of fossil fuel depletion under a range of scenarios.

This outlook is as troubling as it is under-emphasized in its public significance. Fossil fuel is the very source of modern economies and their petrochemical, transport and industrial production systems – and hence also sustain the spatial centers of global civilization: cities and city regions. The modern city, its history, form and growth dynamics, is entirely inconceivable without it. Yet a persuasive agenda to deal with the prospects equitably, humanely and positively still eludes global and local planners alike.

Modern cities have mushroomed on their rich fossil nutrient supply, and especially voracious and dependent are the largest, most rapidly growing urban agglomerations. The very logic of their global rise and regional spread is founded on the availability of powerful, centralized and inexpensive fuels: coal, petroleum and natural gas – yielding fossil urban structures and patterns based on fossil transport, fossil construction machinery and fossil industrial systems and manufacturing processes. Intensive economies and labor markets clustered around the centralized and networked city regions and were anchored by heavy investments in infrastructure: power, transport and communications. This increasingly bolstered cities' primacy over non-urban hinterlands. The new cities of the 19th and 20th centuries – and the very cultures they engendered – were a product of combustion: London exploded with coal-fired power, pre- and post-World War II modern city innovations in the Soviet Union, the United States, Europe and across Asia alike were literally jump-started by the electrifying jolt of the new energy technologies. Los Angeles, as urban system, carrier of cultural content and global paradigm, is the proverbial petroleum city.

It is appropriate to refer to contemporary urban areas as fossil cities. The logic of global urbanization becomes transparent when considering the broad availability of inexpensive power for all urban infrastructures: building construction, lighting, air conditioning, computing and telecommunications and massive freight and human transport systems on surface, sea and air. These new bundles of infrastructure at once link cities globally and drain their regions. As a consequence globalized urban systems are inherently more vulnerable to a serious – and inexorable – decline of global fossil fuel supplies than those that rely more on their local and regional human and land resources.

While some local urban systems theoretically are relatively safe from a terminal fossil fuel shock through their reliance on hydro-electric, nuclear or bio-energetic power, no currently utilized alternative energy source alone can help the vast majority of cities world-wide. Also, the interconnectedness of the global system makes it impossible to seriously contemplate the survivability of regional pockets of self-sufficiency. The only viable option to secure the continuity of urban civilization in this century is a system-wide turn to a broad portfolio of renewable energy sources, based on an overwhelming availability of solar, wind, wave and geothermal energy. The alternative to this path lies in a massive military build-up as it is already being prepared by some leading economies. A global and open escalation of the simmering war over regional fossil resources, currently contained largely in local and regional conflicts is inevitable without a broad and world-wide introduction of renewable energy sources. Cities and city dwellers would bear the brunt of such conflicts.

2.2 Urban Greenhouse Gas Emissions

Global climate change is perhaps the most long-term and devastating effect of the massive, recent and rapid world-wide burning of much of the global Carbon Age heritage: the organic carbon stores, deposited and sequestered, for example, in coal sediment layers over the past 330 million years. The powerful release of stored carbon dioxide and other gases of similar atmospheric impacts destabilizes the planetary climate system through a strengthening of the positive radiation feedback mechanism known as the greenhouse effect. The results of a warming earth surface include rising ocean levels, affected food crops, shifting regional weather patterns and the advance of tropical diseases. There is also the distinctive prospect of catastrophic feedback systems, such as volcanic eruptions in glacial areas, disturbed oceanic methane strata and the thawing of methane-rich tundras.

As primary energy consumers, cities and other urban systems are the largest single sources of CO_2 equivalent greenhouse gas emissions. As hypertrophic urban growth is being fuelled by the fossil-fed economy, the new agglomerations also become the worst offenders in carbon emissions, and the most difficult to retrofit to a renewables-based, zero-emissions behavior.

The protocol resulting from the 1997 Kyoto United Nations Framework Convention on Climate Change had most industrialized nations agree to a five percent cut of carbon-dioxide-equivalent emissions by 2010, although it is widely agreed that a 60% cut is required to actually halt global warming. As a global target based in international development equity principles, a figure of 3.3 tons of carbon dioxide-equivalent emissions per person per year (based on the customary 1990 figures) is also being proposed as a total sustainability measure to reflect the actual oceanic and forest sink capacity of the earth. The United States today produces nearly 26 tons, while India still lies well below this level, at 1.8 tons (see Figure 2). The large imbalance between nations – the world average lies at seven tons – results in the specter of a sizable and complex global carbon market, heralded by some as an effective mechanism to achieve overall reductions and equity and decried by others as a pollution licensing scheme for the rich.

The global need to work towards large-scale savings in fossil energy consumption as well as greenhouse gas (GHG) emissions calls for a systematic integration of renewable energy products, systems and processes in cities and regions – where most human emissions originate. However, many industrialized countries see mitigating action still largely as a challenge of energy efficiency and conservation measures and expect most savings to be made in lowering electricity consumption in residential and commercial buildings and in altered transport and land-use patterns.

In the long run, efficiency and conservation measures have only limited capacity to reduce fossil fuel use and combat global warming given the massive fuel requirements that will continue to grow well into this century, particularly in the developing world. Renewable energy technologies have to be advanced immediately to be available in time for a significant and relevant supplementing and replacement role half a century hence. While cities are increasingly seen as settings for local environmental action and instruments of international and national greenhouse gas abatement, much of the prevailing urban energy agenda is still aimed either at issues of efficiency, or at isolated but highly visible renewable energy technology installations, without a sense of sustained and comprehensive action or an understanding of the relative value of individual agenda items.

2.3   Energy and the Form of the Built Environment

As cultural, economical and technological system the built environment is deeply affected in its form by the nature of its basic fuel supply. Fossil changes in the general urban structure were highlighted by the leaders of the great Modern design movements, from the beginning of the 20th century: *Futurism* in Italy, *Constructivism* in the Soviet Union, *De Stijl* in the Netherlands, *Das Bauhaus* in Germany, the declarations resulting from the *International Modern Architecture Congresses* (CIAM – Les Congrès Internationaux de I'Architecture Moderne, 1928-1956) and the *International Style* that spread from the United States throughout the industrialized world. These movements and the broader societal changes they express were supported by the dawn of the fossil machine age. This new age, an outgrowth of earlier stages of the Industrial Revolution when hydro-power was a main driver of textile mill operations, was boosted by the discovery of coal's potential and the discovery of electricity, giving rise to Frederick Winslow Taylor (1856-1915) and Henry Ford's (1863-1947) ideas about the mechanization of manufacturing and later the increasingly urban and automated global production-consumption systems of the advanced industrial age, boosting global power consumption at an exponential rate.

Indeed, industrialization as powered by electricity, coal-fired steam engines and petroleum combustion motors, meant the rapid growth of cities, driving the search for innovation in urban form. As far as western urban traditions are concerned, a rash of utopian premonitions spread as a result, as revolutionary as the technological changes they were triggered by. Peter Kropotkin's (1842-1921) and other anarcho-syndicalist influences on ideas about ideal communities were a direct outgrowth of this era, as were Sir Ebenezer Howard's (1850-1928) concepts and the rise of the Garden City movement – and the genesis of the Regional Plan Association of New York and its seminal plan of 1929, Frank Lloyd Wright's (1869-1959) Broadacre City (first described in his 1932 work *The Disappearing City*), Ludwig Hilbersheimer's (1885-1967) mass housing concepts, and General Motors' extraordinary 'Futurama' pavilion at the 1933 'Century of Progress' World Exposition in Chicago, anticipating the glorious, then-futuristic car friendly cities to come. Brasilia, Canberra and all of the British New Towns are fossil fuel fantasies come true.

And so are the modern suburbs. Early modern ex-urban subdivisions were developed as aspiring new communities along the tracks of the early electric tramways, while soon thereafter the combustion-engine driven vehicle created a new urban reality altogether: that of the new car suburbs expanding along motorways. The electric mass-media driven promotion of a new fossil-fuel based lifestyle centered around the motor cars and petrochemical products suffused industrial civilization after World War Two, even more so than the heroic visions of humanity's *redemption by machine* had promised before the war, in a new set of aspirations in urban form. And these very much were practical visions indeed: already very early in the 20th century the great electric inventions of the telephone and the elevator opened the door to unprecedented skyscraper cities.

Modern carbon culture arrived in style with the epitomizing work of migrant Bauhaus architects such as Walter Gropius and Mies van der Rohe, and others. Le Corbusier's ideas about the radical modernization of cities are widely regarded as being among the most influential of these in urban terms, brazenly calling for the razing of pre-fossil cities, as exemplified by his *tabula rasa* modernization concepts Plan Voisin, La Ville Radieuse, a famous lone pilot project for a mass housing block, Unite d'Habitation and in some realized projects such as in the Alsatian town of St. Dié. The introduction of this simple and revolutionary technological development doctrine in cities in the United States, United Kingdom and countries around the world, combining the power of the new with a good dose of basic speculative opportunism, brought about urban destruction and blight through wholesale urban renewal projects in what was identified as inner city slum areas: the neighborhoods of the poor and disenfranchised and the pre-fossil relics they inhabited.

The post-war years in Europe saw powerful expansion plans hatched among the spreading webs of electrification that crisscrossed former farmlands: the Greater London strategy, the satellite city concepts, the Stockholm and Copenhagen finger plans gave much vaunted plan form to a wider, world-wide

turn to life in cities, driven by structural shifts and opportunities brought about by
the rise of the fossil fuel economy and its promises.

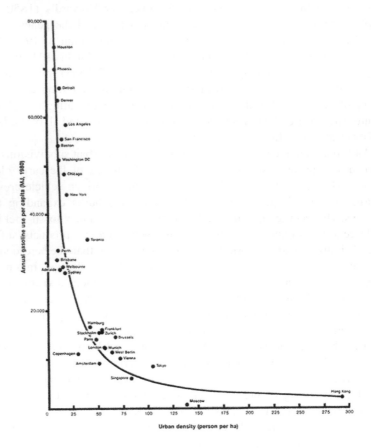

Figure 2   One of Newman and Kenworthy's seminal charts of the 1980s, correlating the density of
international cities with their fuel consumption.

Yet the gleaming new vision of renewed and more orderly, healthier and socially
more equitable cities did not escape the Faustian energy reality of advanced
industrialization. With the burning of fossil fuel arrived sulfur dioxide, nitrous
oxide and a host of other toxic gases as well as carcinogenic airborne combustion
particles, creating the smog and air pollution crises of the 1960s and 1970s.

Today, with the exception of the greenhouse gases themselves, these have been partially conquered as hazards in the industrialized world but still plague most large cities, particularly those of the developing world. Calls for more concentrated and transit-oriented forms of urban development were signs of this declining stage of the fossil-fuel economy when savings in energy conservation and fuel efficiency were identified as the cheapest, fastest and most immediately useful means to limit emissions. Denser cities were shown to be more fuel-efficient (see Figure 2). Car dependence and low-density urban structures that were incapable of sustaining public transport came to be understood as a major hindrance to achieving sustainable urban life. It is clear that in the long run, however, efficiency and conservation measures are not sufficient to halt the powerful world-wide rise in emissions. Massive substitutions of conventional energy technologies will be needed, such as those powering the making, operation, maintenance and upkeep of military machinery, and systems of passenger and freight conveyance with sustainably renewable systems of energy supply and use. In order for these new technologies to be in place in time to be of significant use they need to be widely introduced now. In this sense, transit-oriented development and recent attempts at recreating pre-industrial urban patterns such as 'neo-traditional design' and other approaches of 'new urbanism' are important interim measures but fossil-fuel derived urban form phenomena themselves.

The nature and form of buildings, too, were deeply affected by the development and spread of the revolutionary new fuels and the industrialized manufacturing processes they engendered. As a consequence architecture changed radically in the new fossil age, logically breaking with all earlier traditions. The Modern Movement provided aesthetic and ethical refinement to mass applications in electricity, ubiquitous machinery, air conditioning, industrial steel products, advancing glass technology, mass produced curtain walls, prefabricated building systems and a number of other highly energy-intensive technologies. The new thinking about buildings, highlighted by the International Style, applied to aesthetically refined, ornament-free, skeletal, industrialized and largely corporate structures. It became the new global *aesthetic of the possible* as the advanced fossil age dawned in cities around the world. Buildings became disconnected from their climatic and cultural context due to the end of local resource dependency. An interest in vernacular form and *regionalism* emerged, as a counter-reaction and an attempt to recapture the local language of form: the regional building traditions that had been lost in industrializing change. This movement remained a stylistic artifice since it failed to address the fundamental dynamics of design and development under conditions of unsustainable energy resource practice.

To address the energy issues there are today two kinds of building response. One focuses on regulation- or incentive-driven industry reform along passive energy design, conservation and efficiency lines, and the other seeks to drive more advanced forms of technological and design innovation which are needed to fundamentally transform the way built structures are made. Some countries, most notably among them Germany, pursue a mandated solar energy pricing policy aimed at jump-starting a new generation of buildings and architectural responses. In this vision, the zero-emissions house that functions without fossil energy supply, once the exclusive domain of eccentrics and university research laboratories, would become a living reality for a majority of dwellers. Yet to be addressed, however, remains the question of how to reduce the large amount of energy that is embodied in the building materials themselves, the services contributing to their making, and the energy implications of the very form and design of neighborhoods and cities. Indeed, the energy household of cities is made up by more than buildings and the infrastructures that service these. It is the sum total of all goods and services measurably consumed in a given location.

3 CULTURAL SHIFTS TOWARDS SUSTAINABLE URBAN ENERGY DEVELOPMENT

3.1 Postglobalization: New Energy For Sustainable Regions

Modern globalization, in the general sense of a complex set of global economic, communications and cultural changes, is very much driven by the profoundly fossil energy mode the world operates in. Global supply lines are required for securing oil, coal and natural gas from the limited number of highly productive fields in production: the mining, shipping and processing of the raw material and its world-wide distribution has necessitated a vast network of logistics, military management, security arrangements and diplomatic agendas – as well as specialized economic systems. A majority of current and recent regional and local armed conflicts are resource wars, in part or entirety. The specter of violent global strife over the control of regional and global fossil fuel supplies rises in the short and medium term.

The great 19th and 20th century industrialization and modernization drives accompanied the rise of a globally dominant fossil culture with its specific rules, values and powerful images structuring collective and individual desire. Their unique behavioral patterns are generated by the characteristics of supply and demand in a global fossil fuel fired economy. Seen in this light, the global media, information systems and telecommunications networks play a rather ancillary role in the processes of contemporary globalization: the fundamental *technologies of globalization* are the at once centralized and globally active power generation and petrochemical resource systems tethered to geographically limited and hence geo-strategically critical energy resource and mineral deposits. International trade rules and interpretations of national security are based on and very much support this global system.

The result is a single terrestrial system, rapidly growing and fed by tenuous yet distant, even global supply lines. As a consequence, an increasing number of local urban areas are surrounded by formerly productive but now either suburbanized or relatively impoverished, disconnected rural and semi-rural regions. These new 'globalization hinterlands' are the former supply regions of pre-fossil villages and towns, now increasingly defunct, with their population streaming to the rising, brightly lit and comfortably powered, globally networked urban centers.

The deployment of renewable energy technologies has a potential to help bring about a time of *differentiated globalization*, marked by a distinction between largely local supplies of food and elementary goods on the one hand and the global trade in services on the other. *Postglobalism* as engendered by non-fossil production modes is characterized by a rise of regional economies in support of urban centers, based on regional resources such as productive land for food, biomass and wind energy production. New ways of re-knitting central cities with their regional economies and related spatial structures are already being pursued by a number of communities. These are based on age-old principles of rural urban support economies, boosting the primary sectors of agriculture and forestry: cities around the world are beginning to make concrete links between their renewable energy needs and potential regional resources capable of meeting that need. This movement is also beginning to help spawn new indigenous manufacturing and advanced industry sectors in renewable energy production, supply and services.

On the industrial side current initiatives fall into two categories: that of aiming at technology development on one side and technology use or market uptake on the other. Technology push and supply from the 1970's through the 1990's were limited to a number of limited-scale industry efforts and pockets of largely government-sponsored research and development projects. This history is marked by an absence of a significant market demand, or market-oriented policy push, given the overwhelming subsidies granted generally to the fossil energy sector. However, the international and domestic policy and pricing environment of the early 2000s is changing fast, heralding massive urban technological and practice changes and a natural integration of technology development and markets.

3.2 Energy, Cities and Technological Innovation

Cities face the new challenges largely without national guidance and some seek to go beyond individual technology applications, single structures or limited urban areas. They hope to translate international and national agreements onto the local level, despite the institutional constraints of the inherited sectorially divided systems. Increasingly, urban leaders seek to grapple with the issue of technological innovation, absolute and globally equitable emissions targets, the prospect of urban carbon trading and the pursuit of full integration with mainstream urban management systems.

The most hopeful visions describe entire cities as net renewable energy producers. This idea requires a rethinking of urban-regional alliances as well an adoption of increasingly firm industry promotion practices. The Australian city of Melbourne, as an example, is in the process of investing in renewable energy producers with the dual aim to reduce its fossil fuel dependency and to promote the development of more advanced industries that one day will be capable of competing nationally as well as internationally.

As motivating force compensating action on subsidies and selective pricing, even in the absence of a true deregulation of energy markets, spells a boon for technological innovation. In a technologically advanced renewable economy energy supplies no longer exclusively depend on large, centralized supply models but can unfold in a more diverse and differentiated manner, in keeping with the contemporary culture of convergence. Indeed, emerging conditions are characterized by a certain blurring of conventional distinctions between production and consumption. Traditional appliance and facility users can become net energy generators, for instance through solar systems or zero-emission, renewable-source based hydrogen fuel cells in vehicles, capable of powering homes and external machinery.

Systems convergence dynamics point to a merger of information technology, telecommunications and energy systems. While some electric utilities already lease their grids for information transmission purposes, emerging technology goes much further: future energy systems are ubiquitous and pervasive. Operating on the level of individual units, be they consumer appliances, households, neighborhoods or even city-regions, the long-range energy management paradigm is grid-free, self-sufficient and renewable.

Ubiquitous energy management is introduced here as the notion that in a renewable-energy based economy a myriad of small and medium-scale providers of energy services replaces the system of large-scale centralized ones. This system can operate both at the high end described, but also at low levels of technological sophistication. High-end technology contains vectors from information technology and telecommunications mergers to the blending of these new technologies with energy production and consumption modes residing everywhere, from personal apparel and equipment to cars and facilities. At the lower end of the scale, distributed, low-cost and low-maintenance small hydro-power and solar systems are capable of leveraging access to global information network for small remote communities even in least developed nations.

Another technological dimension of the impending energy revolution is the role the Internet plays in the energy sustainability of cities. The 1994 United Nations Conference on Environment and Development held in Rio de Janeiro has firmly associated the term sustainability with a global action agenda, connoting international processes of working towards sustainability aims, especially in an urban context. The tradition of sister city arrangements was a rudimentary beginning of inter-urban networks in particular, while activist non-governmental organizations such as Greenpeace pioneered the working in global networks as means for local action.

A number of international networks operate today in the area of energy and the sustainability cities, and all are in the process of exploring the best manner in which the nature of the Internet and the world wide web can be exploited to productive ends. Some are listed in the reference section of this paper. It is instructive to remember that the Internet itself, by now a vital global infrastructure, is highly dependent on fossil fuel and needs a strategic action plan to base it on renewable and sustainable energy sources, through the introduction of suitably distributed, even ubiquitous power supply systems.

Finally, there is a number of ways in which the technologically sophisticated management of environmental information such as local, community or point-of-emission accounting data is crucial in the making of policy. Integration of currently available information, modes of visualization and analysis, the massive networking of personal computers – these are all technologies and techniques advancing at national or international levels but remain woefully unavailable or inadequate locally.

3.3 Climate-Stable Development

Local policy environments during the late 20th century were marked by a lack of commitment to planned urban energy practice, due to the investment in existing centralized power arrangements, a pervasive process management culture without clear accountabilities and the absence of local levers, incentives and practical means of allocating or influencing emissions. As global agreements get progressively firm under mounting pressures to resolve the greenhouse challenge, and as nations and trading units such as the European Union begin to establish firmer greenhouse, energy and environment frameworks, the pressure mounts to construct workable arrangements to approach urban emissions and energy practice with.

In a world in which cities emerge as global emissions performers, distributed energy managers and ultimately as credit trading entities, a science and engineering practice emerges that interprets cities and city regions as power systems, resource flow units and point-of-emission entities. The most visible of current community planning models are based on relative improvement targets and selective means of accounting for greenhouse gas (GHG) emissions.

A number of place-based emissions allocation techniques and action approaches are currently in use, ranging from the United States Environmental Protection Agency's systems of support available to states in compiling GHG inventories, producing action plans and staging demonstration projects, to the pragmatically approximating system developed by Ralph Torrie and others for the Cities for Climate Protection™ (CCP) program operated by the International Council for Local Environmental Initiatives (ICLEI), which does not account for all emissions yet is in experimental use in four hundred communities world-wide, to the system based on the behavior of large geographical units – one degree of longitude by one degree of latitude – developed by the Association of American Geographers in a program called *Global Change in Local Places*. Other systems are in development and being promoted as well. There is the potential to apply aspects of the Advanced Local Energy Planning approach developed under the International Energy Agency's (IEA) Building and Community Systems program, the physical model of the economy developed by the Australian Commonwealth Science, Industry and Research Organization, and another ambitious model, a total-flow, end-consumption and value-based accounting system currently being advanced by an IEA initiative known as Solar City (see below), which also aims to compare, evaluate and advance a range of methods.

While most GHG emission sources and mitigation efforts are inherently local, their effects and most easy modes of measurement are global. However, the identification of globally diffused GHG levels carries little practical meaning locally. In order for cities to become active and integral participants of any global action program, they need to opt for one or the other GHG accounting or allocation method that reliably links local practice to global aims. A persuasive argument holds that only systems directly applicable to reliably measuring contributions to climate stability are valid, and more specifically approaches that embrace the aforementioned 3.3 tons of carbon dioxide-equivalent target by 2050.

An ideal method has not yet been arrived at, and in pursuing it, cities and their supporters need to apply certain performance criteria. To be informative, universally acceptable and suitable as a basis for coordinated climate stable practice, the model needs to be comprehensive, precise and accurate. To be action supportive it needs to be pragmatic, affordable and easily replicable across a large number of communities or geographic units. Another challenge lies in avoiding a new information poverty barrier: it could develop were any local carbon trading to be pursued based on highly sophisticated and expensive forms of data gathering – poorer nations have less capability to cope with the technical challenges involved in measuring current emissions performance, let alone reliably monitor its progress over time.

The single most widely promoted method, that deployed by ICLEI's CCP program, is to be introduced by local governments as the first of CCP's five steps to better GHG performance: namely the arrival at an energy and emissions inventory, forecast, reduction target, target achievement plan and active implementation program to achieve measurable GHG reductions. At these relatively early days of the CCP program measurable success in reaching significant implementation goals has been elusive, while the other milestones have been reached to highly varying degrees. It is possible that in future rounds of development, in association with other methods and given a better understanding of the key barriers to success a break-through in implementation can be achieved. As it stands now, the very methodological imperfection of the method and its omission of significant emissions makes the CCP system attractive in practical terms but scientifically difficult to use in globally credible accounting contexts, or even for local improvement monitoring over time, in terms of total performance.

Indeed, there are a number of obstacles to transformed urban energy practice, although none of them are insurmountable in principle. The most formidable is the fact that local governance structures are not usually geared to end-state or long-range planning, and hence the adoption of any form of long-term accountings. The effective allocation of incentives and accountabilities is difficult, too. Notwithstanding the efforts described above there is the persistent question of how to convincingly disentangle resource flows in a local government area not only for carbon accounting but also for scenario modeling purposes, and how to identify agency accountabilities for savings and reduction.

The adoption of broad energy and greenhouse action programs opens up new dimensions for government. It begins to supply new development perspectives and drive regional realignments, the formation of state initiatives and industrial alliances. It can mean a new sense of empowerment and the opportunity for economic strengthening, as well as competitive advantages of a more promotional nature.

Indeed, whatever the difficulties, the benefits are promising. Governments have begun to discover that energy responses can be scaled well to local development conditions and growth options. In newly industrializing countries with interest in maintaining considerable levels of autonomy such as China and India there is a good fit with local traditions and climatic and economic conditions. Local industries get stimulated while environmental agendas are being reinforced. Comprehensive renewable energy practice on the local level directly reinforces broad quality of life aims, as well as healthy water, food and bio-diversity practice.

Besides meeting the aim of responsibility devolution and other local subsidiarity objectives in the search for meeting global greenhouse objectives, another government reform agenda item is being satisfied as well: the call for accountability in public service. Outcome-geared reform and strengthening of local governance in delivering on indexed and measurable performance is a natural byproduct of integrated sustainable energy practice. Local consensus, too, is expected to be forthcoming more easily as agendas become more clearly understood and visible in tangible improvements.

4 SOLAR CITY, A BLUEPRINT FOR INTEGRATED URBAN PLANNING

The challenge for urban governance is unprecedented. In the absence of useful established patterns of practice, a search is under way for new models capable of unifying local government's sectorial concerns, technology imperatives, energy markets as well as global agendas. As a good example of a current attempt to support cities' quest for sustainable energy futures, Solar City represents a new generation of International Energy Agency (IEA) research and development work whereby solar energy techniques are being linked with other renewable energy technologies and means of greenhouse gas emissions reduction and absorption not only in a coherent spatial and social context, but also applied within a finite and community-wide time-line. This program is to work with a number of participating cities in nurturing a Solar City network, supporting local action by augmenting existing efforts.

Since its beginning after the early 1970's oil crisis the IEA has expanded its agenda from international energy security concerns to climate change, and recently moved from serving primarily OECD countries to reaching out to the rest of the world. In its Committee for Energy Research and Technology (CERT) research and development agenda there is a growing emphasis placed on renewable energy, and on expanding from a technology development mission to a community and market push. This program is being developed in co-operation with a number of other renewable energy international implementing agreements within the CERT agenda and is designed to operate in conjunction with related existing industry and community networks.

The mission of this program is to advance an integrated set of renewable energy technologies in cities, with the aim of producing absolute emissions reductions of the extent and time frame required for stabilization of the atmosphere at the sustainable GHG concentration of 3.3 tons of GHG per person per year. The rationale of its aim at absolute and equal targets is based on a belief in equity-based emissions control which implies a reduction for industrialized countries, and scope for a growth supporting increase in many less developed nations.

The IEA conducts well over one hundred research and development projects into a very wide range of relevant energy technologies and management techniques operating under such international 'Implementing Agreements' as Bioenergy, Geothermal, Hydrogen, Hydropower, Photovoltaic Power Supply Systems, Solar Heating and Cooling, Wind Energy, Buildings and Community Systems, Demand Side Management, Energy Technology Systems Analysis, District Heating and Cooling, Energy Storage, Heat Pumping, Heat Transfer and Exchangers, Hybrid and Electric Vehicles as well as a range of other related activities, programs and information centers. Within this extraordinary portfolio many activities offer direct solutions and systems that can be promoted and applied in cities, others have more indirect significance.

4.1 Case: Insulating Urban Development from Fossil Fuel Dependency in China

Solar City operates as a globally coordinated, nationally independent and funded program involving largely OECD countries but also aiming at engaging non-member states such as China. The program banks on the fact that fresh approaches to urban planning are emerging world-wide, based on the realization that the days of relatively problem-free power generation methods as a platform for development are numbered. These fresh approaches challenge national and international institutions to revisit the established local, regional and national patterns of community planning. They also place new demands on multilateral and bilateral aid programs. For example, in China the performance criteria of designing such approaches are clear: it will have to be comprehensive, affordable and replicable. The China Integrated Town Infrastructure and Environment Strategy (CITIES) is currently being advanced by Solar City to help integrate solar and other renewable energy technologies in Chinese towns and cities, along with energy conservation and efficiency measures, with the dual aim of maintaining sustainable emission levels and lower reliance on fossil fuel.

To face this historical challenge, planners in a number of cities, towns and urban regions have begun to link renewable energy objectives to planning and community development aims. As a result, opportunities in infrastructure, built fabric, land use, urban form and regional development are being discovered while efforts to mitigate greenhouse-gas induced global warming have redoubled the resolve behind these initiatives. In this new world of development, China is well placed to become a global leader in innovative, sustainable forms of urban development, domestically and throughout the developing world.

Many aspects of rural decline, unsustainable land use practices, desertification, reduced bio-diversity, water pollution, deforestation, urban and rural poverty and uncontrolled urbanization are either brought about or significantly contributed to by the conditions engendered by centralized, fossil-fuel or nuclear powered electricity production, its enormous costs and serious limitations. Chinese towns and cities that are being developed and built today under the fossil-fuel paradigm will be fundamentally outdated and become unmanageable within this generation, threatening a massive crisis in just a few decades from now, when the conditions underlying contemporary urbanization patterns will have dramatically altered.

Many of China's provincial and smaller town communities hope to take advantage of the opportunities brought about by this new set of circumstances. These not only spark local business and industrial innovational impulses but also deliver renewed higher productivity value to the regional rural economies surrounding the settlement centers. In China, this means that an integrated approach to whole-of-town planning can be embraced, one that rallies institutional, land-use, transport, facility and construction planning around the increasingly central issue of renewable energy management. The rural anchoring potential of this approach is particularly important in a country in which the overwhelming majority of inhabitants are still counted as rural.

4.2   Future Promoter of Urban Energy and Greenhouse Planning

China ranks among the world's three leading energy consumers and producers, and is rapidly rising among the world's top net petroleum importers. Its largely coal-fired economy consumes ten percent of the world's primary energy supply, with an emissions rate to match. The challenge is both national and regional: the great diversity of China's provincial and regional realities makes it necessary to work towards a nationally integrated energy policy with the help of locally meaningful and integrated town and rural development strategies.

Therefore, to focus on China's town development within the framework of national resource sustainability and self-sufficiency strategies makes good sense, to boost local and regional environmental benefits. In 1997, 32% of China's population officially lived in urban areas; this figure is to rise to 45% by 2010. Actual figures become considerably higher when applying criteria that include the de-facto urban but statistically rural component of the population. Energy is significantly consumed in cities and towns, while the very form of energy supply adds a destructive side to growth, rendering it unsustainable both in the short and long term. Traditional electrification schemes based on outdated fossil fuel plants not only wreak damage in air pollution locally, regionally and nationally but also serve to detach rural areas from towns and other forms of urban agglomeration.

This fossil-fuel induced segregation of town and rural context has its mirror in the separation of the sub-disciplines that make up modern town planning. Both are among the root causes of unsustainable development and hamper the relevance and effectiveness of many international aid programs. Sewerage, water supply, power, transport, land use planning, building regulations and agricultural planning are generally managed in isolation from one another, frequently yielding slow and even contradictory outcomes. By contrast, traditional forms of town development depend on the very linkage between these elements and an extremely high efficiency rate. It is telling that modern sustainable approaches to urban management begin to re-establish these in some way. In this new approach waste disposal through recycling and other regional resource flow dynamics begin to inform and enrich the local economy. Local energy strategies and urban management in general are not merely related but actually identical, due to energy's fundamental role in the viability of local economies and ecologies alike.

4.3   An Integrated Urban Energy, Environment and Development Strategy

By not duplicating but enhancing existing and related national and local urban development programs and networks Solar City builds on a growing momentum world-wide, adding valuable intelligence, focus and an unequivocal implementation mission nationally and locally. In seeking coordination it interfaces well with international development aid and loan programs. The program is capable of working as a primary partner in multi-lateral settings of development support, but it also lends itself well to the construction of inter-city and town-regional support networks, with or without affiliation with international NGO or local government networks. World-wide, prospective cities and national partners are positioning themselves in Korea, Japan, Africa, Australia, North America and throughout Europe, at all levels of development and in different size categories and development stages. The participant areas range from entire city regions to individual new town settlement programs.

Solar City aims at working collaboratively in all locally relevant institutional contexts. It focuses on the energy supply and technology side, but embedded in a total town planning and design strategy that also includes institutional arrangements. It promotes a community-wide, rural and urban energy and emissions accounting system as well as performance targets that are linked to urban development and reform initiatives. Finally, it advocates land use strategies that are based on a consideration of urban-rural linkages and value land use and transport investment choices according to their potential contribution to long-range energy and resource self-sufficiency.

The Solar City concept is structured into three thematic areas of inquiry and two general investigation programs to operate across these. They are to be advanced simultaneously, within nationally defined agendas. These are: sustainable-energy focused urban planning strategies; targets, baseline studies and scenario development; and urban energy technology, industry and business development. The two general investigation program streams focus on best practice cases and on learning in action. There three areas of general inquiry are briefly described here

Solar City strategies. This activity is to identify local planning and development approaches that are conducive to the introduction of solar and other renewable energy technologies, within a broadly energy-conscious community development approach. To be addressed are issues of strategy, planning tools, organizational arrangements, legislation and standards, incentive structures, public information and exemplary municipal practice.

By introducing improved ways of adopting solar and other renewable energy technologies the program is to contribute to climate-stable practice in the building and property development industry, land-use planning and infrastructure development. Solar City is also to strengthen local governments' efforts to build enlightened community performance and household preferences.

There are five ways in which better practice can be promoted by cities and towns:

1. direct legislation and standards;
2. the provision of incentives and disincentives;
3. corporate capital asset practice, power purchasing and pricing;
4. institutional reform and improved strategic and general planning practices; and
5. community action development, industry alliances, information and education.

This activity is to investigate each of these in detail and develop advanced means of building improved urban practice approaches, in full partnership with the participating cities.

Targets, baseline studies and development scenarios. The objective of this activity is to introduce, evaluate and enhance suitable approaches that help understand the role of solar and other renewable energy technologies in the broader urban energy context. Means deployed will include absolute climate-stable carbon dioxide-equivalent emissions targets aiming at 2050, introduced in ways that support their quantification and translation into implementation paths along shorter-term milestones. To do so it is important to build a baseline model of the performance of each Solar City in terms of a range of key indicators such as greenhouse gas emissions, renewable and non-renewable energy use, household consumption patterns and transport mode distribution. A general town catchment is to be defined for emissions accounting; the existing energy and emissions situation is to be recorded and assessed; basic global indicators (CO_2-e) are to be established; and urban indicators such as annual emissions output per capita to be introduced to study the behavior of nominal resident, working and visiting populations.

Planning methods based on energy and emissions accounting methods may deploy *backcasting* approaches. This involves the development of alternative urban development growth trajectories, maintaining sustainable CO_2-e emissions and fossil fuel use rate goals for 2050, then 'backcasting' growth milestones for emissions in order to determine workable reduction rates for each milestone period. Scenarios of anticipated emissions reduction rates are important for the determination of alternative sustainable development paths. Suitable scenario approaches are to be developed, for example by using a physical model of the regional economy and vary these to agree with milestones.

Urban renewable energy technology, systems and industry development. The objective of this activity is to work with cities in advancing renewable energy technologies and systems, and to help promote the renewable energy industry, in a way that can serve as a model for the rest of the national urban system. The emphasis is therefore on research and development work into market-led approaches of technology system development and deployment, through pricing, investment, electricity purchasing policies, information, model action and other means.

Optional paths are to be developed, evaluated and implemented, suitable for the informed and broad introduction of solar and other appropriate renewable energy technologies as part of a comprehensive portfolio, for the use of city governments, municipal utilities, businesses, industries and households. Special emphasis will be on micro-generation and distributed low-energy production in buildings, facilities and urban systems. Current, emerging and potentially competing solar and other renewable energy technologies, systems and related urban services are to be assessed for their urban modification and city-wide, systematic introduction in ways that are meaningful to cities' development agendas – physical planning, sustainability objectives, organization, services – and their pursuit of targeted emissions reductions.

Results will include a comprehensive and dynamic portfolio of technology, systems and industry development options, suitable for selective and targeted implementation in general and specific action plans. The focus will be on what city governments in collaboration with industry and constituent urban communities can do to advance the direct use of renewable fuels for industry and transport; the generation of electricity in quantity, such as solar power stations, wind, biomass, geothermal and hydro; and, primarily, on the development and deployment of technology development strategies in industrial and residential consumer-oriented application, such as stand-alone power generators, heat pumps, photovoltaics, solar hot water and solar cooling.

General investigation program in best practice. The objective of this activity is make accessible and apply useful lessons from current and recent related initiatives domestically and world-wide. This objective will be achieved by studying successful practice in integrated urban energy planning, management and projects. The activities include an identification of scope and criteria for evaluation; information gathering and documentation; study and evaluation;

analysis and description; case study development; and communication and dissemination. The scope will encompass technologies, management practices as well as growth strategies. As a point of departure, at least three categories of case studies will be differentiated: comparable cities, urban precincts and settlement projects but also development policies and programs.

General investigation program in learning from action. The objective of this activity is to monitor, analyze and feed back program experience derived from the participating cities. This will help develop a shared understanding of the barriers to, dynamics and impacts of community, institutional, industrial and technological change, with a view towards the planned and targeted, GHG-reductions geared phasing in of solar and other renewable energy sources on an urban and regional scale. This activity will not only be useful to the participants, but of value in the application of lessons and methods across the national urban system.

5 CONCLUSION: RENEWABLE ENERGY RESOURCES FOR SUSTAINABLE URBAN DEVELOPMENT

Because of the very impacts of fossil fuel we live in a rapidly urbanizing world, and cities and city regions are central and fertile settings for effective energy policy, programs and projects. Cities are not only powerful markets for the introduction of renewable energy technologies but also the national and regional seats of political power, and the core settings of cultural discourse and technological innovation. They form the very frameworks for development: local government, planning structures and the powerful civic organizations that are so important in many cultural contexts.

Cities face extraordinary opportunities in the gradual but inexorable change from the risky and costly systems of fossil power reticulation to a world of sustainable, affordable, diverse and ultimately ubiquitous energy management. The hope is one of growing choice in scales of operation and levels of technological sophistication. Fundamental changes in urban power regimes that are in keeping with sustainable development practices promise to revitalize regional and rural development, and boost urban business and technological innovation. And by pursuing energy reform strategies in keeping with globally sustainable greenhouse gas emission levels, local urban leaders can also act globally by helping achieve greater equity and justice in international development.

GLOSSARY

City: Defined for the purpose of this paper as an urbanized area managed and represented by one or several local governments, culturally and communally defined as city, with specific administrative and political boundaries.

City region: The general urbanized, ecological and economic catchment area surrounding and comprising one or several urban nuclei, all or any of which may be defined as city.

Climate-stable practice: A city's climate-stable practice is a practical commitment to lower greenhouse gas emissions by the year 2050 to an amount that is proportionally in keeping with the globally sustainable level deemed to be 3.3 tons per person per year in 1990 levels.

CO_2-e: Carbon-dioxide-equivalent. *CO_2-e:* carbon-dioxide-equivalent. Expresses the presence of all effective greenhouse gases as an amount that would be required in CO_2 to achieve the same effect.

Fossil cities: The majority of modern cities not only depends on the safe supply of fossil fuels, their very existence, form and growth dynamics are explained by the logic of the fossil fuel economy. Fossil cities will be inoperable in a post-fossil era, unless they manage the transition to a renewable energy economy and supply system. The term 'solar city' is used here to polemically emphasize the characteristics of cities in terms of energy base and emissions behavior in contradistinction to those of fossil cities.

Greenhouse gas (GHG): Human activity effected gases that trigger global warming are carbon dioxide (CO_2), methane (CH_4), nitrous oxide (N_2O), chlorofluorocarbons, especially CFC-11, HCFC-22, and CF_4. Their emission levels are in this document expressed as CO_2-e.

Postglobalism, Postglobalization: These expressions are coined here to characterize globalization as a fossil-fuel induced phenomenon, and to denote the post-fossil era as 'postglobalized'. They describe an era in which globalization will be differentiated into local flows of basic food, regional resources and certain manufactured goods, and a time of a reasserted primary industry sector, especially agriculture. Globally it anticipates a time when primarily services and information flows will be global, but basic levels of economic dependency and sufficiency will be defined locally and regionally. As policy of local self-determination and quality of life improvement, postglobalism is already rising in various communities. This is exemplified by the Slow City initiatives of Orvieto, Italy.

Renewable energy: All energy forms that can be replenished without relying on finite sources such as fossil deposits.

Solar city: Denotes for the purpose of this article an urban community embarked on a comprehensive path towards substituting renewable and sustainable forms of energy for fossil and nuclear sources, while aiming at absolute greenhouse emissions targets for its populations in keeping with global sustainability goals. See also 'Climate-stable practice' and 'Fossil city'.

Solar energy: 'Solar' in its wider definition connotes all aspects of energy sources which can be traced to the action of the sun. These include solar thermal, photovoltaic, biomass, bio energy, wind and wave energy (and, strictly speaking, also fossil fuels). In this sense 'solar' describes renewable forms of energy that neither create greenhouse gas emissions nor non-degradable and/or toxic waste. A solar energy economy would exhibit dramatically altered urban development dynamics globally, regionally and locally.

Sustainable energy: Energy sources that do not pollute the atmosphere, water and soils in irreparable or harmful ways.

REFERENCES

Byrne J., Y. D. Wang, H. Lee and J. D. Kim, 1998
  *AN EQUITY- AND SUSTAINABILITY-BASED POLICY RESPONSE
    TO GLOBAL CLIMATE CHANGE.* Energy Policy 26 (4), 335–343.
    [Seminal article on end-user representation of globally sustainable
    greenhouse gas levels.]

Calthorpe, P., 1993
  THE NEXT AMERICAN METROPOLIS: ECOLOGY, COMMUNITY,
    AND THE AMERICAN DREAM. New York: Princeton
    Architectural Press. (Illustrative key work detailing a strategy of
    rebuilding US cities using transit-oriented development principles.]

Droege, P. , 1997
    INTELLIGENT ENVIRONMENTS. Amsterdam; New York: Elsevier.
    (See especially p. 245 ff. Tricia Kaye, Stewart Noble and Wayne
    Slater. *Environmental Information for Intelligent Decisions.*)
    [Scientific compendium on the interaction of information
    technology, telecommunications and human-made spatial
    environments.]

Girardet, H., 1999
    CREATING SUSTAINABLE CITIES. Schumacher Briefings 2. Devon:
    Green Books. [Fundamental agenda for action.]

Howard, E., 1902
    GARDEN CITIES OF TOMORROW. London: Swan Sonnenschein.
    [Classical utopian blueprint to create self-sufficient and sustainable
    communities.]

Kates, R. W., Michael W. Mayfield, Ralph D. Torrie and Brian Witcher, 1998
    *METHODS FOR ESTIMATING GREENHOUSE GASES FROM LOCAL
    PLACES.* Local Environment, Vol. 3, No. 3. Carfax. [A comparison
    of the greenhouse gas emissions accounting systems used by
    International Council on Local Environmental Initiatives – Cities
    for Climate Protection (ICLEI-CCP) and the American
    Geographers – Global Change in Local Places (AAG-GCLP). Ralph
    Torrie is the author of ICLEI-CCP's emissions accounting
    approach.]

Lenzen, M., 1997
    *ENERGY AND GREENHOUSE GAS COSTS OF LIVING FOR
    AUSTRALIA DURING 1993/4.* Energy Vol. 23. No. 6. pp. 487–586.
    Pergamon. [Important paper on the application of end-user
    representation of greenhouse as allocations.]

Newman, P.W.G. and Jeffrey R. Kenworthy Newman, 1987
    GASOLINE CONSUMPTION AND CITIES: A COMPARISON OF
    U.S. CITIES WITH A GLOBAL SURVEY AND SOME
    IMPLICATIONS. Murdoch, W.A. : Murdoch University. [The
    classical study linking car dependent, low-density urban form to
    fuel inefficiency in transport.]

Scheer, H., 1999
SOLARE WELTWIRTSCHAFT. Kunstmann. [Seminal work on the issues surrounding the transformation of the global economy from a fossil to a solar footing.]

Wahlefeld, G., 1999
ZEIT UND MÄDER KOHLE. Feuer und Flamme. Sonne, Mond und Sterne – Kultur und Natur der Energie. Bottrop; Essen: Pomp. ISBN 3–89355–194–8. [Comprehensive catalogue to an exhibition on the nature and impact of coal.]

RELATED WORLD WIDE WEB SITES

citiesnet.uwe.ac.uk – European Sustainable Cities
climatenetwork.org – Climate Action Network (CAN)
climnet.org – Climate Network Europe (CNE)
energie-cites.org – Energie-Cités
eurocities.org – Eurocities, Car Free Cities (CFC)
greencity.dk – European GREEN Cities
iclei.org/co2 – Cities for Climate Protection™/International Council for Local Environmental Initiatives
inforse.org – International Network for Sustainable Energy
sustainable.org – Sustainable Communities Network (SCN)
sustainable-cities.org – European Sustainable Cities Project, Campaign Interactive
resetters.org – RE-Start/RESET European urban energy alliance
solarcity.org – International Energy Agency's Solar City Program
sustainability.org.uk – Action Towards Local Sustainability
unchs.org/scp – Sustainable Cities Program of the United Nations Centre for Human Settlements
wrenuk.co.uk – World Renewable Energy Network

Energy-Efficient Elevators for Tall Buildings

Harri Hakala, Marja-Liisa Siikonen, Tapio Tyni and Jari Ylinen

ABSTRACT

According to elevator traffic studies, the traffic patterns in an office building, such as number of starts, round trips, and number of transported passengers, are repeated quite the same from day to day. In this study, the energy consumption of elevators in tall office buildings is studied by measurements and traffic simulations. A method using elevator load and travel distributions in calculating energy consumption is introduced. The simulation results are verified by measurements in a single-tenant office building. The proportions of savings gained by different drive systems and machinery as well as control systems are compared. A case study for yearly energy consumption in a tall building is presented.

INTRODUCTION

According to the GIBSE guide, elevators consume about 4–7% of the total energy load in an office building [GIBSE]. In Finland, measurements were made in low-rise office buildings with 4–6 floors. In these buildings, the electricity consumption was somewhat lower, about 1–3% of the total electricity load of the building. Most of the energy in prestige office buildings is consumed by the heating and hot water systems, lighting and office equipment, such as computers [Field]. The energy needed for space heating, ventilation, or air conditioning also depends on the outside temperature. Tall buildings vary in height and shape, and in their usage, as well as in elevator layout. In fact, there is no such thing as a typical high-rise building. Consequently, generalized statements about relative elevator energy consumption in tall buildings are difficult to formulate.

Owing to the uniqueness of tall buildings, this article concentrates on two specific office buildings, one in Finland and the other in Australia. The energy consumption of modern traction elevators and control systems is compared with technology that is about ten years old.

ELEVATOR TRAFFIC

Most offices are occupied for about ten hours a day, five days per week. A measurement of the passenger arrival rate in a single-tenant building with 18 floors was made in Finland (Figure 1). The profile shows the numbers of passengers arriving at entrance floors (incoming traffic), travelling to entrance floors and exiting the building (outgoing traffic), and inter-floor traffic where passengers travel between the upper floors. The traffic profile was measured by the elevator control system using load information and photocell signals. In this building, passenger traffic is the most intense during the lunch hour, about 10% of the population in five minutes. In the morning the traffic peak is not as sharp since the working hours are flexible. During normal times, passenger traffic intensity varies from four to 6% of the population in five minutes. The proportion of inter-floor traffic is greater in a single-tenant office building compared to a multi-tenant office building.

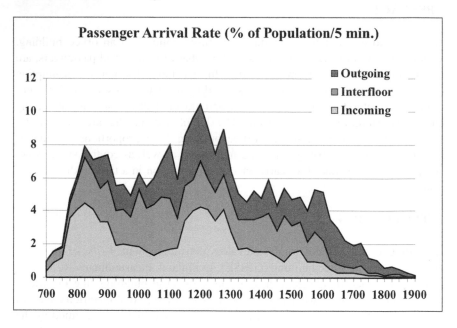

Figure 1   Passenger arrival profile measured in an office building in Finland.

In this building, passengers are served by a group of six elevators with a car size of 16 persons. If the passenger traffic intensity exceeds a certain amount, e.g. 4–6% of the population in five minutes, the number of elevator starts and trips up and down (round trips) tends to reach the limit. The number of starts and round trips of one elevator in half-hour steps can be seen in Figure 2. This measurement was made before the modernization in 1989. In the figure, the number of starts varies from 120 to 160 per hour, and the number of round trips reaches the limit of about 40 per hour. The maximum number of elevator starts within an hour is limited by the elevator performance. A typical measured value for an average cycle time from one elevator start to another is about 20–25 seconds. This time includes the elevator run from one floor to another, door opening and closing times, and passenger transfer times. With a typical cycle time, the maximum number of elevator starts varies from 145 to 180 within an hour.

(a)

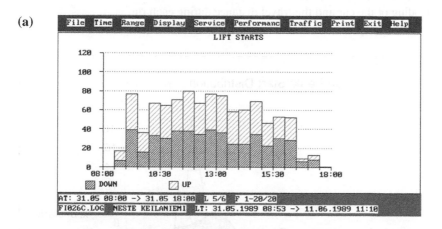

(b)

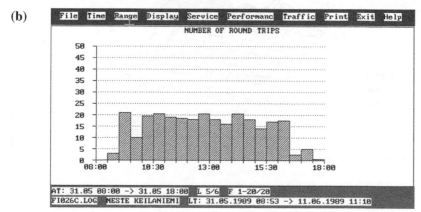

Figure 2 Number of starts (a), and number of round trips (b) of an elevator in a single-tenant office building.

The simulated number of starts with different loads and running distances for the six cars is shown in Figure 3. Figure 3a shows a theoretical up-peak situation, and Figure 3b a typical mixed traffic pattern from 9:00 to 11:00 o'clock, and from 13:00 to 16:00. During mixed traffic, about 25% of the starts are with empty load, and about 25% are for one-floor runs. Most of the measured starts during a day occur with a 10–20% load factor and with a distance of 20–30% of the total. Exceptions are made if the elevator group has an express zone.

ENERGY CALCULATION PRINCIPLE

In the procedure for calculating elevator energy consumption, elevator traffic is simulated for different times of day. The energy consumption is calculated for each elevator run. The direction of travel is considered for each run, as well as the loading of the car and the travel distance during each run.

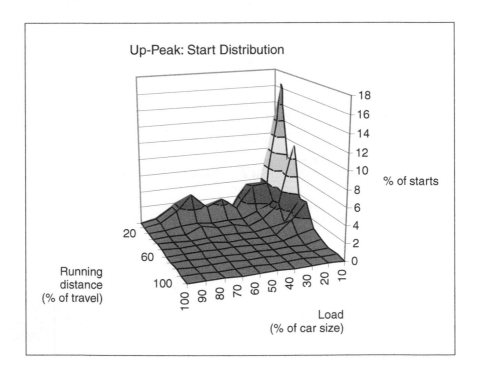

(a)

Figure 3a   Distribution of starts according to passenger loads and running distances during up-peak (3a), and daily mixed traffic (3b).

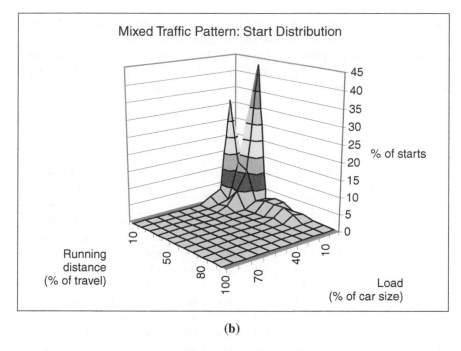

(b)

Figure 3b   Distribution of starts according to passenger loads and running distances during up-peak (3a), and daily mixed traffic (3b).

There are many ways to estimate the energy consumption of an elevator or an elevator group. If only an idea of the magnitude of the energy bill is needed, the following principles can be used:

The simplest way is to use only the motor power and number of starts with an experimental coefficient for different drive types.

Another method is to estimate the lifted mass (E = kmgh), with an experimental coefficient. If the population and height of a building are known, an energy value can be calculated.

When comparing the savings gained by different technologies or traffic control principles, time-step simulation is the only accurate way. This method is used in this study. The simulation method consists of two parts: One-run energy is calculated for each load and travel distance and it is tabulated into a matrix corresponding to number of floors and number of passengers in the car (Tables 1 and 2). For each combination, the energy demand is calculated by an electro-mechanical simulation program (Figure 4). The energy demands are calculated for both directions, but the actual location inside the building has not been taken into account in this study. This causes a small inaccuracy if the balancing is not perfect. For practical purposes this inaccuracy is negligible.

Table 1 Energy consumption in up direction (kWs).

Kg/ Persons	0/0	165 /2	329 /4	494 /6	659 /8	823 /10	988 /12	1152 /14	1317 /16	1482 /18	Distance
Floors											(m)
3	16	30	45	62	80	99	120	143	167	193	11m
6	-18	12	44	78	114	153	193	236	280	327	21m
9	-57	-10	39	91	145	202	261	324	389	457	32m
12	-134	-58	22	106	193	284	380	479	582	690	43m
15	-193	-101	-6	94	198	307	421	539	662	789	54m
18	-252	-145	-33	83	204	330	462	599	741	889	64m
21	-310	-188	-61	71	209	353	503	659	821	988	75m
24	-369	-232	-89	59	214	376	545	719	900	1088	86m
27	-428	-275	-117	48	219	399	586	779	980	1187	96m
30	-486	-319	-144	36	224	422	627	840	1059	1287	107m
33	-545	-362	-172	25	230	445	668	900	1139	1387	118m
36	-604	-406	-200	13	235	468	710	960	1219	1486	128m
39	-662	-449	-228	1	240	491	751	1020	1298	1586	139m
42	-721	-493	-256	-10	245	514	790	1080	1378	1685	150m

Table 2 Energy consumption in down direction (kWs).

Kg/ Persons Floors	0/0	165 /2	329 /4	494 /6	659 /8	823 /10	988 /12	1152 /14	1317 /16	1482 /18	Distance (m)
3	195	171	148	127	108	90	73	58	44	32	11m
6	334	290	248	208	170	134	101	69	39	11	21m
9	468	404	343	284	228	175	124	75	29	-15	32m
12	711	610	512	418	327	240	157	78	2	-70	43m
15	821	699	582	468	359	254	154	58	-34	-122	54m
18	931	789	652	519	391	268	151	38	-70	-173	64m
21	1041	879	721	570	423	282	147	18	-105	-224	75m
24	1151	968	791	620	455	296	144	-1	-141	-276	86m
27	1261	1058	861	671	488	310	140	-21	-177	-327	96m
30	1371	1147	931	722	520	324	137	-41	-213	-378	107m
33	1481	1237	1001	773	552	338	134	-61	-249	-430	118m
36	1591	1327	1071	823	584	352	130	-81	-285	-481	128m
39	1701	1416	1141	874	616	366	127	-101	-320	-532	139m
42	1811	1506	1211	925	648	380	124	-121	-356	-583	150m

Electromechanical simulation is based on an elevator model, which takes into account both the mechanical and electrical parameters of the elevator system. The formulas are quasi-stationary, and no differential equations are used. The model uses the following mechanical parameters:

1. Speed pattern with predetermined top speed, acceleration and jerk.
2. Linear masses of car, counterweight, cables and ropes.
3. Inertia of rotating parts like the traction sheave and other pulleys and the motor.
4. Efficiency of roping, depending on guide-shoe type, number of bends in roping etc.

The electrical part of the model includes the following components:

1. Copper and iron losses of the motor. Temperature rise of half of the maximum allowed has been assumed.
2. Copper and silicon losses in the drive.
3. Power of fans and brakes.

In one simulation step, the required torque is first calculated based on speed, load and acceleration. The corresponding motor current and voltage are then defined and finally the resulting line current. The line current comprises reactive power (small in the case of frequency converter) and true power, which is then used for integrating the energy.

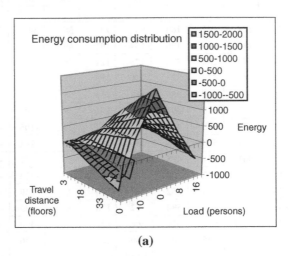

(a)

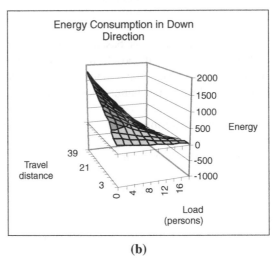

(b)

Figure 4   Energy consumption according to passenger loads and travel distances in up direction (a), and in down direction (b).

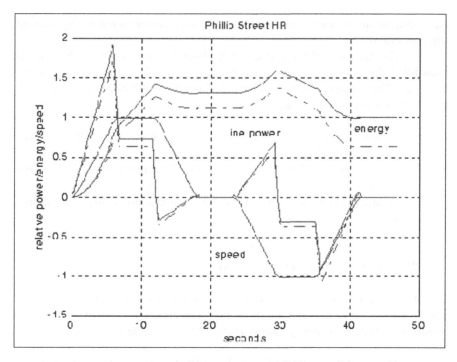

Figure 5   Power and energy consumption, and the speed of an elevator during one run.

An example of one simulation of an up/down cycle is shown in Figure 5. The elevator design specification is 7 m/s and 1 564 kg. The solid line represents conventional technology with a motor efficiency of 80%, and the dotted line is for MX-technology with a motor nominal efficiency of 90%. The picture clearly shows the source of energy savings. Although the peak line power does not decrease more than 10%, the energy of a total run decreases about 35%. The reason is the energy savings accumulate over time in both directions.

EVALUATION OF THE ACCURACY OF SIMULATION

The accuracy of the simulation is compared with measured energy values in the same office building mentioned earlier. The building is an 18-floor office complex with DC-motors with a static converter. Energy consumption was measured for an elevator round trip from the bottom floor to the highest floor of the building with an empty car. The total energy consumption during the round trip was of 124 Wh. Ten round trips were simulated with the same building.

According to simulation results, the average energy consumption during a round trip was 118 Wh.

The accuracy of the simulation is about 5%. It has already been mentioned that the simulation does not take into account the actual location in the hoistway. This causes certain error, which depends on the balancing of the elevator. Other sources of error are:

– Temperature of the motor affects the energy losses, as well the brakes and other wound parts.
– Fans are controlled by thermostats, thus the actual running time is not known.
– Actual friction in the shaft is difficult to estimate, especially when sliding guide shoes are used.

TMS9900 GATM CONTROL SYSTEM

Elevator group performance is traditionally measured in terms of interval, passenger waiting, ride and journey times and transportation capacity. Rarely has attention been paid to energy consumption when allocating landing calls to elevators.

The Optimum Routing Principle (ORP) in TMS9900 GATM (Genetic Algorithm) simultaneously optimizes several targets including energy consumption. In the multi-objective optimization the best routes for elevators to allocate landing calls are selected using a heuristic search method called Genetic Algorithm [Goldberg], [Tyni]. GA imitates 'Mother Nature' in its operation within a computer: route combinations with, e.g. shortest waiting times, are found by developing route alternatives from generation to generation. New generations are created from the selected routes with good properties using inheritance, mutation and crossover. The final combination of routes is a result of processing and evolving thousands of route alternatives. This evolutionary search method is the most powerful tool that has been seen in elevator call allocation problem so far.

Figure 6 shows the principle of the TMS9900 GATM system when optimizing both the passenger waiting times and energy consumption. The group control system uses feedback about the energy consumption regarding the elevator group in its decision making. The target for the service level, e.g. maintain the average waiting times to 20 seconds, is set externally. During the call allocation the GA search method considers both targets when processing each route alternative. The energy consumption in elevators is asymmetric – less energy is consumed if elevators ascend with a smaller load and descend with a greater load. The route alternatives that best satisfy the target waiting time with the least energy consumption are selected and will be carried forwards to the next generation. As a final result, the route combination selected is the one that incurs the least energy consumption, and where the average passenger waiting times stay within the defined limit. During light and normal traffic ORP is able to route the elevator cars via more energy-efficient routes yet maintain good service level.

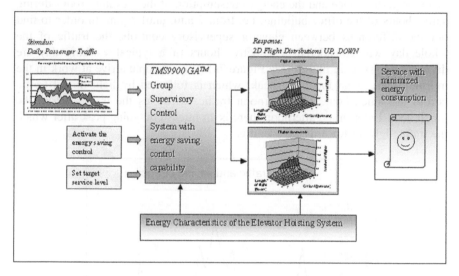

Figure 6 Elevator system with the TMS9900 GATM control system impacted with daily passenger traffic.

CASE 1: PHILIP STREET

A test of energy consumption was made for a high-rise office tower in Sydney, Australia. Totally there are 16 elevators in three groups. The elevator groups serve the floors as follows: the low-rise elevator group serves floors from 9 to 19, the mid-rise elevator group floors 20 to 29, and the high-rise elevator group 30 to 39. Each of the elevator groups has the entrance floor at street level.

The advanced lift traffic simulator (ALTS, Siikonen) was used to estimate the performance and the energy consumption of the elevator groups during active hours of the office building, i.e. from 7 a.m. until 7 p.m. In order to find out the differences between elevator supervisory controls, the traffic of the whole day was considered. The active hours of a typical working day are divided into six main phases (see Figure 7a). When people arrive at work in the morning there is an incoming peak, which is followed by light mixed traffic before the lunch hour. The afternoon traffic resembles the traffic before the lunch hour. Later in the afternoon there is an intense outgoing peak after which the traffic lightens towards the evening.

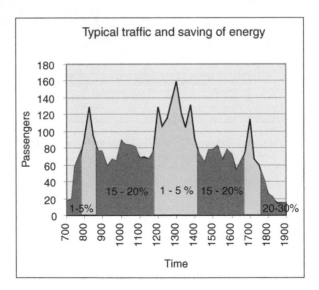

Figure 7a   Energy savings gained with group control systems using ORP and ESP call allocation principles during mixed traffic (7a), and during active hours of a day (7b). The results apply to all the tested elevator groups.

Two different call allocation principles, TMS9000 AI with ESP (Enhanced Spacing Principle) and TMS9900 GA with ORP are compared. The ESP principle is an effective call allocation principle that was used during the last decade in TMS9000 control system. ESP optimizes passenger waiting times in all traffic situations and keeps the elevators evenly spaced in the building.

With ORP principle using the new energy saving control feature, about 15-20% of the consumed energy is saved before and after the lunch hour when the traffic intensity is not very high (7a). According to Figure 7b energy is saved especially during light traffic. In this simulation, waiting times were fixed to stay below 20 seconds with energy saving feature. According to simulation result, for the whole day traffic, about 15% energy saving with ORP is obtained compared to a control where energy consumption is disregarded.

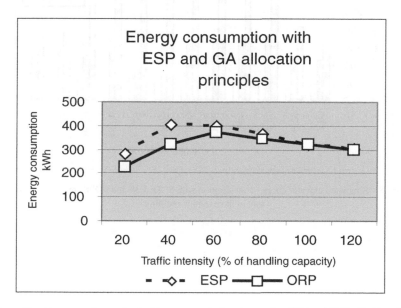

Figure 7b   Energy savings gained with group control systems using ORP and ESP call allocation principles during mixed traffic (7a), and during active hours of a day (7b). The results apply to all the tested elevator groups.

Three different hoisting technologies, the SCD, V3F and MX technologies, are compared by simulating the daily traffic. SCD uses DC motors with static converter, and was used in traction elevators about ten years ago. V3F is a Variable Voltage Variable Frequency converter/inverter drive system that was used in AC elevators during the last decade. MX machinery is also known as EcoDiscTM and is used with the modern TMS9900 GATM control system.

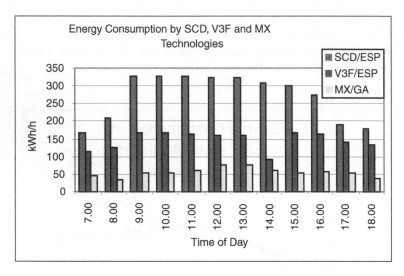

Figure 8   Simulated elevator energy consumption with SCD, V3F and MX technologies during the day in the example building.

According to Figure 8, V3F saves 30–40% energy compared to the SCD technology. MX machinery with the TMS9900 GATM control system consumes 65–75% less energy compared to the SCD technology in this building. If MX technology is compared with V3F drive system, MX saves energy 40–60% in this example case. In the calculations, feedback of the current to the network was taken into account. In this building, elevators with the SCD technology would consume about 700 MWh within a year, and 525 MWh with V3F. With the MX technology the energy consumption is only at minimum 200 MWh within a year.

CONCLUSIONS

Elevators consume 1–7% of the total energy load in a building. In this article, the energy consumption of different elevator technologies was studied. A simulation method to analyze energy consumption was introduced, and the method was verified by a measurement in an office building in Finland. The error with the simulation method stays in the range of 5%.

Using simulation, the effect of different machinery and drive systems, and group control systems were compared. The test was made for a 40-floor high-rise office building in Australia. According to the test case, during a typical day V3F drive system saves energy 30–40% compared to SCD system. With the MX machinery and TMS990 GATM control system, energy savings are huge compared to both systems. About 40–75% less energy is consumed with MX technology compared with the old technology. Energy is saved also by control means, about 10–15% during a day compared to a control where energy consumption is not considered. According to this case study, the MX machinery with TMS9900 GATM control system offers a real green elevator product for the most modern and demanding office buildings.

REFERENCES

CIBSE, 1994
    ENERGY EFFICIENCY GUIDE (26 January 1994 Draft), Section 3.9, 1.

Field, A., 1992
    ENERGY USE IN OFFICE BUILDINGS, Building Research
    Establishment (BRE), IP 20/92, November, 4p.

Fortune, J., 1996
    MODERN DOUBLE-DECK ELEVATOR APPLICATIONS AND
    THEORY, Elevator World, August, pp. 64–69.

Goldberg, D. E., 1989
    GENETIC ALGORITHMS IN SEARCH, Optimisation & Machine
    Learning. Addison-Wesley, Reading.

Hakala, H., 1995
    APPLICATION OF LINEAR MOTORS IN ELEVATOR HOISTING
    MACHINES, Doctorate thesis, Tampere University of
    Technology 1995, Publications nbr 157, ISBN 951–722–327–7.

Siikonen, M.-L., 1993
    ELEVATOR TRAFFIC SIMULATION, Simulation, Vol. 61, No. 4,
    pp. 257–267.

Tyni, T., Ylinen, J., 1999
    METHOD AND APPARATUS FOR ALLOCATING LANDING
    CALLS IN AN ELEVATOR GROUP. United States Patent
    5,932,852, August.

Limits on Energy Efficiency in Office Buildings

Mahadev Raman

1.0 OVERVIEW

Many sophisticated computer tools exist to calculate building energy demand but while they provide good levels of accuracy, they lack the flexibility to quickly compare a large number of scenarios. They also focus primarily on thermal performance for heating and cooling and do not easily allow other energy consuming systems to be compared and optimized.

To assist with the rapid prediction and optimization of building energy consumption at an early stage in the building design process, a spreadsheet-based building energy analysis tool was developed. The spreadsheet calculates energy consumed by HVAC systems, lighting, office equipment, domestic water systems and elevators. It also includes a simple daylighting calculation to allow building façades to be optimized for both thermal and lighting performance.

Many default values are included in order to minimize the amount of input data required. However, the spreadsheet format allows the user to customize the calculations used without the need for programming knowledge.

A further feature of the spreadsheet is that it allows the performance of Building Integrated Photovoltaic panels (BIPV's) to be evaluated.

After briefly describing features of the spreadsheet, this paper presents results obtained in the following applications:

- To chart the history of energy consumption in office buildings over the last 50 years.
- To identify limits to energy conservation in office buildings given current and imminent advances in building technology.
- To explore the likely contribution from BIPV's to offset energy consumption.

2.0 SPREADSHEET DETAILS

2.1 Weather Data

Readily available average climatic data are used. Several sources were found on the internet for the necessary input data including the US National Oceanic & Atmospheric Administration, the Australia Bureau of Meteorology, the World Meteorological Organization and the UK Met Office.

The average daily maximum and minimum temperatures for each month are used to derive daytime and nighttime temperatures for thermal calculations. Similarly, daytime and nighttime relative humidity values are used for latent heat calculations.

Records of bright sunshine hours are used to split each month into a number of equivalent sunny and overcast days. 24 hour average design solar radiation values, from sources such as the Chartered Institution of Building Services Engineers, Guide Book A, are used to derive average radiation intensities for each façade for both sunny and overcast conditions.

Eight thermal calculations are carried out for each building façade for each month as follows:

Sunny	Occupied	Daytime
		Nighttime
	Unoccupied	Daytime
		Nighttime
Overcast	Occupied	Daytime
		Nighttime
	Unoccupied	Daytime
		Nighttime

The results from each calculation are factored by the number of hours in each category to estimate energy consumption for the month. This is significantly more detailed than a manual estimate but not as sophisticated as the typical computational process that uses hourly weather data.

2.2 Building and Occupancy Data

The building is characterized by the length and width of a typical floor, the floor-to-floor height and the total number of floors. Core and perimeter zones can be defined as well as the orientation relative to true North. The thermal properties of glazed and opaque façade elements can be simply defined.

The occupant density, attendance factors, outside air quantities, heat output, hours of operation and days of operation per week may be specified. Lighting and equipment heat gains as well as infiltration levels may also be defined.

2.3 HVAC System Selection

Extensive built-in default values allow the rapid comparison of many system configurations as follows:

- All-air with air-water systems.
- Variable volume with constant volume systems.
- Mixed with displacement systems.
- Fan assisted with static terminal systems.
- Perimeter heating with terminal reheat.
- Control or not of winter humidity level.

Various optional energy saving strategies are also included:

- Exhaust air sensible and latent heat recovery.
- Air and water side 'economizer' for free cooling.
- CO_2 based outside air control.
- Daylight and occupancy sensing for control of artificial lighting.

System pressure losses and equipment efficiency values may be defined by the user.

2.4 Other Systems

The spreadsheet uses the building height and occupancy to estimate energy consumed by elevators. Simple calculations also estimate energy consumed by domestic water systems.

Energy produced by BIPV's is estimated based on collection efficiency and percentage coverage data entered by the user.

2.5 Energy Sources

For each building system, the energy source and unit cost may be defined to estimate operating costs. A calculation is also made of primary energy consumption and equivalent amount of CO_2 released.

2.6 Benchmarking

The spreadsheet technique has not been rigorously benchmarked against other proven techniques or actual measurements although an earlier version, written specifically for a museum in Maine, provided, perhaps by luck, predictions within 10% of actual measured consumption in the first year of operation. Copies of the spreadsheet may be obtained free of cost from the author for further testing and comparative studies.

3.0   THE HISTORY OF ENERGY CONSUMPTION IN OFFICE
BUILDINGS

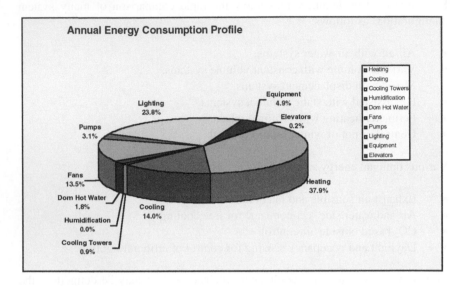

In order to chart the progress of energy conservation measures in office buildings over the last 50 years, a hypothetical building model was analyzed. A 20
story building with a 40m by 40m floor plate and 80m overall height was
assumed. All four façades were assumed to be 50% glazed. New York City
weather records were used.

3.1   Pre-Energy Crisis Building

Prior to the worldwide energy crisis in the early 1970's, energy conservation in
buildings was rarely considered an issue. A typical office building of that era
would include the following features:

- Combination of operable windows and air conditioning.
- Clear single glazing with internal shades.
- Use of generous amounts of energy to solve comfort problems.
- High artificial lighting levels.
- Negligible equipment loads.

Analysis shows that a building of this type would have had an annual energy
consumption of approximately 328 kWh/m^2. The figure below shows the distribution of this energy:

3.2 Post-Energy Crisis Building

The energy crisis precipitated many innovations in building systems, materials and design, many of which were eventually mandated in building energy codes. The features of a typical building of this era are as follows:

- Reduced fresh air supply to occupants.
- Extensive use of dark and reflective glass.
- Increasing use of double glazing.
- Use of 'economizer' free cooling.
- Advent of variable air volume systems.
- Increase in equipment loads with the advent of personal computers.

Despite the increase in equipment loads, the various energy saving measures above had a dramatic impact on energy consumption. In this analysis, a 30% saving was achieved to give an annual value of approximately 230 kWh/m².

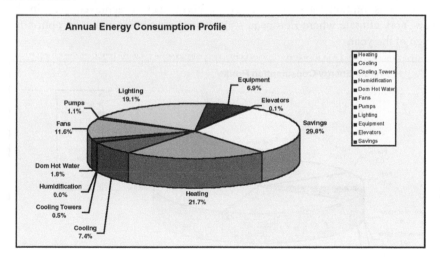

3.3 Contemporary High-Performance Building

Many of the energy saving systems and measures introduced after the energy crisis had the unforeseen side-effect of degrading the working environment. Reduced fresh air levels and variable volume systems in combination with the increased use of artificial materials in the workplace led to poor indoor air quality and problems such as sick building syndrome. The use of low transmission glazing reduced the sense of connection with the outdoor environment.

Modern building designs, particularly since the early 1990's, have begun to address these issues while maintaining good energy efficiency. Common features include the following:

- Increased fresh air supply to occupants.
- Clearer glass with smart coatings and/or physical shading to control heat gain.
- More efficient artificial light sources with better color rendering.
- Better control systems with greater personal control of the immediate environment.
- Further increases in equipment loads.
- Displacement type air-conditioning systems.

The need to improve environmental conditions in the workplace, along with the increase in equipment loads, has slowed down improvements in energy conservation. Nevertheless, analysis shows that contemporary high-performance office buildings use 40% less energy than their pre-energy crisis counterparts. Annual energy consumption rates of under 200 kWh/m^2 can be achieved even in the New York climate where European style natural ventilation is not an option for much of the year.

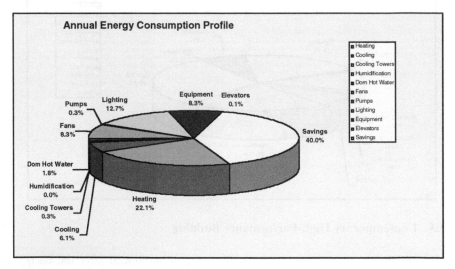

4.0 LIMITS TO ENERGY CONSERVATION

While many innovative systems and design approaches continue to incrementally improve energy conservation, it is theoretically possible to make dramatic improvements with features such as the following:

- Eliminate 'offgassing' materials to allow a reduction in fresh air supply without adverse health effects.

- Improve façade performance with more extensive use of external shading and selective coatings. Possible use of electrochromic glazing products.
- Install daylight collection, distribution and control systems.
- Use more efficient artificial light sources.
- Optimize air-conditioning systems for energy consumption rather than initial cost or space requirements. Use more efficient equipment and drives. Employ heat recovery and more occupancy based controls.
- Use more efficient elevator drives and intelligent control systems.
- Reduce equipment loads by using computers with better power management and low energy flat screen monitors.

Despite the likely need to maintain minimum winter humidity levels in the future, analysis shows that the above measures could theoretically reduce energy consumption to about 90 kWh/m², less than half the current value. The chart below shows the relative energy consumption of different systems in this scenario.

The currently low cost of energy is the major obstacle towards achieving these low levels of consumption as it places severe restrictions on the economic viability of aggressive conservation measures.

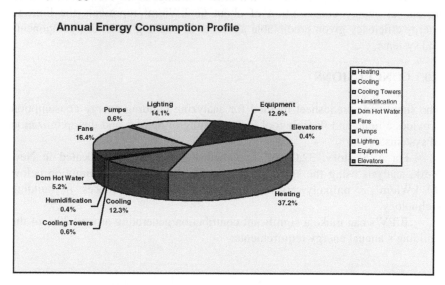

5.0 THE IMPACT OF BUILDING INTEGRATED PHOTOVOLTAICS

As building energy consumption reduces to the level described in 4.0, the potential contribution of BIPV's to the overall energy consumption becomes more significant.

Assuming a 15% PV generation efficiency and an 80% coverage of all opaque building surfaces with PV panels, the figure below demonstrates that 30% of the building's annual energy requirements could be met by BIPV panels.

In the summer months, when electricity demand is at its highest, the PV contribution could be as high as 60%.

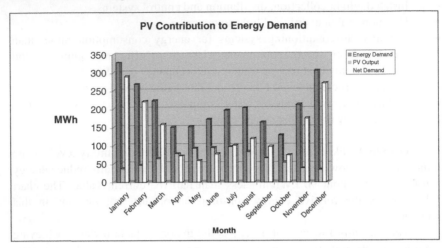

This would suggest that for tall office buildings in the New York context, an annual energy consumption of about 66 kWh/m^2 represents the limit to energy efficiency given predictable advances in building materials, components and systems.

6.0 CONCLUSIONS

The simplified spreadsheet method for analyzing building energy consumption provides a rapid and flexible tool for the testing of options and the optimization of systems.

For a 20 story, 32,000m^2 hypothetical office building located in New York, analysis using the spreadsheet suggests that energy consumption below 90 kWh/m^2 is unlikely given current and probable advances in building technology.

BIPV's can make a significant contribution generating around 30% of the building's annual energy requirements.

BUILDING SERVICE SYSTEMS

An Integrated Design Approach to the Environmental Performance of Buildings

M. Colomban, M. Zobec and M. Kragh

1.1 ABSTRACT

The present paper deals with the complex relationship between architectural intensions, indoor environment, building envelope and environmental systems, energy performance and design process. The development of architecture and the construction industry is discussed with focus on the way design used to be adapted to, and responsive to, local climatic conditions and later on, with commercialisation neglected these aspects and relied on mechanical systems for fully air conditioned buildings. This development resulted in energy consuming buildings with poor occupant comfort. The paper introduces Blue Technology, which is based on integrated design of building envelope and environmental systems, facilitating design of buildings with high levels of occupant comfort and reduced energy usage. Problems and solutions pertaining to transparent building envelopes and the need for integrated design are discussed. Moreover, the financial incentives in terms of tax legislation and potentially increasing productivity are introduced, supplementing the performance related aspects. At the Permasteelisa Headquarters in Italy, a series of full-scale test rooms have been set up in order to test and demonstrate different combinations of façade and environmental system technology. The test room activities are described and a case study is presented as an example of innovative thinking and the potential of an integrated design approach. Finally the recently established architectural Internet portal BuildingEnvelopes.org is introduced. In recognition of the need for dissemination of knowledge and guidance, the portal is dedicated to building envelopes and environmental systems and will constitute an important point of reference for the future.

1.2 BACKGROUND

The first American house built in war-time Java completely bewildered natives there. Instead of building walls of local bamboo, which is closely spaced to keep out rain while admitting light and air, the white man put up solid walls to keep out light and air, and then cut windows in the walls to admit the light and air. Next, he put glass panes in the windows to admit light but keep out the air. Then, he covered the panes with blinds and curtains to keep out the light too (Moore, 1993, adopted from Ken Kerr, 1978).

1.2.1 Building Form and Function

Throughout history, the building envelope of monumental structures has attempted to embody the complex relationships between functional and cultural elements. The history of the building envelope can be described as constant research and development of the following functions:

- Protection: *rain, cold, heat, solar radiation and intrusions*
- Prestige and identity: *dimension, materials and decorations*
- Comfort: *light, ventilation, insulation, perception*

In modern times, buildings have come to represent both corporate and individual identity (much like the palaces and temples of the past). Throughout history; up until the early post WWII years, architectural response to the climate was an integral part of professional architectural education. Unfortunately from the 1960's to the present day, the principles of designing with the climate; or using bio-climatic principles, was almost negated. Architectural institutions no longer focused on teaching fundamentals of building physics and the design of the mechanical system was focused on the suppliers of the mechanical plant. With the advent of mechanical air conditioning, building designers were free to pursue and give precedence to purity of form over human comfort. It was by no coincidence that the suppliers developed most of the HVAC design standards, manuals and software packages used by HVAC engineers. Whilst the research undertaken to develop such standards and systems has facilitated quick development of design, there has been a clear demarcation between the services engineers and architect regarding design responsibility. The inside of the office building became a deep-plan space, artificially both illuminated and ventilated throughout the day. The sealed concrete/steel and glass box was the only option. In urban environments, this of course has some justification due to land prices being at a premium as well as shielding occupants from noise and pollution.

At the same time as the 'sealed glass box' came into popularity, the face of the construction market also changed considerably. Unlike monuments of the past, which were occupied by the owners, buildings began to be seen by owners as a source of financial investment. Construction was beginning to be driven by fundamental market forces and a great deal was undertaken by speculators and developers, who had no intention of occupying the building, but rather saw three fundamental obstacles regarding the façade: cost, timing and warranty. Returns on investments and 'maximum nett lettable area' took preference over occupant comfort.

The designers of HVAC plant are usually commissioned far earlier in the design process than the façade contractor. HVAC designs are therefore carried out based on assumed façade performance parameters rather than adopting an integrated design approach or using what is termed *Blue Technology*.

1.2.2 The Philosophy behind Blue Technology

With the current boom in information systems, the word technology usually conjures up images of technical complexity such as computer chips. On the contrary, *Blue Technology is the understanding of fundamental building physics principles applied in a manner, which enables HVAC and façade to be designed as an integrated, synergetic system rather than individual components.*

Normal glass is almost completely transparent to short wave solar radiation (visible and near infra red) but is a barrier to long wave radiation. As solar radiation strikes the façade, the solar energy passing through the glazing tends to warm up the various internal surfaces by absorption, and these internal surfaces become heat radiators. However, the re-emitted heat is long wave radiation to which glass behaves as a barrier causing the building temperature to rise. This effect is commonly referred to as the 'Greenhouse' effect. Once heat is trapped inside a room, it can be removed by:

- Natural ventilation
- Mechanical ventilation
- Full air conditioning
- Radiant cooling by chilled surfaces and/or building thermal mass

The development of high performance solar coatings on glass has made significant improvements in reducing heat gains yet there is still a down side. In recent years, transparency in architecture has become most desirable. In architectural terms transparency is most associated with maximum natural daylighting. Since more than half of solar radiation is visible light, with high performance glazing, reductions in heat gains result in reductions in natural daylight. Most HVAC systems are also designed on the basis of short duration peak cooling loads and therefore in order to reduce heat gains for a brief peak loads, transparency or the use of natural daylight is limited or even sacrificed.

The problem remains to optimise transparency whilst minimising heat gains and achieving buildings with optimal internal comfort. It is important to note that it is not possible to air condition against direct solar heat gains. Current full air HVAC systems only treat the problem of building conditioning and do little towards preventing the primary problem associated with cooling loads ... solar heat gains. Many alternatives to full air conditioning systems exist (i.e. chilled ceilings), but since the capacities of such systems are limited, reductions in cooling loads must be possible. The obvious solution is not to treat but prevent the problem altogether, starting at the source ... the façade. By excluding or reducing solar gains, the internal environment can be significantly improved and capital, maintenance and operational costs can be reduced.

The main obstacle is not a matter of available technology, but reluctance due to a combination of problems. Research endeavours aim at resolving these as indicated in the following table:

Table 1.1   Transparent building envelopes – problems and solutions

Problems	Solutions
• Clear demarcation between the architect and building services designers. • Conventional building 'practices' in obtaining the expertise of the façade and services engineers at different stages of the project development. • Misunderstanding of building physics principles. • Over reliance on proprietary software, which may not allow the input of parameters other than those related to conventional full air systems. • Lack of adequate design tools, details and cases studies of completed projects	• Promoting an integrated approach to the design of the façade and the environmental system. • Undertaking an R&D test program, which verifies and quantifies the performance of alternate simpler integrated façade and HVAC systems. • Documenting case studies of completed buildings and their performances. • Developing a series of inexpensive design tools, which enable building designers to undertake simple preliminary simulations. • Effectively disseminating and sharing experiences within a global information medium available to the global construction community.

There is a pronounced need for a common language in order to characterise and communicate the performance of innovative systems such as the mechanically ventilated façades

1.3   INTEGRATED BUILDING ENVELOPE AND ENVIRONMENTAL SYSTEM DESIGN

1.3.1   Transparent Building Envelopes

Transparency in architecture has always been desirable and the problem has always been to realise a transparent building envelope without compromising energy performance and indoor climate. For years the development of advanced façade and environmental systems has aimed at creating fully glazed buildings with low energy consumption and high levels of occupant comfort. Ventilated double skin façades reducing solar gains in summer and providing thermal insulation in winter is an example of a technology, which is becoming still more common.

1.3.2 Integrated Design

Intelligent application of advanced façade technology in conjunction with innovative environmental systems results in significant energy savings and – at the same time – improvement of indoor comfort. It has been shown that, when designed carefully, innovative systems do not represent additional initial costs, running costs are lower and energy costs can be reduced by approximately 30 per cent compared with conventional solutions.

> *Successful application of these systems depends closely on the adoption of an integrated design approach from the early, schematic phases of a given project*

Too often the façade design is developed when fundamental decisions, for instance pertaining to the layout of the ventilation system, have already been taken. At this point it can be too late to benefit fully from application of advanced façade solutions. If façade and environmental system are engineered as two parts of the same solution, not only will the performance most likely be superior – both initial and running costs may moreover be reduced significantly.

To this end, there is a need for a change of approach bringing together façade- and M&E engineers during the early design phases. Moreover – and this is a problem we experience frequently these days – there is a pronounced need for a common language in order to characterise and communicate the performance of innovative systems such as the mechanically ventilated façades. For instance, quantities such as U-value and solar factor are not readily applicable when the façade interacts with the ventilation system, and traditional ways of designing HVAC systems may not be adequate when assessing possible application of innovative solutions such as soft-cooling.

1.3.3 Financial Implications of Blue Technology

The emphasis that clients require in particular is the achievement of BEST VALUE.

Value management of the building envelope

In the UK, Government initiatives to secure best value have lead to changes from normal development, making 'cost in use' a key factor in the design. Local authorities must obtain best value when procuring goods and services and must comply with a rigorous regime of performance indicators and efficiency measures. The Authorities need to take into account whole life costing of the service or element.

Both central and local government is thus undergoing significant change. Similar strategic changes are also taking place in the private sector. The purchase decision is moving away from the lowest tendered costs with the focus being more on the cost in use benefits whoever the tenant is going to be. This issue is particularly relevant in the selection of an appropriate façade design.

Tax laws – the UK situation

The tax laws in the United Kingdom are uniquely favourable to technical development in environmental engineering generally and active façades in particular.

UK tax relief is given by way of capital allowance on plant and machinery that writes off the capital value effectively within 8–9 years. It covers for example all heating, ventilation and air conditioning systems, most and in some cases all electrical installations.

The stage is wide open to propose that active façades, which carry air as part of the air conditioning/environmental control system, should be treated as an air duct to the perimeter of the building. This of course will depend on the design of the wall itself (Glanville, 2000).

1.3.4 Comfort and Productivity

Energy-efficient building and office design offers the possibility of significantly increased worker productivity. By improving lighting, heating, and cooling, workers can be made more comfortable and productive. An increase of 1 per cent in productivity can provide savings to a company that exceed its entire energy bill (Romm and Browning, 1998). Efficient design practices are cost-effective just from their energy savings; the resulting productivity gains make them indispensable.

There has always been a consensus that the comfort of the occupants affected their productivity, but until now the hard data proving this have been lacking. The Rocky Mountain Institute (Romm and Browning, 1998) carried out a series of case studies of both new buildings and retrofits of existing ones, all demonstrating correlation between occupant comfort and productivity. The following are examples of the findings:

- Lockheed's engineering development and design facility, which saved nearly US$500,000 a year on energy bills and gained 15 per cent in productivity with a 15 per cent drop in absenteeism.
- West Bend Mutual Insurance's new building, which yielded a 40 per cent reduction in energy consumption per square foot and a 16 per cent increase in claim-processing activity.
- ING Bank's new headquarters, which used one-tenth the energy per square foot of its predecessor, created a positive new image for the bank, and lowered absenteeism by 15 per cent.

However attractive the gains in terms energy-efficiency retrofits for existing buildings, and new buildings designed for energy-efficient performance these gains are tiny compared with the cost of employees, which is greater than the total energy and operating costs of a building. Based on a 1990 US survey of large office buildings, as summarised in the graph below, electricity typically costs US$1.53 per square foot and accounts for 85 per cent of the total energy bill, while repairs and maintenance typically add another US$1.37 per square foot; both contribute to the gross office-space rent of US$21 per square foot. In comparison, office workers cost US$130 per square foot – 72 times as much as the energy costs. Thus an increase of 1 per cent in productivity can nearly offset a company's entire energy cost.

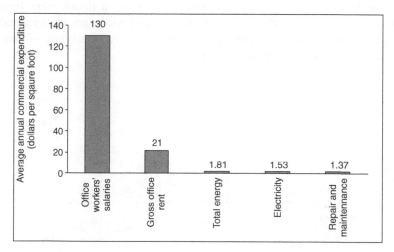

Figure 1.1   Data from Building Owners and Managers Association; Electric Power Research Institute; Statistical Abstract of the United Stated 1991 [Romm and Browning, 1998].

Productivity is measured here in terms of production rate, quality of production, and changes in absenteeism. This can be improved by fewer distractions from eye-strain or poor thermal comfort, and similar factors.

Will just any energy retrofit produce gains in productivity? No, only those designs and actions that improve visual acuity and thermal comfort seem to result in these gains. This speaks directly to the need for good design, a total-quality approach that seeks to improve energy efficiency and improve the quality of workplaces by focusing on the end user – the employee. This is a point that seems to have been forgotten by many designers and building owners (Romm and Browning, 1998).

1.4   PERMASTEELISA'S TEST ACTIVITIES

1.4.1   Test Room Monitoring

At the Permasteelisa Headquarters in Italy, a series of advanced façade solutions have been realised in conjunction with innovative environmental systems. Currently, a total of 10 full-scale test rooms are being continuously monitored in terms of energy consumption and indoor environment and another 4 rooms are in progress. The measurements will enable a direct comparison between different solutions exposed to identical climatic conditions and provide a basis for validation of both simplified and detailed engineering tools.

The building envelope configurations comprise double skin walls (naturally ventilated, mechanically ventilated indoor-indoor and outdoor-outdoor) demonstrating stand-alone systems as well as integration between façade and environmental system. Examples are the Active Wall, a double skin façade ventilated with room return air and the Interactive Wall, a double skin façade, mechanically ventilated with outdoor air by means of micro fans incorporated in the spandrel area. The environmental systems comprise variations of radiant systems as well as displacement ventilation.

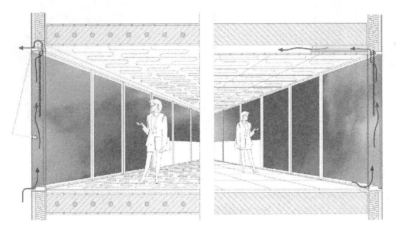

Figure 1.2   Schematic of the Interactive Wall (left) and the Active Wall (right).

For comparison, the innovative systems are installed side-by-side with conventional systems adopting high performance glazing and fancoil cooling and heating.

Figure 1.3   External view of the Permasteelisa test room facilities.

Initially the test rooms are not occupied and no internal loads are simu-lated. They are all kept at the same set point temperature and ventilation air is supplied at a rate corresponding to 2 air changes per hour. The rooms, which have radiant systems are being conditioned mainly by means of these, but the air volume is increased if the capacity of the radiant systems is not adequate, for example during peak load periods. It is important to note that the objective is to monitor combinations of façade and environmental system technology rather than one of the two.

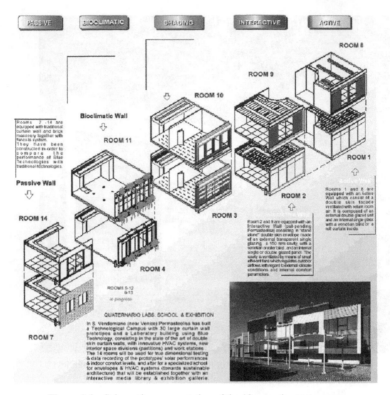

Figure 1.4   Schematic representation of the 10 operating test rooms.

1.4.2   Measured Parameters

Generally speaking, all of the rooms are continuously monitored in terms of energy consumption, ambient temperatures and humidity as well as surface temperatures across the façade (and cavity air temperatures when applicable) and solar radiation transmitted through the façade. Moreover, mobile instruments are available for series of daylight measurements. A meteorological station records the climatic conditions from the roof of the building.

Apart from providing a basis for assessment of system performance and a direct comparison between different solutions exposed to identical climatic conditions, the measurements will yield a basis for validation of simulation tools. The data will prove useful for validation of both existing and future software tools.

1.4.3   Publication of Results

The monitoring/control system has been operating since the summer of year 2000. Preliminary results yield trends, while the system is continuously being modified and improved in terms of both control and monitoring. Since the seasonal variations play an important part in the assessment of façade/HVAC performance, long-term measurement is essential. However, already now, studies of specific climatic situations and pertinent system performance are being carried out in collaboration with the MIT and the results will be published. Apart from publications in journals and at conferences and seminars, the results will be published through the architectural Internet portal www.BuildingEnvelopes.org, which is dedicated to building envelopes and environmental systems.

1.4.4 Preliminary Results

The following graphs show examples of the monitoring output. The specific case is a hot, clear summer day in August and the energy consumption for cooling is compared for two test rooms, both with fully glazed curtain wall: (a) conventional curtain wall and fancoil units (b) Interactive Wall and dynamic beams (cooling/heating by convection and radiation).

The outdoor climate is registered from a meteorological station at the roof of the building. The graph below shows the variation of drybulb temperature, relative humidity, solar irradiance on vertical and illuminance on vertical. The selected day is characterised by outdoor drybulb temperatures between 30 and 35°C and a maximum solar irradiance of 680 W/m2 (on vertical).

These extreme environmental conditions yield a good basis for comparison of two fundamentally different solutions. The following graphs show the cooling energy consumption as recorded for the two rooms. The difference in cooling consumption is due to both façade type and HVAC system.

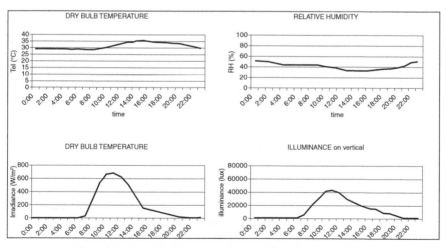

Figure 1.5 Example: hot summer day – outdoor climatic conditions.

Both rooms are conditioned mainly by means of water, with additional background cooling provided by the fresh air supply (two air changes per hour). The room with the Interactive Wall and dynamic beams is consuming between 1400 W and 1750 W, whereas the room with conventional curtain wall and fancoil units is consuming between 1900 W and 2800 W.

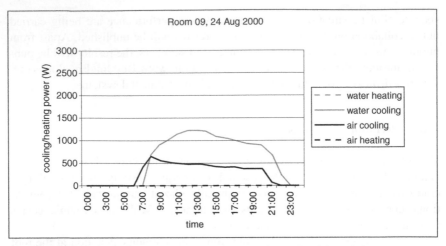

Figure 1.6   Example: hot summer day – Interactive Wall and dynamic beams.

The case demonstrates significant differences in consumption. At peak load, the room with the conventional curtain wall with high performance glazing and internal roller blinds is consuming approximately 60 per cent more than the room with the Interactive Wall.

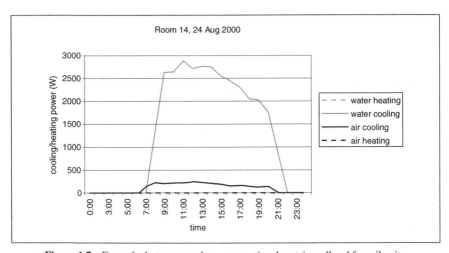

Figure 1.7   Example: hot summer day – conventional curtain wall and fancoil units.

The rooms are being studied in terms of both energy consumption and internal environment. In this regard, it should be noted that with radiant systems the same level of perceived comfort can be obtained with higher ambient temperatures (summer) as long as the operative temperature is acceptable.

1.5 CASE STUDY: THE UCB CENTER, BRUSSELS

1.5.1 Double Skin Façade in Conjunction with Chilled Ceilings

The following case study has been selected to illustrate the potential benefits of careful combination of advanced façade and HVAC technology. This particular case study describes the 'happy marriage' between a mechanically ventilated double skin façade and chilled ceilings as realised in the *UCB Center* by the architects *Assar* (Brussels) with the mechanical and structural engineers *Tractebel*. At a conference in the UCB Center, May 2000, the owner and the engineers involved presented both the initial analyses and the actual building performance the in terms of energy consumption and indoor environment.

Originally there was no intention to use double skin façades and chilled ceilings for the UCB Center. The project was proposed with conventional fancoil units. The directors of the UCB (Union Chimique Belge) wished to have transparent façades. Initially the chilled ceiling concept was considered, but rejected because of its limited capacity (soft-cooling) and the south exposure of the glazed main façade. Even with fancoil units, the thermal balance of the system would be at the limit. The architect's wish to increase the glazed area could not be met without introducing external solar shading, which would be in conflict with the whole design philosophy. These considerations lead to the idea of the double skin façade with solar shading positioned in the façade cavity. This type of façade poses an ideal compromise, offering a smooth, glazed external surface and, at the same time, providing the necessary solar protection. Mechanical ventilation was required in order to extract the solar heat from the façade cavity, but this was not a difficult problem to solve. Because of the solar protection provided by the ventilated double skin façade, the chilled ceiling (soft cooling) technology now became possible. A comparative cost analysis of the alternative solutions was carried out (Marcq & Roba, 2000). Both initial costs and expected running costs were compared. The conclusion of the study was that the solution with double skin/chilled ceiling resulted in better comfort and did not result in higher initial costs, compared with a solution based on conventional façade and fancoil units. The building has been in use for two years, and the expected advantages have all been confirmed (Vervaeck, 2000):

- Transparency, better view to the exterior, which is particularly appreciable because of the nature surrounding the building.
- Thermal comfort, summer and winter, without draughts and fancoil noise.
- Virtually non-existent maintenance of the chilled ceilings.
- Reduction of energy consumption.

1.5.2 Technical Details and Performance

The double skin façade (Active Wall) of the UCB Center is composed of an external double glazed unit, a 143mm deep, mechanically ventilated cavity and a clear single layer internal glazing. Motorised blinds are positioned in the ventilated cavity and controlled depending on the solar irradiance. The airflow rate is 40m³/h per module (width 1.5m).

Figure 1.8   The UCB Center, Brussels, Active Wall and chilled ceilings.

Heating is provided by the supply air, which results in lower installation costs, and means that the glazing can be continued down to floor level (no fancoil units). The ventilation air is re-circulated when the building is not occupied. The temperature of the inlet air is regulated depending on the solar irradiance.

Cooling is provided by means of chilled ceilings operating with water at temperatures between 15°C and 17°C(!). The chilled ceiling is a capillary type in polypropylene, incorporating thermal insulation. The acoustic barrier is horizontal. In order to avoid condensation problems, the ventilation air is dehumidified.

The UCB have reported savings between 12 and 30 per cent on gas and between 39 and 44 per cent on electricity (Caudron, 2000).

Energy savings have been significant. Up to 30 per cent savings on gas, and up to 44 per cent savings on electricity.

The chilled ceiling technology is reducing air movement and increasing occupant comfort. The absence of fancoil units at the façade increases the usable floor area. Utilisation of a static system such as the chilled ceiling leads to better acoustic performance than the dynamic fancoils. Furthermore, the acoustic insulation of the façade is improved due to the extra layer of glazing.

It is important to note that the soft-cooling technology, which leads to energy savings, is enabled by the thermal and solar performance of the ventilated façade. The performance is due to the successful combination of these two elements. If the chilled ceiling is to maintain a comfortable indoor environment, the cooling load cannot exceed 70 W per m2 floor area. In zones with higher cooling loads, such as conference rooms, additional cooling capacity is required.

1.6 WWW.BUILDINGENVELOPES.ORG

1.6.1 Academic and Industry Collaboration

Research in academia has traditionally been focused on basic research, in an effort to lay the foundation for future developments. Apart from some notable exceptions, academic research has been typically government funded, resulting in the demand for tangible results being less than industry carried research that aimed in the development of products and services to reach the market at the shortest possible time. However, is there another role for academia in the new Internet economy?

A main characteristic of the new economy is that information can be disseminated fast and can originate from different sources. Creating an online resource would allow both universities and industry to work together in creating an aggregation of knowledge sharing among those involved in the building industry.

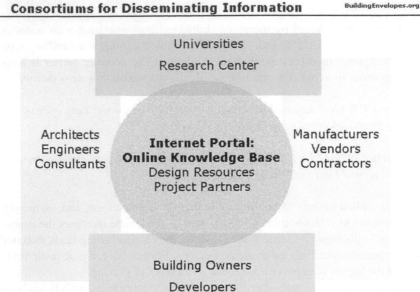

Figure 1.9   Internet portal consortium.

1.6.2   BuildingEnvelopes.org Portal

The result of this academic and industry collaboration is BuildingEnvelopes.org, an online knowledge base for the design and construction of innovative architectural envelopes and environmental control systems. An initial project archive was expanded to a portal because at the start of every new project, designers and owners are faced with two challenges: The *Design Challenge* and the *Information Challenge*. The Design Challenge continually grows as designers try to create state-of-the-art designs but are increasingly challenged by the rapid technical evolution that surpasses their know-how. This Design Challenge is exasperated because there is no single, reliable source of current information on the variety and number of products and systems available. These Design Challenges increase on international projects since local conditions, in terms of regulatory framework, climate conditions, or potential collaborators, are unknown. The second challenge, the Information Challenge, is created because suppliers and manufacturers are constantly challenged with keeping the designers informed about new systems and products. However, catalogues and brochures are unsatisfying because they can become outdated very quickly. This leads to the possibilities of integrating their products and services being limited when suppliers and manufacturers are involved late in the project development.

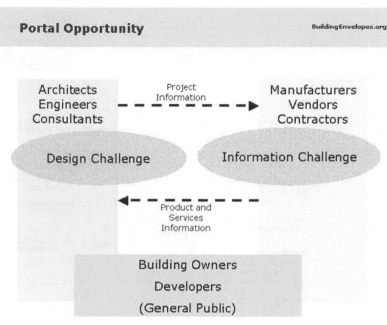

Figure 1.10 Internet portal opportunity.

BuildingEnvelopes.org fills this need by creating a dynamic, up-to-date resource for the building industry to gain knowledge on the most current technologies, products, designs, and methods. BuildingEnvelopes.org provides the framework for the building industry, academic institutions, and research centres to share knowledge and information. This knowledge is provided to owners and designers in order to expand their knowledge about new and innovative potential design solutions, thereby closing the information gap. After providing preliminary information, the portal then leads the owners and designer to the design or product specialists for further advice. It also provides new and inexperienced designers with a resource tool to answer practical, real-world questions.

Worldwide building industry organizations as well as research centres and universities provide information for the portal. Currently this collaboration is between Harvard Design School, ETH Zurich, Lawrence Berkeley National Laboratory, MIT Building Technology Group, the Polytechnic University of Milan, Solar Energy and Building Physics Laboratory (LESO-PB), the University of Michigan, VTT Finland and more than 15 building industry professionals, such as architects, engineers, manufacturers, and consultants.

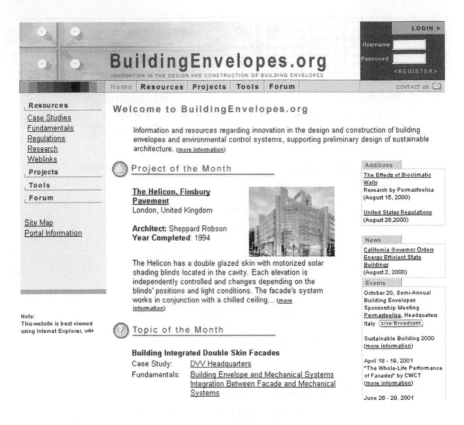

Figure 1.11   BuildingEnvelopes.org home page.

1.7   REFERENCES

Caudron, Y., 2000
> YVAN CAUDRON, TRACTEBEL. SRBII MEETING.
>> *Façades doubles ventilées: Un concept interessant*, Brussels, May 2000.

Glanville, G., 2000
> FINANCIAL IMPLICATIONS OF MODERN ARCHITECTURAL ENVELOPES IN OFFICE BUILDINGS.
>> *Modern Façades for Office Buildings* (in Dutch: Moderne Gevelarchitectuur voor Kantorbouw), Delft Technical University, The Netherlands, October 26, 2000, edited by Renckens, J.

Marcq & Roba, 2000
MARCQ & ROBA CONSULTING ENGINEERS IN COLLABORA-
TION WITH UNIVERSITY OF LIEGE AND CATHOLIC UNIVER-
SITY OF LEUVEN. SRBII MEETING.
Façades doubles ventilées: Un concept interessant, Brussels,
May 2000.
Moore, F., 1993
ENVIRONMENTAL CONTROL SYSTEMS – HEATING,
COOLING, LIGHTING.
(McGraw-Hill Inc.) USA.

Romm, J. J. and Browning, W.D., 1998
GREENING THE BUILDING AND THE BOTTOM LINE –
INCREASING PRODUCTIVITY THROUGH ENERGY-EFFICIENT
DESIGN.
Rocky Mountain Institute, Colorado, USA.

Vervaeck, J., 2000
JACQUES VERVEACK, UCB (UNION CHIMIQUE BELGE). SRBII
MEETING.
Façades doubles ventilées: Un concept interessant, Brussels,
May 2000.

DESIGN CRITERIA AND LOADS

Practical Solution to Reduce the Wind-Induced Response of Tall Buildings

R. J. McNamara, S.E.

ABSTRACT

Adding damping with various energy dissipating devices has become an accepted method to reduce wind-induced vibrations in tall buildings. An example of a 39-story office tower is presented where large projected accelerations generated by the vortex shedding of an adjacent existing 52-story building are reduced by a passive system composed of viscous dampers and a motion amplification system. A description of the damping system and its analytical complexities are discussed. Non-linear analysis of the tower, using time history forcing functions derived from the wind tunnel is presented. Cost data for the damper system is also presented.

INTRODUCTION

The use of energy dissipating devices to reduce building response from dynamic inputs has become an accepted design approach for high-rise buildings. New approaches are continually being developed by designers as evidenced by the varied applications of tuned mass dampers, sloshing dampers, visco-elastic dampers, friction dampers and viscous dampers. Each of these systems has its own idiosyncrasy and which is most appropriate must be evaluated for the particular project under consideration.

This paper presents the results of an investigation of the application of viscous dampers in a high-rise structure located in an urban environment. The structure, a 39-story steel-tube frame was designed using conventional wind engineering methods with code loadings and standard deflection limitations. A model of the tower was tested in a wind tunnel of RWDI facilities in Canada. The building is located within the immediate proximity of a 52-story tower in the center of a coastal downtown urban environment. Wind tunnel results indicated that the structure would experience very high acceleration levels generated by winds coming from a northwestern direction. Detailed investigation into the wind tunnel data indicated that the intense buffeting the tower was experiencing was the result of vortex shedding from the adjacent 52 story existing building. The predicted acceleration levels were double the industry standard for office towers. In order to reduce the projected motion levels, several approaches were investigated and evaluated for cost and planning impact. Tuned mass dampers and sloshing dampers required valuable office space at the top of the tower and proved to be very expensive (although very effective). Viscoelastic dampers were no longer available from US manufacturers. Viscous dampers proved to be the most cost effective and least space intrusive on the office tower. An extensive design program was undertaken with various viscous damper configurations vertically and with many variations of viscous damper properties.

Since the main intent of the damper installation is to reduce accelerations resulting from relatively frequent storms, the viscous dampers need to provide a large force output at very low displacement levels (±1/8"). In order to insure reliability at this small movement and to keep the number and cost of the dampers to a minimum, a motion amplification device was introduced in the design. The motion amplification device was used in one direction of the structure, that being the stiffest with the lowest predicted inter-story displacements.

The introduction of a motion amplification device to amplify inter-story displacements experienced by the damper was essential to the design reliability. The small inter-story movements normally experienced by frequent storms producing annoying accelerations must be amplified to allow the use of an economical viscous damper and to ensure the reliability of the damper force output.

A motion amplification device called a Toggle Brace Damper system (TBD) was tested by Constaintinou, etc. (1998). Their report demonstrates that the TBD system is a very effective mechanism to amplify inter-story motion. However, the efficiency of TBD, as reported by McNamara, Huang and Wan (1999) is highly dependent on various local system design parameters. Careful design of the TBD is extremely important for the proper performance of the damper system. From the above parameter study of TBD system, a total of 60 viscous dampers were used in the 39-story office building to reduce the top floor acceleration into an acceptable range. The viscous dampers in North-South direction use TBD devices. Viscous dampers in the East-West direction use dampers with straight braces. The viscous dampers were then designed for both 100-year return wind and moderate earthquake excitations. (Seismic zone 2, Av = .12 g.)

OFFICE BUILDING STRUCTURAL SYSTEM

The 39-story Office Building consists of three lateral systems at different levels. From the 1st to 7th floors and above 34th floor diagonal bracing is used for the lateral system. Over the remaining of floors the lateral system is a moment frame on the perimeter of building. The typical floor system is composite metal deck with composite joist girders spaced at 10'-0" o.c. Typical floor area is 22,500 square feet. Viscous dampers in E-W direction are straight diagonals placed in two bays of the inner-core on every other floor between 7th floor and 34th floor. The TBD systems are placed in two bays along the N-S direction at the same level as the diagonal dampers. The damper system layout is shown in Figure 1.

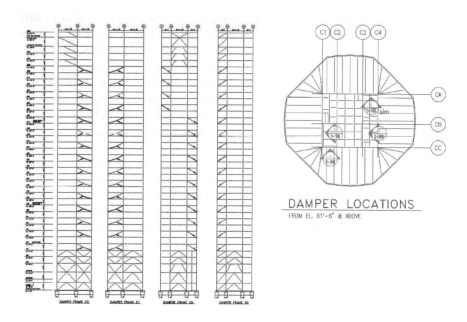

Figure 1   Viscous Damper Elevation and Key Plan.

A static lateral analysis and design was conducted using ETAB6.2. The dynamic response and viscous damper design and the TBD system were analyzed by SAP2000. Simple one story models were used to do parametric studies on the TBD system. The building dynamic properties are tabulated in Table 1.

Table 1   Dynamic Properties of Building for first six modes.

Mode Shape	1	2	3	4	5	6
Period (sec)	5.26	5.00	3.65	1.92	1.82	1.71
Effective Mass (%)	66.1	62.6	81.2	15.3	12.8	8.5
Direction	X (E-W)	Y (N-S)	Rotation	X (E-W)	Y (N-S)	Rotation

Note: Above dynamic properties obtained from ETAB63-D model

Wind tunnel results indicate average story drifts from 7th floor to 34th floor on E-W (X) direction are larger than the (Y) direction. The overall building stiffness in X-direction is less than that on Y-direction. For cost effective design, a TBD system in the Y-direction was used to magnify the story drift. The damper constant (C) was varied throughout the height of the tower. Linear viscous dampers and the TBD system were designed and manufactured by Taylor Devices, Inc. The damper layout is shown in Figure 1. The elevation of the dampers and TBD are shown in Figures 2 and 3. Geometric data for the TBD system is given in Table 2.

The design of the viscous damper system can be conceptualized as the damper system providing a set of loads distributed vertically along the height of the tower. These loads are velocity dependent and are applied to the tower's lateral force resisting system. The damper loads are out of phase with the displacement response of the tower and represent the mechanisms by which the response is reduced. Maximum damper forces occur at response levels of zero displacement and maximum displacement velocity.

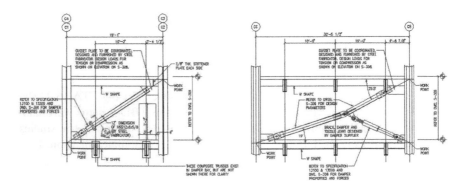

Figure 2   Diagonal Viscous Damper on E–W (X) direction.

Figure 3   Toggle Brace Damper on N–S (Y) direction.

Table 2   Toggle Brace Configuration with Story Height 12'-6" **.

Bay Length (ft)	Low Brace Angle	Upper Brace Angle	Low Brace Length (ft)	Upper Brace Length (ft)	Amp1*	Amp2*
31'-0"	19°	29.5°	24'-0"	9'-5"	2.9	6.1

*: Amp1 and Amp2 are motion amplification factor (δ_c/Δ) and force amplification (FB/FD) respectively.

DESIGN CRITERIA AND STATIC LATERAL LOAD

The design criteria for office building are compliant to BOCA 96 and Massachusetts State Building Code. The lateral structural systems are designed to meet AISC strength requirements and seismic provisions for zone 2B. No force reductions due to the damping increase by viscous dampers was taken into account at this design stage. The design coefficients for the equivalent lateral load of BOCA 96 are tabulated in Table 3. Wind design criteria are for 100-year return for strength and 10-year return for serviceability are also shown.

Table 3   Equivalent Lateral Load Design Parameters for BOCA 96.

Design Wind Load		Design Earthquake Load	
Wind Speed	90 mph	Seismic Zone	2A
Design Category	B	Peak Acceleration (Av)	0.12g
Importance Factor	1	Reduction Factor (R)	4.5
Aspect Ratio of Depth to Width	3	Soil Factor (S3)	1.5
Aspect Ratio of Depth to Width	1	Building Period (Ta)	3.65sec

WIND TUNNEL TEST RESULTS AND WIND TIME HISTORY LOADING

The 39-story office building wind tunnel test was carried on by RWDI, Ontario, Canada. The tests were conducted on a 1:400 scale model in presence of all surrounding buildings within a full-scale radius of 1600ft. The magnitude of simulated wind speed for a 100 year return period was scaled to correspond to a fastest-mile speed of 94mph at 33 ft (10m) above ground in open terrain, which is consistent with the Massachusetts Building Code and ASCE-93 Standard. In order to perform nonlinear time history required for viscous damper design, a specific time series was generated from the high frequency force-balance wind tunnel results. Response comparisons for various wind force time history studies are shown in Table 4 along with wind tunnel predictions.

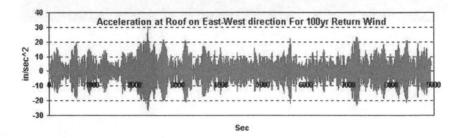

Figure 4   Roof Acceleration Response on (E-W) From Wind Tunnel Test.

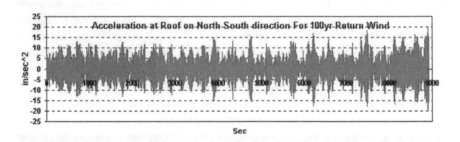

Figure 5   Roof Acceleration Response on (N-S) From Wind Tunnel Test.

Table 4   Response Comparison of Truncated Time Series Data.

	10 yr		100 yr		Wind Tunnel
Building Response	4.2 hrs (45000 data)	5.7mins (1024 data)	3.5 hrs (45000 data)	4.7mins (1024 data)	100yr 1.5%damp.
36th x-Accel. (in/s²)	16.1	14.6	30.6	27.4	NA
y-Accel. (in/s²)	12.2	10.7	20.2	17.9	NA
36th x-Displ. (in)	13.6	12.5	23.1	20.8	18.0
y-Displ. (in)	8.4	6.8	13.3	11.2	13.1
x-Base Shear (kip)	2738	2630	4374	3907	3541
y-Base Shear (kip)	1832	1699	3118	2903	2844
x-Base Moment (kip-in)	6.97×10^6	6.13×10^6	1.19×10^7	1.03×10^7	1.46×10^7
y-Base Moment (kip-in)	1.10×10^7	1.01×10^7	1.86×10^7	1.67×10^7	1.14×10^7

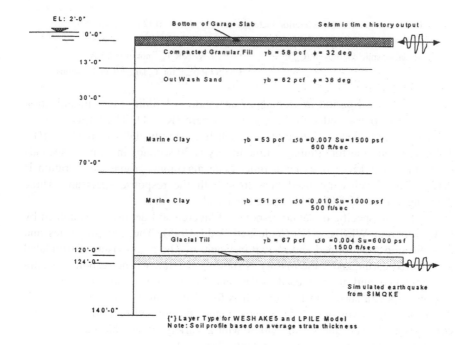

Figure 6   Soil Profile.

EARTHQUAKE ANALYSIS AND DAMPER DESIGN

Once the damper system was designed to reduce wind motion, the response of the system must be investigated under expected earthquake motions. Since no ground motion records are available at this site, ground motions to test the design must be simulated.

The office building is located in Bay Back, Boston, a moderate seismic zone according to the Massachusetts State Building Code. The design peak ground acceleration is 0.12g. Soil profile is shown in Figure 6. Since time history analysis is required for viscous damper design, artificial earthquake time histories are generated by SIMQKE (Vanmarcke 1976). Design spectra factors used here conform to BOCA 96:

Peak velocity-related acceleration factor (Av):   0.12
Site soil profile properties (S):   1.2
Modal seismic coefficient (C_{sm}):   1.2 Av S/R $T_m^{2/3}$ not over 2.5 Av,
                                                 3 Av S / R $T_m^{4/3}$ for T_m larger than 4 seconds

T_m is modal period in seconds of mth mode of building. R is modification factor. No response reduction is considered here (R = 1). The target pseudo-velocity design spectra (in/sec) for SIMQKE is simply defined as C_{sm} $2\pi/T_m$. Total duration time for simulated time history is 20 seconds in which 2 seconds rising time and 15 seconds level time. Code maximum ground acceleration is 0.12g. Three cycles are used here to smooth the response spectrum. Three damping ratios (1%, 2% and 5%) are examined.

The site-specific dynamic response of layered soil deposit is estimated by using program WESHAKE5 (Yule and Wahl 1995). The soil properties and classification are grouped and shown in Figure 6. The comparison of simulated and filtered earthquake time history is shown in Figure 7. The response spectra for simulated and filtered earthquake time history are compared with other response spectra shown in Figure 9. It is found that site-specific period of this building is approximately 2.5 seconds. Figure 8 shows the building response comparison of earthquake simulation with a variety of notable case histories which have different frequency contents. The peak ground acceleration for all time histories is scaled to 0.12g. Typically the earthquake simulations produced forces in the viscous dampers which governed the damper ultimate force capacity. This ultimate force becomes part of the damper design specification.

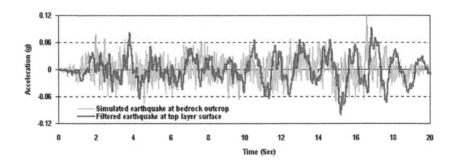

Figure 7   Simulated Earthquake Time History.

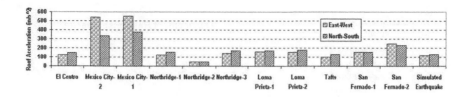

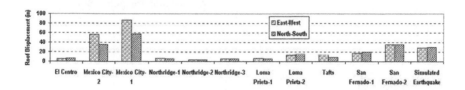

Figure 8   Building Response under Various History Records
(1% Structural damping and 0.12 g peak Acceleration).

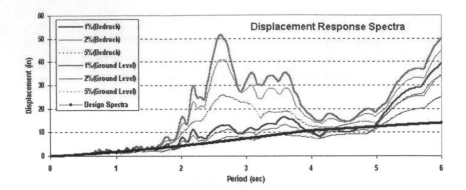

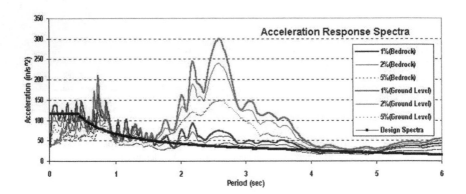

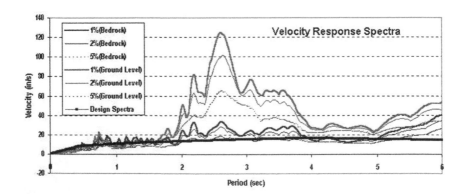

Figure 9   Comparison of Simulated Time History Response Spectra with Design Spectra.

BUILDING RESPONSE REDUCTIONS

High-rise building design is often governed by member stiffness rather than member strength. This is especially true in moderate seismic zones. Under extreme wind conditions, large deflections or story drifts of a building may result in damage of the nonstructural partitions and cladding. Under smaller storms occupant comfort can control the design. The wind tunnel story drift for 100-year return storm is about 1/280 in each direction. With the introduction of sixty viscous dampers, the deflection and minimum story drift index are much improved as demonstrated in Figure 10. The building defection and drift under seismic loading is also greatly improved as shown in Figure 11.

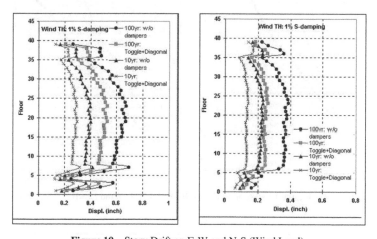

Figure 10   Story Drift on E-W and N-S (Wind Load).

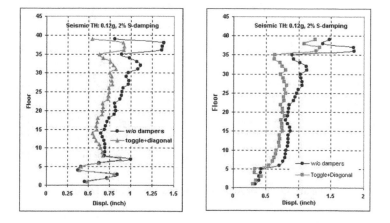

Figure 11   Story Drift on E-W and N-S (Seismic Load).

Humans are sensitive to acceleration and its change rather than building displacement and velocity. Acceptability of motion perception varies widely. In common practice, the suggested peak values range from 10 milli-g to 30 milli-g for a storm with a return period of 10 years (10mg for apartments and 30mg for offices). For this office building, the acceleration at the highest occupied (36th floor) level is predicted to be 41mg. The introduction of the damper system reduced the floor accelerations by approximately 35% as shown in Figure 12 and Figure 13.

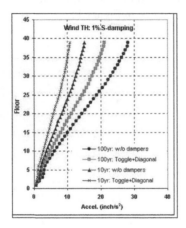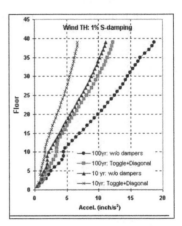

Figure 12   Floor Accel. on E-W and N-S (Wind Load).

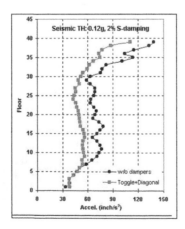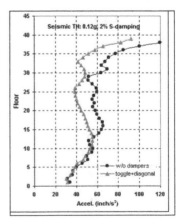

Figure 13   Floor Accel. on E-W and N-S (Seismic Load).

RESULTS SUMMARY

The equivalent static wind and seismic load effects of Massachusetts State Building Code and National Building Code (BOCA 93) on the office building are plotted on Figure 14 and Figure 15. Under wind load, wind tunnel predicts more pressure at 300 feet and above, but diminished quickly on lower floors. In general, the wind load indicates more severe design requirements than that of earthquake loads.

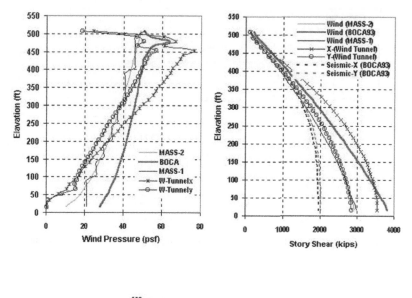

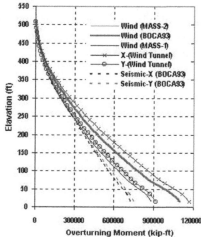

Figure 14   Comparison of Building Behavior for Equivalent Static Lateral Load.

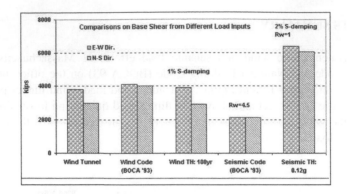

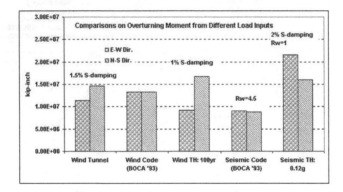

Figure 15   Building base shear and overturning moment comparison for different load condition.

        The effectiveness of viscous dampers on the office building were
summarized in Table 5. As seen from this table, viscous dampers will improve
the building dynamic behavior from 20% to 30%. These dampers give the building an additional inherent damping, equivalent to entire building structural

damping ratio of approximately 3%-4%. This is a significant increase above the assumed 1% structural damping for the tower without the dampers.

COSTS OF DAMPER SYSTEM

The dampers and bracing between the gusset plates shown in Figures 2 and 3 were fabricated and manufactured by Taylor Devices, Tonawanda, NY, USA. The cost of each diagonal damper was approximately $8,000.00 US. The cost of each TBD was $19,000.00 US. The total cost of the system installed was approximately $1,000,000.00 US.

Table 5 Results Summary Table for Time History Analysis.

		Wind Load Condition		Seismic Load Condition	
		E-W (X) Dir.	N-S (Y) Dir.	E-W (X) Dir.	N-S (Y) Dir.
Response without dampers	accel. at 37th Flr.(in/s²)	27.4	17.9	95.1	112.1
	displ. at 37th Flr.(in/s)	20.8	11.2	23.5	26.2
	Base Shear (kips)	3907	2903	6387	6057
Response with dampers	accel. at 37th Flr.(in/s²)	20.6	12.0	76.3	77.8
	displ. at 37th Flr.(in/s)	16.4	7.3	21.9	23.0
	Base Shear (kips)	3172	2038	5852	5246
6th~15th	Max. stroke(in)	0.37	0.77	0.59	1.87
	Max. damper force(k)	113	18	409	81
16th~25th	Max. stroke(in)	0.36	0.82	0.43	2.12
	Max. damper force(k)	108	17	326	80
26th~35th	Max. stroke(in)	0.32	0.80	0.55	2.17
	Max. damper force(k)	60	8	366	66
Overall Damping	Evaluated by energy	1.89%	2.0%	3.56%	3.8%
	Evaluated by accel.	1.94%	3.08%	3.56%	4.58%

Note: 1% and 2% internal modal damping included for wind and seismic condition respectively.

CONCLUSIONS

The viscous damper system with a motion amplification device proved to be a very cost effective method to reduce accelerations resulting from the buffeting by vortex shedding from winds off an adjacent tower. Cost of the installation was less than one million dollars. Other interesting aspects of the design are the following:

1. Current modal analytical approaches can produce erroneous results and should not be used for final design for systems with motion applification devices.
2. Non-linear time history analysis for both wind and seismic effects is required for damper system design.
3. Due consideration of the effects of the local damper forces must be considered. These forces can have a significant impact on the design of the local beams, connections and diaphragms.
4. The stiffness of the damper bracing system has an important role in the overall damper effectiveness. This is especially true for the stiffness of members in a motion amplification system. Large member stiffness' are required to ensure that the response reductions predicted analytically can be achieved in the actual installation.

REFERENCES/BIBLIOGRAPHY

Costantinou, M.C., Tsopelas, P. and Hammel, W., 1998
> TESTING AND MODELING OF AN IMPROVED DAMPER
> > CONFIGURATION FOR STIFF STRUCTURAL SYSTEMS,
> > Technical Report to the Center for Industrial Effectiveness and
> > Taylor Devices, Inc.

McNamara, R. J., Huang, C. D. and Wan, V., 1999
> PARAMETRIC STUDY FOR MOTION AMPLIFICATION DEVICE
> > WITH VISCOUS-DAMPER , Submitted to 2000 ASCE
> > Congress – Philadelphia.

Soong, T. T. and Dargush, G. F., 1996
> PASSIVE ENERGY DISSIPATION SYSTEMS IN STRUCTURAL
> > ENGINEERING, Wiley and Sons, London.

Performance Code Requirements in the Tall Building Environment

Robert E. Solomon
Chief Building Fire Protection Engineer

Brian Hagglund
Civil Engineer

Building risk and hazard from the threat of fire have been largely managed through prescriptive code requirements since the 1800's. Use of fire resistive materials, compartmentation features, and later, installation of automatic sprinkler systems and automatic fire alarm systems have worked together to make tall buildings extremely safe from the effects of a fire. Code prescribed mandates have worked very well to direct architects and engineers towards the necessary level of protection for the building occupants as well as the structure itself.

In the United States, many code enforcement jurisdictions have recognized the effectiveness of these integrated systems. While any number of anecdotal stories or narratives can be used to demonstrate this, perhaps the most compelling argument lies in the evacuation strategies associated with tall buildings. In the majority of circumstances, tall buildings have become a "defend in place occupancy" when the appropriate prescriptive systems and design features are present. In these environments, occupants are normally best served and protected by remaining in a given area provided they are not in close proximity to the initial fire.

The US code system is now moving to incorporate performance design options as an alternative to the long established, time tested, prescriptive design that has dominated fire protection codes and standards for so many years. A performance code, as defined by NFPA is essentially a document that states its goals and provides reference to some, but not all, of the approved methods to achieve those goals.

Numerous NFPA Technical Committees are moving to incorporate this design approach into the codes and standards that they author. This paper will provide a background on the unique issues surrounding high-rise fire protection and how the future solutions to this problem can potentially be handled through application of performance based design.

INTRODUCTION

The use of mandated and legally enforced prescriptive codes has been the primary method used in the US to provide minimum requirements for building design and construction. These codes, normally described as providing a minimum level of protection – to occupants, the structure and the contents in some cases – work to ensure that certain very broad goals are achieved, that buildings owners get what they pay for, and to some extent, to protect engineers and architects from litigation involving their design work.

The fire risk associated with high-rise buildings has been a concern to the fire community since these buildings emerged at the beginning of the century. The special code requirements for these structures reflect the fire experience in high-rise buildings and the need to provide adequate fire protection. The incorporation of performance based design options to the standard prescriptive code requirements is currently being undertaken by NFPA along with the U.S. code system. This paper will provide a background on the unique issues surrounding high-rise fire protection and how the future solutions to this problem can potentially be handled through application of performance based design.

A high-rise building, as defined by Section 3.3.25.6 of NFPA's *Life Safety Code*, 2000 edition is a building greater than 75 ft (23 m) in height where the building height is measured from the lowest level of fire department vehicle access to the floor of the highest occupiable story. However, different definitions may exist in local jurisdictions that may use height or number of stories for their defining criteria. In general, a high-rise structure is a building of seven stories or more. The national fire incident databases provide four categories of high-rise buildings: 7–12 stories, 13–24 stories, 25–49 stories, and 50 stories or more. The four property classes incorporated in high-rise buildings are office, hotel, apartment, and hospitals (or other care for sick facilities). The type of occupancy is important in determining the appropriate fire protection as different types of fires are likely to occur for each property class and the occupant characteristics are quite different for each building. The NFPA Fire Analysis Division utilizes these occupancy divisions and breakdowns to assist in their annual collection and data reports on the US fire problem.

Several key characteristics can be described for high-rise buildings that result in the need for specific fire protection engineering approaches to ensure the life safety of the occupants and the protection of property. The distinguishing features of tall buildings require different and additional design considerations than lower, more conventional buildings. Some of the major issues that influence the fire protection in high-rise buildings are fire department accessibility, egress systems, the effect of natural forces on a fire, increased occupant loads, multiple and mixed use occupancies, and the complexity of internal utility services.

TERMS AND DEFINITIONS

The following terms as used in this paper have the meanings as shown below:

Approved Method. Authoritative procedure used to develop proposed solutions. A more commonly used example of approved methods is contained within prescriptive documents.

Calculation Method.* A description of a system or phenomenon in terms of relationships among elements, permitting study of how some elements vary when other elements are changed. A calculation method normally consists of one or more mathematical relationships, permitting calculation of some elements based on their relationship(s) to other elements. Note that fire science and engineering use "model" as a synonym to calculation method but also for other concepts and elements such as scale models.

Computer Model. A calculation method that is packaged as computer software.

Fire Effects Model. A calculation method that incorporates engineering and scientific principles and applies them in a logical manner to determine possible consequences and extent of physical effects based on an externally specified fire, expressed as heat release rate as a function of time. (Typically referred to as a "fire model", even though it may not model combustion.)

Evacuation/Egress Model. A calculation method used to describe the behavior and movement of people during a fire situation. May be used in combination with a fire effects model to determine whether or not occupants safely escape from a building before being exposed to products of combustion.

Computer Fire Model. A calculation method that is packaged as computer software and used to predict fire behavior.

Design Team. Group of stakeholders including, but not limited to, representatives of the architect, client, and any and all pertinent engineers and other designers. A stakeholder is an individual, or representative of same, having an interest in the successful completion of a project.

Performance-Based Design Approach. A design process whose fire protection solutions are designed to achieve a specified goal for a specified use or application. This process allows performance-based documents to be implemented and ensures that their goals are met.

NOTE: The following describes a performance-based design approach:
a)   Establish fire safety goals.
b)   Evaluate the condition of the occupants, building contents, process equipment, or facility in question with regard to fire protection.
c)   Establish performance objectives and performance criteria.
d)   Identify potential hazards to be protected against.
e)   Define appropriate scenarios.
f)   Select suitable verification methods (e.g., fire models).
g)   Develop trial solutions.
h)   Assess proposed solution.
i)   Document proposed solution along with supplementary information.
j)   Obtain approval of the proposed solution.

Steps a) through f) are also part of the development of a performance-based code or standard. Only steps g) through j) are specific to performance-based design, where the intent is to find a solution for the project. Also, steps c), d), and e) are not necessarily intended to be sequential; they may in fact be concurrent. While the above is presented in a sequential order, the design approach does not necessarily need to begin with step a) and proceed consecutively through step j). Since different stakeholders (e.g., owner, designer, authorities) must be satisfied, some steps of this approach are iterative. Similarly, for performance-based document development, steps a) through e) may or may not be taken sequentially.

Performance-Based Document. A code, standard, or similar document that specifically states its fire safety goals and references approved methods that can be used to demonstrate compliance with its requirements. The document may be phrased as a method for quantifying equivalencies to an existing prescriptive-based document and/or it may identify one or more prescriptive documents as approved solutions. Furthermore, the document allows the use of all solutions that demonstrate compliance using approved methods.

NOTE: A performance-based document may also include separate prescriptive provisions as a parallel, independent approach to meet the performance-based goals and objectives.

Prescriptive-Based Document. A code or standard that prescribes fire protection for a generic use or application. Fire protection is achieved by specifying certain construction characteristics, limiting dimensions, or protection systems without providing a mechanism for how these requirements achieve a desired fire safety goal. Typically these documents do not state their fire safety goals.

NOTE: Many current NFPA codes and standards are not strictly performance-based or prescriptive-based: technically, they can be referred to as prescriptive documents containing some performance provisions. For example, a requirement for a one-hour door sets a measurable performance criterion, going beyond prescription of the door's construction, but does not link the criterion explicitly to a fire safety goal.

Proposed Solution. A fire protection system design intended to achieve the stated fire safety goals and which is expressed in terms that make it possible to assess whether the fire safety goals and objectives have been achieved. If models are used, then the proposed solution should also specify the models and input data employed.

Top-Down. One approach used to develop performance-based provisions. Using this approach, the goals and objectives are developed during document revision processing without consideration of any current prescriptive requirements: a "clean sheet of paper" approach. See "Bottom-up".

Verification. Confirmation that a proposed solution (i.e., candidate design) meets the established fire safety goals. Verification involves several steps. Verification confirms that the building is built as proposed to a design that will achieve the intended level of safety and that the building's ability to achieve the level of safety has been demonstrated by qualified people using the correct methods applied to the correct data.

HISTORY OF FIRE IN HIGH-RISE BUILDINGS

Tall building fires repeatedly show the importance of implementing the existing and proven fire protection technologies. In 1996, high-rise building fires in the US in all occupancies combined had 12,100 reported structure fires that resulted in 64 civilian deaths, 790 civilian injuries, and $69.1 million in direct property damage.

Fires in high-rise buildings are significant events in the fire protection engineering profession. Such fires have often resulted in changes to prescriptive code rules, and have provided the basis for improved technologies to deliver water supplies that are adequate to combat fires in the upper reaches of the building. Automatic sprinkler systems, standpipe systems, and fire alarm systems have all been tailored in their own way to properly function in the unique environment of the tall building. Retroactive automatic sprinkler system regulations have been passed in numerous states and cities in the US in recognition of the need to provide a needed level of protection.

The US, while perhaps having the greatest number of tall buildings in the world, shares in a worldwide fire history where fires in high-rise buildings have resulted in multiple fatalities to both civilians as well as fire suppression personnel.

Table 1 captures a small percentage of fatal fires in high-rise venues (NFPA, 1999).

Table 1   Fatal Fires in High-Rise Buildings.

Date	Building	Location	Number of Fatalities	Floor of Origin/ Total Height (Stories)
25 Mar 11	Asch Building	NY, NY	146	8/10
1 Aug 32	Ritz Tower	NY, NY	8	Sub-basement/42
5 June 46	Hotel Lasalle	Chicago, Il	61	1/22
28 June 63	Astoria Building	Rio De Janeiro, Brazil	7	14/22
7 Dec 67	Time-life	Paris, France	2	8/8
24 Jan 69	Hawthorne House	Chicago, Il	4	36/39
25 Dec 71	Tae Yon Kak Hotel	Seoul, Korea	163	2/21
23 July 73	Avianca Tower	Bogata, Colombia	4	13/36
1 Feb 74	Crefisual Bank Building (Joelma)	Sao Paulo, Brazil	179	12/25
21 Nov 80	Mgm Hotel	Las Vegas, NV	85	1/23
8 Feb 82	Hotel New Japan	Tokyo, Japan	32	9/10
5 Sept 86	Hotel	Kristianstead, Norway	14	1/13
31 Dec 86	Dupont Plaza Hotel	San Juan, Puerto Rico	96	1/20
23 Feb 91	Meridian Plaza	Philadelphia, Pa	3	22/38
20 Nov 96	Office	Hong Kong	40	Basement/16
23 Dec 98	West 60 Th. Street Towers	NY, NY	4	12/40

Some common factors contributing to significant losses in high-rise fires are listed below. These conditions can be eliminated from buildings with the use of adequate, prescriptive based fire protection criteria and performance-based design.

- Lack of automatic detection equipment
- Inadequate/locked/blocked exits
- Inadequately protected vertical and horizontal openings
- Lack of alarm system, poor accessibility of alarm system
- Inadequate water supply for the standpipe system
- Lack of compartmentation
- Lack of automatic sprinkler protection.

NFPA fire investigation reports issued on several of these fires including the ASCH, Hotel LaSalle, Crefisual, MGM Hotel, Dupont Plaza and Meridian Plaza fires could all be narrowed down to some combination of these seven conditions. There is nothing inherently dangerous about high-rise buildings from the point of fire protection and life safety. It is noted in the select cases shown, that several of these fires had their point of origin at or near the ground floor.

Prescriptive Code Provisions and Fire Protection

Building code, life safety code and fire prevention code regulation in the United States have largely grown out of insurance industry regulation and rules. For example, building construction features were managed so as to provide protection to the contents that may have been stored in a warehouse. Protection goals may have simply been to keep weather related events (rain, snow, excessive sun) from damaging or altering the stored content. Beyond weather related damage, damage from fire related events also became a subject of concern for the US insurance industry. In general terms, the initial concerns for such losses focused on the factory and warehouse environment. Single, large, unmanaged fires could easily destroy a building or complex, thereby rendering the local economy, local residents, and entire companies in dire straits.

In a series of related events, the widespread losses that occurred during large, massive, urban conflagrations in key developing cities, including Boston, Chicago, Baltimore and San Francisco, the larger issue of how a fire in one building or structure could impact on an adjacent building, structure, neighborhood, or entire city, soon became an issue for all codes to consider. Code prescribed rules of minimum separation distance to other properties, imposition of selected construction techniques and installation of automatic fire sprinkler systems in many buildings set the stage for formalized rules in codes, standards and even certain zoning regulations.

Multiple story buildings, designed to take advantage of the limited real estate, set back, user needs and space limitations normally found in urban centers, have proven to be an effective use and utilization of these conditions. Code development organizations and insurance interests while initially reluctant to embrace the concept of not only two story structures used to operate factories, had to be brought into the fold of accepting the ever increasing challenge of story 'creep', that is, engineering limitations on building materials and design techniques appeared to be the only forces curtailing the design of taller buildings. With time, these limitations were torn down, and more floors were built on.

Regulatory documents were expanded to require heavy timber construction techniques in certain multiple-story buildings (generally factory and warehouse buildings, five and fewer stories in height), and 'fire proof' construction (generally referred to as fire resistive construction) in buildings taller than five stories in height. In 1920, the US code system incorporated another prescriptive regulation for building construction, the height and area table. Height and area tables, still used in all of the major US codes, impart selected maximum building footprint areas, and maximum building heights, in both a number of stories as well as a vertical, linear dimension. Allowable heights and areas relate directly to allowable building construction types. These combinations are intuitive as fire resistive, building construction types are generally permitted to reach unlimited areas and unlimited heights given the presence of certain other design features. Conversely, selected types of wood construction receive severe limitations with respect to both height and area, even when other positive fire protection attributes are present.

Code regulation trends also started to move towards a method of protecting the occupants of these multiple story buildings. On March 25, 1911, the first documentation of a fire in a US high-rise building (as defined by today's standards) occurred in the 10 story ASCH building in New York City. The primary tenant of this building was the Triangle Shirtwaist Company. The 146 lives lost in this building marked the beginning of the need to control and provide select features that protected not just the building and contents, but also the occupants. This fire also was significant in that the strategies associated with fighting fires in taller buildings were brought to the forefront.

A MOVEMENT TOWARDS PERFORMANCE CODE REQUIREMENTS

A level of performance requirements in US codes and standards was present in certain regulatory standards at the turn of the century. As an example of this, the following statement concerning placement and positioning of automatic sprinklers in buildings is taken verbatim from the 1896 edition of NFPA 13, Standard for the Installation of Sprinkler Systems. "Sprinklers should be located so as to not be shielded by building construction features" (NFPA, 1896). In the 1999 edition of this standard, no less than seven pages of text and accompanying diagrams are necessary to detail how you can achieve the performance goal from the first edition of this standard. The US code system, at least as it relates to fire protection, has come full circle. Minimal, goal oriented text, has been expanded, detailed and otherwise revised to be as thorough as possible.

A new era of design challenges has opened the door to consider other than the tradition of strict, prescriptive approaches to fire protection. Highly specialized industrial facilities, large assembly occupancy buildings, extreme high-rise buildings, and the need to out do other building designs by doing more, doing better and completing projects on time and in a more economical way, have all contributed to the decision to introduce performance-based design. PB designs leverage the ability of engineers and architects to challenge the status quo. This option allows, as a minimum, the following elements to be considered: (NFPA, 2000a)

1. Allow code developers to quantify established prescriptive code requirements. Doing something the same way, for 20, or 30 or even 100 years may not necessarily be the best method or solution.
2. Permits designers to provide highly specialized and innovative designs for aesthetic and effect purposes.
3. Permits designers to filter out excessive or overly conservative design features that may not enhance occupant safety or building protection features.
4. Allow designers to explore techniques, methods and solutions not explicitly covered by the prescriptive regulation.
5. Allows end users (building owners and operators) to modify prescriptive code regulations to provide more than the minimum level of protection that is normally provided in codes and standards.

NFPA embarked on a program in 1995 to begin the development of a system where NFPA Technical Committees could begin establishment of a process to integrate a PB design option in the codes and standards developed in the NFPA code development system (NFOPA, 1995). The first document to complete this process was the 2000 edition of NFPA 101, Life Safety Code (NFPA, 2000; CTBUH, 1999). The process used to arrive at this point followed the outline of the 1995 white paper from the NFPA Board of Directors. In general terms, the method used was one of the top down approach as shown in Figure 1. Establishment of the goals of this particular code (which were actually first specifically refined in 1991) were used to build upon the line of attack to outline the relevant items that would have to be considered in order to arrive at a credible life safety design.

The goals and objectives established in this top down approach can be used to consider all manner of building hazards beyond life safety. For example, this model can be applied to a building's ability to withstand select environmental loads, seismic events, and fire events as they relate the functionality of the building following a severe fire event. In this process, design goals should be established by the legal equivalent of a code. The subset of objectives, established by the design team can establish the level of performance needed during and following the event. In terms of life safety only goals, the building performance criteria, and subsequent objectives, these elements may only have to allow for adequate time for building occupants to relocate to safe areas within the building, or to permit time for occupants to evacuate the building.

The tall building environment fully recognizes the benefit, and the normally accepted practice, of relocating occupants from the floor of origin to areas remote from the fire. This environment also recognizes the importance of maintaining structural integrity during a fire event, and in recognizing the need to complete repair and clean up as soon as possible so as to allow 'business continuity' to the extent possible. Beyond this, the potential for structural collapse during a fire event is intolerable given the likely impact on adjacent and neighboring properties. Objective based PB code designs should always consider the effects of a fire on not only the building of interest, but also on neighboring properties.

The core of arriving at the ability to provide a solution deemed to be acceptable by a governmental regulatory authority (this term is defined by NFPA as the Authority Having Jurisdiction-AHJ) is complex, thus PB designs are unlikely to be applied to typical design, or design build projects. While the NFPA PB Primer establishes the framework for the method to be followed, the collective opinions of the design team must be vetted, refined and codified in a manner such that everyone, including the AHJ, is able to defend the ultimate design solutions, and hence design options that are offered.

Verification of the applied design methods or solutions is expected to come forward in one of four ways, or more likely, some combination of these. The four basic verification methods include:

1. Deterministic. Based on mathematical equations to verify an assumption or phenomenon.
2. Probabilistic. Based on historical loss data.
3. Heuristic. Based on investigation of losses, general problem solving, and judgment.
4. Laboratory Tests. Based on scale tests of materials or assemblies.

While none of these methods should be favored in one manner or another, the use of deterministic and probabilistic methods are likely to have more credibility to obtain recognition of one acceptable solution over another acceptable solution. Deterministic methods encompass the relatively new (25 years) use of computer fire models (Society of Fire Protection Engineers, 1994). A variety of these models can be used to simulate fire growth characteristics, smoke movement and even evacuation and movement times for building occupants. Input for these models as well as interpretation of the output from such models must be carefully scrutinized. Subsets of this category include hand calculations, physics models (which is further divided into 3 categories), and evacuation models.

The subset of 'field models' under the physics model category is viewed by many as the most promising model for simulation of fire effects. This type of simulation evaluates a nearly infinite number of space or compartment volume units and is generally referred to as a computation fluid dynamics (CFD) model. CFD models, while requiring robust computing platform capacity, provide a more thorough analysis of the growth, development, and movement of the fire, as well as its associated products of combustion.

Probabilistic data and tools are essentially used to gauge the occurrence of a given event or a given result once a challenge is applied to a building or structure. Expected value risk models are included in this category and can be used to establish likely outcomes of a given event or scenario. In other words, how likely is it that a given hazard scenario may occur in a given building, and what is the consequence should that hazard scenario occur.

Heuristic methods allow the true creativity (with limits) of the design community to be put fourth. In the US, this has allowed code development organizations to look well beyond our borders and to more formally see how fire protection approaches differ in other countries. In a much broader sense, this area has also brought together various international entities to evaluate key issues that surround the use of PB design options. Notable work includes that of the International Council for Research and Innovation in Building and Construction (CIB, 1998). One method under discussion by CIB involves a whole building approach. In this method, five broad, building categories are identified. Corresponding building attributes are then identified and segmented into three categories. This building performance matrix can then be used to inter-relate the categories and the attributes. Whole building parts are included in Table 2. The related building attributes are in Table 3.

Arguably, it can be defended either way that all of the subcategories impact or relate to the fire protection needs of a building. In the specific case of tall buildings, PB designs must consider the functional needs of the client – which is likely to translate to the number of building occupants. This value alone will begin to drive useable square meters on a given floor. Space taken up by elevator shafts, exit stairs and HVAC and utility shafts must all be considered in the PB design analysis. In these examples, building component parts which improve access (mechanical transport), and which can be used in a time of emergency (exit stairs), must also be evaluated as smoke transport conduits.

Smoke movement in tall buildings is a sometimes-contentious subject. Should vertical smoke movement in a tall building be completely non-existent, or is it reasonable to expect, and even tolerate some 'acceptable' amount of smoke on upper floors provided it does not contain lethal by-products? Current prescriptive code rules in the US provide numerous requirements for enclosing vertical shafts in multiple story buildings, yet these rules do not specify a pass/fail rule for smoke movement in the field. Overall PB design strategies will be able to help quantify such subjective criteria.

Table 2   Whole building parts.

CATEGORY	SUBCATEGORY	RELATED TO FIRE PROTECTION
SPACE	Functional Space	ü
	Building Envelope Space	
STRUCTURE	Sub-structure	
	Super-structure	ü
EXTERNAL ENCLOSURE	Below ground	ü
	Above ground	ü
INTERNAL ENCLOSURE	Vertical	ü
	Horizontal	ü
	Inclined	ü
SERVICES	Plumbing (Water and waste)	
	Heating, Ventilation and Air Conditioning	ü
	Fuel System	
	Electrical system	ü
	Communication system	ü
	Mechanical transport	
	Security and protection	ü
	Fitting	ü

Installation of select, specific systems such as automatic sprinkler systems will all but be insured in the tall building environment. While continuing to be a standing, prescribed system in many codes for the tall building environment, it is nearly inconceivable that a thorough PB design will include an acceptable solution that does not include the installation of an automatic sprinkler system. Nonetheless, PB design recognition will at least open the possibility that someone will at least consider such an option.

As previously stated, PB design is going to be reserved for all but the most challenging and exigent projects. Tomorrow's tall building designers will have more design options available, but issues of reliability, redundancy, predictability, safety factors and conservatism are still being actively debated in the fire protection engineering community.

Table 3 Related building attributes.

CATEGORY	SUBCATEGORY	RELATED TO FIRE PROTECTION
SAFETY	Structural Fire Accident (Safety in Use)	ü ü
HABITABILITY	Structural Serviceability Thermal Comfort Tightness (Water and Air) Air Quality Acoustic Lighting Access Security Condensation Health and Hygiene Functionality Adaptability Aesthetic	 ü ü ü ü
SUSTAINABILITY	Maintainability Durability Economics Decommission Environmental Friendliness	ü ü ü ü

A generally agreed upon approach to help leverage PB design options into the larger picture of tall building design should set or consider six goals. These goals include:

1. Life Safety of Building Occupants
2. Property/Contents Protection
3. Mission Continuity
4. Environmental Consequence of Fire
5. Heritage/Cultural Preservation
6. Fire Suppression Personnel Safety

These elements can begin to set the stage for PB design. The CTBUH as an organization is an obvious venue to share PB design ideas in fields as diverse as steel erection techniques, concrete batching methods and challenges associated with adding larger populations to the tall building environment. PB fire protection engineering design, which is the newest entry into the field of PB design, can contribute to overall goals of allowing larger buildings to continue to be safely built. Innovative and novel design techniques can be safely used to establish reasonable goals and objectives to keep occupants of the tall building environment safe, and to ensure that fire suppression personnel are properly trained, protected and able to safely take any necessary actions to control and supplement automatic fire suppression systems. The end results of a more general movement towards PB design in the fire protection engineering community should contribute to safer buildings, more economical design, and use of improved materials and methods. The added return is also likely to provide improvements to the established status quo of prescriptive code regulations.

REFERENCES/BIBLIOGRAPHY

CIB, 1998
> DEVELOPMENT OF THE CIB PROACTIVE PROGRAM ON PERFORMANCE-BASED BUILDING CODES AND STANDARDS, 1998, International Council for Building Research, Studies and Documentation.

Council on Tall Buildings and Urban Habitat (CTBUH), 1999
> FIRE PROTECTION ISSUES IN THE HIGH RISE ENVIRONMENT, *The Tall Building and the City, The State of the Art for the Millennium* , Proceedings of Conference, 3–4 May 1999, Kuala Lumpur, Malaysia.

National Fire Protection Association (NFPA), 1896
   NFPA 13, National Board of Fire Underwriters – Recommendations and
      Regulations, National Fire Protection Association.

NFPA, 1995
   NFPA'S FUTURE IN PERFORMANCE-BASED CODES AND
      STANDARDS – JULY 1995, National Fire Protection Association.

NFPA, 1999
   INTERNATIONAL LISTING OF FATAL HIGH RISE STRUCTURE
      FIRES 1911–1999: NFPA Fire Analysis Division.

NFPA, 2000a
   NFPA PERFORMANCE-BASED PRIMER-CODES AND
      STANDARDS PREPARATION – 2000 EDITION, National Fire
      Protection Association.

NFPA, 2000
   NFPA 101, SAFETY TO LIFE FROM FIRE IN BUILDINGS AND
      STRUCTURES, 2000 Edition – Chapter 5, National Fire Protection
      Association.

Society of Fire Protection Engineers, 1994
   COMPUTER SOFTWARE DIRECTORY, Society of Fire Protection
      Engineers.

Analysis and Model Techniques for Determining Dynamic Behaviour of Civil Structures with Various Damping Systems

B. Breukelman, T. Haskett, S. Gamble and P. Irwin

1 INTRODUCTION

Controlling vibrations of structures through the use of damping systems continues to increase as structures become lighter and more slender. This includes buildings, towers, vehicular bridges and pedestrian bridges. As a result, the tools used for design of these supplemental damping systems need to take into account the nature of the applied loads to get a more accurate determination of the way the structure will behave with the damping system.

Wind tunnel testing to determine the wind-induced forces on structures has been common for many years. It is common for significant buildings and towers to have wind tunnel testing performed, beginning with the High-Frequency Force-Balance (HFFB) test and occasionally an aeroelastic wind tunnel test is performed to better determine aeroelastic effects such as aerodynamic damping and vortex-induced phenomena. Detailed knowledge of the nature of the wind-induced forces can be used with advanced analysis techniques to accurately describe the behaviour of a supplemental damping system. In addition to wind-induced forces, it may be necessary to consider seismic forces if the structure is located in a region with seismic activity.

To date, damping systems which have been implemented to reduce vibrations of structures have been primarily designed using frequency domain analysis techniques. Beginning with the work of Den-Hartog (1956), many have produced estimates of effective damping of a damped structure and reduction in motion and amplitudes of the damping system. Computing limitations have limited the ability to perform time-domain analysis of the combined structure and damping system. More recently, software commonly available to structural engineers (e.g. SAP2000) has the ability to perform analyses in the time domain. The use of these software packages for time domain analyses is gaining increasing acceptance.

This paper intends to briefly review currently available analysis methods available for analysing the behaviour of a structure with a supplemental damping system. The limitations and advantages of each analysis tool are given. An advanced technique for investigating the behaviour of a structure with a damping system is presented.

2   REVIEW OF ANALYSIS METHODS

Frequency Domain Analysis

The frequency domain type of analysis is computationally efficient, and is based primarily on assumptions about the nature of the forces. For the tuned mass type of damper a simplification of the structure to a simple two-degree-of-freedom (DOF) system is also typically employed. Tuned mass systems would include the well known Tuned Mass Damper (TMD), Tuned Liquid Column Damper (TLCD), Liquid Column Vibration Absorber (LVCA) and the Tuned Liquid Sloshing Damper (TLSD). For wind induced motion it is common to assume that the spectrum of forces can be approximated by a white-noise spectrum. The equations for a 2 DOF system (DenHartog, 1956) can then be used together with the assumption of white noise input forces to arrive at estimates of the building response. (McNamara 1977, Wiesner 1979, Tanaka 1983, Simui and Scanlan 1986.)

The frequency domain method's principle advantage is that it does not require significant computing power. A good approximation of the effective damping provided by a TMD is readily provided. This allows a number of concepts to be quickly investigated, through the use of parameters such as mass ratio, tuning ratio and TMD damping ratio. Another advantage of the frequency domain analysis is the ability to "envelope" the response. Given that there are variations in the frequency content of the input forces, whether wind or seismic, analysis in the frequency domain can ensure that the full range of possible excitation frequencies can be investigated, not just those present in available time histories. This is especially important for seismic events, where time histories of the "design" seismic event are usually synthesized from a code required design spectrum.

Of the limitations of the frequency domain analysis, the primary one is that non-linear effects cannot be easily investigated. For example, off-the-shelf viscous damping units have a force proportional to velocity to the 2nd power. Linear viscous dampers can be produced, but there are additional advantages from using the standard unit, that being the ability to limit motions of a TMD during extreme events. Essentially, due to the higher damping forces produced due to the higher velocities, the TMD becomes overdamped and the amplitudes during the extreme event are reduced compared with those with a linear TMD. Similar non-linearities exist for the TLCD, LCVA and TSLD types of damping systems, some of which are advantageous in design. In addition to limiting the analysis to linear behaviour, analysis in the frequency domain prevents an investigation of more complex coupled TMD-structure interactions.

For many structures, it is sufficient to investigate the seismic behaviour of a structure in the frequency domain. Most code-type calculations find their basis in the frequency domain method combined with a site-specific response spectrum. For more significant structures, it is necessary to investigate non-linear behaviour of structural members. The limitations of analysis in the frequency domain force one to investigate the non-linear (elasto-plastic) behaviour of a seismically loaded structure in the time domain. In considering the design a viscous damping system to reduce wind-induced motion, the forces generated under the design seismic event must also be determined where the structure is located in a seismically active region. As the viscous damping system may have been optimized for wind with its small deflections, the seismic forces generated by the damper can be excessively large if linear viscous damping units are used. A non-linear viscous damper can be utilized, with force being proportional to velocity to a power lower than unity. The total effect of using such a non-linear viscous damper in a structure cannot be readily determined using frequency domain analysis. Similar difficulties arise when visco-elastic dampers are being considered because the non-linearity of the visco-elastic material can be difficult to model in the frequency domain.

Time Domain Analysis

The direct integration of the equations of motion with a variety of input forces for a number of passive damping system types has been reported by several authors (e.g. McNamara, 1977, Kawaguchi et al., 1992, McNamara et al., 2000). Due to the significant computational effort involved, this analysis has generally been used for seismic analyses as the typical seismic event is of much shorter duration than a design wind event. Where a TMD (or TLCD, LCVA, etc.) is being considered, this analysis has in the past only been performed to validate the theoretical basis of the frequency domain method and to verify the selection of parameters determined using the simpler frequency domain analysis. As a result, parametric time domain analyses of the behaviour of a combined building/damping system are not often performed.

Recent advances in computing power and software has given more flexibility to engineers to perform these analyses. Commercially available software packages allow the user to perform analyses in time domain. McNamara et al. (2000) report using this feature to analyse the behaviour of a 39 storey office tower under both wind and seismic loading conditions. However, as commercial software packages compute the response using the fully modelled structure, the number of parametric investigations that can be performed are often still limited by computing time.

For cases where wind-induced motion is being investigated, typically a simulated wind spectrum is used (Kawaguchi et al., 1992). Although the benefit of using measured wind force time histories (for example from a High-Frequency Force-Balance "HFFB" wind tunnel test) as input in the direct integration is understood, to date an efficient method of utilizing all of this data in an efficient integration algorithm has not been developed.

3   ADVANCES IN STRUCTURAL RESPONSE MODELLING

RWDI has developed an efficient means of integrating the response of a structure in the time domain, to both wind and seismic forces. In summary, the numerical integration procedure combines the efficiency of modal analysis with a direct spatial definition of the parts of the damping system. The specific geometry of the damping system, whether it is a TMD or TLCD (or other damping system) is combined with the modal definition of the structure. In this way, coupling of the modes of vibration, torsional effects, etc. can be directly evaluated in the time domain.

Wind-Induced Responses

In analysing the behaviour of a structure to wind-induced forces it is customary to assume that only the fundamental bending and torsional modes are excited by wind forces due to the nature of the wind-force spectrum. This enables us to reduce the description of the building to only the first 3 fully coupled modes of vibration. If necessary, additional modes can be easily modelled. This is a substantial advantage over performing a time domain analysis in the spatial environment with a finite element analysis (FEA) package, where the structural engineer may have a model with thousands of nodes.

This FEA model is still necessary, however, to generate the coupled modes and frequencies for use in this technique. The development of the equations of motion is as follows:

We begin with the basic form of a harmonic oscillatory system,

$$[M]\frac{d^2}{dt^2}\begin{Bmatrix} x_1 \\ y_1 \\ \vdots \\ y_n \\ z_n \\ h \end{Bmatrix} + [C]\frac{d}{dt}\begin{Bmatrix} x_1 \\ y_1 \\ \vdots \\ y_n \\ z_n \\ h \end{Bmatrix} + [K]\begin{Bmatrix} x_1 \\ y_1 \\ \vdots \\ y_n \\ z_n \\ h \end{Bmatrix} = \begin{Bmatrix} F_{1x} \\ F_{1y} \\ \vdots \\ F_{nx} \\ F_{ny} \\ 0 \end{Bmatrix}$$

where the M, C, and K matrices represent the floor-by-floor mass, damping, and stiffness properties of the building/damper system. The position vectors give the x, y, and z deflections of the first floor and are followed by the second floor etc. Similar to what is generally used in a HFFB analysis, a radius of gyration which is characteristic of the topmost floors of the building is multiplied into the rotational component (z) of building motion, to create unit similarity in the position vector. The value h is a single coordinate of damper deflection, and may represent either one principle axis of a TMD, or the column motion of a TLCD etc. For the sake of further discussion we will consider a TMD. The vector of forcing functions is laid out in the same fashion as the position vector.

In this form, significant difficulty is encountered when attempting to define the entries of the C and K matrices, due to our approximation of the large nodal model as a simple chain structure. Therefore, the next step is to separate the M, C, and K matrices into sums of building- and damper-related matrices as

$$
\left(\left[M_{building}\right]+\left[M_{TMD}\right]\right)\frac{d^2}{dt^2}\begin{Bmatrix}x_1\\y_1\\\vdots\\y_n\\z_n\\h\end{Bmatrix}+\left(\left[C_{building}\right]+\left[C_{TMD}\right]\right)\frac{d}{dt}\begin{Bmatrix}x_1\\y_1\\\vdots\\y_n\\z_n\\h\end{Bmatrix}+\left(\left[K_{building}\right]+\left[K_{TMD}\right]\right)\begin{Bmatrix}x_1\\y_1\\\vdots\\y_n\\z_n\\h\end{Bmatrix}=\begin{Bmatrix}F_{1x}\\F_{1y}\\\vdots\\F_{nx}\\F_{ny}\\0\end{Bmatrix}
$$

where all terms coupling the TMD to the building are contained within the subscripted TMD matrices with the TMD properties.

The 3 fundamental eigenvectors of the building (excluding the damping device) are augmented with zeroes in the TMD position, and a new vector is added which is a unit translation of the damping device. It has been found convenient to normalize the magnitude of each column of the eigenvector matrix to unity. We then transform the system into a new coordinate set which is a combination of building modal coordinates and TMD displacement. This is accomplished by the substitution $\bar{x}(t) = \Psi\bar{q}(t)$, where Ψ is the augmented eigenvector matrix, and premultiplication of each term by the transpose of Ψ.

$$
\Psi^T\left(\left[M_{building}\right]+\left[M_{TMD}\right]\right)\Psi\begin{Bmatrix}\ddot{q}_1\\\ddot{q}_2\\\ddot{q}_3\\\ddot{h}\end{Bmatrix}+\Psi^T\left(\left[C_{building}\right]+\left[C_{TMD}\right]\right)\Psi\begin{Bmatrix}\dot{q}_1\\\dot{q}_2\\\dot{q}_3\\\dot{h}\end{Bmatrix}+\Psi^T\left(\left[K_{building}\right]+\left[K_{TMD}\right]\right)\Psi\begin{Bmatrix}q_1\\q_2\\q_3\\h\end{Bmatrix}=\Psi^T\begin{Bmatrix}F_{1x}\\F_{1y}\\\vdots\\F_{nx}\\F_{ny}\\0\end{Bmatrix}
$$

The quantities of generalized mass, damping and stiffness are substituted into the building products, and the system of equations becomes:

$$\left(\left[\begin{array}{c|c}GM & 0 \\ \hline 0 & 0\end{array}\right]+\Psi^T\left[\begin{array}{c|c}M & \bar{m} \\ \hline \bar{m} & m\end{array}\right]\Psi\right)\left\{\begin{array}{c}\ddot{q}_1 \\ \ddot{q}_2 \\ \ddot{q}_3 \\ \ddot{h}\end{array}\right\}+\left(\left[\begin{array}{c|c}GC & 0 \\ \hline 0 & 0\end{array}\right]+\Psi^T\left[\begin{array}{c|c}C & \bar{c} \\ \hline \bar{c}^T & c\end{array}\right]\Psi\right)\left\{\begin{array}{c}\dot{q}_1 \\ \dot{q}_2 \\ \dot{q}_3 \\ \dot{h}\end{array}\right\}+\left(\left[\begin{array}{c|c}GK & 0 \\ \hline 0 & 0\end{array}\right]+\Psi^T\left[\begin{array}{c|c}K & \bar{k} \\ \hline \bar{k}^T & k\end{array}\right]\Psi\right)\left\{\begin{array}{c}\dot{q}_1 \\ \dot{q}_2 \\ \dot{q}_3 \\ \dot{h}\end{array}\right\}=\left\{\begin{array}{c}GF \\ 0\end{array}\right\}$$

Where GM, GC, and GK are diagonal matrices containing the generalised properties of the building system. The constants m, c, and k are properties of the TMD and are chosen according to design objectives. The submatrices M, C, and K, and vectors $\bar{m}$, $\bar{c}$, and $\bar{k}$, are formulated from a free-body diagram of the building-damper interaction and contain the appropriate coupling terms. To date, we have modelled the interaction of both TMD and TLCD systems, where the appropriate terms are determined assuming that the coordinates of damper device motion and building motion are aligned.

The remaining quantities in this system of equations are the generalised forcing functions due to wind excitation. These time series are available from wind tunnel testing which would have been performed to allow a traditional HFFB analysis. Care must be taken when converting forces measured in the wind tunnel to generalised forces that the eigenvectors are scaled as above. These generalised forces are at model scale, and must be subsequently scaled to full size with the appropriate non-dimensional factors.

In this way, we have simultaneously avoided entry of the unavailable or near-singular building C and K matrices, and reduced the number of equations in the system to a much more manageable size – all without any loss of information or simplifying assumptions.

The versatility of this technique is that it now allows for any type of damping device to be simulated within the building. For example, readily available commercial viscous dampers have velocity squared relationships with respect to force. Such a damper may be easily modelled by the addition of a term that uses velocity times the magnitude of the velocity, while simultaneously retaining the linear velocity dependence of modal damping.

For many software packages, it is necessary to change the above second order system of equations into a set of first order equations, i.e. the state-space representation. Numerical simulation of the above system of equations may then proceed with an accurate algorithm like the 4th order Runge-Kutta scheme. This is a relatively quick process for a typical three hour storm at full scale.

Post processing of the building modal state variables may be performed to yield the spatial response of a desired floor, for example the top occupied floor of the building. Unlike the frequency domain approach, statistical measures of the actual TMD displacement or forces may be evaluated with an Extreme Value type of approach, instead of assuming buffeting behaviour with its attendant peak factor characteristics. Figure 1 shows a sample time history of TMD and top occupied floor responses for a strong across-wind response. The degree to which the TMD resonates with the structure is increased for this scenario, and the effective damping is thus greater. Such a phenomenon is not accounted for by a traditional frequency based approach. Additional data, like peak velocity or power dissipation, assists in the physical specification of components to design the damping system.

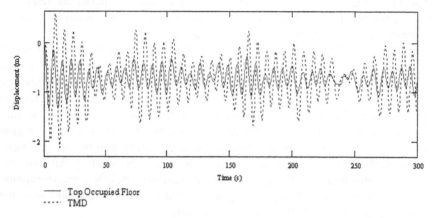

―― Top Occupied Floor
----- TMD

Figure 1 Floor and TMD Motion for an Across-Wind Event.

The numerical simulation may be performed for a range of full scale wind speeds and using data collected in the wind tunnel, can be used to simulate incident wind from multiple sectors around the building. The only change is the input wind tunnel data file, which corresponds to the wind direction. Once all rotational sectors around the building are analysed with the wind tunnel data files, the response of the building has been determined which accounts for the influence of surrounding topography and local structures. Lastly, these results may be combined with meteorological data for the region to determine probabilistic measures of structure and/or TMD response. For example, one can produce within a reasonable degree of effort predicted acceleration levels for a damped structure for a complete range of return periods. This can then be compared to acceleration estimates made by the frequency based HFFB technique to evaluate the effectiveness of a certain set of damping system parameters, such as TMD mass, frequency, damping, etc.

Seismic Responses

When designing a passive damping system to reduce wind-induced motion for a structure in a seismically active region, it may be necessary to evaluate the forces imposed on the structure due to the presence of the damping system. Note that the purpose of the damper is (in our applications) for occupant comfort during wind events, and not to reduce the response of the structure during an earthquake.

The time domain approach proceeds mostly as outlined above. However, it is prudent to make use of a modal model of the building which includes additional modes, either to a reasonable frequency with respect to seismic events for that region, or so as to include most of the mass of the building. In practice, we have found it necessary to include between 10 and 30 modes of the building for a reasonable simulation. Notice that 30 modal coordinates, plus a couple of damper degrees of freedom, still represent a significant computational reduction when compared to the full FE model, yet does not make any substantial sacrifice when macroscopic details are to be investigated.

The second major difference between this and the wind simulation is that of the forcing function. In the wind simulations, all building and damper motion is taken with respect to an inertial frame. For our seismic simulation, without a complete modal model of the building, it is necessary to use a frame of reference that moves with the base of the building. From this frame, we may model excitation forces on each floor by d'Alembert's principle. Our distributed excitation force is then that of ground acceleration, scaled according to the design event, and multiplied by the respective mass of each floor of the building. All post processing of this numerically simulated event must then include a transformation into inertial space, by adding the displacement of the base of the building to the displacement records of all floors of the building.

The principal limitation of this technique is that it simulates the building response, with damping system, to only a single earthquake event, and only from a single incident direction. Many simulations must therefore be run to investigate directional sensitivity, and to attempt to envelope all possible seismic responses.

In an effort to take advantage of the fact that the frequency domain analysis assists with enveloping structural responses, a modification of the National Building Code of Canada (NBCC) frequency domain analysis for seismic events was developed. The NBCC allows for a dynamic analysis to determine the distribution of inter-storey shear loads, which must be scaled to sum to the Code-determined total base shear. The basis of the technique is a summation of modal responses, on a floor-by-floor discretization.

To proceed further in this vein, we must make an assumption. Typically, the structure is assumed to have a modal damping ratio of about 5% against earthquakes in each mode, as per the excitation spectrum in the NBCC. With the addition of a damping device, such as a TMD tuned to the appropriate frequency, we introduce a new mode at a frequency below the fundamental. We make the assumption that the damping ratio of this new mode is also 5%, when in practice it may be 2 to 3 times greater. This is thought to be a conservative assumption, as comparison of damping and restoring forces in the time domain suggests that the restoring forces dominate.

This frequency-based approach with a damping system is begun by first building the system of equations as outlined above for wind, with the exception of the damping matrix. At this point, we are dealing with a system of equations that contains q modal coordinates, and s spatial TMD coordinates for a building with n floors. Solving the new eigenvalue problem produces a set of new natural frequencies and mode shapes for the damped system. We now need to premultiply the second generation eigenvector solution matrix by the original set to create a composite matrix that relates spatial building floor displacements with these new modal coordinates.

The lateral floor loads for each frequency of vibration are determined by

$$P_{ij} = m_i \frac{\sum\limits_{k=1}^{n} m_k \Phi_{kj}}{\sum\limits_{r=1}^{n} m_r \Phi_{rj}^2} \Phi_{ij} \omega_j S_{v(j)}$$

where i represents the building floor, j the response frequency, ω the natural circular frequency of the jth mode, Φ the composite eigenvector matrix, and $S_{v(j)}$ the Code spectral velocity for frequency j. Due to the closely spaced building frequencies at a moderate level of damping, it has been found prudent to combine the modal floor loads with the Complete Quadratic Combination (CQC) technique, as opposed to the simpler Square Root of the Sum of Squares (SRSS) method. Figure 2 illustrates a sample of floor-by-floor shear loads induced by a Code seismic event. Good agreement has been reached with the worst-case time domain simulation, in which numerous local ground acceleration records were applied from multiple directions to the same building.

Comparison of Interstorey Shear Force
Lateral Force on a Single Axis

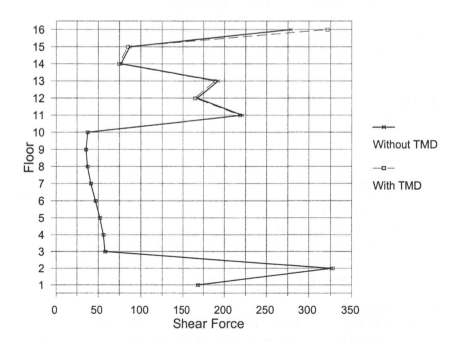

Figure 2 Distribution of Interstorey Shear Forces by Dynamic Analysis.

REFERENCES

Chang, C. C. and Hsu, C. T., 1998
  CONTROL PERFORMANCE OF LIQUID COLUMN VIBRATION
    ABSORBERS, *Engineering Structures*, Vol. 20, No. 7, (Great
    Britain: Elsevier Science Ltd.) pp. 580–586.

Den Hartog, J. P., 1956
  MECHANICAL VIBRATIONS, 4th edition, McGraw-Hill, New York.

Hitchcock, P. A., Kwok, K. C. S., Watkins, R. D. and Samali, B., 1997
  CHARACTERISTICS OF LIQUID COLUMN VIBRATION
    ABSORBERS (LCVA) – I, *Engineering Structures*, Vol. 19, No. 2,
    (Great Britain: Elsevier Science Ltd.) pp. 126–134.

Kawaguchi, A., Teramura, A. and Omote, Y., 1992
  TIME HISTORY RESPONSE OF A TALL BUILDING WITH A
    TUNED MASS DAMPER UNDER WIND FORCE, *Journal of
    Wind Engineering and Industrial Aerodynamics*, pp. 1949–1960.

Luft, R. W., 1979
  OPTIMAL TUNED MASS DAMPERS FOR BUILDINGS, *Journal of the
    Structural Division*, ASCE, pp. 2766–2772.

McNamara, R. J., 1977
  TUNED MASS DAMPERS FOR BUILDINGS, *Journal of the Structural
    Division*, ASCE.

McNamara, R. J., Huang, C. D. and Wan, V., 2000
  VISCOUS-DAMPER WITH MOTION AMPLIFICATION DEVICE
    FOR HIGH RISE BUILDING APPLICATIONS, Proceedings
    ASCE Structures Congress, Philadelphia, May 8–10.

Simiu, E. and Scanlon, R. H., 1986
  WIND EFFECTS ON STRUCTURES, 2nd ed., John Wiley, New York.

Tanaka, H. and Mak, C. Y., 1983
  EFFECT OF TUNED MASS DAMPERS ON WIND INDUCED
    RESPONSE OF TALL BUILDINGS, *Journal of Wind Engineering
    and Industrial Aerodynamics*, pp. 357–368.

Wiesner, K. B., 1979
  TUNED MASS DAMPERS TO REDUCE BUILDING MOTION, ASCE
    Annual Convention and Exposition, Boston, Reprint 3510, April 2–6.

Recent Applications of Damping Systems for Wind Response

Peter A. Irwin and Brian Breukelman

1 INTRODUCTION

A traditional solution to the problem of excessive motion of tall buildings under wind action is to add more structure so as to stiffen the building and increase the mass. However, another approach is to supplement the damping of the structure. Some of the first buildings where this approach was taken were in North America. An early example is the World Trade Center towers, New York, which were designed in the 1960's with viscoelastic dampers distributed at many locations in the structural system. In the 1970's the Citicorp building in New York, which has a single Tuned Mass Damper, weighing 400 tons, at the top was constructed. The Columbia Seafirst Center in Seattle, with distributed viscoelastic dampers, followed in 1984. There have been a few other applications (e.g. John Hancock tower, Boston, as a retro-fit, 28 State Street, Boston also as a retro-fit) but not many. Bearing in mind the large number of tall buildings constructed since 1960, there have thus been relatively few in North America, compared with Japan say, where the concept of adding damping has been applied for reducing wind response. However, as we view the scene in the year 2000, there does appear to be a renewed surge in activity. At the time of writing there are now at least six tall buildings under construction with dampers incorporated as part of the design to resist wind excitation. These buildings are in Chicago, New York, San Francisco, Boston and Vancouver. The author's firm RWDI became involved with these projects initially through wind tunnel testing models of each building and, in several cases, was assigned the task of designing the damping systems.

2 DESIGN ISSUES

Two approaches have been used to add damping to buildings to reduce wind excitation. The first is to add many sources of damping throughout the structure. Both hydraulic dampers and visco-elastic type dampers have been used in this type of application, and so have multiple small Tuned Sloshing Liquid Dampers. The second approach is to concentrate the damping mechanism at one or two locations, usually near the top of the building. The Tuned Mass Damper (TMD), the Tuned Liquid Column Damper (TLCD) and the large scale Tuned Sloshing Liquid Damper (TSLD) fall into this category.

Distributed Systems. The analysis of structures with various distributed types of dampers is covered in a number of references, e.g. Soong and Dargush (1997) and this paper is not the place to re-iterate the theories involved. However, a full analysis can be complex and it is helpful to have a simple approximate method of estimating required damper properties for preliminary design purposes. For a highly distributed system with small local damping forces, the increment, ζ, in overall damping ratio due to the added dampers is given by

$$\zeta = \left(\sum_{j=1}^{N} C_j \phi_{sj}^2 \right) / \left(2\omega M_G \right) \tag{1}$$

where N = total number of added dampers, C_j = viscous damping constant of the added dampers at the jth level in the tower in terms of the differential velocity between the floors, φ_{sj} is the difference in modal deflection shape between the floors spanned by the dampers at the jth level, ω = circular natural frequency for the particular mode of vibration being considered and M_G = generalised mass of the building for the mode of vibration.

   If large dampers are added at only a few locations then the tower's modes of vibration become appreciably altered and Equation 1 will overestimate the amount of damping that will be added because the structure starts to deform around the dampers and they become less effective. As an illustration, if a viscous damping system is installed half way up a building with pure shear flexibility only, and the damping system, with damping constant C_0, spans one floor only, then the damping ratio increase ζ due to the damper can be shown, using methods similar to those employed by Carne (1980), to be approximately given by

$$\zeta = \pi C \lambda^2 / \left(1 + \pi^2 C^2 \lambda^2 \right) \tag{2}$$

where $C = C_0 / \sqrt{Gm}$, G = shear constant, m = mass per unit height, and λ = ratio of floor-to-floor height to twice the building height. Note: C_0 = (floor damping force) / (relative velocity of floors). As the damping constant C_0 is increased from zero, ζ initially increases, but it reaches a maximum value of $\zeta = \lambda/2$ at $C = 1/\pi\lambda$.

Further increases in damping constant will actually be counter productive since ζ will only decrease due to the excessive restraint imposed by the damper. For a 50 storey simple shear building, for example, the increment in damping ratio can be no more than 0.005 according to Equation 2. Since real buildings flex and twist as well as shear, the actual maximum possible damping may be even less than this in reality. So generally it is necessary to add viscous or visco-elastic damping systems to many floors for them to work efficiently. To fully evaluate the effectiveness of a given system of viscous type dampers (linear or non-linear) it is best to undertake a full time history simulation, using wind tunnel derived time histories of the input wind forces and a full structural model incorporating the dampers, such as described by McNamara and Huang (2000).

Since typical floor to floor deflections of interest are only one or two millimetres, it may be necessary to use a mechanical linkage system that will magnify the damper stroke for a given floor to floor deflection, as used on the 111 Huntington St. tower in Boston (MacNamara and Huang, 2000). Also, the exponent in the damping force versus velocity relationship can be important. Available hydraulic dampers obey a force versus velocity relationship of the form force $\propto$ (velocity)$^\alpha$. The classic, strictly viscous, damper has $\alpha = 1.0$. However, there are advantages to using exponents less than 1.0 in distributed damping systems for motion reduction purposes. To avoid the dampers putting excessive forces into the structure locally in full design winds or under seismic actions an exponent α less than 1.0 is desirable. Typically a value of about 0.7 is found to work well.

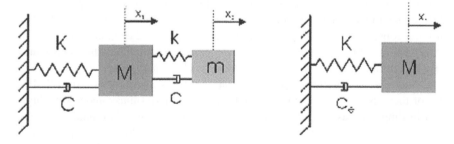

Figure 1   Conceptual Model of a Tuned Mass Damper and System with Same Effective Viscous Damping.

Tuned Mass Dampers. The principle of the Tuned Mass Damper is illustrated in Figure 1. The main mass, M, in Figure 1 represents the generalised mass of the building in its fundamental mode in a particular direction and the spring connecting the main mass to ground represents the stiffness of the building. The viscous damper connecting the main mass to ground represents the inherent damping of the building. The secondary mass, m, is the mass of the TMD, which is connected to the main mass via a suspension system, the stiffness and damping of which is represented in Figure 1 by the spring and viscous damper between the main and secondary masses. In practice the stiffness of the "spring" connecting the secondary mass to the main mass can be the stiffness arising from pendulum action when the TMD mass is hung from cables attached to the building structure.

The equations of motion of the TMD were laid out by Den-Hartog (1956) and have been re-iterated in many references. Therefore they will not be described here. The TMD is most effective in damping motion of the main mass when optimally tuned. The optimum frequency ratio, optimum TMD damping ratio, and resulting maximum dynamic amplification factor for the case where the damping of the main mass is zero are given respectively by

$$f_2 / f_1 = \frac{1}{1+\mu} \ , \quad \zeta_{opt} = \sqrt{\frac{3\mu}{8(1+\mu)}} \ , \quad \frac{x_{1max}}{x_{static}} = \sqrt{1+\frac{2}{\mu}} \tag{3}$$

where f_1 and f_2 are the natural frequencies of the main mass and secondary mass respectively, μ is the mass ratio m/M, x_{1max} is the maximum deflection of the main mass under the sinusoidal loading and x_{static} is the main mass deflection under the action of a static force equal to the amplitude of the sinusoidal force. In reality the wind loading of the structure is not sinusoidal but has a broad band of frequencies. Also, of course, the structural damping is not zero. However, most towers are lightly damped and the above simple relationships can be useful for arriving at preliminary estimates of optimal tuning and TMD damping. Weisner (1979) has published computations done for the Citicorp building TMD for various assumed inherent damping values for the building.

A useful concept is that of effective viscous damping. The effective viscous damping provided by a TMD depends on the type of loading being applied to the main mass. The definition of effective viscous damping is that it is the viscous damping ratio that would give the same root-mean-square building response as the TMD under the selected type of loading. In Figure 1 this is represented by the damping constant C_e in the right hand diagram. Since wind excitation is typically broad band in nature, one type of loading frequently used in the definition is white noise excitation, i.e. assuming equal excitation at all frequencies.

There are a number of practical considerations in the design of a TMD. One of these is needed to limit the motions of the TMD mass under very high wind loading such as will occur in the design storm or under ultimate load conditions. One way of doing this is to use a nonlinear hydraulic damper in the TMD. By employing a damper with an exponent $\alpha = 2$, say, rather than 1.0 or less, the motions of the TMD mass can be greatly reduced under very high wind loading conditions or under strong seismic excitation. A further safeguard against excessive TMD motion is to install hydraulic buffers around the mass. When the mass comes into contact with the buffers high velocities are quickly reduced.

To properly simulate the response of a TMD with nonlinear dampers and with hydraulic buffers it is best to undertake time history simulations of the tower response in both wind and earthquake loading, (Breukelman, *et al.*, 2001). These simulations can be used not only to evaluate the motions of the TMD and tower but also to determine the maximum forces the TMD will impart locally to the tower structure.

Tuned mass dampers can in principle be readily converted to be an active system by incorporating sensors and feedback systems that can drive the TMD mass to produce more effective damping than is possible in a purely passive mode. As a result a larger effective damping can be obtained from a given mass.

This approach has been used in several commercially available ready-to-install systems. The TMD is thus made more efficient, a benefit to be weighed against the increased cost, complexity and maintenance requirements that are entailed with an active system.

Liquid Column Dampers and Sloshing Dampers. Tuned liquid column dampers are in many ways similar to tuned mass dampers. The difference is that the mass is now water (or some other liquid). The detailed equations of motion for a TLCD have been worked out by several researchers (e.g. Xu *et al*, 1992). The damper is essentially a tank in the shape of a U, i.e. it has two vertical columns connected by a horizontal passage and filled up to a certain level with water. Within the horizontal passage there may be screens or a partially closed sluice gate. The TLCD is mounted near the top of the building and when the building moves the inertia of the water causes the water to oscillate into and out of the columns, travelling in the passage between them. The columns of water have their own natural period of oscillation which is determined purely by the geometry of the tank. If this natural period is close to that of the building's period then the water motions become substantial. Thus the building's kinetic energy is transferred to the water. However, as the water moves past the screens or partially opens the sluice gate in the horizontal portion of the tank the drag of these obstacles to the flow dissipates the energy of the motion and results in damping.

An even simpler liquid type damper is a rectangular tank filled to a certain level with water (Isyumov *et al.,* 1995). In this case the tank's natural period of wave oscillation is approximately matched to the building period by appropriate geometric design of the tank. If screens and baffles are placed in the tank then dissipation of the waves takes place and the result is again that the tank behaves like a TMD. However, analysis indicates that a sloshing water tank does not make as efficient use of the water mass as a TLCD.

3   RECENT APPLICATIONS

Simple Pendulum Damper Application. Recently RWDI designed a TMD for the 67 storey Park Tower, Chicago, Figure 2, in cooperation with architects Lucien Lagrange and Associates and structural engineers Chris Stefanos and Associates and is currently in the process of commissioning it. The developers are the Hyatt Development Corporation. The wind tunnel tests showed that with the initially planned structural system the accelerations would be above the desired values for a residential building. The 10 year return period acceleration was predicted to be in the range of 26 to 30 milli-g at 2% damping ratio and a target of 15 milli-g had been set. The higher than normal accelerations were primarily due to wind flows off the John Hancock Tower nearby. After extensive investigations into various structural or shape change solutions, the decision was made to add damping. From the developer's point of view the damping system had to be economical, require little maintenance and avoid compromising valuable floor space. The design selected consisted of a simple pendulum damper

Figure 2   Photo of the Park Tower under construction.

mounted under the mansard roof of the tower, an area being used for mechanical equipment. With some minor changes in the geometry of the mansard, there was enough space to accommodate a simple pendulum TMD. Figure 3 illustrates the design. It consists of a 300 T mass block slung from cables with lengths adjustable up to 34.5 ft. This mass represented approximately 1.4% of the building's generalised mass in the first mode of vibration. The simplicity of the design minimises the need for maintenance and also kept the cost low. The only components in need of maintenance will be the hydraulic dampers and present day hydraulic dampers can be manufactured to have very low maintenance.

Features of the TMD are that it has a tuning frame which can be moved up and down and clamped on the cables to allow the natural period of the pendulum to be adjusted. The damper constants can be adjusted. The dampers are nonlinear with a force proportional to velocity squared so as to prevent excessive mass motions during extreme wind events. The mass is connected to an anti-yaw device to prevent rotations about a vertical axis. Below the mass there is a bumper ring connected to hydraulic buffers to prevent travel beyond the hydraulic cylinder length. The main hydraulic dampers of the TMD are sloped from their floor mountings up to the TMD mass. This was found to be advantageous in shortening the stroke required of the dampers which reduced cost. Installation of the TMD is in progress at the time of writing. The building frequencies have been measured using accelerometers mounted in the building and by recording motions caused by ambient winds. The measured sway frequencies were within 10%–20% of computed modes. The results of the measurements will allow the correct pendulum length to be set. The predicted 10-year acceleration for the building with the TMD in operation is 15 milli-g.

Nested Pendulum Damper Application. In some situations the height available in the building is insufficient to allow a simple pendulum TMD to fit. In such a case, at the cost of a little more complexity in design, a nested pendulum design can be used. RWDI has recently completed such a TMD design of mass 600 T for another tall North American residential tower. The nested TMD design is illustrated in Figure 4. The total vertical space occupied by the damper, which has a natural period of about 6 seconds, is less than 25 ft. The cost, including installation, is anticipated to be about $2.5 million US.

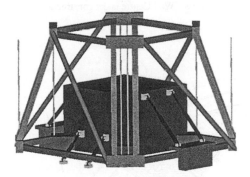

Figure 3   Nested Pendulum Design.

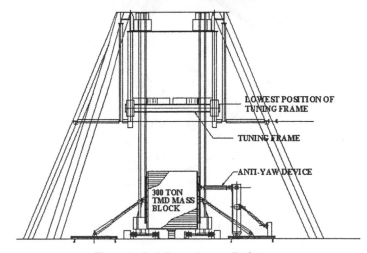

Figure 4   Park Tower Damper Design.

Figure 5   Wall Centre under construction.

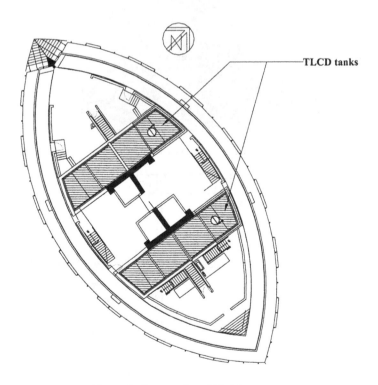

Figure 6   Cross-section of Wall Centre.

Tuned Liquid Column Damper Application. The Wall Centre is a 48 storey residential tower currently under construction in Vancouver, BC, Figure 5. Its cross-section is illustrated in Figure 6. From wind tunnel tests predicted 10 year accelerations were in the range 28 to 40 milli-g, depending on the structural systems being explored by the structural designers Glotman Simpson Engineers. The effects of shape changes were also examined in conjunction with the architects, Busby and Associates, and have been described by Irwin (1999). However, in the end, when all things were considered by the owners, the Wall Financial Corporation, and the design team, it was decided that the approach of increasing the damping was preferable. In discussions with the owner it transpired that a damper using water could serve a dual purpose by also providing a large supply of water high up in the tower for fire suppression. Therefore a TLCD was designed. Initially a sloshing water damper was considered but the TLCD was found preferable due to its greater efficiency in using the available water mass. The TLCD design turned out to be a remarkably economical solution in this case, especially considering the saved cost of having to install a high capacity water pump and emergency generator in the base of the building, as initially required by fire officials. The total mass required was in the order of 600 T which corresponds to a large volume of water. However, sufficient space was available. Also a helpful factor was that the motions of the tower were primarily in one direction only. Therefore only one direction needed to be damped which simplified the design.

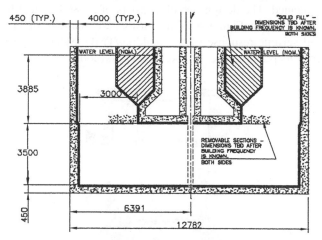

Figure 7 Tuned Liquid Column Damper design for Wall Centre.

Figure 7 illustrates the TLCD design which consisted of two identical U-shaped concrete tanks. Since the building was concrete it was relatively easy to incorporate the tanks into the design and to construct them as a simple addition to the main structure. The locations of the two TLCDs in the tower cross-section can be seen in Figure 6. The dimensions of the TLCDs were initially worked out using analytical methods. However, since a number of assumptions are inherent in these methods a scale model was also constructed at RWDI's Guelph, Ontario, laboratories. The model behaviour gave good confirmation of the theoretical results.

The tower is currently under construction and its natural frequencies have been checked by vibration tests. Commissioning will consist of adjustments to the sluice gate to obtain the correct TLCD damping and adjustments to the water level in the columns to obtain the desired natural frequency. The design incorporates a system for sealing off the top of a vertical column, pressurizing it and then releasing the seal so as to set the TLCD into oscillation. From the decay of oscillations the TLCD's internal damping can be measured. The predicted 10 year accelerations with the TLCDs operational are 16 milli-g, within 1 milli-g of the target.

Tuned Mass Damper as a Building Feature. The view of many owners is that the presence of a special damping device in the building is not something that they necessarily want widely broadcast. In most cases it is tucked away out of view. However, the architects, C. Y. Lee and Associates, and owners of the 101 storey Taipei Financial Centre, have taken the route of making the RWDI designed TMD a feature of the building. A special space has been allocated for it near the top of the building and people will be able to walk around it and view it from a variety of angles. It will be brightly coloured and special lighting effects are planned. The design, which consists of an 800 T steel ball slung on cables is illustrated in Figure 8. Time history simulations of both the wind response and seismic response of the building/TMD system have been undertaken to verify its performance, Breukelman *et al.* (2001).

Figure 8   Taipei Financial Center TMD rendering.

REFERENCES

Breukelman, B., Haskett, T. C., Gamble, S. L. and Irwin, P. A., 2001
ANALYSIS AND MODEL TECHNIQUES FOR DETERMINING
DYNAMIC BEHAVIOUR OF CIVIL STRUCTURES WITH
VARIOUS DAMPING SYSTEMS. Proceedings, Council on Tall
Buildings and Urban Habitat, 6th World Congress, Melbourne,
Australia, Feb. 26–Mar. 2.

Carne, T. G., 1980
GUY CABLE DESIGN AND DAMPING FOR VERTICAL AXIS WIND
TURBINES, Sandia National Laboratories Report No.
SAND80-2669 Albequerque, New Mexico.

Den Hartog, J. P., 1956
MECHANICAL VIBRATIONS, 4th edition, McGraw-Hill, New York.

Irwin, P. A., Breukelman, B., Williams, C. J. and Hunter, M. A., 1998
SHAPING AND ORIENTING TALL BUILDINGS FOR WIND,
Proceedings, Structural Engineers World Congress, San Francisco,
July 19–23, CD ROM, Elsevier.

Isyumov, N., Breukelman, B., Fediw, A. A. and Montpellier, P., 1995
INFLUENCE OF A TUNED SLOSHING WATER DAMPER ON THE
WIND-INDUCED RESPONSE OF A TALL OFFICE TOWER,
Proceedings, ASCE Structures Congress, Boston.

McNamara, R. and Huang, C.D., 2000
VISCOUS DAMPER WITH MOTION AMPLIFICATION DEVICE FOR
HIGH-RISE BUILDING APPLICATION, Proceedings, ASCE
Structures Congress, Philadelphia, May 8–10.

Soong, T. T. and Dargush, D. F., 1997
PASSIVE ENERGY DISSIPATION SYSTEMS IN STRUCTURAL
ENGINEERING, John Wiley & Sons.

Wiesner, K. B., 1979
TUNED MASS DAMPERS TO REDUCE BUILDING MOTION, ASCE
ANNUAL CONVENTION AND EXPOSITION, Boston, Preprint
3510 April 2–6.

Xu, Y. L., Samali, B. and Kwok, K. C. S., 1992
CONTROL OF ALONG WIND RESPONSE OF STRUCTURES BY
MASS AND LIQUID DAMPERS, *J. of Engrg Mech.*, ASCE,
118(1), pp. 20–39.

Geologic and Seismic Hazards in the Urban Habitat

Marshall Lew and Farzad Naeim

1.0 ABSTRACT

Strong ground shaking presents the greatest hazard to property and personal safety in the urban habitat in the seismically active regions of the world. However, there are geologic and seismic hazards that threaten a smaller population in the urban habitat. These hazards include the effects of surface fault rupture, liquefaction, and landsliding. With proper precautions, the exposure to these hazards could be avoidable.

2.0 INTRODUCTION

As the world population continues to grow at an ever-increasing rate, the teeming masses of many major metropolitan cities are swelling at an even greater rate. Many of these cities, such as Tokyo, Auckland, Mexico City, Manila, Santiago, Los Angeles, San Francisco, Athens, Istanbul, Tehran, and Vancouver, are located along or near the boundaries of major tectonic plates where damaging earthquakes have already been experienced or are expected to occur. However, rapid urbanisation and unchecked growth of many of these cities and surrounding areas have proceeded without considering or exploring the geologic and seismic hazards that may exist beyond the ground shaking.

Recent earthquakes occurring through the end of the last millennium illustrate that the recognition of geologic and seismic hazards in the urban habitat is still not a reality, even in the most technologically advanced countries of the world. However, bold attempts are being made in some jurisdictions to identify land areas where such hazards may exist and to require that the public be informed about the inherent risks of occupying and ownership of properties within such areas. Such efforts are not without opposition, as such important issues as maintaining economic value and use of property are weighed versus public safety.

The most recent major earthquakes in California, Turkey, and Taiwan illustrate very clearly the need to address the geologic and seismic hazards associated with earthquakes other than the strong ground shaking. These hazards include surface fault rupture, liquefaction, and landsliding. Although the first of these hazards may be limited to just the immediate near-fault area, the second and third hazards may be significant in even distant locations to the fault rupture zone. The economic and personal losses associated with these hazards can be very high.

3.0 SURFACE FAULT RUPTURE

When an earthquake fault ruptures, it sometimes extends to the ground surface and results in differential ground movement and permanent deformation across the fault plane. Strike-slip faults would primarily result in permanent horizontal displacements across the surface rupture. Normal or thrust faults would result in primarily permanent vertical displacements across the surface rupture. These fault displacements can range from a few centimetres to as much as 10 meters in a large earthquake. Some examples of surface fault rupture and resulting damage to structures are shown in Figures 1 and 2.

These figures illustrate the destructive effects of surface fault rupture. Structures and utility lifelines set across a fault would be subject to large deformations if a surface fault rupture were to occur. Figure 1 shows the effect of thrust faulting on a public street in the 1971 San Fernando earthquake; such ruptures can disrupt vital surface transportation lifelines such as vehicular and rail transportation systems. Figure 2 shows the effects of about 8 meters of vertical thrust faulting on the Shihkang Dam in central Taiwan from the 1999 Chi-Chi earthquake; the thrust faulting caused the southern portion of the dam to be raised relative to the northern portion. Figures 3 and 4 show the effects of thrust faulting through buildings in Turkey and Taiwan. Structures obviously cannot accommodate significant displacements associated with fault rupture.

Figure 1   Surface thrust faulting in 1971 San Fernando Earthquake. Steinbrugge Collection, Earthquake Engineering Research Center, University of California, Berkeley.

Figure 2   Thrust faulting through Shihkang Dam in 1999 Chi-Chi earthquake in Taiwan.

Figure 3   Fault rupture under building in Golcuk, Turkey (1999). Izmit Collection, Earthquake Engineering Research Center, University of California, Berkeley.

Figure 4   Thrust faulting through building in 1999 Chi-Chi, Taiwan earthquake.

4.0   LIQUEFACTION

Liquefaction in saturated soils can occur during an earthquake if the strong ground shaking causes a loss of strength in the soil. Liquefaction usually requires the presence of shallow ground water and loose sandy soils. The effects of liquefaction are dramatically illustrated in Figures 5 and 6. As shown in Figure 5, liquefaction-induced settlement and lateral spreading occurred in loose filled-in areas of San Francisco Bay; structures, as well as underground water pipes vital to fire-fighting activities, were heavily damaged in the 1906 San Francisco earthquake.

Figure 5   Settlement and lateral spreading in San Francisco after the 1906 earthquake. Loma Prieta Collectiion, Earthquake Engineering Research Center, University of California, Berkeley.

Figure 6 shows liquefaction-induced bearing capacity failures of residential buildings founded on shallow spread foundations in reclaimed landfill in Niigata, Japan in 1964.

Figure 6   Liquefaction-induced loss of bearing in 1964 Niigata, Japan, Earthquake. Steinbrugge Collection, Earthquake Engineering Research Center, University of California, Berkeley.

Figure 7 shows an example of loss of bearing capacity due to liquefaction in the Turkish city of Adapazari. Figure 8 shows a similar occurrence in Taiwan.

| **Figure 7**   Liquefaction-induced bearing capacity failure of building in Turkey. Izmit Collection, Earthquake Engineering Research Center, University of California, Berkeley. | **Figure 8**   Liquefaction-induced tilting of buildings (bearing capacity failure) in 1999 Chi-Chi, Taiwan earthquake. Photograph by Dr. Sampson C. Huang |

The effects of liquefaction can range from subtle to very dramatic. If liquefaction is limited to a few isolated thin layers of soil below the water table, only minor settlement of structures supported on the ground may occur. If the structures are supported on deep foundations extending to deeper non-liquefiable soils, the effects on the structures may be minimal or non-existent. However, if the liquefiable soils are very thick, greater settlement could occur. If the ground slopes even slightly, there is a possibility that lateral spreading could occur. Liquefaction can cause significant financial losses and pose a threat to life safety.

5.0 LANDSLIDING

Landsliding can occur during earthquakes when seismic strong ground motions impart additional impetus to driving forces of slopes that may only be marginally stable under static conditions. Recent slope failures during earthquakes are shown in Figures 9, 10 and 11.

Figure 9   Central area of Turnagain landslide area in Anchorage, Alaska from April 10, 1964 earthquake. Steinbrugge Collection, Earthquake Engineering Research Center, University of California, Berkeley.

Figure 10   Nigawa landslide in Kobe, Japan. Kobe Geotechnical Collection, Earthquake Engineering Research Center, University of California, Berkeley.

Figure 11   Fourth Avenue landslide area in Anchorage, Alaska from April 10, 1964 earthquake. Steinbrugge Collection, Earthquake Engineering Research Center, University of California, Berkeley.

Just recently on January 13, 2001, a massive mudslide was triggered by a magnitude 7.6 earthquake that was centered about 110 kilometers south-southwest of San Miguel, El Salvador in Central America. The mudslide buried a portion of the Las Colinas neighborhood in Santa Tecla, a suburb of the capital of San Salvador. About 500 homes were buried by the slide debris.

Man-made development on or near slopes can create conditions that exacerbate the stability of natural slopes. Some of these man-made effects include steepening of slopes, removal of stabilising soil or rock masses, addition of driving forces (such as structures), increases in moisture content of slope materials, and changes in site drainage that may accelerate erosion and slope deterioration.

6.0   EVALUATION OF SEISMIC HAZARDS

With ever-increasing damaging earthquakes occurring throughout the world, the seismic hazards related to surface fault rupture, liquefaction and landsliding require much more attention than presently given. With massive urbanization and explosive population growth in many of the world's major metropolitan areas in seismically prone areas, development quite naturally extends into areas that were previously not considered desirable. Many of these areas are subject to the hazards discussed in this paper.

It appears that some form of land-use management is the only practical means to control the development of lands that may be subjected to seismic hazards. Besides the technical details, which may require extensive and quite costly investigation and evaluation, the political and economic forces in control in many areas of the world may prevent or greatly inhibit the use of land-use management. The issues of property rights, reduced land values, and disclosure of hazard can be difficult obstacles to overcome, especially if there is no immediate or even long-term economic relief.

In the United States, the State of California has been the pioneer in protecting public safety from the effects of strong earthquakes. The Alquist-Priolo Earthquake Fault Zoning Act was passed in 1972 to mitigate the hazard of surface faulting to structures for human occupancy (Division of Mines and Geology, 1997). This state law was a direct result of the 1971 San Fernando Earthquake, which was associated with extensive surface fault ruptures that damaged numerous homes, commercial buildings, and other structures. The Alquist-Priolo Earthquake Fault Zoning Act's main purpose is to prevent the construction of buildings used for human occupancy on the surface trace of active faults. The law requires the State Geologist to establish regulatory zones (known as Earthquake Fault Zones) around the surface traces of active faults and to issue appropriate maps. The maps are distributed to all affected cities and counties, and to state agencies for their use in planning and controlling new or renewed construction. A portion of the Earthquake Fault Zone in the city of Palmdale, north of Los Angeles, is shown in Figure 12; the figure shows zones related to the San Andreas fault. Local agencies must regulate most development projects within the zones. Projects include all land divisions and most structures for human occupancy. Before a project can be permitted, cities and counties must require a geologic investigation to demonstrate that proposed buildings will not be constructed across active faults. A licensed geologist must prepare an evaluation and written report of a specific site. If an active fault is found, a structure for human occupancy cannot be placed over the trace of the fault and must be set back from the fault (generally 50 feet). The law also requires disclosure that a property within an Earthquake Fault Zone is within such a zone at the time of sale or property transfer.

Single family wood-frame and steel-frame dwellings up to two stories not part of a development of four units or more are exempt from the law in California. However, local agencies can be more restrictive than state law requires. Thus it is possible that some new dwellings in California can be constructed within zones that could be subject to fault rupture. However, the existence of the Earthquake Fault Zone maps does provide notice to potential builders that their construction may be subject to seismic risk.

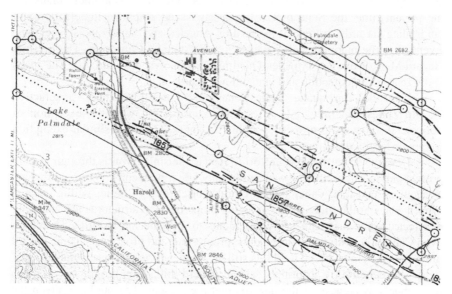

Figure 12   Portion of Alquist-Priolo Earthquake Fault Zone map for Palmdale area, north of Los Angeles. The zones are for branches of the San Andreas fault. (Division of Mines and Geology, 1979).

The 1972 Alquist-Priolo Earthquake Fault Zoning Act only applied to surface fault rupture and not other seismic hazards. It was only after the 1989 Loma Prieta earthquake in Northern California that the State of California enacted the Seismic Hazards Mapping Act which became law in 1991 (California Division of Mines and Geology, 1997). The purpose of this act is to protect public safety from the effects of strong ground motion, liquefaction, landslides, or other ground failure, and other hazards caused by earthquakes. The law requires that new construction within areas determined to be potentially subject to certain seismic hazards have the geologic and soil conditions investigated for those hazards. If those hazards exist at the development site, appropriate mitigation measures are to be incorporated into the development plans. At the present time, the State of California has issued maps for liquefaction and landsliding hazards, primarily in the Los Angeles and San Francisco regions. Eventually, the mapping will cover the entire state of California. The Seismic Hazard Map of the Los Angeles Quadrangle is shown in Figure 13. The maps show zones of liquefaction and landsliding hazard.

Like the Earthquake Fault Zones, the Landsliding and Liquefaction Hazard Zones generally only affect new construction with the exception of notice of disclosure for properties within the zones. There is a similar exemption for single family dwellings.

Despite California's best effort to manage seismic hazards through laws to protect public safety, the laws currently only affect new construction with some exemptions possible. Other than providing disclosure of being within seismic hazard zones at the time of sale, existing properties and structures are generally unaffected since there are no requirements to mitigate the hazards. Therefore, even California, with some of the most progressive and proactive seismic hazard mitigation programs in the world, is still at significant risk from seismic hazards.

One benefit of California's seismic hazard mitigation programs is to raise the public's awareness of the hazards that are present. The public is given the opportunity to evaluate for themselves and perhaps have an opportunity to limit their risks by relocating to a better location or to mitigate the problem. The California seismic hazard mitigation program would be a good model for other seismically active areas to consider for implementation.

Implementation of a seismic hazard mitigation program will require a large commitment of funding to carry out the fundamental research to identify the types of hazards that could affect a region. The susceptibility of the ground or landforms to failure will need to be determined on a rational basis, preferably based upon some uniform level of acceptable risk. There will need to be proper enforcement and acceptance of building codes, land use restrictions, and good construction practices to have the desired result of improved seismic safety.

Figure 13 Seismic Hazard Zone map for the Los Angeles quadrangle. Landsliding hazard zones (light shading) and Liquefaction hazard zones (dark shading) are shown. (California Division of Mines and Geology, 1999).

7.0 CONCLUSIONS

Seismic hazards will exist wherever earthquakes can occur. Besides the effects of strong ground shaking, earthquakes can also cause damage and injury from fault rupture, liquefaction, and landsliding. In addition, there are other seismic-related phenomena, such as tsunamis (sea waves) and seiches (waves in enclosed bodies of water), that could also result in severe damage and loss of life. To limit losses from these effects, evaluations are needed to identify those areas within the urban habitat that may be subjected to these hazards. However, the public will to undertake these evaluations may not be strong, unless mobilized by knowledge and outrage; however, experience has shown that calls for such measures usually only occur immediately after a major earthquake catastrophe and the interest drops off faster than the rate of the aftershocks.

The economic costs of identifying the sources of geologic and seismic hazards can be large and large public sector funding will most likely be needed to accomplish the required research. Once the hazards are identified, new construction can be protected by proper mitigation techniques. Strong enforcement of the building codes and proper construction techniques will also be needed to provide assurance that the mitigation techniques will be effective in the new construction. However, existing construction will likely continue to pose a threat for severe economic losses and life safety concerns as it will be unlikely that there will be sufficient capital or will to deal with the complex social and economic issues.

8.0 ACKNOWLEDGEMENTS

The authors are members of the Los Angeles Tall Buildings Structural Design Council, which was formed to provide a forum for the discussion of issues relating to the design of tall buildings. The support and encouragement of the Los Angeles Council to investigate, research and analyse structural behaviour in earthquakes is most greatly appreciated.

9.0 REFERENCES

California Division of Mines and Geology, 1997
    FAULT-RUPTURE HAZARD ZONES IN CALIFORNIA. DMG Special
        Publication 42.

California Division of Mines and Geology, 1997
    GUIDELINES FOR EVALUATING AND MITIGATING SEISMIC
        HAZARDS IN CALIFORNIA. DMG Special Publication 117.

California Division of Mines and Geology, 1999
    OFFICIAL MAP OF SEISMIC HAZARD ZONES, LOS
        ANGELESQUADRANGLE.

California Division of Mines and Geology, 1979
    REVISED OFFICIAL MAP, SPECIAL STUDIES ZONES, PALMDALE
        QUADRANGLE.

Built Environment Hazards in the Urban Habitat

Farzad Naeim and Marshall Lew

1.0 ABSTRACT

As disastrous as major earthquakes are, earthquake disasters and building performance provide valuable insight to design professionals, owners, and public officials for development of planning guidelines and building design and construction mitigations that may be utilized to minimize death and destruction of the urban habitat caused by future seismic events. It is imperative that we do not ignore the recent lessons learned and press on to implement needed changes to current societal, design, and construction practices. This paper provides a basic understanding of the major earthquake performance issues of importance worldwide, the lessons we have learned during the recent past, and the promising paths to a less risk-prone future.

2.0 INTRODUCTION

As disastrous as major earthquakes are, they teach us important lessons that we as earthquake engineers and public officials may utilize to minimize death and destruction caused by future events. We need to optimise our learning process so that we learn new lessons from each earthquake rather than re-learning old lessons. We owe it to the people in general and to our clients in particular, to eliminate unnecessary death and devastation. The major known structural issues causing devastation of the built environment include:

- Soft and weak story building configurations
- Poor material quality and workmanship
- Poor structural detailing
- Unaccounted for torsional response
- Weak column strong floor configurations
- Lack of adequate load paths
- Influence of "non-structural" components on building response
- Lack of adequate correlation between the building as analysed and designed and the building as built.

These issues are explored in more detail in the following sections of this paper.

3.0   SOFT/WEAK STORIES ARE DEADLY

Probably among all urban habitat structural problems the soft/weak story failures have been responsible for more deaths and destruction than any other. A pure soft-story problem exists when the stiffness of one floor (most often the first floor) is significantly less than that of adjacent floors. The soft-story then displaces substantially more than the floors on the top of it aggravating the P-Δ loads and leading to instability and failure of the soft story. A pure weak story problem exists when the strength of the weak floor (most often the first floor) is significantly less than the strength of adjacent floors. Here, plastic hinge formation or brittle failure is concentrated at this floor causing damage and sometimes total collapse of the weak floor. Pure cases of soft-story or weak-story failures are rare and generally the same floor is both soft and weak therefore justifying the use of the term soft/weak story instead. The soft/weak story problem is commonly magnified by regional response where due to a solid back wall at the opposite end of the row of soft/weak columns the displacement and forces on the soft/weak row is amplified.

Collapse potential of soft/weak stories is not a new phenomenon. This type of failure was documented during the 1906 San Francisco earthquake. The seismic design codes, however, were belated in addressing this issue and the first U.S. code provisions related to this problem appeared in the late 1970s and early 1980s. However, quite often, the open-front one or two story single-family residential buildings which are most prone to this type of failure generally fell outside the jurisdiction of the codes embodying soft/weak story safeguards. Nevertheless, the soft/weak stories have consistently claimed a large number of lives and have caused serious destruction during many of the 20th century earthquakes. From Turkey to Taiwan and from California to India, the mark of soft/weak story failures can be seen on the face of many collapsed structures.

4.0   MATERIAL QUALITY AND WORKMANSHIP ARE CRITICAL

No matter how good a design is the resulting performance of the structure can be poor if less than sufficient attention is paid to the quality of the material used and the degree of care and workmanship afforded to the building during its construction. A rather humorous example of this issue is provided by the failure of the first author's backyard masonry block walls during the 1994 Northridge earthquake. The author who had examined the simple design drawings for the block walls and had checked the mix and compressive strength of the concrete delivered to his house was surprised to learn that the block walls have totally collapsed out of plane during the earthquake. However, an inspection of the collapsed walls showed that the contractor had consistently poured the concrete in one cell and placed the reinforcement in the adjacent empty cell, thereby for all practical purposes creating an unreinforced masonry block wall.

Concrete quality problems significantly contributed to the size of disasters experienced during the 1999 Turkey and Taiwan earthquakes. In Turkey, the dominant problem was use of very large aggregates and unwashed salty gravel in concrete. In Taiwan the major problem appeared to be the small size of aggregate and/or lack of sufficient amount of aggregates mixed with the cement. This generally resulted in powdering of concrete after a few cycles of shaking. Similar concrete quality problems have been observed in the aftermath of the 1985 Mexico, 1988 Armenia, and the 1990 northern Iran earthquakes among others. As far as steel construction is concerned, poor welding practices and improperly prepared heat affected zones are considered to be major factors in pre-mature failure of steel moment resistant frame connections during the 1994 Northridge, USA and 1995 Kobe, Japan earthquakes. Simply stated, the importance of quality assurance measures during the construction process cannot be overemphasized.

5.0 POOR DETAILING CAN BE DISASTROUS

Earthquakes have consistently proven that poor concrete detailing and lack of adequate concrete confinement can lead to disasters. Despite this very clear message, such practices remain prevalent worldwide. Examples of failures due to lack of adequate beam column joint confinement were abundant after the 1999 Turkey and Taiwan earthquakes. However, the fact that poor detailing and lack of adequate confinement can result in non-ductile and sudden failures has been clearly known at least since the 1971 San Fernando earthquake. As a result, most modern seismic design codes worldwide require a minimum amount of structural member and joint confinement to ensure ductile behaviour. The problem is, that in many places around the globe these code recommendations are routinely ignored or not enforced. Further complicating the picture, is the large stock pile of existing non-ductile concrete buildings built prior to mid 1970s which are still in existence and occupied. For example, in the Los Angeles county alone more than 2,000 of such buildings are in operation mostly as offices and warehouses. Addressing the issue of non-ductile concrete buildings in urban areas should be given a high priority in earthquake hazard mitigation efforts worldwide.

6.0 POOR CONFIGURATIONS AND UNANTICIPATED TORSIONAL RESPONSE CAN BE DISASTEROUS

We talked about the soft/weak story problem in a previous section of this paper where we mentioned that unaccounted for torsional response can seriously aggravate that problem. Here, we provide a few examples. Simple application of structural theory suggests that concurrent lines of resistance provide no torsional stiffness. This is simply because torsional resistance requires a torsional arm and concurrent lines by definition provide no such arm. Therefore, a building with a triangular plan configuration which derives its lateral stiffness from structural walls or frames located exclusively at the perimeter of the structure provides little to no torsional resistance. If concurrent lines of resistance provide no torsional stiffness, then it follows that nearly concurrent lines provide very little stiffness. Prevalence of two-dimensional structural analysis, however, has resulted in a very widespread and unfortunate desertion of this simple principle.

Unfortunately, failures of multi-story buildings during earthquakes due to ignoring of this simple principle are numerous.

In Taiwan for example, in the city of Taipei in an area that experienced less than 10%g peak ground acceleration, one of the very few buildings that was seriously damaged was a semi- triangular shaped building built on a corner lot. In the city of Taichung, a multi-story residential building collapsed and totally overturned in the middle of a residential district where all other buildings surrounding it remained practically undamaged. Many of the surviving buildings in the vicinity of this collapsed building featured undesirable configuration attributes such as open fronts and soft/weak first floors. What apparently made the difference here was the semi-rectangular plans of the surviving buildings and the semi-triangular plan of the collapsed one. Although, poor detailing and poor concrete quality definitely also contributed to this collapse. Similar poor performance of torsionally irregular buildings have been reported in many other events including the 1985 Mexico earthquake.

7.0 STRONG COLUMN WEAK GIRDER DESIGN AND PREVENTING PUNCHING SHEAR FAILURES ARE VITAL

Arguably the failure of buildings due to collapse of columns which are weaker than the girders or floor slabs they support is second only to the soft/weak story configuration in producing deadly results during earthquakes. As a matter of fact, the latter configuration problem may be classified as a sub-category of the former one.

Pancaking of several floors due to this problem was observed in abundance during the 1985 Mexico earthquake. Similar, and equally drastic and deadly failures could be observed during both the Turkish and Taiwaneese earthquakes of 1999. During the 1994 Northridge USA earthquake a hospital floor collapsed in this manner. Fortunately, however, no lives were lost in that incident.

8.0 NO SEISMIC RESISTANCE WITHOUT ADEQUATE LOAD PATHS

Lack of adequate positive load paths can be disastrous. An early proof of this was provided by the collapse of the elevator/stairway towers of the Olive View hospital during the 1971 San Fernando earthquake. In this case, the forces from these towers were not properly dragged into the main building which was supposed to provide the lateral resistance for the towers. As a result these towers overturned. Yet a better illustration was provided by the performance of a parking structure during the 1994 Northridge earthquake. In this case, floor diaphragms and columns collapsed while the shear walls which were supposedly the lateral load resisting elements of the building remained in large part undamaged. Apparently, there was not sufficient load path to transfer earthquake generated forces from the diaphragms to the walls. Although, large vertical accelerations have also been mentioned as a cause of this failure.

9.0 COMPONENTS NOT DESIGNED FOR LATERAL RESISTANCE CANNOT BE IGNORED

Non-lateral load resisting components, such as gravity columns must be designed to sustain the level of deformation induced by earthquake ground motions. Perhaps the best illustration of this issue was provided by the collapse of a newly constructed parking structure on the campus of California State University, Northridge (CSUN) during the 1994 Northridge earthquake. The ductile concrete moment frames of this structure exhibited an amazing level of ductility. However, because the interior gravity columns could not withstand the lateral displacements they collapsed and brought the building down notwithstanding the exemplary ductile behaviour of the lateral system.

10.0 STRUCTURAL ANALYSIS SHOULD REFLECT THE REALITY OF THE BUILDING

Application of complicated structural analysis techniques such as sophisticated nonlinear analysis methods is becoming increasingly widespread. While this in itself is a positive trend, it should not lead to forgetting the cardinal fact that no analysis, no matter how sophisticated, can compensate for sound construction practices and reasonable engineering judgment. If the way building is constructed has no or little resemblance to how the building has been analysed, fancy analysis is just an exercise in futility. This obvious fact is commonly, and with increasing frequency, being ignored. In addition, more critical than whether the building has been designed according to proper and adequate seismic code provisions or postulated earthquake ground motions, is whether the building has been constructed according to the visions of the design engineer and whether the codes are being enforced in the field. Simply stated, the earthquakes neither look at our engineering calculations or read our governing seismic design codes, they merely pay attention to the status queue of the building as it stands at the time of the earthquake. This almost tautological assertion is often ignored.

11.0 CONCLUSIONS

The important lessons learned from old and new earthquakes were reviewed. It is unfortunate that these lessons which have been learned many times over have to be re-learned following each major earthquake. It is hoped that this paper contributes to the general understanding of how critical the understanding of these issues are and results in a coordinated course of action to reduce the damaged caused by them during the future earthquakes. Let us learn only new lessons from future earthquakes and avoid re-learning the old lessons anew.

12.0   ACKNOWLEDGEMENTS

The authors are members of the Los Angeles Tall Buildings Structural Design Council, which was formed to provide a forum for the discussion of issues relating to the design of tall buildings. The support and encouragement of the Los Angeles Council to investigate, research and analyse structural behaviour in earthquakes is most greatly appreciated.

13.0   REFERENCES

Naeim, F. and Lew, M., 2000
    THE 1999 EARTHQUAKE DISASTERS WORLDWIDE: HOW MANY
        TIMES DO WE HAVE TO RE-LEARN THE FUNDAMENTALS
        OF SEISMIC ENGINEERING? *Struct. Design Tall Build.* 9,
        161–182.

DESIGN CRITERIA AND LOADS

New CTBUH Monograph on Building Motion, Perception and Mitigation

N. Isyumov and T. Tschanz

ABSTRACT

This paper provides an overview of the new Monograph which deals with motions experienced by tall buildings and their impact on design. Of various types of building motions wind-induced vibrations are of greatest concern to designers since they can occur relatively frequently and are difficult to control and mitigate. The Monograph also treats floor vibrations caused by pedestrian traffic, in recognition of their importance for the satisfactory performance of buildings. The new Monograph includes recent findings of controlled "moving room" experiments and deals with the response of human to motions in actual buildings. This includes results by researchers from North America, Japan, Australia, Europe and other parts of the world. Methodologies for predicting wind-induced accelerations of tall buildings are presented and criteria used for their evaluation are discussed. Measures for reducing the wind-induced dynamic motions of tall buildings are discussed. These include the effects of altering the dynamic properties of tall buildings, notably their effective damping, as well as possible modifications to their aerodynamic characteristics, in order to reduce the wind-induced dynamic forces.

Finally, while significant progress has been made since the 1980 edition of this Monograph, many questions about how to predict, evaluate and if necessary mitigate wind-induced and other types of building motion still remain. While the need to limit motions of tall buildings is universally recognized there is still incomplete consensus on how to best gauge acceptable performance. As a result, different points of view are presented to express the diversity of practice in different parts of the world.

1.0 OVERVIEW

Motions can become perceptible and potentially unacceptable in situations where the response of buildings and other occupied structures to the action of external loads has a significant dynamic or oscillatory component. Oscillatory motions or vibrations occur in situations where structures resonate under the action of external loads such as seismic ground accelerations, wind action and various man-induced excitations. The latter include explosions, sonic booms, vehicular, rail and pedestrian traffic, operating machinery, etc. Of these, wind-induced vibrations tend to have the greatest influence on the performance of tall buildings. While there can be substantial oscillatory motions of a tall building during a strong earthquake, such motions are of short duration and the primary concern of occupants is the integrity of the building rather than motion perception and discomfort. Wind-induced motions of tall buildings can persist for hours and can become perceptible and possibly annoying to its occupants without causing structural distress. Some man-induced motions such as floor vibrations due to pedestrian traffic can also occur frequently and be annoying. Generally, the public does not expect buildings nor their components to move and noticeable vibrations which persist are usually judged to be indications of inferior quality.

Wind-induced motions of tall buildings are not new phenomena and reports of the motion of early skyscrapers abound in the literature. The following quotation from Cushman Coyle's writing in the American Architect in 1929 is good advice for both past and current designers of tall buildings:

> *"In the case of high buildings, the frame must be designed to resist wind pressures with sufficient stiffness to keep the vibration caused by wind within limits that inspire the occupants with confidence in the strength of the structure."*

Few building codes, past or present, provide designers with the necessary insight for judging exactly what limits should be placed on the magnitude of vibrations occurring with different recurrence rates and for spaces intended for different activities. It is generally accepted that large motion amplitudes may still be acceptable if they occur rarely and/or the activity intended for the area is a casual one. The public may not be surprised if perceivable motions are experienced in the viewing gallery of a tall flexible structure. The same public will have different expectations, however, for office or residential space in tall buildings. Consequently, attention to wind-induced drift and concern for perceptible motions and potential occupant dissatisfaction have become important considerations in the design of tall buildings. These concerns are being pushed to the limit by continuing construction trends towards super-tall and ultra-tall buildings.

The requirements to limit the wind-induced drift and horizontal accelerations constrains the design of most tall buildings, including those in seismic areas. The wind-induced drift of buildings is limited to assure an acceptable performance of the building envelope and interior finishes. Limits on wind-induced accelerations are imposed in order to assure the comfort of occupants and their confidence in the integrity and quality of the building. While there is considerable experience in setting appropriate limits on the total or the inter-storey drift of a tall building, much less is known about how building motions are perceived by occupants and under what circumstances they can become objectionable.

The human perception and acceptance of tall building motions depends on both psychological and physiological considerations. Both tend to be highly subjective and therefore difficult to quantify, except in statistical terms. The acceptability of wind-induced building motions is most commonly judged by the magnitude and the recurrence rate of horizontal accelerations which determine the body forces experienced by occupants. The emerging feed-back from the performance of actual tall buildings furthermore indicates that the "habitability" or the total occupant comfort is also influenced by other factors. Particularly important are audio and visual cues which can accentuate motion sensation and heighten the irritation of occupants. While the literature is full with information on the effects of shock and vibrations on humans, most attention has been given to frequencies well above those of interest for tall buildings, whose fundamental natural frequencies of vibration are typically in the range of .1 to .3 Hz (typical periods of 3 to 10 seconds). Higher modes of vibration are rarely excited by wind action. One area with a similar frequency range is the experience with the effects of ship motions on passengers. Unfortunately however, that motion is primarily vertical and of a pitching nature rather than horizontal as experienced in buildings. As a result, it is difficult to transfer the experience with ship motions to the evaluation of horizontal motions in tall buildings.

No specific recommendations on acceptable motions of tall buildings were included in the previous 1980 edition of Monograph 13. Also there are no requirements to control building motions in any of the US Codes, including Standard 7 of the American Society of Civil Engineers. Some codes of practice do contain limits for wind-induced motions of tall buildings. The National Building Code of Canada, since its 1975 edition, contains procedures for evaluating peak wind-induced accelerations and recommendations for evaluating the acceptability of motions. It suggests that the peak horizontal acceleration at the top of a building, predicted for a return period of 10 years, should be limited to 30 and 10 milli-g for office and residential occupancies respectively. Guidelines for the evaluation of the response of occupants of buildings and offshore structures in the frequency range of 0.063 to 1 Hz appear in the International Standard ISO 6897, published in 1984. Detailed requirements have been published by the Architectural Institute of Japan in its "Guidelines for the Evaluation of Habitability to Building Vibration". Nevertheless, a full understanding of the effects of motion on the habitability of tall buildings has yet not fully emerged.

While the physiological effects of horizontal motions are now more clearly understood, largely as a result of systematic "moving room" experiments, their psychological effects and the influence of other factors, such as wind-induced noise and visually apparent motions, require further study. One common trend which has arisen is the realization that many currently recognized "problem" buildings experience wind-induced motions which contain an appreciable torsional component. In addition to increasing the resultant horizontal acceleration and therefore the body forces experienced by occupants, particularly those near the building corners, torsional motions can be more apparent visually to occupants along the building perimeter with sight lines to other buildings and outside reference points. The readily apparent lateral swinging of the horizon, which appears due to torsional motions, is an important prompt of building motions, which otherwise may have gone unnoticed. As a result, some criteria suggested in the literature and included in this Monograph impose limits on the wind-induced torsional velocity.

Finally, while significant progress has been made since the 1980 edition of this Monograph, many questions remain about how to predict, evaluate and if necessary mitigate wind-induced and other types of building motions. While the need to limit motions of tall buildings is universally recognized, there is still no consensus on the criteria for gauging acceptable performance. This can only be achieved through further research and feed-back from subjective studies carried out in full scale.

2.0 CONTENTS OF NEW MONOGRAPH

A Table of Contents of the current draft of Monograph 13, entitled "Motion Perception, Tolerance and Mitigation" is included in Attachment 1 to this paper. As seen from the Table of Contents, an attempt has been made to reflect the practice followed in different parts of the world. This has not been an easy undertaking and the inclusion of a sufficiently wide cross-section of different approaches has largely contributed to the delays in finalizing this Monograph. Our current expectations are that a final draft of the Monograph will be established on a web-site later this year, to be announced in the CTBUH Times. Printing will follow once it has been viewed and commented on by tall building designers in the technical community.

3.0 RETROSPECTIVE AND FUTURE DIRECTIONS

The ingredients of current practice largely rely on the results of controlled "moving room" experiments, subjective feed-back from building occupants and the judgement of designers and building owners. To insist that wind-induced building motions should be below occupant perception would invariably require special measures and the commitment of additional resources with a corresponding impact on cost. This should not be done lightly and without careful consideration of alternatives. It is important, therefore, for designers to be familiar with both the physiological and psychological consequences of excessive motions and with both good and bad experiences of the past.

Initial information on how to judge the acceptability of building motions, acquired in support of the design of major tall buildings, such as the World Trade Center Towers in New York and the Sears and John Hancock Towers in Chicago, have become corner-stones of acceptable practice in North America and world-wide. Peter Chen and Leslie Robertson through systematic experiments found that the threshold of human perception to horizontal motions was a peak acceleration of about 3 and 4 milli-g respectively for the 2 and 10 percentiles of their subject. Parallel work by Fazlur Khan and Dick Parmalee confirmed these perception thresholds and furthermore, suggested that peak acceler-ations in excess of 20 milli-g would be disturbing. Hansen, Reed and Vanmarcke were the first to provide subjective information on the response of a substantial number of occupants of tall buildings. Based on their findings, they concluded that both the perception of motion and its judged severity were due to the total effect of body forces and various perceived sensory cues. These included wind-induced noise and such visual cues as the movement of fixtures and the "swinging of the horizon" due to torsional motions. Furthermore, their survey of owners and developers suggested that complaints from a relatively small portion of the population of occupants can be sufficient to trigger concerns and to lead to a potential loss of rentability. They recommended that the onset of unacceptable motions should be taken as the acceleration level which would be considered unacceptable to 2% of occupants of the top 1/3 of the building. Their specific recommendation was to limit the predicted 6-year return period rms acceleration at the top of the building to 5 milli-g. This, in turn, corresponds to mean hourly peak acceleration of approximately 18 to 19 milli-g. This is remarkably consistent with earlier recommendations by Khan and Parmalee. Davenport used the perception threshold data of Robertson and Chen and the subjective feed-back from actual building occupants, provided by Hansen, Reed and Vanmarcke to suggest criteria for acceptable peak wind-induced accelera-tions for office buildings. The accelerations judged acceptable in Davenport's Criteria were allowed to increase with decreasing recurrence rate. Results of these early studies are reflected in the practices of different parts of the world and are expected to influence targets for acceptable performance in the future.

While "moving room" experiments have been extremely effective in para-metrically examining the response of humans to horizontal motions, including the initial perception of motion, the onset of nausea or "sea-sickness" and the effects of motion on balance, task performance and motor functions in general, such experiments unfortunately have not captured the influence of additional psychological factors, present in real buildings. Most occupants do not expect buildings to move, consequently noticeable motions can cause anxiety and raise suspicions about the quality and integrity of the building. These concerns have been recognized since the birth of the skyscraper. These are elusive questions and can only be settled through subjective feed-back from the occupants of actual tall buildings. It is furthermore important that such feed-back be obtained not only from buildings which have proven to be troublesome, but also buildings which perform well and are recognized as quality structures. This we believe should be one of the main targets of the next edition of this Monograph.

ACKNOWLEDGEMENT

The authors would like to express their appreciation to all contributors of the present Monograph as listed in Attachment 1. Thanks are also extended to Professor Takeshi Goto, Vice-Chairman of Committee 36 both for his technical contributions to the Monograph and his membership of the Editorial Group.

ATTACHMENT 1

Monograph 13

**Motion Perception, Tolerance and Mitigation
2000 Draft**

**Council on Tall Buildings and Urban Habitat
Committee 36**

CONTRIBUTORS

D.E. Allen
D. Boggs
M. Franco
T. Goto
J. Huancheng
P. Irwin
N. Isyumov
A. Kareem
J. Kilpatrick
W. Melbourne
T.M. Murray
E. Nielsen
G. Oosterhout
T. Ohkuma
K. Shioya
Y. Tamura
T. Tschanz

Editorial Group
   Nicholas Isyumov Chairman
   Takeshi Goto   Vice-Chairman
   Tony Tschanz   Editor

Table of Contents – Motion Perception, Tolerance, and Mitigation in Tall Buildings

Aerodynamic Solutions to Minimize the Wind-Induced Dynamic Response of Tall Buildings

W. H. Melbourne and J. C. K. Cheung

ABSTRACT

Background to the generation of perceptible accelerations in tall buildings is described with the conclusion that these tend to be dominated by the cross-wind response to wake (vortex) excitation. A sensitivity analysis demonstrates the dependence on wind speed, damping, building massiveness and the cross-wind force spectrum. The remainder of the paper gives guidelines to the way in which the shape or configuration of a building may be directed to achieve reductions in the cross-wind force spectrum and hence reductions in acceleration response to meet occupancy comfort criteria.

1 INTRODUCTION

The design of many tall buildings has become increasingly determined by the requirement to keep motion from wind action to within acceptable levels for human occupation. The most significant response in this respect is the cross-wind response driven by the wake (vortex) excitation mechanism. There are a number of ways of shaping or configuring a tall building to reduce the cross-wind excitation, such that it can be hypothesized that there should never be any need to incorporate auxiliary damping into a tall building which has been appropriately shaped. This paper will discuss the parameter sensitivity of tall buildings to cross-wind response and develop some guidelines for the shaping of tall buildings to avoid excessive response to wind action.

2 PARAMETER SENSITIVITY AND ACCELERATION CRITERIA

The more technical aspects of the mechanisms of tall building response to wind action and acceleration criteria have been covered in several papers by Melbourne et al (1988, 1992, 1998). For the purpose of this paper the essential background information will be summarised.

2.1   Response and excitation mechanisms

The response of tall buildings to wind action can be conveniently separated into along-wind and cross-wind motion in relation to two distinctly separate excitation mechanisms. The along-wind response is made up of a mean component and two types of fluctuating component, a low frequency response to background turbulence and a resonant response component. For the cross-wind response the mean component is usually small and the fluctuating components again have background and resonant response components. In terms of generating accelerations felt by building occupants it is only the resonant components which are significant, because the first cantilever bending modes have significantly higher frequencies than the main motions generated directly by the background turbulence. In the case of a tall building in a turbulent wind environment the resonant component of the along-wind response is usually relatively small which is the primary reason why along-wind response is rarely a problem in acceleration terms. The cross-wind resonant component can, however, be quite dominant particularly when the building is operating near the peak of the cross-wind force spectrum, only at lower Reduced Velocities does the background component tend to become significant. The dominance of the cross-wind resonant component in response to the relatively large amount of energy available from the vortex excitation is the main reason why acceleration levels tend to be dominated by the cross-wind response of the building. It follows then that shaping and configuring a tall building to reduce response acceleration levels essentially has to target reductions in vortex excitation.

2.2   Acceleration criteria

Acceleration criteria to achieve occupancy comfort in tall buildings have received somewhat varied attention over the past 30 years. The pioneering work of Chen and Robertson (1973) gave valuable information about human perception of sinusoidal excitation as a function of frequency, and they suggested using an occupancy sensitivity quotient which defines the ratio of the tolerable amplitudes of motions to the threshold motion for half the population. The work by Reed (1971) gave the first full-scale evaluation of occupants' responses to accelerations on two buildings and the first criteria in terms of standard deviation of acceleration for a return period for the frequencies of those buildings. Irwin, in a series of papers, further studied the responses of humans to sinusoidal acceleration over a range of frequencies. Through a number of unpublished full-scale studies of crane operators and building occupants he was primarily responsible for the standard deviation acceleration criteria in ISO 6897. He focused these on tall buildings in Irwin (1986). In North America some use appears to be made of an unreferenced peak acceleration criterion of 20 milli-g once in 10 years, with no reference to frequency.

Based on the above earlier studies and some full scale experience Melbourne and Cheung (1988) commenced the development of frequency-dependent criteria for peak accelerations rather than sinusoidal or normally distributed oscillations. This resulted in a single expression for peak acceleration for occupancy comfort in a building with dependence on the building frequency, and return period under consideration, and which allowed a peak factor dependent on the operating reduced frequency to be taken into account. This criterion or expression for peak acceleration which should not be exceeded to achieve acceptable occupancy comfort was

$$\hat{y} + \sqrt{2 \ln nT} \left(0.68 + \frac{\ln R}{5} \right) \exp(-3.65 - 0.41 \ln n) \tag{1}$$

where
n = building resonant frequency
T = duration of experience of acceleration (usually put at 600 seconds)
R = return period in years.

2.3 Determination of cross-wind response

One of the simplest ways of evaluating the cross-wind response, involving all the important parameters in the process of resonant response to wake excitation, is to use a mode-generalized force spectrum approach proposed by Saunders and Melbourne (1975). The method makes use of measured cross-wind displacement spectra to give a mode-generalized force spectrum (for the first mode) of

$$S_F(n) = \frac{(2\pi n_0)^4 m^2 S_y(n)}{H^2(n)} \tag{2}$$

where
$S_y(n)$ = spectrum of cross-wind displacement at top of building
n_0 = first-mode frequency
m = modal mass
$H^2(n)$ = mechanical admittance; $1/\{[1 - (n/n_0)^2]^2 + 4\zeta^2 (n/n_0)^2\}$
ζ = critical damping ratio

For a linear mode, and if excitation by low frequencies is small and the structural damping low so that the excitation bandwidth is large compared with the resonant bandwidth, the standard deviation of displacement at the top of the building may be approximated by

$$\sigma y = \left| \frac{\pi n_0 S_F(n)}{(2\pi n_0)^4 m^2 4\zeta} \right|^{1/2} \tag{3}$$

and the standard deviation of acceleration is given by

$$\sigma_{\ddot{y}} = \sigma y (2\pi n_0)^2 \tag{4}$$

The Cross-Wind Force Spectrum may be expressed in coefficient form by

$$C_{FS} = \frac{n_0 S_F(n)}{(\frac{1}{2}\rho \overline{V}_h^2 bh)^2} \tag{5}$$

where
h = building height
b = building width normal to wind direction
V_h = mean wind speed at top of building.

Then in terms of this force spectrum coefficient the standard deviation of acceleration becomes

$$\sigma_{\ddot{y}} = \frac{\rho \overline{V}_h^2 bh}{4m} \sqrt{\frac{\pi CFS}{\zeta}} \tag{6}$$

For a parallel sided rectangular building, depth d, average building density ρs, and a linear mode, the modal mass is given by

$$m = \tfrac{1}{3}\rho_s bdh \tag{7}$$

and the peak acceleration at the top of the building due to cross-wind response is given by

$$\hat{\ddot{y}} = \frac{3}{4} \frac{g\rho \overline{V}_h^2}{\rho_s d} \sqrt{\frac{\pi CFS}{\zeta}} \tag{8}$$

where
g = peak factor = $\sqrt{2 \ln nT}$

Corrections for mode shape, complex motion and background component have been discussed by Melbourne and Palmer (1992) and Melbourne and Cheung (1999), and it is noted from the latter reference that for low values of Reduced Velocity, in particular, it is important to use values of Cross-Wind Force Spectra for the resonant component only.

2.4 Parameter sensitivity

Inspection of Equation (8) shows the obvious dependence of cross-wind response on wind speed, building density, the cross-wind force spectrum and damping, but less obvious is the dependence on building shape. Firstly the acceleration is not, as one might intuitively think, directly dependent on height or aspect ratio but rather on planform size. Indirectly height is involved because wind speed is a function of height. Hence relatively tall slender buildings will have higher accelerations than low squat buildings, but the important relative parameters in the denominator are planform size and average density, i.e. massiveness. Secondly the value of the Cross-Wind Force Spectrum is a function of Reduced Velocity (V/nb) and the shape of the building. This relationship is quite complex and it is in the sensitivity of the Cross-Wind Force Spectrum to building shape that the key to developing tall buildings with reduced cross-wind response really lies.

3 CONTROLLING THE CROSS-WIND FORCE SPECTRUM

With some understanding of the aerodynamic excitation mechanisms involved it is a short step to formulate some guidelines which will direct the shape of a tall building towards reduction of the Cross-Wind Force Spectrum.

The primary cause of the cross-wind excitation is the vortex shedding process. The strength of the cross-wind excitation depends primarily on the strength of the shed vortices, the height correlation of the vortex structures and the location of streamwise surfaces on which cross-wind acting pressure fields can develop with sufficient energy at the natural frequency of the building. For buildings only the first two cantilever bending modes are significant, but this does not follow for towers where sufficient energy may be generated to excite higher modes.

Whilst far from rigorously proven it is known that sharp edged structures will shed stronger and more coherent vortices than curved surfaces. Tapering the cross-section will broaden the frequency bandwidth of excitation and lower the peak, as will increased turbulence in most cases. Tapering also reduces massiveness. Bleeding flow through the body will weaken the vortex structure and tend also to broaden the bandwidth. With these few observations it can be simply concluded that guidelines for changes in building shape to achieve reductions in the Cross-Wind Force Spectrum are as follows:

1. Square or rectangular buildings be made more circular, i.e. by cutting or rounding corners or adding features in the middle of sides.
2. Buildings be tapered or stepped back with increasing height.
3. Porosity be introduced across the building, particularly over the upper levels and near sharp corners.

4 EXAMPLES OF TALL BUILDING SHAPES WHICH REDUCE CROSS-WIND RESPONSE

All examples given in this section are based on data obtained from wind tunnel measurements using linear mode aeroelastic models tested in a properly scaled boundary layer model of the natural wind, that is with correctly scaled mean velocity and turbulence intensity profiles with height and spectral distribution. In this respect the measurements all relate to very tall buildings, that is between 300 and 500m, as identified by the turbulence intensity values quoted for the flow at the top of building height.

The benchmark for comparison is the square sectioned parallel sided tall building. Plots of the cross-wind force spectrum coefficient for a square sectioned building, with height to width ratio of 9:1, are given in Figure 1 as a function of Reduced Velocity for two turbulence intensity conditions, in suburban and urban terrain. The effect of increasing turbulence in reducing the peak and broadening the width of the cross-wind force spectrum is clearly evident.

The second example, shown in Figure 1, is for a tapered square tower with chamfered corners. This takes two steps to lower and broaden the width of the forcing spectrum. Broadening the vortex shedding frequency range reduces the correlation of the vortices shed into the wake, and the chamfered corners start towards a section with the characteristics of a circular section which has significantly lower cross-wind excitation; conjectured to be because of weakening of the shed vorticity as well as some reduction of wake correlation.

The third and fourth examples, shown in Figure 1, have the basis of a parallel square section but to which additions have been added to the centre of each side along with two levels of corner chamfer. This has transformed the square section into an aerodynamically circular section as evidenced by the shift of the peak of the spectrum from a reduced velocity of 10 to between 6 and 7 (for a circular cylinder the value would be about 5). The reduction of the cross-wind force spectrum achieved by these two modifications is more than an order of magnitude over much of the Reduced Velocity range, which indicates that cross-wind response will be reduced by around a factor of three.

The building geometries of the examples given in Figure 1 are defined as follows:

(i) Square, $h/b = 9$
(ii) Square, tapered, chamfered,
$b_{base} = 1.25\ b_{top}$ (i.e. at 3/4 height)
$h/b_{ref} = 9.6$
45° corner chamfer $= 0.2\ b$ to $0.2\ b$

(iii) and (iv) Square modified,

h/b = 9

rectangular addition at centre of each side, step out 0.075b, width 0.25b

45° corner chamfer (iii) = 0.1b to 0.1b

45° corner chamfer (iv) = 0.2b to 0.2b

5 EXAMPLE OF REDUCTION IN ACCELERATION RESPONSE

Using the data in Figure 1 an example evaluation of the acceleration for a parallel sided square section building and the fourth example, which has additions to the centre of each side and chamfered corners, will be given as follows:

Building parameters –

Height, h	=	400m
width, b	=	45m
Frequency, n	=	0.12 Hz
Building density, ρ_S	=	250 kg m^{-3}
Damping, ζ	=	0.01
mean wind speed at top of building, $\overline{V}$	=	32 ms^{-1}
peak factor, g	=	3.0
Reduced velocity, V_r	≈	6.0
for square building, C_{FS}	=	0.01
for square building with additions to the centre of each side and chamfered corners, C_{FS}	=	0.001

Using Equation 8 the peak acceleration for the two buildings evaluate to 44 milli-g and 14 milli-g respectively. Relative to the criterion evaluated from Equation 1, for a 5 year return period, of 19 milli-g, the parallel square building is clearly well outside the criterion and the building with additions and chamfered corners is well inside the criterion. It is of interest to note that tall buildings in Australia which tend to be more dense and more highly damped (due to reinforced concrete construction) a typical evaluation of the above for ρ_S = 300 kg m^{-3} and ζ = 0.015 would be 30 milli-g and 10 milli-g respectively, and whilst response is reduced, the conclusions relative to the criterion remain the same.

6 CONCLUSIONS

A summary of the mechanisms of the response of tall buildings to wind action have been given along with a simple method of calculating the cross-wind acceleration response and a demonstration of the dependence on wind speed, damping, building massiveness and cross-wind force spectrum. From this examples have been given with respect to the geometric shapes of tall buildings that will reduce the cross-wind force spectrum and the acceleration response to meet occupancy comfort criteria.

7 ACKNOWLEDGEMENTS

The assistance of J. Hick in the preparation of models and H. Fricke in the preparation of the paper, and the support of the Australian Research Committee is gratefully acknowledged.

8 REFERENCES

Chen, P. W. and Robertson, K. E., 1973
> HUMAN PERCEPTION THRESHOLDS TO HORIZONTAL MOTION, *Journal of the Structural Division, ASCE,* Vol. 98, pp. 1681–1695.

Irwin, A., 1988
> MOTION IN TALL BUILDINGS, *Proceedings of the 3rd International Conference on Tall Buildings, Second Century of the Skyscraper, Chicago*, Council on Tall Buildings and Urban Habitat, Bethlehem, Pa.

Melbourne, W. H., 1998
> COMFORT CRITERIA FOR WIND-INDUCED MOTION IN STRUCTURES, *Structural Engineering International*, Vol. 8, No.1, pp. 40–44.

Melbourne, W. H. and Cheung, J. C. K., 1988
> DESIGNING FOR SERVICEABLE ACCELERATIONS IN TALL BUILDINGS, *Proceedings of the 4th International Conference on Tall Buildings,* Hong Kong and Shanghai, pp. 148–155.

Melbourne, W. H. and Cheung, J. C. K., 1999
> RESONANT AND BACKGROUND CROSS-WIND RESPONSE COMPONENTS IN TURBULENT FLOW, *Proceedings 10th International Conference on Wind Engineering*, Copenhagen, pp. 1531–1536, Publ AA Balkema/Rotterdam/Brookfield.

Melbourne, W. H. and Palmer, T. R., 1992
> ACCELERATIONS AND COMFORT CRITERIA FOR BUILDINGS UNDERGOING COMPLEX MOTIONS, *Journal of Wind Engineering and Industrial Aerodynamics*, Vol. 41, pp. 105–116.

Reed, J. W., 1971
> WIND INDUCED MOTION AND HUMAN COMFORT, *Research Report 71–42, Massachusetts Institute of Technology*, Cambridge, Mass.

Saunders, J. W. and Melbourne, W. H., 1975
TALL RECTANGULAR BUILDING RESPONSE TO CROSS-WIND EXCITATION, *Proceedings of the 4th International Conference on Wind Effects on Buildings and Structures*, London, Cambridge University Press, pp. 369–380.

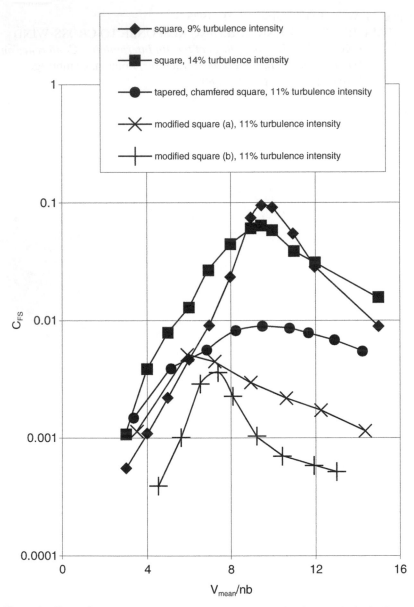

Figure 1 Cross-wind Generalised Force Spectra as a function of Reduced Velocity for Square, Modified Square and Tapered Chamfered Square Buildings, height/width ≈ 9, freestream longitudinal turbulence intensity at top = 9% to 14% (as noted on Figure).

The Behaviour of Simple Non-Linear Tuned Mass Dampers

Barry J. Vickery, Jon K. Galsworthy and Rafik Gerges

1.0 INTRODUCTION

Tuned mass dampers have been used to add damping to structural and mechanical systems for many years and the theory of their design and behaviour when subjected to sinusoidal forcing has been elegantly described by Den Hartog (1956). The optimization of these dampers with linear dashpots and springs and for sinusoidal excitation is well treated by Den Hartog but considerably less attention has been paid to non-linear devices on systems subject to the random loads imposed by the wind acting on tall buildings and other slender structures such as towers and chimneys.

The present paper deals with the behaviour of non-linear T.M.D.'s subject to random Gaussian excitation. Attention is limited to two simple non-linear forms that occur commonly in practice. These two forms are dry-friction or constant force dampers and "velocity squared" or V^2 damping associated with, for example, flow through an orifice or flow in a rough pipe. Although there is some mathematical development, this is not the essential part of the paper. The prime intention of the paper is to examine the more general characteristics of these non-linear dampers with a view to assessing the advantages and disadvantages of each compared to linear T.M.D.'s.

The characteristics of linear T.M.D.'s are reviewed briefly in Section 2 which includes sample design charts for linear systems. Linearization of the simple friction and V^2 T.M.D.'s is treated in Section 3. The characteristics of non-linear as opposed to linear T.M.D.'s is also addressed in Section 3. Section 4 presents some numerical results and designs for non-linear T.M.D.'s for a fictitious building.

2.0 LINEAR T.M.D.'S

A linear T.M.D. is shown schematically in Fig. 1. The values of M, K and C define the primary system ie the building or tower in question while the values of m, k and c define the T.M.D. The value of the primary mass is a function of the mass and mass distribution of the building, the mode shape under consideration and the modal deflection at the point of attachment of the T.M.D.

For a simple translational mode, the value of M is given by

$$M = \int m(z)\phi^2(z)dz / \phi^2(z_a)$$

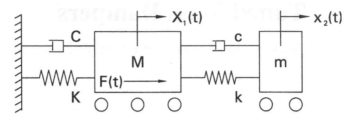

Figure 1 Definition sketch of a S.D. of F. system with an added T.M.D.

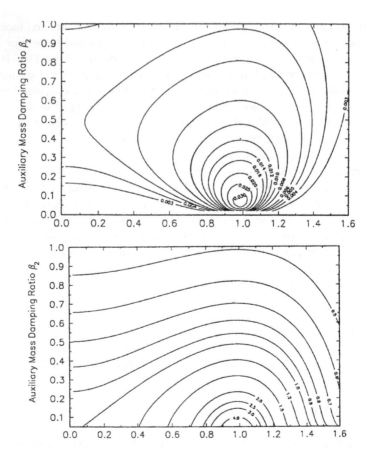

Figure 2 Values of βe (upper) and σ_r/σ_1 (lower) for $\mu = 0.02$, $0 < \beta_2 < 1$ and $0 < f < 1.6$.

where $\phi(z_a)$ = modal deflection at height z

     $m(z)$ = mass per unit height

       z_a = height of attachment

The value of M is typically about 30% of the actual mass of the building if the point of attachment is at or near the top of the building. The behaviour of the T.M.D. can be expressed in terms of the following dimensionless parameters

μ = mass ratio = m/M; β_2 = damping of T.M.D. = $c/2m\Omega_o$
f = tuning ratio = $\omega\Omega_o/\Omega_o$; g = forcing ratio = ω/Ω_o
β_1 = damping of primary system = $C/2M\Omega_o$

where

Ω_o = $\sqrt{K/M}$; $\omega_o = \sqrt{k/m}$
ω = angular frequency of sinusoidal force acting on the primary system
$F(t)$ = $F_o e^{i\omega t}$

The displacements of the system can be expressed as;

$$X_1(t) = \frac{F_o}{K}H_1(i\omega)e^{i\omega t}$$

$$X_r(t) = X_2(t) - X_1(t) = \frac{F_o}{k}H_r(i\omega)e^{i\omega t}$$

The performance of the T.M.D. is best measured by the amount of damping that it adds to the primary structure. This damping is not easily defined but a satisfactory approach is to define the added damping as that which, when added to the primary system, will result in a response of this S.D.F. system equal to the actual response of the 2 D.O.F. system. If the exciting force is random and the spectrum does not vary strongly in the vicinity of Ω_o then the added damping, β_e, is given as;

$$\beta_e = \left[\frac{4}{\pi}\int_O^\infty |H_1(g)|^2 dg\right]^{-1} - \beta 1$$

β_e is a function of β_1, β_1, μ and f but is only weakly dependent on β_1 and the value of β_e computed with $\beta_1 = 0$ is an adequate approximation. For this special case;

$|H_1(g)|^2 = [(f^2 - g^2)^2 + (2\beta_2 g)^2]/F(\mu, \beta_2, g)$
$|H_R(g)|^2 = g^4/F(\mu, \beta_2, g)$
$F(\mu, \beta_2, g) = [(f^2 - g^2)(1 - g^2) - \mu f^2 g^2]^2 + (2\beta_2 g)^2[1 - g^2 - \mu g^2]^2$

The second parameter that is required is the magnitude of the X_r compared to X_1. The ratio of the rms values (σ_r and σ_1) is given by,

$$\sigma_r/\sigma_1 = [\int|H_r(g)|^2 dg/\int|H_1(g)|^2 dg]^{1/2}$$

The dependency of β_e and σ_r/σ_1 on β_2 and f is shown in Fig. 2 (Vickery et al, 1982) for a mass ratio (μ) of 0.02. The existence of an optimum design to achieve maximum β_e is apparent. The optimum tuning ratio for frequency is slightly lower than unity and the optimum value of β_2 is close to 0.07. In the case of a non-linear damper for which β_2 varies with amplitude, it will be possible only to obtain maximum β_e at some chosen amplitude. The performance for optimum frequency tuning is shown in Fig. 3 for mass ratios of 0.01, 0.02, 0.04 and 0.08.

3.0 LINEARIZATION

Performance diagrams such as those shown in Figs. 2 and 3 can be employed in the designs of linear T.M.D.'s but to employ them for non-linear (friction or V^2) devices requires that the latter be linearized. This can be achieved by defining an equivalent linear β_2 for which the energy dissipation rate matches that of the non-linear device. It is assumed that the non-linearities are such that;

(i)   individual cycles are sinusoidal in form with a slowly varying amplitude, $a(t)$

(ii) the distribution of $a(t)$ follows the Rayleigh form associated with a narrow-band Gaussian process.

Both the above assumptions are reasonable and should lead to a linear model which is adequate for preliminary design of a non-linear T.M.D. Final design however should employ a time domain analysis using a true non-linear model of the system that will reproduce the correct probability distribution.

The energy dissipated per cycle of amplitude "a" is given by,

$$E(a) = \pi C_e \omega a^2; \qquad F_D = C_e v$$

$$E(a) = \frac{8}{3}C\omega^2 a^3; \qquad F_D = Cv^2$$

$$E(a) = ; \qquad F_D = F_r$$

$$F_D = \text{damping force}$$

The average dissipated energy per cycle is given by $\overline{E} = \int_0^\infty E(a)p(a)da$; where $p(a)$ is the probability density function defined as $p(a) = a/\sigma^2 \exp(-a^2/2\sigma^2)$.

Evaluation of this integral leads to the following results;

$$\overline{E} \text{ (linear)} = 2\pi c\omega\sigma^2$$
$$\overline{E} (V^2) = 4\sqrt{2}\pi c\omega^2\sigma^3$$
$$\overline{E} \text{ (friction)} = 2\sqrt{2\pi}F_r\sigma)$$

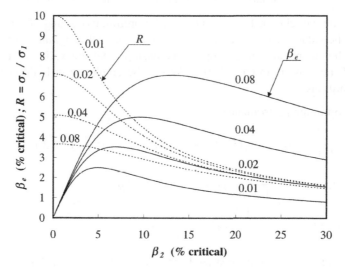

Figure 3 Values of βe and $R = \sigma_r/\sigma_1$ for $f = 1/(1 + \mu)$ and $\mu = 0.01, 0.02, 0.04$ and 0.08.

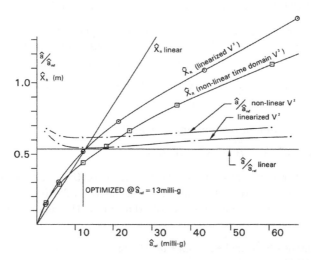

Figure 4 Response of a building with a V^2 T.M.D. with $\mu = 0.094$.

The equivalent linear values of β_2 for the two non-linear devices are then;

$$\beta_2 = \sqrt{\frac{2}{\pi}} \frac{c\sigma_r}{m} \frac{\omega_o}{\Omega_o}; \quad F_D = cV^2$$

$$\beta_2 = \frac{1}{\sqrt{2\pi}} \frac{F_r}{K\sigma_r} \frac{\omega_o}{\Omega_o};$$

σ_r = rms relative displacement , F_r = constant friction force

The above relationships can be used for preliminary sizing of a non-linear T.M.D. The values of β_2 can be used in conjunction with diagrams such as Figs. 2 and 3 but there are some simple relationships that are valid for the small values of the mass ratio ($\mu < 0.10$) likely to be employed in buildings. These relationships are;

(i)   for optimum performance

$$f = \frac{1}{1+\mu}, \beta_2 = \sqrt{\mu}/2, \beta_e = \sqrt{\mu}/4, R = \sigma_r/\sigma_1 = 1/\sqrt{2\mu}$$

(ii)   for $\beta_2 \rightarrow 0; f = \frac{1}{1+\mu}; \beta_e = \beta_2, R = 1/\sqrt{\mu}$

(iii)   for $\beta_2 \rightarrow \infty ; f = \frac{1}{1+\mu}; \beta_e = \frac{\mu}{4\beta_2}, R = \frac{1}{2\beta_2}$

The adequacy of the linearized approach is illustrated in Fig. 4 in which the results of the linearized approach are compared with those obtained from a time-domain non-linear simulation. The results in Fig. 4 are for an actual damper with the following properties;

mass = 250,000 kg ; $\mu = 0.0094$; $F_D = CV^2$; $\beta_o = 0.01; f_o = 0.16$ Hz

The damper is designed to have an optimum performance when the peak acceleration (without the T.M.D.) is 13 milli-g and is expected to reduce the acceleration by at least 35%. For optimum performance we have;

$f = 1/1 + \mu = 0.99 , \beta_2 = \sqrt{\mu}/2 = 0.0485 , \beta_e = \sqrt{\mu}/4 = 0.0242$
$R = \sigma_R/\sigma_1 = 1\sqrt{2\mu} = 7.29$

The acceleration of the building is given by; $\hat{a} = 13/\sqrt{1 + \beta_e/\beta_o}$ milli-g

$\hat{x}_1 = 0.128/\sqrt{1 + \beta_e/\beta_o}$ M = 0.069 m

The peak relative motion is 0.069 × R or 0.503m and the root-mean-square value is 0.503/3.6 = 0.140m. The peak factor of 3.6 is consistent with a Gaussian process with a sample length of 30 minutes. The required value of c is then given by;

$\beta_2 = 0.0485 - \sqrt{\pi/2} \bullet c\sigma_R/m \bullet \omega_o/\Omega_o;$ or
$C = 110,000 \, N/(m/s)^2$

This value of C was confirmed by the true non-linear time domain simulation. Fig. 4 shows the predicted reductions in the peak acceleration as a function of $\hat{a}_{ref}$, the peak acceleration in the absence of the T.M.D. Also shown are the peak displacements for the T.M.D. Results are presented for,

(i)   a non-linear ($F_D = CV^2$) system predicted using the linearized theory,
(ii)   a non-linear system predicted by time domain simulation; and
(iii) a linear system optimized for $\mu = 0.0094$.

There are a number of features of Fig. 4 that deserve comment.

(i)   The V^2 system compares favourably with the optimized linear system for reference accelerations from the perception level to 60 milli-g. The acceleration of 13 milli-g corresponds to a return period of about 10 years while for a return period of 1000 years the reference acceleration is 45 milli-g.

(ii)   The linearized predictions of the peak accelerations are about 10% lower than those from the simulation.

(iii) The predicted motions of the T.M.D. given by the linearized model are about 20% higher than those given by the simulation.

(iv)   At the 1000-year return period the motions of the T.M.D. in the V^2 system are some 50% lower than those in an optimized linear T.M.D.

4.0   PRELIMINARY DESIGN OF A V^2 T.M.D.

The building under consideration has a natural frequency of 0.20 Hz, a damping of 1% of critical, a height of 200m, a cross-section of 30m $\times$ 30m, a bulk density of 240 kg/m^3 and a mode shape $\phi(z) = (z/H)^{1.23}$. For a 10-year return period, the tip acceleration is 20 milli-g and it is desired to reduce this to a maximum of 12 milli-g. the T.M.D. will be installed at or very near $z = H$. For $R = 1000$ years, the tip acceleration without additional damping is 70 milli-g and the damper must accommodate this motion.

The chosen design is a simple "U-tube" damper as illustrated in Fig. 5. Dampers of this general type have been discussed by Sakai et al (1989). The aimed reduction in acceleration is from 20 to 12 milli-g but an allowance should be made for inaccuracies in the design method and the inaccuracies of the theoretical and experimental data. Somewhat arbitrarily, the target acceleration will be chosen as 10 milli-g, thus allowing for a 20% error in the design process. For an optimum damper, the added damping is $\sqrt{\mu/4}$ or 0.0354 for $\mu = 0.02$, this would reduce the acceleration by a factor of $1/\sqrt{1 + 0.354/.01}$ for or 0.47 (i.e., to 9.4 milli-g) and thus provide an adequate margin on the target value of 12 milli-g.

The exact analysis of a U-tube damper is complex but a simple one-dimensional analysis enables the parameters x_r, m, c and k for the system in Fig. 1 to be evaluated. The results of such an analysis are;

$$m \cong w\,B\,D\,L; \cong (L_e/L)x_o$$

$$c \text{ (where } F_D = c\dot{x}_r^2) = C_L\frac{\rho}{2}A_o(L/L_e)^2$$

$$k = 2(A_o^2/A_R)\,(L/L_e)\,(y + K_a/H) \text{ both reservoirs sealed}$$
$$ = 2(A_o^2/A_R)\,(L/L_e)\,(y + K_a/2H) \text{ one reservoir sealed}$$

where; $\dot{x}_o$ = average flow velocity over A_O; γ = specific weight of water; K_a = adiabatic bulk modulus of air; $L_e = \int_1^2 \dfrac{A_O}{A_S}ds$; A_S = area of stream tube for one-dimensional analysis; C_L = loss coefficient of screen, orifices, etc.

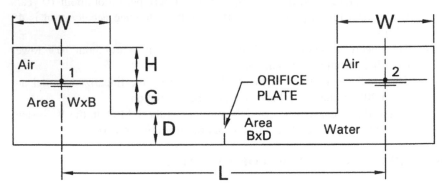

Figure 5   Arrangement of a "U-tube" V^2 T.M.D.

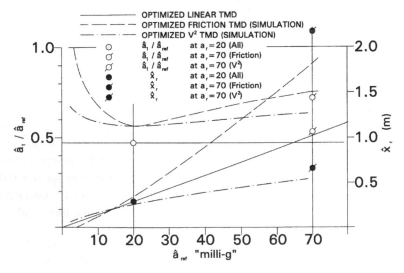

Figure 6   Performance of linear, friction and V^2 T.M.D.'s with $\mu = 0.02$.

For $\mu = 2\%$, $\sigma LBW = 250{,}000$ or $LBW = 250$ m^3. The value of β_2 is given by;

$$\beta_2 = \frac{\sqrt{\mu}}{2} = 0.0707 = \sqrt{2/\pi}\,(c\sigma_R/m)(\omega_0/\Omega_0)$$

$$\omega_0/\Omega_0 = 1/1 + \mu = 0.98,\ \sigma_R = \sigma_R/R,\ R = \sqrt{2/\pi} = 5$$

The value of σ_1 can be computed from the peak acceleration of 9.4 milli-g; for a peak factor of 3.6 the value of σ_1 is $9.4 \times 9.81/1000/3.6/\Omega_0^2 = 0.016$ and hence $\sigma_R = 5\sigma_1$ 0.081m. Using this value for σ_R the required value of C is 279,000 N/(m/s)^2 and hence $\Sigma C_L = 62$ for an assumed value of $L/L_e = 0.95$. The value of k must be chosen to yield $f_0 = 0.20/1.02 = 1/2\pi\sqrt{k/m}$.

There are of course a variety of designs that will satisfy the requirements but that chosen is as follows;

$D = 1.0$m , $W = 3.0$m , $L = 25.0$m , $B = 10.0$m,
$H = 0.90$m , $G = 0.90$m   one reservoir sealed, one vented

The orifice plate might be a plate with 48mm diameter holes at 100mm centres and supported on a frame to resist the imposed loads. Final tuning of the system after on-site measurements are comparatively simple. The frequency tuning is accomplished by increasing or decreasing H, i.e. by adding or removing water. Damping tuning can be accomplished by blocking off or opening out a selected number of orifices.

Dealing with the 1000-year accelerations is not a problem since the motions of the water are very small compared to the available free space of about 1.0m. Design details will not be covered in any depth but it is worthy of note that the primary source of stiffness is the "air spring" in the sealed reservoir. This spring provides 87% of the system stiffness and hence small surface waves will not markedly influence the performance. Even so it would be prudent to use vertical baffle plates in the end reservoirs to ensure that resonant waves will not develop.

The use of turning vanes to link the tube to the reservoir is also a consideration. These would lead to an even distribution of the flow into the reservoir. The selection of the orifice plate (or alternately screens) is important since the available data is from steady flow rather than oscillating flows. The steady flow condition will be approached if $a >> d$ where a is the amplitude of the fluid motion and d the hole diameter (or bar size in a grid). In the present instance, a typical value of a (2σ say) is 0.16m and the value of a/d is 6.7 which should ensure that the orifice plate performance is similar to that under steady flow conditions but tests would be advisable to examine the performance for a range of a/d. For $R = 1000$ years the computed value of σ_R is 0.18m and the peak value $\hat{x}_R$ is 0.648m. This corresponds to a maximum rise in the reservoir of 0.205m and a maximum flow velocity in the tube of 0.78 m/s. These two values define the design pressure in the sealed reservoir, $\Delta p_{max} = K_a \times 0.205/H = 27.5$ kPa and the pressure drop across the orifice plate, $p = C_L \rho V^2 = 18.9$ kPa. This could be spread across many orifice plates or screens rather than concentrated at the single orifice plate of this example.

The performance of the above damper is shown in Fig. 6 together with that for an optimized linear T.M.D. and an optimized friction damper with the optimum value of F_r being 5560N for $\mu = 0.02$. The curves shown in Fig. 6 were determined using a non-linear time domain simulation but also shown are the points predicted using the linearized theory. These predictions are shown for $\hat{a}_{ref}$ (the peak acceleration in the system without a T.M.D.) equal to 20 milli-g and 70 milli-g. The linearized theory closely predicts the optimum values of C and F_r and provides reasonable estimates of the reduction in acceleration and the motions of the T.M.D.

A problem with the friction damper are the large T.M.D. motions at excitation levels substantially higher than that at the design point. The amplitude reduction also falls off quite rapidly away from the design point. The friction damper is more suited to situations where the design point corresponds to maximum excitation eg, when used to reduce vortex-induced motions in the vicinity of the critical speed.

5.0   CONCLUSION

The linearization of V^2 and friction T.M.D.'s provides a simple and adequate approach to optimizing design and predicting performance but time domain non-linear simulation should be employed to refine performance predictions.

The V^2 damper significantly reduces motions of the T.M.D. with little sacrifice in performance compared to the optimum linear T.M.D.

The friction damper performs well in the vicinity of the design point but results in large T.M.D. motions at higher levels of excitation.

6.0   REFERENCES

den Hartog, J. P., 1956
   MECHANICAL VIBRATIONS, 4th edition, (New York, McGraw-Hill Book Company).

Sakai, F., Takeda, S. and Tamaki, T., 1989
   TUNED LIQUID COLUMN DAMPER – NEW TYPE DEVICE FOR SUPPRESSION OF BUILDING VIBRATION, *Proceedings of Int'l Conference on High-Rise Buildings, Nanjing, China.*

Vickery, B. J., Isyumov, N. and Davenport, A. G., 1982
   THE ROLE OF DAMPING, MASS AND STIFFNESS IN REDUCTION OF WIND EFFECTS ON STRUCTURES, *Proceedings of the 5th Colloquium on Industrial Aerodynamics*, Aachen, Germany.

DESIGN CRITERIA AND LOADS

Earthquakes and Tall Buildings – a Review

Graham L. Hutchinson, Nelson T. K. Lam and John L. Wilson

ABSTRACT

This paper has provided an overview of both the philosophy and methods used in the design and analysis of building structures to earthquake ground shaking. In particular, the following topics and issues were discussed : (a) seismic hazard and soil response modelling, (b) aseismic design methods, (c) design and detailing of structural systems and (d) future directions for seismic design.

KEYWORDS

seismic hazard, soil response modelling, aseismic design methods, structural systems, passive damping, ductility, earthquake engineering, tall buildings.

1   INTRODUCTION AND BASIC DESIGN PHILOSOPHY

In seismically active areas of the world design engineers are faced with a dilemma of being required to design for an earthquake event which has a high probability of not occurring during the life of the facility. The engineer has many choices between the bounds of designing the structure to remain elastic during a rare event and totally ignoring the hazard thus possibly leaving themselves open to charges of negligence.

To overcome this problem the concept of a dual design philosophy was introduced. This philosophy involved designing for a damageability limit state (an event that has, say, a 50% chance of being exceeded during the design life of the structure) and an ultimate limit state event (where there is a 10% chance of exceedance). The design philosophy associated with the ultimate event is based on the preservation of life. Following a number of significant earthquakes in the past 15 years which caused extensive damage and costly dislocations to the function of large cities this philosophy has been revised. A new philosophy incorporating performance based design concepts has been developed and is gaining acceptance for inclusion in modern building standards (SEAOC, 2000; FEMA273, 1997).

The philosophy involves selecting a performance level for a given return period which is dependent on the importance of the facility. Some performance levels include: (a) immediate occupancy level (facility remains fully operational after the event) (b) life safety level (structural drift is controlled so that damage is limited and lives are not endangered) (c) collapse prevention level (structural drift is at the ultimate limit). For a typical facility these three performance levels would be associated with a 50%, 10% and 2% probability of exceedance during a facility life of 50 years (i.e. return periods of 75,500 and 2500 years respectively).

This paper is of an overview nature and therefore it is not possible to discuss each point in detail. However, some key references are provided for further reading. Both the philosophy and methods used in the design and analysis of building structures to earthquake ground shaking are reviewed. In particular, the following topics and issues are introduced: (a) seismic hazard and soil response modelling, (b) aseismic design methods, (c) design and detailing of structural systems and (d) future directions for seismic design.

2   HAZARD MODELLING AND SOIL EFFECTS

2.1   Response Spectrum Modelling

The design response spectrum for rock sites specified in many codes of practice is based on the well known normalised response spectrum model developed by Newmark and Hall (1982). This model uses the peak ground acceleration (PGA) or peak ground velocity (PGV) to scale the spectrum to reflect the seismicity of the site. These response spectrum models have been principally derived from empirical information and experience in the interplate regions of California. Their application in intraplate regions such as Australia has been assumed to be conservative.

Traditional response spectrum models, including the Newmark and Hall model, assume a unique normalised response spectrum for rock sites, which implies that the spectral property of the earthquake depends solely on the site conditions and not on the earthquake source and path. In contrast, Uniform Hazard Spectra (UHS) developed in the United States display significant regional variations in the shape of the response spectrum across the country for similar site conditions (Frankel *et al.*, 1995). The UHS possesses a higher frequency content in the intraplate regions of eastern north America (ENA) than the interplate regions of western north America (WNA). The contrasts between the WNA and ENA response spectra have often been considered to be representative of the contrasting conditions between intraplate and interplate regions around the world. The different frequency contents observed in the response spectra of these regions is believed to be the result of the different wave transmission in the bedrock and the different stress drops associated with the different faulting mechanisms.

The derivation of a regional response spectrum from first principles requires the analysis of a large number of strong motion accelerograms representative of the whole range of source, path and site conditions. In low seismicity intraplate areas such as Australia and ENA, insufficient strong motion accelerograms have been recorded to satisfactorily undertake such an empirical study.

Seismic loading provisions in contemporary standards have typically been developed from the following three step procedure:

(a) Step One: the level of seismic activity within a defined area is represented by design earthquake magnitudes for different mean recurrence intervals (MRI). Relationships between the earthquake magnitude and the ground motion parameters of engineering interests are then established. The earthquake magnitude is typically based on the measurement of very long period earthquake waves (e.g. Surface Wave Magnitude is defined in accordance with the amplitude of the 22 second period wave measured by a seismometer). By contrast, ground motion parameters such as the peak ground acceleration (PGA), peak ground velocity (PGV) or response spectral accelerations (RS_a) are associated with much shorter period waves. The accuracy of such "long to short" period extrapolation relies heavily on whether the frequency content of the earthquake has been correctly represented.

(b) Step Two: response spectra representing different levels of seismicity and site conditions are developed from one or more of these ground motion parameters (e.g. PGA, PGV or both). A response spectrum defined solely in accordance with a single parameter, although becoming outdated, is still adopted by the majority of earthquake loading standards around the world. It should be noted that the response spectrum defined in accordance with the PGA can accumulate errors as it is a "short to long" period extrapolation which is the reverse of Step One. In regions of high seismicity where strong motion data is abundant, more elaborate dual parameter response spectrum models have been developed to model the average frequency contents of earthquakes occurring in the area of interest (IBC, 2000).

(c) Step Three: the design loads obtained from the elastic spectra developed in Step Two are reduced by what is known as a "structural response factor" (R_f), or a load reduction factor, to account for the overstrength and ductility of the building. The R_f factor essentially extrapolates the initial elastic response behaviour of the building to the ultimate inelastic response behaviour. Such transition is typically associated with a significant lengthening of the structure natural period resulting from ductile yielding. Thus, the R_f procedure can be considered a further "short to long" period extrapolation which is associated with further uncertainties.

2.2   Ground Motion Modelling

The deterministic and stochastic ground motion modelling methods, which are the two main methods used for the generation of earthquake accelerograms, are reviewed below.

One of the earliest deterministic methods used to generate artificial accelerograms was based on the point shear dislocation theory (Aki, 1968). The method uses the seismic moment and an assumed fault-slip function to characterize the earthquake source and wave theory to model the transmission of the generated seismic shear waves to a site through a homogeneous half space. The shortcomings of this method are that the assumed slip function simplistically represents the source and does not account for random irregularities of the rupture surface, whilst the half space model does not account for reflections, refractions and scattering of waves in a typical heterogeneous medium.

Sophisticated deterministic simulation methods have since been developed to generate more realistic accelerograms. The two most popular contemporary simulation methods are the empirical Green's function (EGF) method and the ray-theory method (Atkinson and Somerville, 1994; Irikura, 1986; Beresnev and Atkinson, 1997; Irikura and Kamae, 1994).

The generation of intraplate earthquake accelerograms must consider random variabilities which can be accounted for using stochastic methods. With the advent of the computer and Fast Fourier Transform algorithms, stochastic methods based on frequency domain analysis have become popular. The stochastic procedure typically consists of a deterministic Target Fourier amplitude spectrum defining the frequency content and a set of random phase angles defining the phase arrivals (Vanmarcke, 1977). An amplitude function is used to modulate the accelerogram to a realistic duration. The well known Kanai-Tajimi filter has been developed within a stochastic framework to generate artificial accelerograms. More elaborate Fourier spectrum models have been developed by McGuire (1978) and Trifunac (1976; 1989) using earthquake magnitude, source distance and site classifications as the controlling parameters.

A seismological model developed by Boore, Atkinson and other investigators (Boore, 1983; Boore and Atkinson, 1987; Atkinson & Boore, 1995 and 1998; Hanks and McGuire,1981), identifies the important factors affecting the properties of the earthquake ground motion and distills these factors into a few key parameters. The model is generic, simple to use, and appears most suitable for the seismic hazard modelling of low seismicity areas where details of the potential earthquake sources are generally unknown. The Fourier spectrum specified in the model is expressed as the product of a source factor, a geometrical attenuation factor, a whole path attenuation factor, and factors representing effects near the surface. Interestingly, the source factor has been found to be consistent with the Fourier transform of the shear waves predicted by the point shear dislocation theory (Beresnev and Atkinson, 1997). The source factor can be combined with the crustal factors developed for generic crusts by Boore and Joyner (1997) to predict the Fourier spectrum of the transmitted seismic waves. Ground motion parameters can be obtained from the specified Fourier amplitude spectrum either using random vibration theory (Hanks and McGuire, 1981) or by generating synthetic accelerograms. Such parameters predicted by the model generally provide a good match with field observations, particularly following recent modifications to the original source function by (Atkinson, 1993; Atkinson and Silva, 1997; Atkinson and Boore, 1995 and 1998) . A recent comparative study revealed that the seismological model produces results which are comparable to the previously described deterministic ray-theory method (Atkinson and Somerville, 1994). A further review on ground motion modelling can be found in Somerville (2000).

A generic response spectrum model, termed the "Component Attenuation Model" (CAM), has been developed recently based on the stochastic simulation of the seismological model described above. CAM is particularly suited to engineering applications in low and moderate seismicity regions where empirical earthquake data is scarce (Lam *et al,* 2000a–2000c).

2.3 Soil Effects

The modification of seismic shear waves by soil overlying bedrock is dominated by a number of mechanisms including: (a) period dependent amplification associated with the conservation of energy as seismic shear waves propagate through a steep shear wave velocity gradient (b) attenuation of high frequency wave components caused by energy loss along the wave travel path, and (c) multiple reflections of waves trapped in the soil medium causing resonance.

Shear wave analyses of soft soil columns has shown narrow band periodic waves arising from resonance of seismic waves trapped within the soil column which results in a noticeable amplification at a distinct "site period" in the soil response spectrum. Such resonance effects are most pronounced when the soil is underlain by a distinct soil-rock interface with a very high impedance contrast. Geology of this type is commonly found in delta areas where young unconsolidated alluvium, or artificial fill, have been deposited on top of much older, harder, sedimentary rocks or on crystalline metamorphic or volcanic rocks. The dominant period of the soil surface motion is directly related to the time taken for the incident (or reflected) wave-front to propagate through the soil layer. Consequently, the site natural period (at which the peak of the soil response spectrum is located) correlates very well with the soil depth. This analytical observation is supported by numerous field recording of earthquake tremors on the soil surface and in instrumented boreholes (Lam and Wilson, 1999; Cheng, 2000).

Analytical models have been developed to model the modification effects soils have upon the transmitted seismic shear waves (Romo, 2000). These analytical methods tend to be complex and more suited to detailed site specific studies. In contrast, simplified manual methods have been developed for general seismic hazard analyses and for rapid seismic performance assessment of ordinary facilities.

The most popular manual method is the code response spectral amplification model which scales the soil response spectrum from a bedrock response spectrum (IBC, 2000; SAA; 1993). Such code models were developed typically by statistical analyses of large volumes of empirical data. Contemporary empirical models typically parameterize site dynamic properties based on the shear wave velocity averaged over a certain depth below the soil surface (Borcherdt, 1994a and b; Crouse 1996). Thus, information related to thickness of the soil layers and sub-soil discontinuities, which induce resonance in the soil, are usually not parameterized. In empirical modelling, measurements taken on soil columns with different depths and layering conditions are usually averaged so that dynamic amplification associated with periodicity is diffused into a period band rather than focused on distinct periods.

Thus, some codes provide amplification of the soil response spectrum in the low period range even for soft deep soil columns in which low period (high frequency) waves are expected to be attenuated. Similarly, the high period range of the soil response spectrum is also amplified for shallow soil columns, contrary to expectations. Despite the conservatism associated with the averaging, the code model has been widely accepted for practical applications. Further, the diffused periodicity assumption seems justified in the modelling of deep sedimentary deposits which possess a large number of layers with different properties.

An alternative "Frame Analogy Soil Amplification" (FASA) calculation procedure which parameterizes soil depth and addresses soil resonance directly has been proposed and is being further developed (Lam *et al*, 1997 and 2000).

3 DESIGN METHODS

3.1 Force-Based Method

The response spectra described in previous sections were based on a linear elastic response analysis of a single-degree-of-freedom system structure. Inelastic responses are instead typically characterised by the system ductility demand. The Newmark and Hall (1982): Equal Displacement, Equal Energy and Equal Acceleration relationships have been most widely used to predict kinematic ductility demand (m) of lumped mass SDOF systems assuming linear elasto-perfectly plastic (LEPP) behaviour. The value of m is defined by:

$$R = \mu \text{ for Equal Displacement)} \tag{1}$$
$$R = (2\mu - 1) \text{ (for Equal Energy)} \tag{2}$$
$$R = 1 \text{ (for Equal Acceleration)} \tag{3}$$

where R is the ratio of the elastic strength demand to the yield strength of a LEPP SDOF system.

The above procedure, though popular, was mainly derived from analyses of single displacement pulses and the very limited earthquake data available at the time. Recent studies employing a very extensive database have greatly simplified the prediction of inelastic responses from elastic models (Miranda, 1993; Lam, 1996 and 1998). The mean system ductility demand was found to depend on the ratio of the natural period of the structure to the site natural period. The ductility demand model recently proposed by Priestley (1995) has incorporated the dominant site period T_g in the relationship:

$$R = 1 + (\mu - 1) \, T/(1.5T_g) < \mu \tag{4}$$

Very large system ductility demands are clearly evident for buildings with a low natural period of for buildings founded on soft soil sites with a large site period. This equation appears consistent with the extensive studies carried out by Miranda (1996) and by Lam (1998), although a slight modification in the form of equation (5) is considered justified.

$$R = 1 + (\mu - 1) \, T/T_g < \mu \tag{5}$$

3.2 Displacement-Based Methods

New analysis and design procedures have been developed for the performance-based seismic analysis, design and rehabilitation of buildings (Fajfar and Krawinkler, 1998; Priestley, 2000). These procedures fall into the broad category of displacement-based (DB) methodologies which include the Displacement Coefficient Method, the Direct Displacement Method and the Capacity Spectrum Method (FEMA273, 1997; Whittaker, 1998; Priestley, 1995; Edwards, 1999). Displacement response spectrum models which predict displacement demand for a given magnitude and distance have also been developed (Lam, 2000d; Boomer and Elnashai, 1999). All these methods assume that the structure dynamic response is dominated by the first mode and hence can be represented by an equivalent single-degree-of-freedom (SDOF) system (Calvi and Pavese, 1995).

3.2.1 Displacement Coefficient Method. The Displacement Coefficient Method is based on a comparison of the displacement demand with the displacement capacity at the top of the structure. The displacement capacity is typically estimated from the static push-over analysis and failure is deemed to occur when the performance limits based on the strength capacity or the deformation capacity are exceeded in a member or a joint. The peak displacement demand recommending FEMA273 is based on the product of a series of coefficients with the spectral displacement of an elastic SDOF system. The coefficients account for the modal participation factor, second order effects, different hysteretic shapes and the increased displacement of short period systems. The spectral displacement is based on the effective natural period of the system assuming a lateral stiffness equal to the secant stiffness calculated at a base shear force equal to 60% of the yield strength.

3.2.2 Direct Displacement Method. This method is based on the substitute structure procedure where a structure is reduced to a SDOF system (Priestely, 1995). The iterative procedure comprises the following steps:

(a) Estimate the yield displacement and displacement capacity of the system and expresses in terms of the displacement ductility.
(b) Find the equivalent damping of the system based on the ductility factor.
(c) Estimate the required effective natural period (T_e) based on the design displacement response spectra.
(d) Estimate the effective stiffness (K_e) corresponding to T_e .
(e) Estimate the required strength based on K_e and the ultimate displacement capacity.

3.2.3 Capacity Spectrum Method. The Capacity Spectrum Method is the seismic evaluation method recommended for concrete buildings in ATC40 (1998). Application of the Capacity Spectrum technique requires both the demand response spectrum and structural capacity curves be plotted together in a spectral acceleration versus spectral displacement format. The structural capacity curve is developed from a push-over analysis of the structure and converted to a SDOF system using the first mode properties. An ensemble of demand response spectrum curves are plotted for a range of damping ratios. The Capacity Spectrum Method involves superimposing the capacity spectrum curve on top of the capacity demand curves and determining the point at which the two curves intercept with the same effective damping. The effective damping for the system can be estimated from the ductility factor in a similar fashion to the method cited in Section 3.2.2.

These simplified non-linear static procedures are very useful for a rapid and direct assessment of the seismic performance of all structures. For structures dominated by higher mode effects, non-linear dynamic procedures using multi-degree-of-freedom are recommended to be used in conjunction with the simplified methods.

4 DESIGN AND DETAILING OF STRUCTURAL SYSTEMS

4.1 Initial Design Considerations

The major factors which determine the performance of a structure subjected to earthquake induced loading are: a) the configuration, b) the material characteristics, c) the structural framing system and d) the collapse mechanism of the structure.

4.1.1 Structure Configuration. The configuration of a structure can significantly affect its performance under earthquake excitation. For simplicity, structured configuration is considered as either plan configuration or vertical configuration. Experience around the world has clearly shown that structures which have irregular configuration in either plan or elevation suffer greater damage in earthquakes than structures having a regular configuration. It is highly desirable that at the preliminary design stage structures should be regularly configured in both plan and elevation in terms of the distribution of stiffness, strength and mass with clear and direct load paths for the transmission of seismic forces from roof level to the foundation. Examples of undesirable features include; torsional irregularity where the centres of mass and stiffness/strength do not coincide and soft storeys where one particular storey maybe significantly weaker than the adjoining storeys. Further, the foundation should be configured and designed so that the whole building is excited in a uniform way by the seismic motion.

4.1.2 Material Characteristics. In earthquake resistant design it is necessary to have an understanding of the post-yield behaviour of the materials being used. To some extent the choice of structural material may be governed by the economics of the framing system of the structure. In terms of earthquake resistance, the desirable features which structural materials should possess include: high ductility, high strength to weight ratio, homogeneity, orthotropy and ease in making full strength connections. The earthquake resisting characteristics of some common building materials are listed below:

(a) Steel is a building material possessing most of the above characteristics, and if well detailed generally performs well under earthquake loading.

(b) Reinforced concrete also possesses most of the desirable characteristics provided it is well detailed.

(c) Prestressed concrete and precast concrete units are inherently strong, however overall performance under earthquake loading is limited by the strength and detailing of the connections.

(d) Masonry which is unreinforced and badly detailed performs poorly under earthquake induced loading as it lacks most of the features listed above. Performance can be improved by reinforcing the masonry with steel reinforcement, strengthening the connections and providing sufficient cross walls to both reduce the out of plane effects and increase the shear capacity of the building.

4.1.3 Structural Framing Systems. The overall performance of a structure subjected to earthquake induced loading is not only dependent on the configuration and structural materials chosen, but also on the structural framing systems selected. Different structural framing systems possess different energy absorption capabilities i.e. different overall ductility values. Earthquake codes of practice provide guidance on the level of ductility that can be developed in various framing systems. Low values of R_f reflect the poor energy absorption characteristics of the material and framing system concerned. The four major types of structural systems used in buildings are briefly summarised below. Paulay and Priestley (1992), ATC40 (1998) and Whittaker (1998), FEMA273 (1997) contain extensive materials on the earthquake design and performance of reinforced concrete and structural steel systems respectively.

(a) *Bearing Wall Systems:* Typically these systems are either structural wall systems, such as reinforced concrete lift shafts, or concentrically braced frames as are often used in steel construction. Normally these systems are designed to carry both vertical and lateral loading.

(b) *Building Frame Systems:* Essentially these systems are specifically designed to carry lateral loading and alternative load paths are available for the vertical loads. The fact that these systems are carrying limited vertical load means that, if they fail under earthquake induced loading, the major vertical load paths are still intact. Consequently they are considered to have a greater energy absorption capability (and hence a greater R_f factor) than an equivalent bearing wall system.

(c) *Moment Resisting Frame System:* A moment resisting frame system has rigid connections between its beams and columns and resists lateral loading through sway action. The amount of energy that can be absorbed by such a system is highly dependent on the detailing of the members and connections. The energy absorption capability of special moment resisting frames (SMRF) is greater than for intermediate moment resisting frames (IMRF) which in turn are greater than for ordinary moment resisting frames (OMRF). This variation is reflected in the R_f factors which range from 4 to 8. Although the design earthquake forces in an SMRF are relatively lower than either the IMRF or the OMRF, the specifications for the design and construction detailing are considerably more demanding. The 1994 Northridge earthquake demonstrated that special attention is needed when detailing the SMRF steel connections so that brittle fractures are avoided.

(d) *Dual Structural Systems*: A dual structural system is a combination of a moment resisting frame with either a building frame system or a bearing wall system. The larger R_f factors associated with various dual systems reflect the alternative load paths available to resist the earthquake induced loads and also the higher energy absorption capabilities.

4.2 Capacity Design Method

For the structure to survive the ultimate limit state earthquake event, a rational yielding mechanism should be designed and detailed to dissipate the seismic energy. The "capacity design" method which was developed in New Zealand requires the designer to identify a rational yielding mechanism and provide over-strengths in other members to avoid brittle and premature failure. The method can be likened to a chain, in which each link represents a possible failure mode. In traditional strength design the structure fails at the weakest link, whether that mode is ductile or brittle, is not explicitly considered. In contrast, with "capacity design" the structure is explicitly designed and detailed so that the chosen failure mechanism develops (i.e. the ductile link) to dissipate the seismic energy. A detailed description of the capacity design procedure is provided in Park (1986), Paulay (1988) and Paulay and Priestley (1992).

4.3 Non-structural Components

Earthquake induced motions in a building may endanger the continuous functioning of the equipment and the safe keeping of the contents. The consequences of seismic damage to such items can exceed the damage to the building structure itself. For example, the uninterrupted functioning of medical and communication equipment in a lifeline facility can be vital for the protection and saving of lives. Hence their seismic vulnerability is a major concern. Further, falling debris from building facades, false ceilings and architectural appendages can cause severe injuries and deaths both within and outside the building. The overturning of containers of hazardous materials could also create a widespread safety hazard whilst invaluable damage may be caused to the contents of museums.

Non-structural components may be supported at different locations in the building and cover a large area (e.g. partitions, facade walls, false ceilings, pipes and cables *etc*). It is necessary to check that such components can tolerate the deformations imposed in the building structure caused by the earthquake. Other components can be termed "isolated objects" which are either fastened to the building (e.g. mechanical/electrical equipment, architectural appendages *etc*) or free-standing (e.g. containers, mobile equipment, objects, storage racks, book shelves *etc.*). Seismic assessment of such "isolated objects" is mainly concerned with overturning or failure of the fastenings. The effects of floor amplifications must be accounted for in the seismic assessment of non-structural components (Wilson and Lam, 1994; Lam, 1998). It should be noted that certain objects possess significant reserve capacity to rock without overturning. An estimate of such rocking capacity has been described in the literature (Lam, 1995).

Non-structural components need to be installed, and secured, with due considerations to their vulnerability and consequence of failure. The vulnerability aspect considers the likelihood of component failure whereas the consequence of failures consider the effect of component failure. The Federal Emergency Management Agency Publication 273 (FEMA273, 1997) provides a state-of-the-art code overview of the seismic design and analysis of non-structural components.

5 FUTURE DIRECTIONS

Base isolation and passive energy dissipation are emerging technologies which enhance the performance of the building. This is accomplished in base-isolated structures by adding damping and changing the natural period of the building. In contrast, passive energy dissipation devices only introduce additional damping to the structure. The primary use of energy dissipation devices is to reduce earthquake displacement of the structure, and consequently reduce the seismically induced forces in the structure in the elastic range. A recent publication by Buckle (2000) provides an excellent review on this topic.

For most applications, energy dissipation devices provide an alternative approach to conventional stiffening and strengthening schemes, and should result in improved seismic performance levels. In addition, these energy dissipation devices can also be used to control the building response due to small earthquakes, wind and other form of dynamic loading.

6 CONCLUSIONS

This paper has provided an overview of both the philosophy and methods used in the design and analysis of building structures to earthquake ground shaking. In particular, the following topics and issues were discussed:

(a) seismic hazard and soil response modelling,
(b) aseismic design methods,
(c) design and detailing of structural systems and (d) future directions for seismic design.

7 ACKNOWLEDGEMENTS

The assistance given by Adeline Sze in the preparation of the paper is gratefully acknowledged.

8 REFERENCES

Aki, K., 1968
SEISMIC DISPLACEMENTS NEAR A FAULT, *Journal of Geophysical Research,* **73**(6), pp. 5359–5376.

ATC40, 1998
SEISMIC EVALUATION AND RETROFIT OF CONCRETE BRIDGES, Applied Technology Council, U.S.A.

Atkinson, G. M. and Boore, D. M., 1995
GROUND-MOTION RELATIONS FOR EASTERN NORTH AMERICA, *Bulletin of the Seismological Society of America,* **85**, pp. 17–30.

Atkinson, G. M. and Boore, D. M., 1998
EVALUATION OF MODELS FOR EARTHQUAKE SOURCE SPECTRA IN EASTERN NORTH AMERICA, *Bulletin of the Seismological Society of America,* **88**, pp. 917–934.

Atkinson, G., 1993
EARTHQUAKE SOURCE SPECTRA IN EASTERN NORTH AMERICA, *Bulletin of the Seismological Society of America,* **83**, pp. 1778–1798.

Atkinson, G. and Silva, W. 1997
AN EMPIRICAL STUDY OF EARTHQUAKE SOURCE SPECTRA FOR CALIFORNIAN EARTHQUAKES, *Bulletin of the Seismological Society of America,* **87**, pp. 97–113.

Atkinson, G. and Somerville, P., 1994
CALIBRATION OF TIME HISTORY SIMULATION METHODS, *Bulletin of the Seismological Society of America,* **84**, pp. 400–414.

Beresnev, I. A. and Atkinson, G. M., 1997
MODELLING FINITE-FAULT RADIATION FROM THE ω^N SPECTRUM, *Bulletin of the Seismological Society of America,* **87**(1), pp. 67–84.

Bolt, B.A., 1995
    INTRAPLATE SEISMICITY AND ZONATION, *Proceedings of the Fifth Pacific Conference on Earthquake Engineering,* Melbourne, Australia, **1**, pp. 1–11.

Bommer, J. J. and Elnashai, A.S., 1999
    DISPLACEMENT SPECTRA FOR SEISMIC DESIGN. *Journal of Earthquake Engineering,* **3**(1), pp. 1–32.

Boore, D. B., Joyner, W. B. and Fumal, T. E., 1997
    EQUATIONS FOR ESTIMATING HORIZONTAL RESPONSE SPECTRA AND PEAK ACCELERATION FOR WESTERN NORTH AMERICAN EARTHQUAKES: A SUMMARY OF RECENT WORK, *Seismological Research Letters,* **68**, pp. 128–153.

Boore, D. M. and Atkinson, G., 1987
    STOCHASTIC PREDICTION OF GROUND MOTION AND SPECTRAL RESPONSE PARAMETERS AT HARD-ROCK SITES IN EASTERN NORTH AMERICA, *Bulletin of the Seismological Society of America* ,**73**, pp. 1865–1894.

Boore, D. M. and Joyner, W. B., 1997
    SITE AMPLIFICATIONS FOR GENERIC ROCK SITES, *Bulletin of the Seismological Society of America,* **87**(2), pp. 327–341.

Boore, D. M., 1983
    STOCHASTIC SIMULATION OF HIGH-FREQUENCY GROUND MOTIONS BASED ON SEISMOLOGICAL MODEL OF THE RADIATED SPECTRA, *Bulletin of the Seismological Society of America,* **73**(6), pp. 1865–1894.

Borcherdt, R. D. 1994
    ESTIMATES OF SITE-DEPENDENT RESPONSE SPECTRA FOR DESIGN (METHODOLOGY AND JUSTIFICATION), Earthquake Spectra, **10**, pp. 617–653.

Borcherdt, R. D. 1994
    NEW DEVELOPMENTS IN ESTIMATING SITE EFFECTS ON GROUND MOTION, *Proceedings of ATC-35 seminar on New Developments in Earthquake Ground Motion Estimation and Implications for Engineering Design Practice,* Applied Technology Council, paper no.10, pp. 10.1–10.44.

Buckle, I. G., 2000
PASSIVE CONTROL OF STRUCTURES FOR SEISMIC LOADS, *Bulletin of the New Zealand National Society for Earthquake Engineering (special issue on the 12th World Conference on Earthquake Engineering, Auckland,2000)*, **33(3)**, pp. 209–221.

Bruneau, M., Uang, C. M., Whittaker, A., 1998
DUCTILE DESIGN OF STEEL STRUCTURES, McGraw Hill.

Calvi, G. M. and Pavese A., 1995
DISPLACEMENT BASED DESIGN OF BUILDING STRUCTURES. *Proceedings of the 5th SECED Conference : European Seismic Design Practice, Research and Application.* Chester, U.K., pp. 127–132.

Cheng, M. W. K., Lam, N. T. K., Wilson, J. L., Balendra, T., Koh, C. G. and Lim, T. K. 2000
REPONSE SPECTRUM MODELLING OF SOIL RESONANCE AT KATONG PARK (SINGAPORE) BY FIELD MEASUREMENTS AND SHEAR WAVE ANALYSES, *Joint Report: Department of Civil & Environmental Engineering, The University of Melbourne; Department of Civil Engineering, The National University of Singapore; and Meterological Service Singapore.*

Chopra, A. K., 1995
DYNAMICS OF STRUCTURES, THEORY AND APPLICATIONS TO EARTHQUAKE ENGINEERING. Pub. Prentice-Hall, NJ.

Crouse, C. B. and McGuire, J. W, 1996
SITE RESPONSE STUDIES FOR PURPOSE OF REVISING NEHRP SEISMIC PROVISIONS, Earthquake Spectra, 12(3), pp. 407–439.

Edwards, M., Lam, N. T. K., Wilson, J. L. and Hutchinson, G. L.,1999
THE PREDICTION OF EARTHQUAKE INDUCED DISPLACEMENT DEMAND OF BUILDINGS IN AUSTRALIA AN INTEGRATED APPROACH, *Proceedings of the Annual Technical Conference for the New Zealand National Society for Earthquake Engineering,* Rotorua, pp. 43–50.

Fajfar, P. and Krawinkler, H. (eds.) 1998
SEISMIC DESIGN METHODOLOGIES FOR THE NEXT GENERATION OF CODES. Pub. A.A. Balkema, Brookfield, VT.

Federal Emergency Management Agency,1997
FEMA273, 1997: NEHRP PROVISIONS FOR THE SEISMIC
REHABILITATION OF BUILDINGS – GUIDELINES, FEMA,
Washington, D.C., U.S.A.

Frankel, A., Thenhaus, P., Perkins, D. and Leyendecker, E. V.,1995
GROUND MOTION MAPPING – PAST, PRESENT AND FUTURE
(1995), Applied Technology Council publication ATC-35-1,
pp. 7.1–7.4.

Hanks, T. C. and McGuire, R. K., 1981
THE CHARACTER OF HIGH-FREQUENCY STRONG GROUND
MOTION, *Bulletin of the Seismological Society of America,* **71***(6),*
pp. 2071–2095.

International Code Council, 2000
INTERNATIONAL BUILDING CODE 2000, U.S.A.

Irikura, K. and Kamae, K., 1994
SIMULATION OF STRONG GROUND MOTION BASED ON
FRACTAL COMPOSITE FAULTING MODEL AND EMPIRICAL
GREEN'S FUNCTION, *Proceedings of the 9th Japan Earthquake
Engineering Symposium,* Tokyo.

Irikura, K., 1986
PREDICTION OF STRONG ACCELERATION MOTIONS USING
EMPIRICAL GREEN'S FUNCTION, *Proceedings of the Seventh
Japan Earthquake Engineering Symposium,* pp. 151–156.

Lam, N. T. K., Wilson, J. L. and Hutchinson, G. L., 2000
THE MODELLING OF INTRAPLATE SEISMIC HAZARD BASED ON
DISPLACEMENT, *Proceedings of the 12th World Conference of
Earthquake Engineering,* Auckland, paper no. 1933 (OS1-T4).

Lam, N. T. K., Wilson, J. L., Doherty, K. and Griffith, M., 1998
HORIZONTAL SEISMIC FORCES ON RIGID COMPONENTS WITHIN
MULTI-STOREY BUILDINGS, *Proceedings of the Australasian
Structural Engineering Conference,* Auckland, pp. 721–726.

Lam, N. T. K., Wilson, J. L. and Hutchinson, G. L., 1997
INTRODUCTION TO A NEW PROCEDURE TO CONSTRUCT SITE
RESPONSE SPECTRUM, Proceedings of the15th ACMSM,
Melbourne, pp. 345–350.

Lam, N. T. K., Wilson, J. L. and Hutchinson, G. L., 1995
TIME-HISTORY ANALYSIS FOR ROCKING OF RIGID OBJECTS
SUBJECTED TO BASE-EXCITATION, Proceedings of the 14th
ACMSM, Hobart, Vol. 1, pp. 284–289.

Lam, N. T. K., Wilson, J. L. and Hutchinson, G. L., 1996
BUILDING DUCTILITY DEMAND: INTERPLATE VERSUS
INTRAPLATE EARTHQUAKES, Earthquake Engineering and
Structural Dynamics, **25**, pp. 965–985.

Lam, N. T. K., Wilson, J. L. and Hutchinson, G. L., 1998
THE DUCTILITY REDUCTION FACTOR IN THE SEISMIC DESIGN
OF BUILDINGS, *Earthquake Engineering and Structural
Dynamics*, **27**, pp. 749–769.

Lam, N. T. K., Wilson, J. L. and Hutchinson, G. L., 2000
GENERATION OF SYNTHETIC EARTHQUAKE ACCELEROGRAMS
USING SEISMOLOGICAL MODELLING : A REVIEW, *Journal
of Earthquake Engineering*, **4**(3), pp. 1–34.

Lam, N. T. K., Wilson, J. L., Chandler, A. M. and Hutchinson, G. L., 2000
RESPONSE SPECTRAL ATTENUATION RELATIONSHIPS FOR
ROCK SITES DERIVED FROM THE COMPONENT
ATTENUATION MODE, *Earthquake Engineering and Structural
Dynamics*, **29** (In Press).

Lam, N. T. K., Wilson, J. L., Chandler, A. M. and Hutchinson, G. L., 2000
RESPONSE SPECTRUM MODELLING FOR ROCK SITES IN LOW
AND MODERATE SEISMICITY REGIONS COMBINING
VELOCITY, DISPLACEMENT AND ACCELERATION
PREDICTIONS, *Earthquake Engineering and Structural Dynamics*,
29 (In Press).

Lam, N. T. K., Wilson, J. L., Edwards, M., Cheng, M. and Hutchinson, G., 2000
DISPLACEMENT ASSESSMENT IN THE MELBOURNE
METROPOLITAN AREA ACCOUNTING FOR SOIL
RESONANCE, *Proceedings of the Australian Earthquake
Engineering Society Seminar*, Tasmania, November 2000.

Lam, N. T. K. and Wilson, J. L., 1999
ESTIMATION OF THE SITE NATURAL PERIOD FROM BOREHOLE
RECORDS, Australian Journal of Structural Engineering, SE1(3),
pp. 179–199.

McGuire, R. K.,1978
A SIMPLE MODEL FOR ESTIMATING FOURIER AMPLITUDE
SPECTRA OF HORIZONTAL GROUND ACCELERATION,
Bulletin of the Seismological Society of America, **68***(3),*
pp. 803–822.

Mirander, E., 1993
EVALUATION OF SITE-DEPENDENT INELASTIC SEISMIC DESIGN
SPECTRA, Journal of Structural Engineering, **119**(5),
pp. 1319–1338.

New Zealand Stand NZS 4203:1994
CODE OF PRACTICE FOR GENERAL STRUCTURAL DESIGN AND
DESIGN LOADINGS FOR BUILDINGS. Standards New Zealand.

Newmark, N. M. and Hall, W. J., 1982
EARTHQUAKE SPECTRA AND DESIGN, *Earthquake Engineering
Research Institute Monograph Series.*

Park, R., 1986
DUCTILE DESIGN APPROACH FOR REINFORCED CONCRETE
FRAMES, Earthquake Spectra **2** (3): pp. 565–619.

Paulay, T. and Priestley, M. J. N., 1992
SEISMIC DESIGN OF REINFORCED CONCRETE AND MASONRY
BUILDINGS, John Wiley & Sons INC.

Paulay, T., 1988
SEISMIC DESIGN IN REINFORCED CONCRETE; THE STATE OF
THE ART IN NEW ZEALAND, Bulletin of the NZNSEE **21** (3),
pp. 208–232.

Priestley, M. J. N., 2000
PERFORMANCE BASED SEISMIC DESIGN, *Bulletin of the New
Zealand National Society for Earthquake Engineering (special issue
on the 12th World Conference on Earthquake Engineering,
Auckland, 2000),* **33(3),** pp. 325–346.

Priestley, M. J. N., 1995
DISPLACEMENT-BASED SEISMIC ASSESSMENT OF EXISTING
REINFORCED CONCRETE BUILDINGS, *Proceedings of the
Fifth Pacific Conference of Earthquake Engineering,* Melbourne,
pp. 225–244.

Romo, M. P., Mendoza, M. J. and Garcia, S.R., 2000
GEOTECHNICAL FACTORS IN SEISMIC DESIGN OF
FOUNDATION: STATE-OF-THE-ART REPORT, *Bulletin of the
New Zealand National Society for Earthquake Engineering (special
issue on the 12th World Conference on Earthquake Engineering,
Auckland, 2000)*, **33(3)**, pp. 347–370.

Somerville, P., 2000
SEISMIC HAZARD EVALUATION, *Bulletin of the New Zealand
National Society for Earthquake Engineering (special issue on the
12th World Conference on Earthquake Engineering, Auckland,
2000)*, **33(3)**, pp. 371–386.

Structural Engineers Association of California (SEAOC): Vision 2000
Committee, 2000
PERFORMANCE BASED SEISMIC ENGINEERING OF BUILDINGS.
J. Soulages, ed. 2 Volumes. SEAOC, Sacramento, CA.

Trifunac, M. D., 1976
PRELIMINARY EMPIRICAL MODEL FOR SCALING FOURIER
AMPLITUDE SPECTRA OF STRONG GROUND
ACCELERATION IN TERMS OF EARTHQUAKE
MAGNITUDE, SOURCE-TO-STATION DISTANCE AND
RECORDING SITE CONDITIONS, *Bulletin of the Seismological
Society of America*, **66*(4)***, pp. 1343–1373.

Trifunac, M. D., 1989
DEPENDENCE OF FOURIER SPECTRUM AMPLITUDES OF
RECORDED EARTHQUAKE ACCELERATIONS ON
MAGNITUDE, LOCAL SOIL CONDITIONS AND ON DEPTH
OF SEDIMENTS, *Earthquake Engineering and Structural
Dynamics*, **18**, pp. 999–1016.

Vanmarcke, E. H., 1977
CHAPTER 8 OF SEISMIC RISK AND ENGINEERING DECISIONS,
(editor: Lomnitz, C. and Rosenblueth, E.), Elsevier Publishing Co.

Whittaker, A., Constantinou, M. and Tsopelas, P., 1998
DISPLACEMENT ESTIMATES FOR PERFORMANCE-BASED
SEISMIC DESIGN. *Journal of Structural Engineering, ASCE*,
124(8), pp. 905–912.

Wilson, J. L. and Lam, N. T. K., 1994
HORIZONTAL SEISMIC FORCES ON MECHANICAL AND
ELECTRICAL COMPONENTS MOUNTED WITHIN BUILDING
STRUCTURES in Proceedings of the Australasian Structural
Engineering Conference Sydney, Australia, vol. 1, pp. 327–336.

The Tallest Concrete Building in Shanghai, China – Plaza 66

Richard L. Tomasetti, Dennis C. K. Poon and Ling-en Hsaio

1.0 INTRODUCTION

Plaza 66 is the latest addition to the skyline of Shanghai, China; with a height of 281.5 meters, it's the tallest concrete building in the city (see Figure 1). A 62-story tower, adjacent five-story retail podium and three level below-grade parking area form a three million square foot mixed-use commercial development (see Figure 2). This project embodies the challenges of building in a seismic and typhoon area with poor foundation conditions and materials of relatively limited strength, and illustrates solutions developed to meet those challenges.

Figure 1   City with Plaza 66.

2.0   DESCRIPTION OF THE PROJECT

The most visible feature of the project is the 62-story concrete tower, topped by a 36.5 m (120 ft.) tall steel-framed lantern that will glow on the skyline. Total building height is 281.5 m (923 feet). Adjacent to the tower is a 60,000 m² (600,000 ft²) retail podium (see Figures 3 and 4). The retail podium includes three feature spaces: an atrium with a boat shaped skylight, an arc-shaped sky lit galleria, and a column-free rotunda six stories tall (see Figure 5).

Under the retail podium, a three-story underground parking structure for 9,000 cars and 1,500 bicycles encompasses the entire project footprint. The underground structure is enclosed by a slurry wall built with special construction bracing techniques.

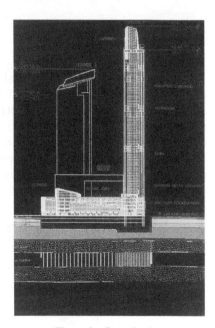

Figure 2   Aerial Overall Plaza 66.   **Figure 3**   Cross Section.

3.0 TOWER STRUCTURAL SYSTEMS

3.1 Lateral Load Resisting System

The key to designing the tallest concrete building in Shanghai was its lateral system. The challenge of resisting lateral loads from typhoon winds was compounded by the limited strength of available materials to resist the forces generated. Modern concrete high-rise towers often use concrete strengths exceeding 80 MPa cube strength (10 ksi cylinder strength), but locally only strengths up to 50 MPa (6 ksi) were available. Steel reinforcing with fy = 410 MPa (60 ksi) is typically used today for main reinforcing, but locally available reinforcing was limited to about 335 MPa (48 ksi).

A further challenge was a local building code that limited the allowable drift of the building due to wind loads to height/800, whereas height/500 is commonly used for high-rise towers in other areas. Based on wind studies and analyses of building accelerations, Thornton-Tomasetti Engineers was able to convince a panel of local building authority experts over several meetings that an allowable drift of height/650 was acceptable for this project.

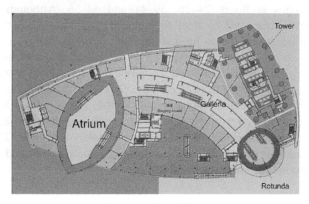

Figure 4 Overall Plan.

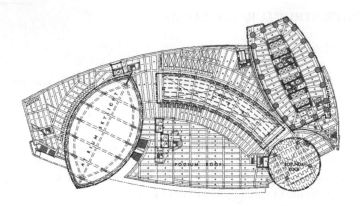

Figure 5   Roof Structural Plan.

A crucial innovation for the Plaza 66 tower was the use of a concrete core with perforated concrete outriggers. The concrete outriggers are located at three mechanical zones at floors 24–26, 39–41 and 54–56. Although steel truss outriggers have been used in steel framed and composite towers, the use of concrete outriggers has been limited because typically they are solid wall elements that cannot readily be penetrated. In the case of Plaza 66, an innovative two-story outrigger system addressed this problem. Each outrigger wall has four large openings, two on each floor (see Figure 6).

In this arrangement, the top and bottom members of the outrigger provide the tension and compression force couple required of the outrigger while the central member at the middle floor transfers the shear. The result is four large planned openings that allow mechanical ducts, piping and pedestrians to circulate around the core (see Figure 7).

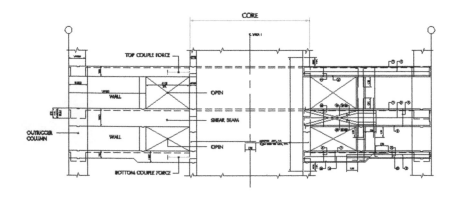

Figure 6   Outrigger Elevation.

Figure 7   Outrigger Photo.

Six paired sets of outriggers extend from the long faces of the core, enabling all longitudinal perimeter columns to efficiently contribute to the lateral load resisting system (see Figure 8). Augmenting the outrigger system is a perimeter frame to resist torsional modes of building motion. Perimeter beams are 1250 mm (49") deep by 975 mm (38") wide, between perimeter columns spaced about 9 meters (30 feet) on center. The perimeter frame meets the requirements of a special seismic moment frame.

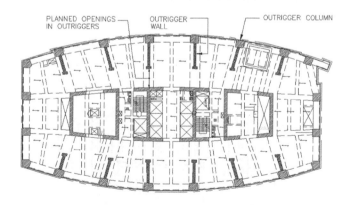

Figure 8   Typical Plan.

3.2   Foundation System

Soil conditions in Shanghai are not conducive to supporting the heavy loads generated by tall buildings. Layers of sand and clay alternate to a great depth. Bedrock is beyond reach. Therefore, this tower is supported on a dense sand-bearing stratum about 90 meters (300 feet) below grade. End bearing concrete bored piles 800mm (2' 7") in diameter, 80 meters (263 feet) long were used. The piles extend 9 meters (30 feet) into the sand bearing strata (see Figure 3) to create effective embedment.

Complicating the challenge of resisting overturning moments, the tower was located near the perimeter of the site. A special cellular mat, bearing on the concrete bored piles, was designed to resist tower gravity and overturning forces (see Figure 9). Although the tower is 58.5 meters (193 feet) long and 34 meters (111 feet) wide at its widest point, the mat is considerably larger, 80 meters (253 feet) long and 55 meters (180 feet) wide to resist the overturning and minimize differential settlement.

Differential settlement is a major concern in Shanghai due to highly compressible clay layers, long piles and varied loading between towers and podiums. Initial geotechnical studies predicted tower settlement to exceed 1000 mm (40 inches). But after studying the actual behavior of other towers nearby, predicted settlement of the tower was revised to 280 mm (11") with anticipated differential settlement between the tower and the podium of up to 230 mm (9"). Through the use of the load-spreading cellular mat, actual performance to date has been consistent with these revised settlements, and differential settlement is only about 80% of predictions. The structure was designed to accept this differential settlement through use of delayed concreting at pour strips.

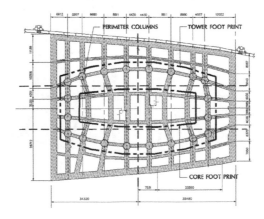

Figure 9   Mat Plan.

The three-story below-grade foundation walls, almost totally under water, also used an innovative approach, slurry wall construction with concrete cross-lot bracing. The slurry wall in the podium area is 800 mm (32") thick and 25 meters (82 feet) deep. In the tower area it is 1 m (39 inches) thick and 33 m (108 feet) deep. During construction, three planes of temporary bracing were placed in the podium and four planes of temporary bracing were used in the tower area. Bracing was constructed from top down, with bracing struts cast against earth as excavation proceeded downward. Bracing was removed from the bottom up, with each level removed as its corresponding basement floor immediately below was completed.

3.3 Floor Construction

The typical floor footprint represents a truncated ellipse (see Figure 10) 58.5 m (192 feet) long and up to 34 m (111 feet) wide. The perimeter frame discussed for the lateral system extends through the podium. In addition, beams at 4.5 m (15 feet) on center span between the core and perimeter frame. Beams on column lines participate in the lateral resistance of the tower and are 1200 mm (48") deep. Other beams are 450 mm (18") deep, including the 125 mm (5") slab.

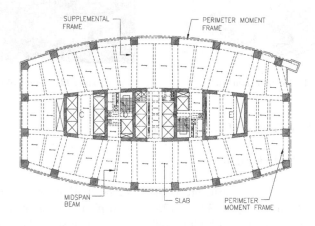

Figure 10   Typical Framing Plan.

3.4   Roof Top Lantern Structural System

The 36.5-meter (120 feet) high lantern structure is a hollow irregular square in plan. It surrounds and supports the mechanical cooling towers. The lantern consists of perimeter horizontal trussed rings (see Figure 11) and vertical trussed column pairs. Lateral resistance is provided by vertical bracing parallel to the perimeter, creating an open-topped trussed "box". Architectural lighting completes this luminescent structure that crowns Plaza 66.

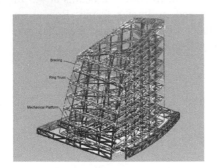

Figure 11   Isometric Lantern.   **Figure 12**   Atrium Skylight.

4.0   PODIUM STRUCTURAL SYSTEMS

4.1   Atrium Roof Skylight

The atrium roof skylight structure consists of tapered concrete portal frames spanning the 67-m (220') by 38 m (125') atrium plan. The architect conceived the atrium configuration as a hollow, reflecting the inverse of the solid form of the tower (see Figures 12 and 5). The tapered frame members that span the space vary in depth, shallow at their column bases and beam mid spans, and deeper at beam/column "knees". The sculptural effect at the columns is further emphasized by using elongated octagons in cross section.

4.2 Galleria Skylight

The curved gallery skylight utilizes a structural frame system similar to the atrium skylight but with smaller spans, as the galleria is 66 m (217') by 17 m (54') (see Figure 5).

4.3 Rotunda Roof

The drama of the rotunda structure is enhanced by a roof 28 m (93 feet) in diameter, bearing on only four column supports and shifted off the rotunda's center. The four columns form a square in plan, with two of the columns 3 meters (10 feet) in from the circumference and the other two 9 meters (30 feet) in.

The rotunda roof cantilevers varying distances past the four support columns to reach the circumference, creating a special sense of entry for the Plaza 66 project (see Figure 5).

5.0 CONCLUSION

Plaza 66 is a unique project that overcame challenges of poor soil conditions, limited material strengths, and a conservative building code to create a tower that soars above the city, providing a new landmark for Shanghai. Innovative design approaches included creating two-story outriggers with a system of cross openings. Plaza 66 stands out as an example of designing and building a tower under difficult conditions and achieving economical structural solutions without compromising aesthetic design integrity

6.0 CREDITS

The Authors would like to thank Len Joseph, I. Paul Lew and Yi Zhu, Project Manager, for their tireless work during design and construction.

7.0 REFERENCES

ASCE 7-90 "Minimum Design Loads" for Buildings and Other Structures, 1990, New York, NY, American Society of Civil Engineers.

ASCE 7-95 "Minimum Design Loads" for Buildings and Other Structures, 1995, New York, NY, American Society of Civil Engineers.

A Report on Africa

Isaac. G. Wanjohi

I have obliged a request to stand in for Mr. Syd Parsons, the CTBUH Vice Chairman for the African region, and speak about Tall Buildings and Urban Habitat in Africa. This is a difficult subject for anyone to speak with authority, particularly one who had not made prior preparations. This is so because Africa is a very large continent with not only diverse cultural heritages, climatic features and economic characteristics but also different state administrations and governments. There are currently over 40 sovereign states in Africa with hardly any effective professional communication and dialogue among them.

In the context of urban habitat some of our cities in Africa such as Cairo in Egypt, Tripoli in Libya, Casablanca in Morocco, Tunis in Tunisia, Addis Ababa in Ethiopia and Khartoum in Sudan are among the earliest cities in the World. With the Pyramid structures in Egypt, Africa held the world record of tall buildings for over 4500 years. Other urban areas in the hinterland of central Africa are among the youngest cities in the world, which include capital cities such as Kinshasa in Congo, Dar-es-Salaam in Tanzania, Kampala in Uganda, Lilongwe in Malawi, Harare in Zimbabwe, Lusaka in Zambia, Kigali in Rwanda, Bujumbura in Burundi, Gaborone in Botswana, Maseru in Lesotho, Mbabane in Swaziland. Yet others, like Johannesburg, and other cities on the coast of Africa such as the capital cities of Kisimayu in Somalia, Asmara in Eritria, Accra in Ghana, Lagos in Nigeria, Dakar in Senegal, Maputo in Mozambique etc. are of the same age as other important cities in the world which were established in the seventeenth and eightieth centuries.

These cities have had varied growth and exhibit different urban characteristics with respect to tall buildings and urban habitat. Though I have visited many cities in Africa, I have not carried out any serious study, but I have made some observations. Today the African urban centers are growing at the rate of 5 to 7.5% per year and the population in the urban areas is making huge demands in employment, housing, water supply, sanitation and sewerage, transportation, communication and energy, particularly electricity.

The housing problem in the cities is greatly constrained by the affordability factor because only a few people are able to afford decent housing. Land may be available in abundance, but the cost of serviced land is beyond reach of most families. Government agencies often have little to do with unplanned urban development and therefore most city dwellers are struggling to live on their own. The majority of urban dwellers therefore live in informal shelters often without any form of services within reasonable reach.

Thus there is pressure on the little serviced land available within the Central Business District (CBD) of every city. Developers guided by Town Planning and Building regulations and by-laws in terms of zoning, ground coverage, building lines, plot boundary distances and plot ratios on land use and densities thereof tend to make maximum use of the available land. The result is construction of multi-story buildings in the CBD and the immediate neighborhood. The zoning and plot ratios are based on constraints in services, including the fire fighting capacity of the relevant urban Authority. It is too risky to have a building so tall that the local fire fighting brigade can not externally reach its top if it catches fire.

In operating tall buildings, a matter of consideration among others is the power to move the lifts. The failure of electricity power is common in many countries and unless alternative electrical power supply is on standby, it is a nightmare to work within such a building. So far tall buildings are generally for offices and hotels, though 4 stories high blocks for residential flats are very common. Up to this level, laws in most countries do not impose the requirement of a lift on the developers.

Currently there are only a few buildings over 20 stories high in most African cities. For example, in Nairobi the tallest building is the Central Bank-owned Times Tower, which is 35 stories and 127.5160 m tall. It was completed in 2000. The next building is the Telfosta Towers comprising 3 towers of 29, 27 and 26 stories. The height up to the tip of the steel mast mounted on top is 150 m. This building was also completed in 1998. The third tallest building is the 28-story, 105 m tall Kenya International Conference Center (KICC), which was completed in 1969. In Kampala, the 25-story Workers House building, which is currently under construction, will be tallest and the 20-story Crested Towers completed in 1965 is the second tallest building. In Dar-es-Salaam, the 21-story Mafuta House currently under construction will be the tallest building, surpassing the 18-story PPF Tower. I am not too familiar with Johannesburg and Cairo, where it is possible that buildings taller than the Times Tower could be found.

So far Africa has had many earthquake tremors but has not had a serious earth quake disaster. It however should be borne in mind that a volcanic fault called the Great Rift Valley runs from Central Africa through Sudan and Egypt to the River Jordan valley in the Middle East. In most countries earthquake forces are taken into consideration in the design of all buildings above 3 stories high though the applied earthquake forces differ from one country to another.

The common building material is reinforced concrete. But the structural engineer should keep an eye on the source of the concrete making materials, particularly the steel. It can be inferior in terms of strength assumed in the design as the country where it is imported from may not be using the same strength production standards and design codes as the country where the material is used.

Because of funding constraints and other local logistic difficulties in most cities, water is scarce, and even when available may not be up to WHO wholesome quality standard. Consequently as an alternative source of drinking water, a big industry of bottled water has emerged and is flourishing in the cities for those who can afford. Too many people live under constant threat of contracting waterborne diseases such as cholera.

COUNTRY REPORTS

Asia Report

Raymond Bates

INTRODUCTION

This is a report of my personal impression of property development in Asia. Any misrepresentation or omissions are mine. I will be concentrating on China and the Special Administrative Region of Hong Kong. This report will focus on four areas:

(a) The Asia economic crisis and its impact on development;
(b) The demographics of the region;
(c) "Bread and Butter" tall buildings; and
(d) The "darker side of construction".

The Asian Region streaches from India and Pakistan in the East, to Korea and Japan in the West. China is in the North and the Philippines and Indonesia flank the Southern boundary. China and India alone account for over 2 billion of the world's population.

ASIAN ECONOMIC CRISIS

In mid to late 1997, the Asian economic crisis evolved. The effects are still with us. It moved like a falling pack of cards through Korea, Japan, Indonesia, Malaysia, Singapore and Hong Kong. China and India were not really affected.

Looking at the Asia from a property market perspective, the region can be divided north and south. The south is stagnant, many construction sites are vacant and negative equity is widespread. In Indonesia, Philippines, Malaysia, Vietnam, Burma, Laos and Cambodia, development is stationary or sluggish. There is little investment and a large number of projects have come to a stop. Some countries have papered over the economic cracks, but investors fear the cracks may return. While Singapore has been impacted by the events in Indonesia and Malaysia, there is some movement in the office and residential markets.

By contrast, North Asia is on the move. Korea is doing quite well. The property market was never oversupplied, and there is a shortage of offices. While the economy of Japan has not recovered, there is both demand and investment in the office and residential sectors, although this is not really translating into new build. Turning to China, there has been an upsurge in office demand in Beijing, and rents have increased by some 30% in the past year on the back of W.T.O. However, there is oversupply and many projects are mothballed.

In Shanghai, there is a large oversupply, but rents are now increasing. These cities are, however, overshadowed by the huge volume of development taking place elsewhere in China. This is due to the underlying demographics of the country.

DEMOGRAPHICS OF CHANGE

I will use China to illustrate the demographics that are leading to change throughout the Region. First, there is the move from the rural to the urban areas. Secondly, there is growth in the population. You have heard elsewhere in this Congress how the population in Hong Kong is increasing at one million people a decade. Third, there is, as elsewhere, a reduction in household sizes requiring more homes for the same population.

These dynamics create the pressure for urban development, and in turn, the use of tall buildings. Since the late 1970's and early 80's, it has been the policy to concentrate much of the development in some 250 coastal towns and cities. A second area of development is along the three great rivers, the Pearl River, the Yangsze River and the Yellow River. In the last year or so, there has been an attempt to develop a third area, the North West Region of China. This third area is an ambitious target, and I suspect it may take decades to realize.

There is considerable investment in road and rail infrastructure projects, including the Gwanjau line 2 and Nanjing underground railways. Water projects are proceeding, as are major urban planning studies such as Shandung and Ji Nan.

BREAD AND BUTTER TALL BUILDINGS

Urban development creates the demand for tall buildings. Egotism and economics drive the creation of the very tall buildings, however, most tall buildings are in the 12–20 storeys range, or in the case of Hong Kong 30–60 storeys. I will use Hong Kong, and in particular, my own organization, the Hong Kong Housing Authority, as an example for what I will call the "Bread and Butter" Tall Buildings.

The Hong Kong Housing Authority provides homes for over 3 million people in 1,800 buildings over 20 storeys. 250,000 flats are to be produced between 1999/2000, and 2003/2004 at a cost of US$17 billion. This five year production programme comprises 181 projects and 572 buildings over 40 storeys. Over the past two years, 297 forty-storey residential buildings had been completed. This year 85,000 flats will be completed. This is just one organization in one city.

This level of activity is taking place in cities all over China and Asia on a much larger scale, with 12–16 storey buildings the norm. These are the "Bread and Butter" Tall Buildings. The focus of this Congress is on the state of the art of tall buildings. As with Formula I car racing, the technology moves on to the family car. So to with Tall Buildings, the mega structures provide the future technology for the "Bread and Butter" buildings.

THE DARKER SIDE OF CONSTRUCTION

People do not like washing their dirty laundry in public. An earthquake will often reveal poor design or shoddy construction. The quality of Bread and Butter Tall Buildings throughout Asia is variable. They do not have the concentration of expertise focused on the mega structures. As an example over the past three years, a number of piling problems have been detected in both the public and private sectors in Hong Kong. All were detected before occupation. This has led to a comprehensive review of the construction industry by government known as the "Henry Tang Report".

It follows the footsteps of the Australian Commission's of the early 90's, the United Kingdom Latham and Eagan Report and the recent Singapore "Construction 21". The Hong Kong Report points to changes needed in the structure and the culture of the industry. It builds on the work of the Hong Kong Housing Authority Report "Quality Housing – Partnering for Change". It faces up to the darker side of construction and the issues that must be dealt with in order to ensure an efficient and effective industry.

I see that Asia must face a number of challenges in the future. These are to ensure that buildings are designed for flexibility and durability, and of particular importance to construct buildings in a way that ensures their quality.

A Report on Oceania

Henry J. Cowan

Until the mid-19th century few buildings in Australia were higher than two storeys. This changed after the Gold Rush of 1851. By 1880 the population of Melbourne had increased from 20000 to more than half a million. It had become the third largest city of the British Empire, and the headquarters of most Australian banks and important commercial enterprises. But even with this increase in population density few buildings exceeded four stories, because it was considered that people would not climb more stairs.

In 1884 Wilhelm Prell proposed to erect another 4-storey building at 15 Queen Street when he received a visit from a Vice-President of Otis who had been sent to Australia to get business for the American company. He convinced Prell that if he put up a six-storey building with an elevator, the upper two storeys would yield as much rent as the ground floor. The success of this building caused Prell to erect three nine-storey elevator buildings, and several more followed.

Melbourne was then not far behind Chicago, where a newspaper editorial writer coined the term skyscraper for buildings ten storeys or more in height. In 1891 Chicago had nine skyscrapers.

The demand for efficient elevators was met by the installation of a system of high-pressure water pipes under the city street surfaces. In 1887 Melbourne became the world's fourth city to build a public hydraulic power system, used for hoists and wool presses, as well as elevators. Since this power was available at the turn of a tap, there was a rapid increase in the number of elevator buildings.

At the turn of the century Sydney again became Australia's leading city. As its building heights increased, it also installed a hydraulic power system. Culwalla Chambers was built in 1912 to a height of 170 feet, and this led to a demand for a limitation on the height of buildings. The fire brigade's turntables could only reach 150 feet, so that the top floors were out of their reach. The growing environmental movement argued that tall buildings ruined the city as a place to live in; but the main problem was that the building, as constructed in 1912, was very ugly. The City Council passed a resolution limiting building heights to 150 feet, which remained law for 43 years, and was soon adopted in Melbourne. This limitation was not as restrictive as it might now seem, because the short era of post-World-War I prosperity was followed by the Great Depression, and then by World-War II, and there was really no pressing demand for taller buildings.

The height limitation was removed in 1955 when commercial construction resumed after the War. The first building to exceed 150 feet was the AMP Building at Circular Quay in Sydney, a 31-storey steel-framed building, 117 m high, completed in 1961. Its architects were Peddle Thorp and Walker, and the structural engineer was John Rankine, a former Chairman of the Council on Tall Buildings. His design introduced the basic principles followed by most subsequent tall buildings: the use of a long-span floor structure to obtain a column-free interior, stiff connections for rigid-frame action, and transfer trusses at plantroom level.

Many more high-rise buildings were erected in Sydney and Melbourne in the 1960s and 70s. New height records were set in 1968 by the 40-storey concrete-framed Australia Square Tower and in 1977 by the 60-storey concrete-framed MLC Centre, both designed by architect Harry Seidler and consulting engineers Miller, Milston Ferris. At 228 m the latter remained the tallest building outside North and South America for many years. In 1985 the Rialto Tower, a 60-storey concrete office building, 242 m high, was completed in Melbourne, and this is still the tallest building in Australasia and Oceania. Its architects were Perrott Lyon Matheson, and its engineer was W. L. Meinhardt.

Of the ten tallest buildings in the region, 5 are at present in Melbourne, 3 in Sydney and 2 in Perth. There has been no significant increase in height since 1977, just as there has been no significant increase in height in the United States since the 1930s when the Empire State Building was built. But there has been in other countries, particularly in Asia. So now there are only two Australian buildings among the one hundred of the world's tallest.

There is a clear case for tall buildings in Australia, as in other countries of the New World. They make the Central Business District more efficient, and more interesting architecturally. The first tall building was built in Central Sydney in 1960. Recent tall buildings are not as tall as in America, because the population density is lower. The maximum height is in Australia determined by economic factors, and not by technical considerations. This economic limitation on height is over-ruled only when an all-powerful company or a politician not restrained by democratic procedures wants to set a new record. We could build much higher if we wanted to do so.

Adelaide, the capital of South Australia, entered the tall building field early, but then decided that they were out of character with its Georgian traditions, and its tallest building today is the Santos Building, a 32-storey structure originally built for the State Bank.

The presently tallest buildings in Western Australia and in Queensland both date from 1988. The R&I Tower in Perth by architects Cameron Chisholm and Nicol and engineers Ove Arup has a 52-storey concrete frame, 214 m high. Central Plaza One in Brisbane, by architects Peddle Thorp and engineers Maunsell, also has a concrete frame, but it is only 44 storeys, 174 m in height. All three are office buildings.

New Zealand is part of my territory, and there are some significant tall buildings in Auckland. The tallest is the 38-storey ANZ Building, constructed in 1989. Oceania is also part of my territory. It is a very large part of the Earth's surface, and consists mostly of small islands, which need, and have only small buildings.

There has been a steady move of population to the warmer climate of the North, and I think the most interesting future developments in tall building design may eventuate in Brisbane, the capital city of resource-rich Queensland.

Contributors

Abdelaziz Attia, Abdalla – Ain Shams University, Cairo, Egypt
Awang, Azman bin Haji – Universiti Tecknologi Malaysia, Skudai Johor, Malaysia
Baker, William – Skidmore, Owings & Merrill, Chicago, Illinois, USA
Bates, Raymond – The Hong Kong Housing Authority, Hong Kong, China
Blutstein, Harry – Environment Protection Authority Victoria, Melbourne, Victoria, Australia
Bressi, Rocco – Bovis Lend Lease, Sydney, Australia
Breukelman, Brian – Rowan Williams Davies & Irwin, Inc., Guelph, Ontario, Canada
Brogan, Stuart – Belt Collins Australia, Brisbane, Australia
Castillo, Phil – Murphy/Jahn, Chicago, Illinois, USA
Chan, Pun Chung – Government of Hong Kong Special Admin., Hong Kong, China
Chancellor, Peter – Connell Mott Macdonald, Melbourne, Australia
Cheung, J. C. K. – Monash University, Clayton, Victoria, Australia
Colomban, M. – Permasteelisa, Spain
Cowan, Henry J. – Mosman, Australia
Dean, Brian – Connell Wagner, South Melbourne, Victoria, Australia
Droege, Peter – University of Sydney, Sydney, Australia
Duncan, Rod – Department of Infrastructure, Melbourne, Victoria, Australia
Ellyard, Peter – Preferred Futures, Melbourne, Australia
Emery, David – Connell Mott Macdonald, Australia
Facioni, Robert – Hyder Consulting (Australia) Pty. Ltd., St. Leonards, Australia
Galsworthy, Jon K. – Boundary Layer Wind Tunnel, London, Ontario, Canada
Gamble, S. – Rowan Williams Davies & Irwin, Inc., Guelph, Ontario
Gerges, Rafik – Boundary Layer Wind Tunnel, London, Ontario, Canada
Hagglund, Brian – National Fire Protection Association, Quincy, Massachusetts, USA
Hakala, Harri – Kone Elevators, Helsinki, Finland
Hart, Gary – Hart-Weidlinger, Santa Monica, California, USA
Haskett, T. – Rowan Williams Davies & Irwin, Inc., Geulph, Ontario
Haysler, Mike – Hyder Consulting (Australia) Pty. Ltd., St. Leonards, Australia
Hsaio, Ling-en – The Thornton-Tomasetti Group, Inc., New York, USA
Hutchinson, Graham L. – The University of Melbourne, Victoria, Australia
Ingenhoven, Christoph – Ingenhoven Overdiek und Partner, Dusseldorf, Germany
Irwin, Peter A. – Rowan Williams Davies & Irwin, Inc., Geulph, Ontario, Canada
Isyumov, N. – The University of Western Ontario, London, Ontario, Canada
Jerome, P. – Department of Infrastructure, Melbourne, Victoria, Australia

Keating, Richard – Keating/Khang, Santa Monica, California, USA
Khattab, Omar – Kuwait University, Salmiyah, Kuwait
Kimm, Jong Soung – SAC International, Korea
Konvitz, Josef – O.E.C.D., Paris, France
Kragh, M. – Permasteelisa, Spain
Lam, Nelson T. K. – The University of Melbourne, Victoria, Australia
Landry, Charles – Comedia, Neas Stroud, Glos., United Kingdom
Lew, Marshall – Law Gibb Group, Los Angeles, California, USA
Martin, Owen – Connell Mott MacDonald, Neutral Bay, Australia
McNamara, R. J. – McNamara/Salvia Inc., Boston, Massachusetts, USA
Melbourne, W. H. – Monash University, Clayton, Victoria, Australia
Michael, Duncan – Ove Arup & Partners, London, England
Monheim, Rolf – University Bayreuth, Bayreuth, Germany
Naeim, Farzad – John A. Martin & Associates, Los Angeles, Calif., USA
Neilson, Lyndsay – Department of Infrastructure, Melbourne, Vic., Australia
Newman, Peter – Murdoch University, Perth, Western Australia
Poon, Dennis C. K. – The Thornton-Tomasetti Group, Inc., New York, USA
Pran, Peter – NBBJ, Seattle, Washington, USA
Prieto, Robert – Parsons Brinckerhoff, New York, USA
Raman, Mahadev – Ove Arup and Partners Consulting Engrs, New York, USA
Rodger, Allen – Allan Rodger Consulting, Carlton, Australia
Schipporeit, George – Illinois Institute of Technology, Chicago, Illinois, USA
Schroff, Ro – Callison Architecture, Inc., Seattle, Washington, USA
Siikonen, Marja-Liisa – Kone Corporation, Helsinki, Finland
Skibniewski, Miroslaw J. – Purdue University, West Lafayette, Indiana, USA
Solomon, Robert E. – National Fire Protection Association, Quincy, Massachusetts, USA
Stevens, L. K. – The University of Melbourne, Parkville, Victoria, Australia
Tabart, John – Docklands Authority, Melbourne, Victoria, Australia
Tomasetti, Richard L. – The Thornton-Tomasetti Group, Inc., New York, USA
Troy, Patrick – The Australian National University, Canberra, Australia
Tschanz, T. – Skilling Ward Magnusson Barkshire, Seattle, Washington, USA
Tyni, Tapio – Kone Elevators, Helsinki, Finland
Vickery, Barry J. – Boundary Layer Wind Tunnel, London, Ontario, Canada
Wanjohi, Isaac G. – Wanjohi Consulting Engineers, Nairobi, Kenya
Weise-v. Ofen, Irene – IFHP, Essen, Germany
Whitzman, Carolyn – Safe Cities Collaborative, Toronto, Canada
Wilson, John L. – Lehigh University, Bethlehem, USA
Wong, Hazel W. S. – WSW Architects, Dubai, United Arab Emirates
Wong, Raymond W. M. – Division of Building Science & Technology, City University of Hong Kong
Ylinen, Jari – Kone Elevators, Helsinki, Finland
Zobec, M. – Permasteelisa, Spain

UNITS

In the table below are given conversion factors for commonly used units. The numerical values have been rounded off to the values shown. The British (Imperial) System of units is the same as the American System except where noted. Le Système International d'Unités (abbreviated "SI") is the name formally given in 1960 to the system of units partly derived from, and replacing, the old metric system.

SI	American	Old Metric
	Length	
1 mm	0.03937 in.	1 mm
1 m	3.28083 ft	1 m
	1.093613 yd	
1 km	0.62137 mile	1 km
	Area	
1 mm^2	0.00155 in.^2	1 mm^2
1 m^2	10.76392 ft^2	1 m^2
	1.19599 yd^2	
1 km^2	247.1043 acres	1 km^2
1 hectare	2.471 acres[1]	1 hectare
	Volume	
1 cm^3	0.061023 in.^3	1 cc
		1 ml
1 m^3	35.3147 ft^3	1 m^3
	1.30795 yd^3	
	$264.172 \text{ gal}^{[2]}$ liquid	
	Velocity	
1 m/sec	3.28084 ft/sec	1 m/sec
1 km/hr	0.62137 miles/hr	1 km/hr
	Acceleration	
1 m/sec^2	3.28084 ft/sec^2	1 m/sec^2
	Mass	
1 g	0.035274 oz	1 g
1 kg	2.2046216 lb[3]	1 kg
	Density	
1 kg/m^3	0.062428 lb/ft^3	1 kg/m^3

SI	American	Old Metric
	Force, Weight	
1 N	0.224809 lbf	0.101972 kgf
1 kN	0.1124045 tons[4]	
1 MN	224.809 kips	
1 kN/m	0.06853 kips/ft	
1 kN/m^2	20.9 lbf/ft^2	
	Torque, Bending Moment	
1 N-m	0.73756 lbf-ft	0.101972 kgf-m
1 kN-m	0.73756 kip-ft	101.972 kgf-m
	Pressure, Stress	
1 N/m^2 = 1 Pa	0.000145038 psi	0.101972 kgf/m^2
1 kN/m^2 = 1 kPa	20.8855 psf	
1 MN/m^2 = 1 MPa	0.145038 ksi	
	Viscosity (Dynamic)	
1 N-sec/m^2	0.0208854 lbf-sec/ft^2	0.101972 kgf-sec/m^2
	Viscosity (Kinematic)	
1 m^2/sec	10.7639 ft^2/sec	1 m^2/sec
	Energy, Work	
1 J = 1 N-m	0.737562 lbf-ft	0.00027778 w-hr
1 MJ	0.37251 hp-hr	0.27778 kw-hr
	Power	
1 W = 1 J/sec	0.737562 lbf ft/sec	1 w
1 kW	1.34102 hp	1 kw
	Temperature	
K = 273.15 + °C	°F = (°C × 1.8) + 32	°C = (°F − 32)/1.8
K = 273.15 + 5/9(°F − 32)		
K = 273.15 + 5/9(°R − 491.69)		

(1) Hectare as an alternative for km^2 is restricted to land and water areas.
(2) 1 m^3 = 219.9693 Imperial gallons.
(3) 1 kg = 0.068522 slugs.
(4) 1 American ton = 2000 lb.   1kN = 0.1003612 Imperial ton.   1 Imperial ton = 2240 lb.

Abbreviations for Units

Btu	British Thermal Unit	kW	kilowatt
°C	degree Celsius (centigrade)	lb	pound
cc	cubic centimeters	lbf	pound force
cm	centimeter	lb_m	pound mass
°F	degree Fahrenheit	MJ	megajoule
ft	foot	MPa	megapascal
g	gram	m	meter
gal	gallon	ml	milliliter
hp	horsepower	mm	millimeter
hr	hour	MN	meganewton
Imp	British Imperial	N	newton
in.	inch	oz	ounce
J	joule	Pa	pascal
K	kelvin	psf	pounds per square foot
kg	kilogram	psi	pounds per square inch
kgf	kilogram-force	°R	degree Rankine
kip	1000 pound force	sec	second
km	kilometer	slug	14.594 kg
kN	kilonewton	U_o	heat transfer coefficient
kPa	kilopascal	W	watt
ksi	kips per square inch	yd	yard

Building Index

Italic page numbers refer to illustrations

Name Index

Subject Index

Printed and bound by CPI Group (UK) Ltd, Croydon, CR0 4YY

01/11/2024

01782621-0015